Nursery Production and Orchard Management

Volume 1: Allied Fields

The Editor

Dr Wasakha Singh Dhillon has significantly contributed in the field of education, research, extension and research management in horticultural sciences. He did his graduation in agriculture, masters and doctorate in horticulture from PAU, Ludhiana. He worked on various research management positions as Director, Fruit Research Station Gangian; Director, Punjab Horticultural Post-harvest Technology Centre, PAU campus; Assistant Director General Horticulture in ICAR, New Delhi, and after retirement working as Consultant NHB, GOI. He guided 12 M Sc and 7 Ph D students. He has written 9 books namely : Viticulture in north-west India; Impact of climate change on fruit crops; Fruit tree physiology; Fruit production in India; Doubling farmers income through horticulture; Horticulture for food and environment security; Plantation crops in India; Post-harvest management of Fruits and plantation crops (Two Volumes *i.e.* Basic concepts, and Applied concepts. His technological pursuits led to the development of 10 fruit varieties, 1 root-stock and 38 technologies. Dr Dhillon published 152 original research papers in national and international journals of repute, 25 book chapters, 4 review papers, 20 full proceeding papers, 60 technical articles, and 1 technical bulletin. He attended 60 national and international seminars. He has been nominated as member QRT's of Directorate of Cropping Systems, Modipuram, Meerut, and Central Citrus Research Institute, Nagpur by ICAR. He was also member of RAC and IMC of 12 ICAR Institutes. He was member of Peer Review Teams for accreditation of BSKKV, Agricultural University, Dapoli, and State Agricultural University of Jharkhand at Ranchi. He has been awarded with prestigious Fellowships by Indian Academy of Horticultural Sciences (IAHS); Confederation of Horticulture Association of India (CHAI); and Society for Horticultural Research and Development. CHAI also awarded Dr Dhillon with Life Time Achievement Award-2024, and Life Time Achievement Award in Sports by PAU. He was also awarded with Dr JC Anand Gold Medal, and Shri GL Chadha Memorial Gold Medal by IAHS; Meritorious Teacher Award, S GS Nihalsingwala Award, Letter of Appreciation for Agriculture Development by PAU; Distinguished Scientist Award by High-tech Horticultural Society. Apart from research, academics and management, Dr Dhillon was an outstanding athlete of PAU during his student days. He was best athlete of the College, University and Ludhiana district, and won numerous medals like Gold, Silver and Bronze at College, University, District, State, and Open National level competitions. For recognition of achievements in sports he was honored with College and University colors, and merit certificates.

Nursery Production and Orchard Management

Volume 1: Allied Fields

– Editor –

W.S. Dhillon

Former Assistant Director General, Horticulture, ICAR, New Delhi

Former Director, Punjab Horticultural Post-harvest Technology Centre

Former Director PAU-Fruit Research Station, Gangian

Former Professor of Horticulture, PAU, Ludhiana

2026

Daya Publishing House®

A Division of

Astral International Pvt. Ltd.

New Delhi – 110 002

ISBN 978-93-5919-844-6

Published by : **Daya Publishing House®**
A Division of
Astral International Pvt. Ltd.
– ISO 9001:2015 Certified Company –
4736/23, Ground Floor, Ansari Road
Darya Ganj, New Delhi-110 002
Ph. 011-43549197, 8130496929
e-mail: info@astralint.com
Website: www.astralint.com

Printed at : **Replika Press Pvt Ltd**

This volume of the book is dedicated to

Late Prof. (Dr.) K.L. Chadha, Padam Shri Awardee

(Father of Golden Revolution in Horticulture in India)

Former National Professor-ICAR, DDG Hort.-ICAR, Commisioner Horticulture-GOI, MD-National Horticulture Board-GOI, and Director-IIHR-Bangaluru

He played very significant role to shape me professionally

डॉ. हिमांशु पाठक
Dr. Himanshu Pathak
सचिव (डेयर) एवं महानिदेशक (आईसीएआर)
Secretary (DARE) &
Director General (ICAR)

भारत सरकार
कृषि अनुसंधान और शिक्षा विभाग एवं
भारतीय कृषि अनुसंधान परिषद
कृषि एवं किसान कल्याण मंत्रालय, कृषि भवन, नई दिल्ली–110 001

GOVERNMENT OF INDIA
DEPARTMENT OF AGRICULTURAL RESEARCH AND EDUCATION (DARE)
AND
INDIAN COUNCIL OF AGRICULTURAL RESEARCH (ICAR)
MINISTRY OF AGRICULTURE AND FARMERS WELFARE
Krishi Bhavan, New Delhi 110 001
Tel: 23382629 / 23386711 Fax: 91-11-23384773
E-mail: dg.icar@nic.in

Foreword

The horticulture have great potential due to varied agro-climatic conditions prevailed in the country that are suitable to grow a variety of horticultural crops ranging from temperate to sub-tropical to arid to tropical and tropical humid climates. India ranks second in the world, producing 352 million tonnes of horticultural produce. The horticulture sector has great economic significance and contributing 33 per cent to agricultural GDP that is having 18.2 per cent share in the overall GDP of the country. Among the horticultural crops, fruits played a significant role and are providing and ensuring nutritional security to the large population of the country. Fruits are vital component of daily diet because it contains all the essential nutrients, minerals, vitamins, fiber and amino acids required to maintain a good health. Presently, India is producing 112.62 million metric tonnes of fruits (2023-24) from 7.04 million hectares of area (NHB). In spite of producing large quantities of fruits, India is still lacking in per capita consumption that is required for nutritious diet recommended by the World Health Organization(WHO). To fulfil the minimum dietary requirement per head per day, the country require to produce more than 150 million tonnes of fruits every year. To achieve this mission concerted efforts by the scientists, Government agencies, policy support and fruit growers are being made to enhance the productivity per unit land area through the production of quality planting material, and manipulations in canopy and orchard management practices.

Considering the increasing demand of healthy and clean planting material, the nursery production is heralded as sunrise for raising of healthy orchards and improving economic status of fruit growers. For the successful cultivation of fruits, the first and foremost area is to ensure healthy and disease free true to type planting material produced with right method of propagation. On the other hand the management of the orchard should be in such a way that the fruit trees give higher yield of quality fruits in successive years for sustainable economic returns.

In recent years, the concept of fruit growing has totally been changed with lot of advances made in the areas of nursery production, canopy architecture and manipulation of tree growth to produce high quality fruits. Apart from these, the allied areas such as nutrient and weed management, soil and water conservation, drip irrigation, plant protection measures, incompatibility of scion and rootstock, inter-cropping in orchards, high density planting, mechanised grafting, horti-silviculture systems, natural fruit growing, suitable machinery for orchards, nursery accreditation and quarantine, biotechnological and tissue culture techniques, artificial intelligence and drones are also playing very important role in nursery production, canopy and orchard management. The first volume of this book exclusively deals with these areas for the successful and economically sustainable fruit production.

The book on "Nursery Production and Orchard Management (Volume 1: Allied Areas) is being published and edited by Prof. (Dr.) WS Dhillon, Former Assistant Director General Hort., ICAR, New Delhi. It consists of chapters on different allied areas will be useful to the students, teachers, extension workers and policy planners. The chapters are contributed by highly experienced scientists is highly praiseworthy.

Himanshu Pathak

Preface

Incorporation of fruit cultivation in horticulture farming systems is a cornerstone of global agriculture, providing nutrition, economic development, and ecological sustainability. At the centre of this endeavour lie two critical components: nursery production of fruit plants and orchard management. Nursery production mainly deals with propagation, production and supply of high-quality, disease free planting material. Nursery production is a science and art of propagating fruit plants under controlled conditions and best agro-horticultural practices until they are ready for transplantation. It serves as the foundation of orchard establishment by ensuring uniformity, vigour, and genetic fidelity in planting material. Poor planting materials usually lead to low quality yield. Whereas, orchard management encompasses the long-term care and to maintain proper health of fruit trees in the field. In unison, these high precision practices decides the success of fruit production systems.

India is the second largest producer of fruits in the world. The average productivity of fruits in India is, however, low as compared to many developed countries. The main reasons for low productivity are small orchards, low yielding traditional varieties, poor orchard efficiency and inadequate technological upgradation and adoption by the growers. The high-density planting technique represents a contemporary approach to fruit cultivation that involves planting trees densely, resulting in smaller or dwarf trees with modified canopies. Effective orchard management systems impact fruit quality, output, and development by using best practices in orchard culture, which also preserve moisture, reduce weed competition, and increase nutrient availability.

Intercropping in orchard systems offers significant environmental, economic, and ecological benefits, enhancing sustainability and resilience. By integrating crops such as legumes, grasses, and aromatic plants into orchards, farmers can improve soil health, promote biodiversity, and reduce pest and disease pressures. Intercropping enhances natural soil fertility through nitrogen fixation, increases microbial activity, and boosts carbon sequestration. It also provides habitats for beneficial organisms, supporting natural pest control and reducing reliance on chemical inputs.

Water is one of Earth's most vital resources, essential for all forms of life and crucial for human civilization. The planet's water resources include oceans, rivers, lakes, glaciers, groundwater, and atmospheric moisture. The declining per capita of fresh water availability especially in the developing countries is causing a serious threat. Therefore, there is need to go for proper planning, development, management and optimal utilization of water resources. It is of paramount importance for food, environmental, social and economic security of the country.

India's diverse climate and topography contribute to its rich fruit diversity and have potential for feeding the growing global population. Despite the tremendous potential, challenges like the abiotic and biotic stresses due to climate change during the last two decades have resulted in productivity decline of fruit crops. Optimizing nutrient management is essential for improving both the yield and quality of fruit.

New frontiers in horticulture are playing significant role in improving the efficiency of operations, time saving and reducing labour dependence. These technologies like small machinery, mechanized grafting, IT and AI tools, drones, horti-silviculture systems and natural farming are becoming more realistic and helpful to improve fruit orchards productivity and improvement in quality traits. These practices will have significant role in improving environment, ecological balance and sustainability in production systems. The first volume of the book solely deals with the allied fields that are essential for the successful cultivation of fruit crops.

I humbly acknowledge the contribution of contributors, who are experts on their own specialized fields that helped me getting this book (Volume: 1) in the present shape. No professional and academic journey is complete without constructive feedback and comments from the readers of this book. I urge upon the scholars and readers to share their critical views on wasakhasingh@yahoo.com regarding the coverage of chapters in this book so that it would value for the improvement in future editions.

W.S. Dhillon

Contents

Chapter 1

Graft Incompatibility in Fruit Crops

Rajbir Singh Boora and Daljinder Singh

PAU–Regional Fruit Research Station, Bahadurgarh–147021, Patiala, Punjab

Grafting is the asexual mean of plant propagation that dates to antiquity. The fruit tree propagation using grafting dates back to 1560 BC in China, approximately 500 BC in Mediterranean region (the Talmudic-Hellenistic times) and Roman era (Mudge *et al.*, 2009; Tsutsui and Notaguchi, 2017). References of grafting are found in literature of Theophrastus (Einarson antd Link, 1976), Virgil and Shakespeare's Richard II and The Winter's Tale. Grafting is uniting the two or more plant tissues to generate a single plant *i.e.* rootstock (for providing base, root system) and scion (as shoot or above ground portion). Grafting is the commercial method of plant propagation, especially fruit crops and has numerous utilizations in horticulture, such as true-to-type plant propagation, integrating the promising traits of the scion and/or rootstock to create a single plant, reducing juvenile phase and enhancing precocity, accelerating growth of progeny in crop improvement programmes, obtaining desired growth habit, canopy management, changing of cultivar on existing root system (top working), repairing trees with bark injury and virus indexing in some cases (Hartmann *et al.*, 1990). Most of the fruit crops are heterozygous in nature, hence becomes mandatory to propagate through vegetative means to preserve the hereditary integrity of a particular cultivar/variety.

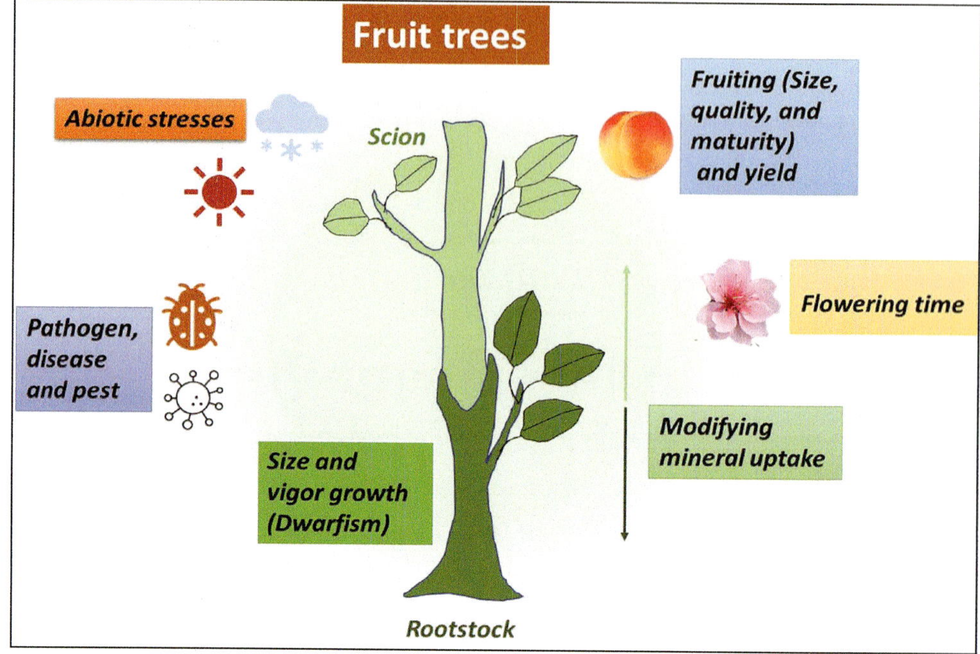

Figure 1.1: Application of Grafting in Controlling different Horticultural Traits (Habibi *et al.*, 2022).

For majority of fruit crops, plant propagation via grafting is applied commercially which are difficult to propagate by cutting or air layering. Further, many superior varieties/cultivars lack strong root system, resistance to soil borne pathogens, nematodes, insects and tolerance to abiotic stress, so by using rootstock with strong root system and resistance to biotic and abiotic factors improves the vigor, fruit yield and quality of a particular scion cultivar. Grafting is commercially applied in plant propagation of various temperate, tropical and subtropical fruit trees like mangoes, citrus, apple, pear, peach, plum, apricots, walnuts, cherries, guava, sapota, *etc.* across the globe.

Branches or roots of the same species can graft in nature (Janick, 2009) or synthetically where the tissues of two different plants are joined to form a single plant. In natural conditions, root grafting happens when nearby plants' roots physically touch and cross one another. Whereas shoot or stem grafting takes place above ground when shoots come in direct contact in dense stand get injured by friction between them by air movement and tissues fuse with each other (Pravdin, Nekrasov, 1953). For holly, lime, oak, beech, Scots pine, and Jack pine, natural grafting is general (Pravdin, Nekrasov, 1953; Janick, 2009). Root grafting in nature can have positive impact on natural bionetwork by supporting weaker trees, enhancing resistance to fire and wind, and nutrient and moisture distribution through root grafts, (Bormann, 1966; Loehle, Jones, 1990). Natural grafting can influence natural ecosystem in positive way through sharing water and nutrients via root grafts (Bormann 1966), by supporting weaker tree and promote fire and wind resistance (Loehle, Jones, 1990).

Grafts can be classified into two broad terms *i.e.* homograft's and heterografts. In case of homograft's, the scion partner and rootstock belong to same species, whereas in lateral class, the grafting partners belong to different species. Several investigations reported that grafting is more successful between homograft's and growth and functioning of homograft's is much stable as due to higher genetic and anatomical similarities in grafting partners (Kovalishin, 1966; Yakovleva, 1974). Heterografts experience incompatibility due to anatomical, morphological and physiological differences between graft components, however successful heterografts can be more beneficial than homograft's. Grafting a flood sensitive *Annona* species onto *A. glabra* improves flood tolerance (Nunez Elisea, 1999).

It's a common fact that a certain combination of grafting partners needs to have some taxonomic affinity in order to be compatible and successful. In other words, a given graft combination's success is inversely correlated with the taxonomic separation of the stock and scion. (Gainza *et al,* 2015). Andrews and Marquez (1993) opined the following hierarchy to determine the theoretical success of graft combinations: intraspecific > intra-clonal > intraspecific > interspecific > intrageneric > intergeneric > intrafamilial.

Within a species, grafting is frequently successful. Similar to how a peach variety can be successfully grafted onto another peach variety that is used as rootstock and this is true for many other fruit crops like mango, guava, apple *etc.* grafting between species is also successful. For example, most species in the Citrus genera can successfully be grafted on each other, and this is done on a commercial basis for plant propagation. Numerous almond cultivars can be commercially grafted using peach and plum rootstocks (Lordan *et al.,* 2019), apricot (Reig *et al.,* 2018), European and Japanese plum (Reig *et al.,* 2019) varieties. Reciprocal interspecific grafts are not successful always. For example, successful graft combination between 'Marianna' plum on peach can be obtained, whereas reverse combination fails to develop normal graft union and dies soon after grafting. Graft combination between different genera can be successful within the same family. Successful and commercial application of this can be noticed in rutaceae and rosaceae families. Trifoliate orange can be successfully used as rootstock for orange, quince for pear and loquat. However, reciprocal grafting of quince on pear is not successful. Reports on successful interfamilial grafting in fruit crops have not been documented (Hartmann *et al.,* 2011).

Sequence of Events during Healing of Grafting

Before throwing insight into graft incompatibility in details, let's understand the cascade of events that transpire while graft union heals and disruption in these events at any stage can lead to graft incompatibility. Hartmann *et al.* (2002) reviewed these events as described below:

1. Sliced scion tissue with meristematic activity is introduced into close, safe contact with stock tissue that has been cut similarly so that their cambial regions are near each other and form a callus bridge. Following close contact between the components of the stock and scion, new parenchymatous cells proliferate from both, forming the callus tissue that quickly entwined and fill the space between the two partners.
2. A incessant cambial association between the rootstock and scion is formed when new cambial cells differentiate from the freshly produced callus. Some xylem and phloem may differentiate before the vascular cambium connects across the callus bridge. Usually, the xylem is the first differentiated tissue to bridge the graft union, and then the phloem, which repairs wounds.
3. As the final event of graft union formation, the freshly formed cambial layer in the callus bridge starts normal cambial activity, forming new vascular tissues. The vascular connections between the grafting partners are thus made possible by the development of newly formed xylem and phloem. Majority of researchers consider this as basic requirement of successful graft (Moore and Walker, 1981 a,b; Yeoman 1984; Tiedemann, 1989).

Definitions

Regardless of the long history and recognition of graft incompatibility, there isn't a particular definition that is accepted unanimously. Due to fact that the reason of incompatibility varies and is dependent on several factors. Among the basic predicaments in describing graft incompatibility is distinguishing it from graft failure caused by various factors, such as unfavorable environmental conditions, incompetence of the grafter, variations in the growth stages of the stock and scion stick, *etc.* However incompatibility between rootstock and scion is one reason for graft failure (Andrews and Marquez, 1993).

Garner (1979) defined it as either the grafted plant's inability to grow strong or premature death after grafting, when the grafting failure could be ascribed to dissimilarity between grafting partners. Nevertheless, it was not made clear by this definition whether these variations are pathological, developmental, anatomical, or metabolic. Graft compatibility, according to Hartmann (1990), is the capacity of two distinct plants to form a fruitful graft union and grow into a single plant in a satisfactory manner; incompatibility is simply the inability to accomplish these goals. This definition leads to confusion because it fails to distinguish between graft incompatibility and graft failure; incompatibility is just the opposite of compatibility.

The definition put forth by Mosse (1962) is based on anatomy and describes what happens when normal vascular tissues do not develop at the graft union. Instead, partly lignified ray tissue replaces the missing vascular tissue, causing cambial and vascular continuity to be interrupted and ultimately leading to mechanical breakage. According to this description, incompatibility is the incapacity to reassemble a vascular continuity that is both structurally sound and operational.

Based upon physiological characteristic, Feucht (1988) defined graft incompatibility as occurrence of early senescence resulting from physiological and biochemical processes, excluding failures caused by a diseased partner. Santamour (1988sca) concurred and proposed that graft failure resulting from microbial infection should be excluded from this.

All the above definitions may not be universally accepted. Andrew and Marquez (1993) suggested the following definition: failure of a stock/scion combination to form a strong union and to remain healthy due to cellular, physiological intolerance consequent upon metabolic, developmental and anatomical variations.

Types

Angles (1937) classified graft incompatibility based on differences in symptoms, however this system of classification is inadequate as it does not differentiate graft failures due to adverse ecological circumstances, pathological contamination or inappropriate graft technique from those originated as a result of anatomical, or physiological factors. Herrero (1956) categorized it into 4 classes: a) unsuccessful bud development, b) graft failure due to viral infections, c) mechanical blockage at graft interface and d) aberrant structures at union linked with unbalanced starch buildup.

Table 1.1: Types of Graft Incompatibility in Fruit Crops and their Causes

Species	*Type of Incompatibility*	*Causes*	*References*
Apple	Anatomical flaws	Vascular discontinuity	Warmund *et al.,* 1993
Apricot/plum	Anatomical flaws	Weak graft union formation	Errea *et al.,* 1994a
Cherries	Anatomical flaws	Poor phloem development and/or weak union	Errea *et al.,* 1994a
Apricot/plum	Anatomical flaws	Bark and wood discontinuity at the graft union	Reig *et al.,* 2018
Apricot	Localized	Differences in the phenol content between tissues above and below graft union	Usenik *et al.,* 2006
Pear/quince	Localized	Lower lignification, disruption of vascular continuity and cambium	Gur, 1968; Musacchi *et al.,* 2000; Ciobotari *et al.,* 2010

Species	*Type of Incompatibility*	*Causes*	*References*
Grapevine	Localized	Accumulation of phenolic compounds at graft interface	Sarooshi *et al.*, 1982; Canas *et al.*, 2015; Assuncao *et al.*, 2016; Assuncao *et al.*, 2019
Kiwifruit	Localized	Differences in genetic affinity coefficients	Li *et al.*, 2021
Litchi	Localized	Yellow leaves and lower superoxide dismutase, peroxidase and polyphenol oxidase activities	Chen *et al.*, 2017; Chen *et al.*, 2016
Olive	Localized	Problems in differentiation of cambium and vascular systems at graft interface	Azimi *et al.*, 2016
European and Japanese plums on peach/ almond hybrids	Localized	Prune brown line disease symptoms	Reig *et al.*, 2019
Sweet cherry	Translocated	Peroxidase activity	Guclu and Koyuncu (2012)
Peach/plum	Translocated	Phloem degeneration and carbohydrate remobilization limitation.	Zarrouk *et al.*, 2010; Amri *et al.*, 2021
Pear	Pathogen induced	Disruption of graft union by phytoplasma	Hudina *et al.*, 2014
Citrus	Pathogen induced	Quick decline by production of viral protein	Moren *et al.*, 2008
Walnut	Pathogen induced	Blackline and death of scion	Mircetich and Rowhani (1984)
Apple	Pathogen induced	Apple union necrosis and decline	Stouffer and Uyemoto (1976)

Habibi *et al.*, 2022.

Scaramuzzi (1955, 1959) classified incompatibility in fruit plants into three groups based on the response of commonly compatible graft intermediate: a) where incompatibility is overcome by insertion of an intermediate between two incompatible graft partners; b) where intermediate fail to prevent incompatibility and c) whereby an intermediate causes two typically compatible partners to become incompatible. Translocated and localized incompatibilities are the two broad terms Mosse (1962) used to describe it, depending on the underlying causes of incompatibility.

Translocated Incompatibility

This type of incompatibility is observed during first year as a stoppage of plant growth, defoliation and yellowing of the leaves. (Herrero, 1951; Mosse, 1962). This kind of incompatibility in peach/plum combinations has been attributed to biochemical and functional changes at the site of graft union that cause a blockade of carbohydrates at the scion, above the graft union (Moing and Carde, 1988; Moing *et al.,* 1990). Other notable features of this type of incompatibility include:

- ☆ Usual vascular linkage at the union, though occasionally scion expansion might cause the bark tissues to become compressed.
- ☆ Starch buildup above union and its nearly total absence below the graft junction.
- ☆ The primary indication, phloem disintegration.
- ☆ Variable performance of reciprocal graft combinations.
- ☆ Effects are noticed in the early phases of growth post grafting.
- ☆ Blockages at the union rarely result in graft failure.

Translocated graft incompatibility is primarily caused by two factors: (a) the translocation of a causal factor or factors from one graft partner to another, and (b) the inability of a commonly compatible intermediate to metabolize it. For example, a number of scions on almond and peach trees on the plum rootstock "Marianna 2624" exhibit phloem degeneration and the inability of a jointly compatible inter-stock to prevent graft incompatibility. A toxin factor produced by incompatible almond scion is transferred to the graft union's phloem and causes a toxic reaction with the rootstock. Another example is the peach known as "Hale's Early," which grows on a plum rootstock known as "Myrobalan B." This peach leaf metabolite appears to be overcome by reciprocal grafting (Herrero, 1955). The characteristics of this type of pairing are starch accumulation above graft union, phloem degeneration, and hyperplasia.

The basic processes underlying translocated graft incompatibility are a) a causal factor(s) is translocated from one graft partner to other and b) a mutually compatible inter-stock fails to metabolize it. For instance, several almond and peach scions on 'Marianna 2624' plum rootstock which are characterized by phloem degeneration and failure of mutually compatible intermediate to overcome incompatibility. Incompatible almond scion produce a toxin factor which is translocated to phloem of graft union, producing a toxic reaction with rootstock. Another example is that of 'Hale's Early' peach on 'Myrobalan B' plum rootstock, which appear to be due to metabolite from peach leaves that is overcome by reciprocal graft (Herrero, 1955). Other characters of this combination are hyperplasia, phloem degeneration and, starch accumulation above graft union. Herrero and Tabuenca, 1969 grafted peach on cotyledonary seedlings of Myrobalan B plums and reported successful growth of grafted plants even for 13 years of grafting without signs of incompatibility, whereas the similar scion stock combination showed symptoms of graft incompatibility when mature tissues were used for propagation. On the basis of this, they suggested absence or no translocation of the factor responsible for incompatibility from juvenile seedling tissues.

Localized Incompatibility

The signs of localized incompatibility manifest later on during development as observed in cherry and peach/plum combinations, the presence of certain biochemical changes across graft union may result in slight or delayed compatibility (Treuther and Feucht, 1991). Anatomical irregularities at the stock/scion graft interface, along with asymmetrical cambial and vascular continuity and bad xylem/ phloem connections at union, are characteristics of this type of incompatibility (Errea *et al.,* 2001; Koepke and Dhingra, 2013). These irregularities may break after years of orchard establishment. Localized incompatibility arises at the graft interface and necessitates proper connections between stock and scion vascular tissues. This type of incompatibility can be overcome by inserting a mutually compatible inter-stock (Mosse 1962). Notable characteristics of localized incompatibility are mentioned below:

- ✰ The breakage at the site of grafting primarily due to necrosis of cambium layer and discontinuous vascular connectivity.
- ✰ Reverse scion/stock grafting fail to overcome this type of incompatibility responses.
- ✰ The root system starves gradually with slow development of external signs relative to severity to degree of vascular discontinuity at the graft interface.

Rather than phloem development, corky bark along with necrosis take place, resulting in gradual starch gradients as opposed to sharp ones as in translocated incompatibility. Localized incompatibility can be seen, for instance, in pears grafted directly onto quince rootstock (Bartlett). 'Old Home' pear inter-stock was added to fully resolve this incompatibility (Mosse, 1962).

Symptoms

The symptoms of graft incompatibility vary according to species involved in grafting. The internal symptoms appear first as compared to external. Internally among vascular tissues, xylem bundles are less affected as compared to phloem. Symptoms include disintegration of cortex and phloem tissues (Breen, 1974); abnormal axillary xylem parenchymatous tissues (Copes, 1980a); cortex cell necrosis and absence of axillary phloem parenchyma (Copes, 1980b); enhanced activity of peroxidase enzyme in scion as well as rootstock, accumulation of polyphenols and lignin at graft union which obstruct vascular connection development and translocation of metabolites (Deloire and Hebant, 1982). Increased levels of carbohydrates such as polysaccharides, starch and sugar alcohols above graft union (in scion) and high concentrations of inorganic nutrients below graft union (in rootstock) have been reported due to poor vascular conductivity and reduced translocation of nutrients and metabolites across graft union (Breen, 1975; Salesses and Al Kai, 1985).

Study conducted by Breen and Muraoka (1975) disproved the widely accepted belief that unbalanced starch distribution causes graft incompatibility by reporting starch distribution as a result of vascular discontinuity. Both homo and hetero-grafts showed increased metabolite deposition from the phloem at graft interface; this phenomenon could be attributed to either an extracellular movement across the graft union or increased metabolism as a result of cell differentiation and growth. Enhanced hexose content have the potential to impact the polyphenol pathway, leading to elevated endogenous polyphenol levels that may be destructive to graft compatibility (Treutter *et al.,* 1985). This may also be attributed to vascular discontinuity in incompatible grafts (Andrews and Marquez, 1993).

In woody perennials, the incompatibility symptoms include excessive suberization and bark overgrowth due to increased sieve cell production and less tracheid formation, increased tannin buildup indicated by abnormally dark-stained bark and a wavy annual growth rings formation (Copes, 1980a). The visible signs include delayed bud break, irregular leaf morphology and early abscission, decreased vegetative growth, die back of shoots, poor plant vigor and premature death (Hartmann *et al.,* 1990). Loss of turgidity, reduced transpiration during day time (Schmid *et al., 1988*), and scion leaves show discoloration attributed to discontinuous vascular connectivity at graft interface (Breen, 1975) as seen in the girdled plants. Graft incompatibility symptoms include overgrowth at the graft interface. Swellings often occur above graft union with vascular discontinuity, overgrowth also occur simply due to genetic differences in relative growth rates between stock and scion. For example Bottleneck and inverted bottleneck symptoms and overgrowth at graft interface as observed in *Zyzyphus* species (Figures 1.2–1.4) and variation in growth rates of grafting partners as seen in Jamun (*Syzygium cuminii*) (Figure 1.5).

Mechanism

The wound healing molecular, biochemical and physiological processes establish the graft union. Rogers and Beakbane (1957) reviewed historical insight on hypothesis of graft incompatibility mechanisms. Including 1) an active 'principle' lacking in scion necessary for function of sieve tubes, 2) substances needed by one partner which are absent or inactive in the other counterpart, 3) substances are present and transportable across graft union but lack activator/enzyme in the other partner, 4) the ability to reabsorb the necrotic layer.

Cell to Cell Recognition and Communication

Cellular contact between both the graft partners is mandatory for the establishment of symplastic and apoplastic translocation system. The main factor required for successful graft establishment in fruit and vegetable crops is formation

Figure 1.2: Bottleneck Type Graft Incompatibility in *Zyzyphus* Species.

Figure 1.3: Inverted Bottleneck Type Graft Incompatibility in *Zyzyphus* Species.

Figure 1.4: Accumulation of Metabolites at Graft Interface in *Zyzyphus* Species.

Figure 1.5: Elephant Foot Type Abnormal Growth at Graft Interface in *Syzygium cuminii*.

of symplastic cell connections which enable the transfer of compounds between graft partners (Hartmann *et al.,* 2002). With increasing information it has become evident that cell recognition and direct cellular communication between the callus of scion and rootstock is crucial for successful and functional grafting (Pina *et al.,* 2009, 2012; Yin *et al.,* 2012).

Plasmodesmata are diverse and dynamic structures for symplastic cell communication and play important role in cellular pathways in the graft bridge. They symplastically connect the scion and rootstock (Strasburger, 1985) permitting mutual metabolic interaction in the cells at graft interface. This requires movement of some compound from both sides of union to the other diffusing into cell walls in order to obtain a firm cohesion between rootstock and scion (Yeomann and Brown, 1976).

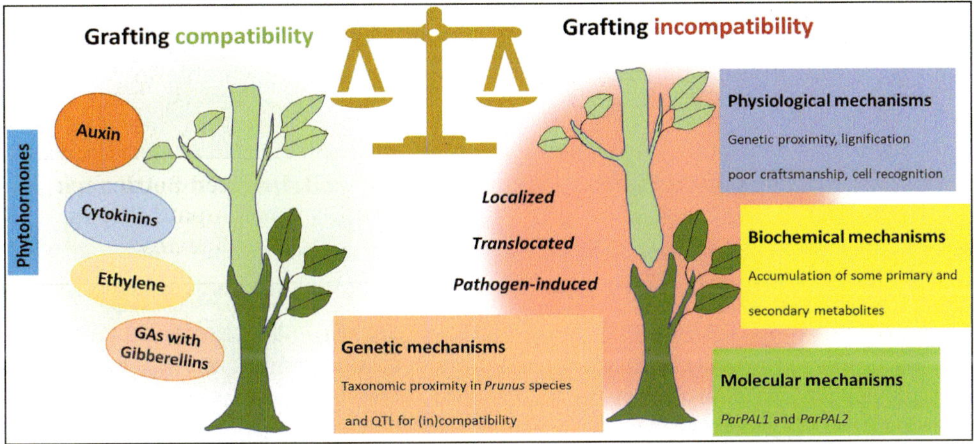

Figure 1.6: General Mechanism of Grafting (Habibi *et al.,* 2022).

Regarding the events leading to formation of secondary plasmodesmata begin with the attachment of smooth domains of the Endoplasmic Reticulum to the plasmalemma determining the location and shape of the developing plasmodesma strands. After that the process of invagination of the plasmalemma occurs around the ER cisternae that are interpreted as a consequence of wall material secretion by a plasmalemma/Golgi vesicles fusion process. The membranes of fusing Golgi vesicles form the plasmalemma that lines the plasmodesma connections (Kollmann and Glockmann, 1991). In order to form continuous secondary plasmodesmata, an exact cooperation of both cell partners is required since insufficient coordination between adjacent cells leads to formation of mismatching, half plasmodesmata through the wall cell only in one cell partner. In graft interface of incompatible heterografts, discontinuous half plasmodesmata have been observed in the graft union of different types of cells using species-specific cell markers in order to identify the graft interface (Kolmann *et al.,* 1985). These reports imply that plasmodesmata may lead to graft failure due to misalignment of the graft partners, although they may not be unique to graft incompatibility.

Jeffree and Yeoman (1983) described a cell recognition mechanism in which opposing cells of the graft partner touch, cell wall undergo dissolution, holes in the cell wall appear, plasmalemmas contact and plasmodesmata form. Cell wall dissolution is mediated by cell wall degrading enzymes released upon cellular contact. After formation of plasmodesmata, the plasma lemmas release proteins which determine compatibility or incompatibility. However they failed to describe the nature and role of these proteins. McCully, 1983 and Moore. 1983a challenged this mechanism from anatomical and physiological point of view, respectively. They opined that visual evidences of plasmalemmas in contact were unsatisfactory and many reports of deposition of cell wall material compared to cell wall dissolution. The necrotic cellular layer present at graft interface constitute initial mechanical barrier against cellular communication and chemical signaling. Moreover, it cannot explain why some compatible grafts become incompatible in reciprocal combinations.

Parkinson *et al.* (1987) suggested cell recognition mediated by enzymatic release of oligosaccharides from opposing cells. Cell wall derived oligosaccharides are necessary for cell wall synthesis. Sharon and Lis (1989) stressed upon the importance of speed with which cell wall recognition occur and suggested that studies related with cell communication should consider rate measurements.

Investigations on formation of plasmodesmata have shown variation in the development of interspecific plasmodesmata between scion and rootstock, suggesting the role of cell recognition and functional coordination between two different genotypes at the graft interface for graft formation (Kollmann and Glockmann, 1985 and Kollmann *et al.,* 1985) with special attention given to establishment and modification of primary plasmodesmata as in the novo formation or modification of secondary plasmodesmata (Ehlers and Kollmann, 2001). Insufficient coordination between adjacent cells at graft junction has been reported to formation of mismatching, half plasmodesmata in incompatible combinations. Late graft rejection in fruit trees has been reported to be associated with limited plasmodesmal coupling at the time of grafting within one of the partner (Pina *et al.,* 2009).

The graft incompatibility in grafted P*runus* spp. plants may be attributed to low intercellular transportation or a less number of functional plasmodesmata in either rootstock or scion callus tissues or between both (Pina *et al.,* 2012). The *de novo* formation of many plasmodesmata between the grafting partners with high compatibility, particularly in globe artichoke-wild cardoon combination have been reported by Trinchera *et al.* (2013). The gene expression differences suggest the involvement of plasmodesmata in grafting success. The expression levels of plasmodesmata-located protein PDLP1A, AT5G43980 and AT2G41870 were elevated at early stages of development in *Arabidopsis* homograft's (Yin *et al.,* 2012).

The protein PDLP1A targets the plasmodesmata and involve in plasmodesmal trafficking (Thomas *et al.,* 2008). These results confirm the role of plasmodesmata in cell to cell communication in graft union development (Pina *et al.,* 2017).

Role of Growth Hormones

Scion and rootstock associations are influenced by growth regulators and may be involved in the graft incompatibility also. Pectin secretion to start graft partner adhesion, de-differentiated callus cell formation, cellular junction development (plasmodesmata), cell division beginning in the cambium, cortex, and pith cells near vascular tissues are all mediated by plant growth regulators (Nanda and Melnyk, 2018). Plant hormones like auxin, ethylene, cytokinin, gibberellin, abscisic acid and jasmonic acid have emerging roles in the regulation of several crucial physiological processes happening at the graft union (Sharma and Zheng, 2019).

Callus proliferation and differentiation into vascular tissues can be modulated by auxins. Callus tissues surrounding vascular bundles are supposed to be receptor of auxins released by served vascular tissues causing the callus differentiation into xylem and phloem (Sachs 1981). Stoddard and McCully (1983) evaluated influence of organ (source of growth regulators) excision on grafting in autografts of *Coleus.* The removal of leaves, shoots or stem segment didn't affected speed of cohesion, appearance of vascular tissues and necrotic cell layer and ratio of xylem to phloem. When auxin sources such as shoot tips and leaves were removed, the formation of callus and further differentiation into vascular strands were reduced at the base of scion suggesting the development of callus and vascular connectivity dependent upon accumulation of auxins at the scion base in intact grafts.

Recently, the understanding of hormonal signaling manners in graft communication during wound healing and vascular regeneration has increased. The scion –rootstock communication is mainly affected by endogenous factors especially plant growth hormones (Aloni *et al.,* 2010). In normal development, auxin is synthesized in shoot tips and actively translocated basipetally to other plant parts affecting various physiological processes such as root development, morphology and functioning (Yin *et al.,* 2012; Spiegelman *et al.,* 2015). Auxin transportation from scion to rootstock was reported to enhance the formation of successful grafts in cactus (Shimomura and Fujihara, 1977). The main cause of graft incompatibility in *Cucurbita* spp. was reported due to hormonal imbalance, mainly auxin and ethylene in root system following the establishment of graft connectivity (Aloni *et al.,* 2008).

During the cell division proximal to vascular strands, auxin as a morphogenic substance is released from vascular tissues (Nanda and Melnyk, 2018). It has been reported that low concentrations of auxin transform callus into phloem in several plant species, while a high level induces both xylem and phloem development (Aloni, 2021). Auxin is translocated to wound via PIN transporters and triggers differentiation of vascular tissues (Bishop *et al.,* 2011).

Investigation at genomic level have identified genes with defined functions in either auxin influx or efflux differentially expressed during graft union development. Several researchers have reported importance of auxins at initial graft union formation in *Arabidopsis* (Yin *et al.*, 2012; Wang *et al.*, 2017). Cookson *et al.* (2013) identified gene encoding auxin influx carrier at 3 days after grafting at graft interface of grapevine. Auxin promotes the expression of specific transcription factors that affect pith cell proliferation (Sharma and Zheng, 2019). The downstream auxin signaling for normal phloem reconnection in root stock during graft union formation is reported by (Zhai *et al.*, 2021). Transcriptome analysis showed that a series of auxin response genes show changes in steady state transcript accumulation during the grafting process (Yin *et al.*, 2012). Transcriptome compatible autografts and incompatible heterografts of litchi showed that auxin mediated gene expression related to wound response and signal transduction pathways likely play a key role in graft healing process (Chen *et al.*, 2017).

The important role of auxin was reported in the analysis of graft compatibility in citrus, as the expression of nine genes in the auxin pathway were upregulated and three were downregulated in the compatible combinations relative to transcripts in the incompatible group (He *et al.*, 2018). Auxin is also reported to work in conjugation with other growth regulators during the process of graft union formation (Nanda and Melnyk, 2018).

Activation of the ethylene and jasmonic acid biosynthesis pathways one day after grafting confirmed that plants utilize these wound induced programmes for graft union formation (Aloni *et al.*, 2010). The role of ethylene in response to wounding is well established and is reported to promote cell expansion, callus formation and cell division (Zhai *et al.*, 2021). The gene expression in response to wound healing is also regulated by ethylene (Nanda and Melnyk, 2018).

Cell proliferation at graft union is induced by cytokinin's during the process of wound healing. The analysis of cells at the graft union detected higher levels of zeatin riboside coincident with establishment of the graft union in pecan (Mo *et al.*, 2018). Cytokinin's can stimulate the regeneration process of vascular tissues during the healing of stem wounds and sieve tubes, and along with auxins enhance vascular differentiation and increase the phloem/xylem ratio (Sharma and Zheng, 2019). In addition CKs can promote vascular cell growth, development of vascular bundles and division of cambium tissues. (Nieminen *et al.*, 2008). It is reported that exogenous CK application including benzyl adenine (BA) or kinetin (Ki) significantly increase callus formation and improved the graft union formation in grapevine (Kose and Guleryuz, 2006). Furthermore CK regulates the expression of some key genes related to vascular tissue development (Hirose *et at.*, 2008). For instance CK regulated the activity of Lonesome Highway (LHW) gene that is responsible for development of stele cells and formation of protoxylem (Dettmer *et al.*, 2009).

Gibberellins (GAs) have demonstrated role in vascular growth, cambium activity, xylem expansion and xylem fiber differentiation as well as plant secondary growth (Sharma and Zheng, 2019). GA affects gene expression to facilitate the development of xylem tissue and stem growth. GA is mobile across the graft union and co-ordinates normal xylogenesis, formation of vascular bundles as well as controlling cambium activity, xylem fibre differentiation and reunion of cortex in the joining of scion and rootstock (Martinez *et al.*, 2016). A role of ABA in establishment of the graft union has not been identified. ABA does induce differentiating xylem and because ABA has an inhibitory effect towards wound healing, reducing of ABA synthesis or signal transduction may underlie graft union establishment (Ikeuchi *et al.*, 2017). Based on its known roles Jasmonates likely has a function in promoting cambium and vascular formation during grafting although it has been described as not necessary. A definitive role of JA in the establishment of a graft union has not been described (Nanda and Melnyk, 2018).

Biochemical Mechanisms

A number of primary and secondary metabolites are linked to graft incompatibility (Amri *et al.*, 2021) yet investigations on role of primary metabolites in graft compatibility/incompatibility in fruit crops is scanty (Loupit and Cookson, 2020). Moing *et al.* (1990) studied carbon partitioning (variation in sugar and starch accumulation) in peach/plum graft combination as marker to determine graft incompatibility during the early stages of graft union formation. Differences in the primary metabolite content were observed between the surrounding woody tissues and the graft interface in an analysis of grapevine unions. After 28 days of grafting, the union's starch concentration was lower than that of the surrounding woody tissues, but its glucose content was higher. At the graft interface, total proteins, γ-amino butyric acid, and glutamine were found in higher amounts than in the surrounding tissues while Arginine, histidine, lysine, phenylalanine, threonine, and tyrosine were in lower concentrations. The variation in the concentrations of metabolites may be attributed to callus formation at graft interface (Prodhomme *et al.*, 2019). In peach/plums graft combination, the levels of soluble proteins in peach scion was lower in all organs in incompatible grafts. Similar pattern was found in scion for asparagine, aspartate, and glutamate concentrations indicating nitrogen starvation.

Graft incompatibility in fruit crops is influenced by certain secondary metabolites. Nevertheless, graft failure is not always caused by the presence of particular secondary metabolites; these are only seen in response to particular inter-generic graft combinations. The role of phenolic compound accumulation in graft incompatibility is reported in various investigations (Pina *et al.*, 2017). Table 1.2 presents the some secondary metabolites accumulation at graft union of incompatible scion/rootstock combination in fruit crops.

Prunasin is cyanogenic glycoside and well-known example of a secondary metabolite involved in graft incompatibility in pear/quince combination. This compound is present in quince rootstock but absent in pear scion. The prunasin present in quince rootstock translocate into the pear scion phloem and pear tissue breakdown it by hydrolyzing with glycosidase and toxic hydrocyanic acid (cyanide) is released, which causes cell death or damages xylem, and phloem at the graft interface (Gur A, 1968). However, cyanogenic glycosides are present in only few genera of woody plants and this phenomenon cannot be applied universally as cause of graft failure/incompatibility.

Table 1.2: Secondary Metabolites Involved in Incompatible Graft Combinations in Fruit Crops

Compound	*Graft Combination*	*Time after Grafting*	*References*
Arbutin	Pear	4 years	Hudina *et al.*, 2014
4-hydroxyphenyl acetic acid	Olive	1 year	Azimi *et al.*, 2016
Catechin	Apricot	1 year	Usenik *et al.*, 2006
Catechin	Pear/quince	2 years	Musacchi *et al.*, 2000
Catechin	Pear	4 years	Hudina *et al.*, 2014
Catechin	Grapevine	1 month	Canas *et al.*, 2015
Catechin	Grapevine	End of rooting stage	Assuncao *et al.*, 2019
Epicatechin	Pear/quince	2 years	Musacchi *et al.*, 2000
Epicatechin	Grapevine	3 months	Assuncao *et al.*, 2016
Ferulic acid	Olive	1 year	Azimi *et al.*, 2016
Ferulic acid	Grapevine	3 months	Assuncao *et al.*, 2016
Gallic acid	Grapevine	1 month	Canas *et al.*, 2015
Gallic acid	Grapevine	End of rooting stage	Assuncao *et al.*, 2019
p-coumaric acid	Apricot	1 year	Usenik *et al.*, 2006
Prunasin	Pear/quince	5 years	Gur A, 1968
Sinapic acid	Grapevine	1 month	Canas *et al.*, 2015
Sinapic acid	Grapevine	End of rooting stage	Assuncao *et al.*, 2019

Source: Habibi *et al.*, 2022.

Phenolic compounds are unique class of plant metabolites that are essential to higher plants. They take part in many different metabolic pathways and contribute to the structure of plants. (Vermerris and Nicholson, 2006; Lattanzio *et al.*, 2008). Plants produce these substances in reaction to physical harm or in response to a variety of biotic and abiotic stress conditions. (Treutter, 2006). Various studies have reported involvement of phenolic compounds with both localized graft incompatibility in cherry, pear, apricot (Guclu and Koyuncu, 2012; Machado *et al.*, 2017; Pereira *et al.*, 2017) and translocated type in peach/plum (Amri *et al.*, 2021). These compounds function as inhibitory factor for callus differentiation, formation of new vascular tissues and cell wall lignification's.

Auxins are involved in callus differentiation, formation of new vascular bundles and lignification's, the auxin transport is adversely affected by high phenolic levels above the graft union and lead to graft failure/incompatibility (Koepke and Dhingra, 2013). The phenolic compound metabolism is regulated by Phenylalanine ammonia-lyase (PAL) and polyphenol oxidase (PPO). Phenolic compounds undergo fast oxidation and polymerization after interaction with endogenous oxidative enzymes. When there is graft incompatibility, the integrity of the vacuolar membrane is lost, which modifies lipid peroxidation, allowing polyphenols to leak out of the vacuolar compartment into the cytoplasm. Phenols that are liberated from the vacuole and enter the cytosol are oxidized by peroxidases and polyphenol oxidases (Errea, 1998). Grafting may not work at all as a result of interactions between proteins and the quinones and polymers produced by these reactions. The polymerized toxic compounds such as polymeric melanin's and quinines are produced by oxidation of phenolic compounds by PPO and POX. (Amri *et al.*, 2021). Pereira *et al.*, 2017 reported incompatibility in peach and apricot attributed to high prunasin concentrations and increased PAL enzyme activity in rootstock. In addition, Pina and Errea (2008) suggested involvement of protein UDP-glucose pyro phosphorylase in graft incompatibility. In this respect, it is interesting to mention the importance of peroxidase and polyphenol oxidases in grafting incompatibility in trees.

Higher peroxidase activity at the graft interface was linked to graft incompatibility in *Prunus* sp. (Rodrigues *et al.*, 2002) and peach–plum combinations (Zarrouk *et al.*, 2010). Peroxidase enzyme activity is linked with callus differentiation into xylem tissues (Sterjiades *et al.*, 1993) as well as with cell wall lignification (Harkin and Obst, 1973). Irregular vascular connectivity after grafting may be caused by differences in peroxidase isozyme banding signature between grafting partners (Santamour, 1992). A steep gradient is established at the union line in Prunus combinations because the phenolic compounds produced by heterospecific Prunus graft partners differ both quantitatively and qualitatively (Geibel and Feucht, 1991; Errea *et al.*, 1994b).

Among the polyphenols, proanthocyanidins, also known as condensed tannins, are compounds that are abundant in nature. They are composed of oligomers and polymers of elementary flavan-3-ol units. After lignin's, they are thought to be the second most abundant group (He *et al.*, 2008). In addition to their involvement in typical wounding and stress processes, these polyphenols can also be important in the formation of graft unions through their protein-precipitating properties, inhibition of auxin transport, and influence on lignification processes (Lockard and Schneider, 1981; Hagerman, 2012). All of these factors can negatively impact the quality of a grafting union.

Quantitative variation in the phenolic compound profiles at the graft union suggests that it as a conserved marker of graft incompatibility based on the large number of incompatible scion–rootstock combinations. Musacchi *et al.* (2000) opined that compounds that are typically present in both the scion and the rootstock and that also accumulate at the graft interface may be better suited as markers of graft incompatibility. Several lines of research have examined the impact of grafting on the content of phenolic compounds in graft unions characterized by varying degrees of compatibility. These studies have found that phenol accumulation affects the relationships between the scion and rootstock. Increased synthesis of phenylpropanoid metabolites is a well-established phenomenon in graft-incompatible combinations from various species, including eucalyptus (de Cooman *et al.*, 1996), grapevine (Canas *et al.*, 2015), pear (Musacchi *et al.*, 2000; Hudina *et al.*, 2014), apricot (Errea *et al.*, 1992; Usenik *et al.*, 2006; Pina and Errea, 2008a), plum (Rodrigues *et al.*, 2002; Zarrouk *et al.*, 2010), cherry (Gebhardt and Feucht, 1982), plum (Rodrigues *et al.*, 2002; Hudina *et al.*, 2014) and loquat (Mng'omba *et al.*, 2008). According to Errea *et al.* (2001), the presence of phenolic compounds is typically linked to small cells in incompatible combinations that fail to differentiate and form successful unions.

Additionally, a number of phenolics, such as proanthocyanidins and derivatives of cinnamic acid, have been suggested as cross-linking and cell-wall-stabilizing compounds in relation to the mechanical stability of the union seen in combinations of *Prunus* and *Pyrus* (Pizzi and Cameron, 1986). Similarly, incompatible Arabidopsis–tomato grafts' scion–rootstock interface was made of a dark red-enriched substance that resembled polyphenols in reaction to grafting. This material seemed to have limited, or at least contributed to, the formation of a vascular bridge between the two species, which is a common incompatibility response (Flaishman *et al.*, 2008). When considered collectively, the majority of authors have stated that a notable build-up of (+)-catechin above the graft union may serve as a biochemical indicator of graft incompatibility (Errea, 1998; de Cooman *et al.*, 1996).

Some investigations have shown that monophenols activate IAA oxidases, leading to the accumulation of naringenin-7-glucoside (prunin) above the graft union of incompatible *P. avium–P cerasus* grafting's (Treutter *et al.*, 1985, 1986) and lead to the destruction of auxin (Feucht *et al.*, 1983).

Molecular and Genetic Mechanisms

Graft incompatibility is linked to the genetic makeup of the scion and rootstock. The genetic foundation of genotype compatibility has been determined by genome-wide quantitative trait loci (QTL) mapping. It is difficult to genotype graft incompatibility because large populations are needed to provide sufficient statistical power for inference (Loupit and Ciikson, 2020). A major factor contributing to graft incompatibility is alterations in the expression of genes encoding secondary metabolism-related enzymes.

Genes differentially Expressed during Graft Union Formation

Most of the molecular mechanisms underlying the establishment of graft unions remain unclear. The worldwide alterations in gene expression that occur during the graft union development process have received increased attention in recent years. The mechanisms underlying the establishment of graft unions in different plants have been clarified by gene expression investigations. The genes that are differently expressed at the graft interface in homograft's are most likely the same genes that are activated during the wound-healing processes. Cookson *et al.* (2013) reported upregulation of many genes responsible for cell-wall synthesis, wound responses, secondary metabolism, hormone signaling, phloem and xylem development from 3 to 28 DAG in homograft's of grapevine. The reactivation of stem development in the spring and the process of graft union formation may be related in some ways, as evidenced by the overlap of genes differentially expressed at the scion–rootstock interface and the modulation of gene expression throughout time. Most of the genes that were differently regulated in the graft interface were unique to the graft union establishment. Thus, many genes specific to the graft interface were found to be linked to wound healing and defense, including those that encode MYB-like 102, germin-like proteins, peroxidases, and chitinases. (Cookson *et al.,* 2013).

Furthermore, the establishment of the graft union caused the transcription factors involved in callus maintenance to express differently in the graft interface compared to the rootstock tissue, indicating that callus formation necessitates extensive transcriptional control. The lateral boundary domain 4 (LBD4) gene, in particular, was discovered to be activated at 3 DAG and may be connected to the development and maintenance of non-differentiated callus cells (Fan *et al.,* 2012; Ikeuchi *et al.,* 2013).

Cookson *et al.* (2014) found numerous genes induced by grafting in two distinct grapevine genotypes (heterografts) in a subsequent study that compared the changes in wound-like gene expression in homograft's during the first month following grafting. Generally speaking, heterografting caused a great deal of genes involved in plant defense and stress responses to express differently. When heterografting was done instead of homo-grafting, oxidative stress responses and pathogenesis-related proteins were quickly induced. This was followed by the induction of numerous additional genes related to plant stress responses.

Genes encoding oxidative stress-related enzymes, such as glutathione S-transferases, ascorbate oxidases, polyphenol oxidases, and peroxidases, were also discovered during this early developmental stage and may be connected to the induction of an oxidative burst at the graft interface. The functional category polyamine oxidase was enriched by 7 and 14 DAG and this enzyme is essential for oxidative burst, which causes programmed cell death (Yoda *et al.,* 2006). Towards the end of graft union formation (14–28 DAG), repression of protease kinases and prolonged induction of numerous genes involved in secondary metabolism enzymes (cell-wall precursors, lignin, and flavanol biosynthesis) were also noted (Cookson *et al.,* 2014). This may indicate that an immune-type response may be triggered by the cells at the graft interface detecting the presence of an intraspecific or interspecific heterograft.

Genes differentially Expressed between Compatible and Incompatible Graft Combinations

It is well known that the likelihood of a successful graft union increases with the degree of botanical relatedness between the plants (Hartmann *et al.,* 2002). Graft incompatibility has been extensively documented, but its underlying molecular causes are not well understood (reviewed by Hartmann *et al.,* 2002; Pina and Errea, 2005; Goldschmidt, 2014). According to recent research, molecular changes may also contribute to the development of distinct behaviors between compatible and incompatible combinations during graft formation, in addition to physiological and anatomical changes (Aloni *et al.,* 2008; Pina and Errea, 2008a,b; Nocito *et al.,* 2010; dos Santos Pereira *et al.,* 2014; Irisarri *et al.,* 2015).

In vitro callus unions have been linked to the induction of phenylalanine ammonia lyase (PAL) due to genotype incompatibility. When compared to both homograft controls, incompatible heterografts (apricot–plum) further enhanced PAL expression (Pina and Errea, 2008a). In the genus Prunus, two isoforms of PAL have been identified (PAL1 and PAL2); the PAL1 isoform is more influential in the process of scion–rootstock incompatibility (dos Santos Pereira *et al.,* 2014). Prunus spp. were also found to harbor UDP-glucose pyro phosphorylase (UGPase), a potentially relevant protein in relation to the graft compatibility/ incompatibility responses (Pina and Errea, 2008b). UGPase mRNA transcript and enzyme activity levels were found to be lower in the rootstock from incompatible heterografts (apricot–plum) at early stages of graft development compared with the corresponding autografts. This protein is involved in sucrose metabolism and the biosynthesis of cell-wall polysaccharides (Pina and Errea, 2008b).

Reactive oxygen species (ROS) generation is one of the main and common reactions to wounding in higher plants (Slesak *et al.,* 2008). Therefore, managing oxidative stress may be crucial to the success of grafts. While plants' antioxidant defense systems would be activated by stress conditions like grafting, there are signs that in incompatible scion–rootstock interfaces, either the amount of ROS is higher or the detoxification system is less effective (Aloni *et al.,* 2008; Irisarri *et al.,* 2015). For example, ineffective grafts exhibit cellular stress indicators like elevated H_2O_2 accumulation, superoxide radicals (O^{2-}), and elevated 2-thiobarbituric acid reactive metabolite levels (Aloni *et al.,* 2008; Nocito *et al.,* 2010).

In a study, Nocito *et al.* (2010) found that, following a seven-day co-culture period, the activity of five antioxidant enzymes–ascorbate peroxidase (APX), dehydroascorbate reductase (DHR), glutathione reductase (GR), SOD, and catalase (CAT)–was higher in incompatible pear-quince heterografts compared to compatible pear-pear autografts. Moreover, particular gene isoforms linked to oxidative stress were found in pear–quince and pear–pear heterografts. In vitro callus heterografts that were compatible or incompatible showed different expression patterns for certain genes. Comparing the incompatible heterograft to the compatible graft, the transcript levels of six antioxidant genes (SOD1, SOD3,

APX3, APX6, CAT1, and CAT3) were repressed at 10 DAG (Irisarri *et al.,* 2015). These findings implied that in the incompatible heterograft, there was less oxidative protection as the graft union developed. Some have suggested that the symptoms of oxidative stress could be an indirect reaction to the imbalance in auxins (Aloni *et al.,* 2010; Goldschmidt, 2014).

Prediction

An early and accurate prediction of graft incompatibility has great significance because incompatible combinations could be avoidable while compatible ones could be selected (Petkou *et al.,* 2004). The information about graft compatibility/ incompatibility is essential before releasing grafted fruit trees for establishing commercial orchards especially with regard to new cultivars and rootstocks where compatibility has not been extensively tested. Graft incompatibility may sometimes only be observed years after grafting hence an early prediction of the compatibility or incompatibility in fruit trees would be extremely useful.

The complexity of incompatibility and the mechanism behind the reactions have been studied in several ways." *In vitro* pear and quince combinations (Moore, 1984), or between callus cultures of many different Prunus species, peroxidase activity and the production of phenolic compounds in Prunus (Rodrigues *et al.,* 2001) and pear- quince grafting (Musacchi *et al.,* 2000) and the analysis of cyanogenic glycosides in some incompatible combinations. In this case histological techniques could be used for visualization of the first stage of graft development including plasmodesmata formation, callus organization and program cell death through X-ray tomography (Irisarri *et al.,* 2021).

The development of molecular markers associate with compatibility would also be a great advantage for selection of rootstock and scion combination. Different metabolic pathways may be used to make predictions about graft incompatibility such as genes associated with the phenylpropanoid pathway, oxidated stress and defense response as the end product of gene expression (Tedesco *et al.,* 2021). Translocated incompatibility may be evaluated using chlorophyll meter (SPAD) where low spade value may indicate defects in carbohydrate transportation due to incompatibility. Examination of isozymes variants may be applied to evaluate compatibility. In sweet cherry peroxidase isozymes variants and activity match well with compatibility and may be able to predict long term incompatibility (Guclu and Koyuncu, 2012). Tissue culture techniques including *in vitro* shoot tip grafting and callus grafting are useful for identifying graft incompatibility (Loupit and Cookson, 2020). Additionally proteomics molecular markers and gene expression associated with specific metabolites are emerging methods for predicting graft incompatibility.

In vitro Techniques

As the physiological and cellular responses of *in vitro* graft systems mimic those that occur *in vivo*, the use of tissue culture techniques may be an interesting approach to predict compatible and incompatible combinations. The previous studies have shown that *in vitro* graft systems have closely matched to that of *in vivo* in fruit crops (Richardson *et al.*, 1996; Errea *et al.*, 2001; Espen *et al.*, 2005). These tissue culture techniques include *in vitro* shoot tip grafting (micrografting), grafting of an explanted stem segment *in vitro* and callus grafting. Advances in micro grafting methods for studying graft union formation have been particularly done in *Arabidopsis*, as reported by Yin *et al.* (2012) and facilitated sample analysis and yielded higher success rate then previous methods and could be applied to other dicotyledonous plants.

Histological Studies

Light and electron microscopy have given detailed information related to graft union formation in pear quince grafts (Ermel *et al.*, 1997) and apricot (Errea *et al.*, 2001). Scanning electron microscopy can be used effectively in anatomical studies on graft union formation without prior sample preparation, saving time and avoiding possible after effects. The development of confocal bio-imaging tools constitutes a valuable non-invasive method to study plasmodesma permeability in living tissues with high temporal and spatial resolution (Martens *et al.*, 2004). Photo-activation an photo-bleaching techniques have been applied to study cell to cell communication via plasmodesmata in *in vitro* grown callus. It was demonstrated that callus cells in compatible and incompatible grafts varied in their ability to establish simplistic connections (Pina *et al.*, 2012).

a) **Chlorophyll fluorescence imaging:** the activation of repair mechanism after grafting place a high metabolic demand on the grafting area to supply carbohydrates, synthesize new molecules or increase anti-oxidative enzyme activity (Pina and Errea, 2005; Aloni *et al.*, 2008). Many of these processes can be supported by photosynthetic activity. Information on changes in the efficiency of photochemistry can be obtained by measuring and visualizing the yield of the chlorophyll a through images that reveal spatial- temporal changes in the grafted plants using imaging fluorometers and has been validated in grafted melon plants (Calatayud *et al.*, 2013).

b) **Electrophoresis method:** Isoforms of enzymes separated by electrophoresis where one of the earliest *in vitro* methods used for prediction of grafting incompatibility. Santamour *et al.* (1986) reported that isozyme analysis of scion and rootstock could be used to determine incompatibility prior to grafting. Past research with other plant species showed that analysis of isozymes, especially peroxidases and protein spectra between rootstock and scion could be used to predict intra specific compatibility or incompatibility (Fernandez Garcia *et al.*, 2004; Pedersen, 2006). By comparison of protein profiles of graft combinations to predict graft incompatibility using SDS-PAGE was studied in Prunus species (Huang *et al.*, 1984; Schmid and Feucht, 1985) and in *V. vinifera* (Masa, 1989).

c) **Phenol analysis:** The presence of phenols was generally associated with small cells in compatible combinations which did not form successful unions. When apricot is grafted on other Prunus species like peach, plum and peach x almond, graft incompatibility frequently occurs (Errea *et al.*, 2001). The activity of IAA oxidase and transport of IAA transport can be altered naturally occurring substances such as phenols. The quantitative and qualitative variation in the phenol pattern between two graft partners can imply metabolic disfunctions at the graft union. Higher concentrations of catechin and epicatechin were found in quince incompatible cultivars before the appearance of visible incompatibility symptoms (Musacchi *et al.*, 2000). In less compatible apricot combinations higher level of flavanols, catechin and epicatechin were found (Errea *et al.*, 2000). Phenol analysis can be used as early sign for predicting the graft incompatibility especially when new cultivars/rootstocks are used for grafting.

d) **X-ray tomography:** The cellular events at the graft interface have been well characterized by histological studies in various woody plants and give beautiful and detailed images of the graft interface but only in one plane. Till date graft union morphology has been studied with magnetic resonance imaging (MRI) in pine trees (Leszezynski *et al.*, 2000) and in grapevine (Bahar *et al.*, 2010) but these MRI studies are of comparatively low resolution and only present 2D virtual sections of the graft union.

X-ray tomography is a least invasive structural imaging method that allows 3D reconstruction of scan object. X-ray tomography has been applied to the study of anatomy of stem, wood, samples of trees and recently to the study of vessel dimensions and inter vessel connections in grapevine (Brodersen *et al.*, 2011). After different tests and optimization of the scanning parameters the method was applied to young vines with varying degree of graft success in order to understand how grapevine tissues and structures organize in response to grafting and this could open new avenues to access graft compatibility in woody plants.

REFERENCES

Aloni R, 2021. Phloem and Xylem Differentiation. In: *Vascular Differentiation and Plant Hormones*. 1st Ed. Switzerland AG: Springer. Pp: 101–130.

Aloni B, Karni L, Deventurero G, Levin, Cohen R, Katzir N, Kapulnik Y, 2008. Physiological and biochemical changes at the rootstock-scion interface in graft combinations between *Cucurbita* rootstocks and a melon scion. *J Hort Sci Biotech* **83**(6): 777-783. https: //doi.org/10.1080/14620316.2008.11512460

Aloni R, Cohen R, Karni L, Aktas H and Edelstein M, 2010. Hormonal signaling in rootstock-scion interactions. *Sci Hort* **127**: 119-126.

Amri R, Font i Forcada C, Gimenez R, Pina A and Moreno MA, 2021. Biochemical characterization and differential expression of PAL genes associated with translocated peach/plum graft-incompatibility. *Front Plant Sci.* **12**: 1-14.

Andrews PK., Serrano Marquez C, 1993. Graft incompatibility. In: Janick, J. (Ed.), Horticultural Reviews Volume 15. John Wiley and Sons, Inc., Oxford, UK, pp. 183–232. https: //doi.org/10.1002/9780470650547.ch5.

Argles GK, 1937. A review of the literature on stock-scion incompatibility in fruit trees, with particular reference to pome and stone fruits. *Tech. Commun. Imp. Bur. Fruit Prod.* 9.

Assunçao M, Canas S, Cruz S, Brazao J, Zanol GC and Eiras-Dias JE, 2016. Graft compatibility of *Vitis* spp.: the role of phenolic acids and flavanols. *Sci. Hort.* **207**: 140–145. https: //doi.org/10.1016/j.scienta.2016.05.020 (Amsterdam).

Assunçao M, Santos C, Brazao J, Eiras-Dias JE and Fevereiro P, 2019. Understanding the molecular mechanisms underlying graft success in grapevine. *BMC Plant Biol* **19**: 1–17. https: //doi.org/10.1186/s12870-019-1967-8.

Azimi M, Ozkaya MT, Colgecen H and Buyukkartal HN, 2016. Analysis of phenolic compounds for determination of cambium differentiation and tracheal elements in olive graft combinations. *J Exp Biol Agric Sci.* **4**: 714–720.

Bahar E, Korkutal I, Carbonneau A and Akcay G, 2010. Using magnetic resonance imaging technique (MRI) to investigate graft connection and its relation to reddening discoloration in grape leaves. *J. Food Agric. Environ.*, **8**: 293-297.

Bishopp A, Help H, El-Showk S, Weijers D, Scheres B, Friml J, Benková E, Mähönen AP and Helariutta Y, 2011. A mutually inhibitory interaction between auxin and cytokinin specifies vascular pattern in roots. *Curr Biol.* **21**(11): 917-926. doi: 10.1016/j.cub.2011.04.017.

Bormann FH, 1966. Structure, function and ecological significance of root grafts in *Pinus strobus L. Ecol. Monographs* **36**(1): 1-26.

Breen PJ, 1974. Cyanogenic glycosides and graft incompatibility between peach and plum. *J. Am. Soc. Hort. Sci.* **99**: 412-415.

Breen PJ and Muraoka T, 1975. Seasonal nutrient levels and peach/plum incompatibility. *J. Am. Soc. Hort. Sci.* **100**: 339-342.

Brodersen CR, Lee EF, Choat B, Jansen S, Phillips RJ, Shackel KA, McElrone AJ and Matthews MA, 2011. Automated analysis of three-dimensional xylem networks using high-resolution computed tomography. *New Phytol* **191**: 1168–1179.

Calatayud A, San Bautista A, Pascual B, Maroto JV and López-Galarza S, 2013. Use of chlorophyll fluorescence imaging as diagnostic technique to predict compatibility in melon graft. *Sci Hort* **149**: 13–18.

Canas S, Assunção M, Brazão J, Zanol G and Eiras-Dias JE, 2015. Phenolic compounds involved in grafting incompatibility of *Vitis* spp: development and validation of an analytical method for their quantification. *Phytochemical Analysis* **26**: 1–7.

Chen Z, Zhao J, Qin Y and Hu G, 2016. Study on the graft compatibility between 'Jingganghongnuo' and other litchi cultivars. *Sci Hort* **199**: 56–62.

Chen Z, Zhao J, Hu F Qin Y, Wang X and Hu G, 2017. Transcriptome changes between compatible and incompatible graft combination of *Litchi chinensis* by digital gene expression profile. *Sci Rep.* **7**: 1–12.

Ciobotari G, Brinza M, Morariu A and Gradinariu G, 2010. Graft incompatibility influence on assimilating pigments and soluble ugars amount of some pear (*Pyrus sativa*) cultivars. *Not Bot Hort Agrobot Cluj Napoca.* **38**: 187–192.

Cookson SJ, Clemente Moreno MJ, Hevin C, Nyamba Mendome LZ, Delrot S, Trossat-Magnin, C and Ollat N, 2013. Graft union formation in grapevine induces transcriptional changes related to cell wall modification, wounding, hormone signalling, and secondary metabolism. *J. Exp. Bot.* **64**: 2997–3008.

Cookson SJ, Clemente-Moreno MJ, Hevin C, Nyamba-Mendome LZ, Delrot S, Magnin N, Trossat-Magnin C and Ollat N, 2014. Heterografting with nonself rootstocks induces genes involved in stress responses at the graft interface when compared with auto-grafted controls. *J. Exp. Bot.* **65**: 2473–2481.

Copes DL, 1980a. Anatomical symptoms of graft incompatibility in *Pinus monticola* and P. *ponderosa. Silvae Genet.* **29**: 77-82.

Copes DL, 1980b. Development of internal graft incompatibility symptoms in *Pinus radiata* D. Don. *New Zealand J. For. Sci.* **10**: 367-380.

de Cooman L, Everaert E, Curir P and Dolci M, 1996. The possible role of phenolics in incompatibility expression in *Eucalyptus gunnii* micrografts. *Phytochemical Analysis* **7:** 92–96.

Deloire A and Hebant C, 1982. Peroxidase activity and lignification at the interface between stock and scion of compatible and incompatible grafts of *Capsicum* on *Lycopersicum. Ann. Bot.* **49**: 887-891.

Dettmer J, Elo A and Helariutta Y, 2009. Hormone interactions during vascular development. *Plant Mol Biol.* **69** (4): 347-360.

dos Santos Pereira I, da Silva Messias R, Diniz Campos A, Errea P and Pina A, 2014. Growth characteristics and phenylalanine ammonia-lyase activity in peach grafted on different *Prunus* spp. *Biologia Plantarum* **58**: 114–120.

Ehlers K and Kollmann R, 2001. Primary and secondary plasmodesmata: structure, origin, and functioning. *Protoplasma* **216**: 1–30. https: //doi.org/10.1007/ BF02680127.

Einarson B and Link GKK, 1976. Theophrastus de causis plantarum. Volume1. Harvard University Press, Cambridge.

Ermel FF, Poëssel JL, Faurobert M and Catessons AM, 1997. Early scion/stock junction in compatible and incompatible pear/pear and pear/quince grafts: a histocytological study. *Annals of Botany* **79**: 505–515. C q

Errea P, Felipe A and Herrero M, 1994a. Graft establishment between compatible and incompatible *Prunus* spp. *J Exp Bot.* **45**: 393–401.

Errea P, 1998. Implications of phenolic compounds in graft incompatibility in fruit tree species. *Sci Hort* **74**: 195–205.

Errea P, Gutmann M and Feucht W, 2000. Physiological implications of falvan-3-ols in apricot-rootstock combinations. *Adv Hort Sci* **14** (3): 126–134.

Errea P, Garay L and Marín JA, 2001. Early detection of graft incompatibility in apricot (*Prunus armeniaca*) using *in vitro* techniques. Physiologia Plantarum **112**: 135-141.

Errea P, Treutter D and Feucht W, 1992. Specificity of individual flavan-3-ols interfering with the grafting stress of apricots. *Angewandte Botanik* **66**: 21–24.

Errea P, Treutter D and Feucht W, 1994b. Characterization of flavanol type-polyphenols in apricot cultivar and rootstocks. *Adv Hort Sci* **8**: 165–169.

Espen L, Cocucci M and Sacchi GA, 2005. Differentiation and functional connection of vascularelements in compatible and incompatible pear/quince internode micrografts. *Tree Physio* **25**: 1419–1425.

Fan MZ, Xu CY, Xu K and Hu Y, 2012. Lateral organ boundaries domain transcription factors direct callus formation in *Arabidopsis* regeneration. *Cell Research* **22**: 1169–1180.

Fernandez-Garcia N, Carvajal M and Olmos E, 2004. Graft union formation in tomato plants: peroxidase and catalase involvement. *Ann. Bot.*, **93** (1): 53–60.

Feucht W, 1988. Graft incompatibility of tree crops: An overview of the present scientific status. *Acta Hort.* **227**: 33-41.

Flaishman MA, Loginovski K, Golobowich S and Lev-Yadun S, 2008. *Arabidopsis thaliana* as a model system for graft union development in homografts and heterografts. *J Plant Grow Regul* **27**: 231–239.

Gainza F, Opazo I, Mu˜noz C, 2015. Graft incompatibility in plants: metabolic changes during formation and establishment of the rootstock/scion union with emphasis on Prunus species. *Chil. J. Agric. Res.* **75**: 28–34. https: //doi. org/10.4067/S0718- 58392015000300004.

Gebhardt K and Feucht W, 1982. Polyphenol changes at the union of *Prunus avium/ Prunus cerasus* grafts. *J Hort Sci* **57**: 253–258.

Geibel M and Feucht W, 1991. Flavonoid 5-glucosides from *Prunus cerasus* bark and their characteristic weak glycosidic bonding. *Phytochemistry* **30**: 1519–1521.

Goldschmidt EE, 2014. Plant grafting, new mechanisms, evolutionary implications. *Front Pl Sci* **5**: 7-27.

Guclu SF, Koyuncu F, 2012. A method for prediction of graft incompatibility in sweet cherry. *Not Bot Horti Agrobot Cluj Napoca* **40**: 243–246.

Gur A, 1968. The role of the cyanogensic glycoside of the quince in the incompatibility between pear cultivars and quince rootstocks. *Hort Res* **8**: 113–134.

Habibi F, Liu T, Folta K and Sarkhosh, 2022. Physiological, biochemical and molecular aspects of grafting in fruit trees. *Hort Res* **9:** 1-18 doi: https: //doi.org/10.1093/ hr/uhac032.

Hagerman AE, 2012, Fifty years of polyphenol-protein complexes. In: Cheynier, V., Sarni- Manchado, P. and Quideau, S. (eds) *Recent Advances in Polyphenol Research*, Vol. 3. Wiley, Oxford, Pp. 71–98.

Harkin JM and Obst TR, 1973. Lignification in trees: indication of exclusive peroxidase participation. *Sci* **180**: 296–297.

Hartmann HT, Kester DE, Geneve, 1991. *Hartmann and Kester's Plant Propagation Principles and Practices, Upper Saddle River* (*Nueva Jersey, Estados Unidos*). 8th ed. Nueva Jersey, Estados Unidos, US State: Prentice Hall; 2011: 915.

Hartmann HT, DE Kester and Davies FT, 1990. Theoretical aspects of grafting and budding. p. 305-348. In: Plant propagation. 5th ed. Prentice Hall, Englewood Cliffs, NJ.

Hartmann HT, Kester DE, Davies FT and Geneve RL, 2002. Plant Propagation. Principles and Practices, seventh ed. Prentice Hall, Upper Saddle River, NJ, 849 pp. ISBN 0-13-679235-9.

He W, Wang Y, Chen Q, Sun B, Tang HR, Pan DM and Wang XR, 2018. Dissection of the Mechanism for Compatible and Incompatible Graft Combinations of *Citrus grandis* (L.) Osbeck ('Hongmian Miyou'). *Int J Mol Sci.* **19**(2): 505. doi: 10.3390/ ijms19020505. PMID: 29419732; PMCID: PMC5855727.

He F, Pan QH, Shi Y and Duan CQ, 2008. Biosynthesis and genetic regulation of proanthocyanidins in plants. *Molecules* **13**: 2674–2703.

Herrero J, 1956. La compatibiliM entre les sujets porte-greffes et les varietes fruitieres. Congr. Pomol. Inter., Namur.

Herrero J, 1951. Studies of compatible and incompatible graft combinations with special reference to hardy fruit trees. *J. Pomol. Hort. Sci.* **26**: 186-237.

Herrero J, 1955. Incompatibilidad entre patron e injerto. 1. Comportamiento de algunas combinaciones recfprocas. *An. Est. Expt. Aula Dei.* **4**: 149-166.

Herrero J and Tabuenca MC, 1969. Incompatibilidad entre patron e injerto. X. Comportamiento de la combinacion melocotonero/mirobal8.n injertado en estado cotiledonar. *Annu. Est. Expt. Aula Dei.* **10**: 937-945.

Hirose N, Takei K, Kuroha T *et al.,* 2008. Regulation of cytokinin biosynthesis, compartmentalization and translocation. *J Exp Bot.* **59**: 75–83.

Huang FH, Tasai S and Rom RC, 1984. An electrophoresis method for watersoluble protein of *Prunus. Hort Sci* **19**: 242-243.

Hudina A, Orazen MP, Jakopic J and Stampar F, 2014. The phenolic content and their involvement in the graft incompatibility process of various pear rootstocks (*Pyrus communis* L.). *J Pl Physiol* **171**: 76–84.

Ikeuchi M, Iwase A, Rymen B *et al.*, 2017. Wounding triggers callus formation via dynamic hormonal and transcriptional changes. *Plant Physiol.* **175**: 1158–1174.

Ikeuchi M, Sugimoto K and Iwase1 A, 2013. Plant callus: mechanisms of induction and repression. *Plant Cell* **25**: 3159–3173.

Irisarri P, Binczycki P, Errea P, Martens HJ and Pina A, 2015. Oxidative stress associated with rootstock–scion interactions in pear/quince combinations during early stages of graft development. *J Pl Physiol* **176**: 25–35.

Irisarri P, Errea P and Pina A, 2021. Physiological and molecular characterization of new apricot cultivars grafted on different prunus rootstocks. *Agron* **11**: 1–18. https: //doi.org/10.3390/agronomy11081464.

Janick J, 2009. Horticultural reviews V.35. New Jersey: John Wiley and Sons, Inc., Pp: 531

Jeffree CE and Yeoman MM, 1983. Development of intercellular connections between opposing cells in a graft union. *New Phytol.* **93**: 491–509.

Koepke T and Dhingra A, 2013. Rootstock scion somatogenetic interactions in perennial composite plants. *Pl Cell Rep* **32**: 1321-1337.

Kollmann R and Glockmann C, 1991. Studies on graft unions. III. On the mechanism of secondary formation of plasmodesmata at the graft interface. *Protoplasma* **165**: 71–85.

Kollmann R, Yang S and Glockmann C, 1985. Studies on graft unions. II. Continuous and half plasmodesmata in different regions of the graft interface. *Protoplasma* **126**: 19–29.

Kose C, Guleryuz M, 2006. Effects of auxins and cytokinins on graft union of grapevine (*Vitis vinifera*). *New Zea J Crop Hortic Sci.* **34**: 145–150.

Kovalishin FP, 1966. Mezchvidovye I mezchrodovye privivki hvoynyh porod. Lesohozyay stvennaya informaciya. *Moskva*: CBNTIleshos, **4**: 1-2.

Lattanzio V, Kroon PA, Quideau S and Treutter D, 2008. Plant phenolics – secondary metabolites with diverse functions. In: Daayf, F. and Lattanzio, V. (eds) *Recent Advances in Polyphenol Research*, Vol. 1. Wiley, Oxford, pp. 1–24.

Leszczynski R, Byczkowski B, Jurga S and Korszun S, 2000. NMR microimaging studies of the union between stock and scion. *Appl. Magn. Reson.*, **18**: 147–153.

Li D, Han F, Liu X *et al.*, 2021. Localized graft incompatibility in kiwifruit: analysis of homografts and heterografts with different rootstock and scion combinations. *Sci Hort* **283**: 1–7.

Lockard RG and Schneider GW, 1981. Stock and scion relationships and the dwarfing mechanism in apple. *Hort Rev* **3**: 315–375.

Loehle C and Jones R, 1990. Adaptive significance of root grafting in trees. *Functional Ecol.* **4**(2): 268-271.

Lordan J *et al.,* 2019. Horticultural performance of 'Marinada' and 'Vairo' almond cultivars grown on a genetically diverse set of rootstocks. *Sci Hort.* **256**: 1–12.

Loupit G and Cookson SJ, 2020. Identifying Molecular Markers of Successful Graft Union Formation and Compatibility. *Front Pl Sci* **11**. 610352.

Machado BD, Magro M, Rufato L *et al.,* 2017. Graft compatibility between European pear cultivars and east Malling 'C' rootstock. *Rev Bras Frutic* **39**: 1–9.

Martens HJ, Hansen M and Schulz A, 2004. Caged probes: a novel tool in studying symplasmic transport in plant tissues. *Protoplasma* **223**: 63–66.

Martinez C, Espinosa-Ruiz A and Prat S, 2016. Gibberellins and plant vegetative growth. *Annu Plant Rev.* **49**: 285–322.

Masa A, 1989. Biochemical affinity between the *Vitis vinifera* L. cv. Albarino and several rootstocks. *Connaissance de la vigne et du vin,* **23** (4): 207–214 (French, abstr. English).

McCully ME, 1983. Structural aspects of graft development. In: R. Moore (ed.). Vegetative Compatibility Responses in Plants. Baylor University Press, Waco, TX. Pp. 71-88.

Mircetich S and Rowhani A, 1984. The relationship of cherry leafroll virus and blackline disease of English walnut trees. *Phytopathology.* **74**: 423–8.

Mng'omba SA, du Toit ES and Akinnifesi FK, 2008. The relationship between graft incompatibility and phenols in *Uapaca kirkiana* Muell Arg. *Sci Hort* **117**: 212–218.

Mo Z, Feng G, Su W *et al.,* 2018. Transcriptomic analysis provides insights into grafting union development in pecan (*Carya illinoinensis*). *Genes.* **9**: 71.

Moing A, Carbonne F and Gaudillere J, 1990. Growth and carbon partitioning in compatible and incompatible peach/plum grafts. *Physiol Plant.* **79**: 540–546.

Moing A and Salesses G, 1988. Peach/plum graft incompatibility: structural, physiological and genetic aspects. *Acta Hort.* **227**: 57-62.

Moing A and Carde JP, 1988. Growth, cambial activity and phloem structure in compatible and incompatible peach/plum grafts. *Tree Physiol* **4**: 347-359.

Moore R, 1984. A model for graft compatibility incompatibility in higher plants. *Am J Bot* **71**: 751–758.

Moore R, 1983a. Physiological aspects of graft formation. In: R. Moore (ed.), Vegetative Compatibility Responses in Plants. Baylor University Press, Waco, TX. Pp. 89-105.

Moore R and Walker DB, 1981b. Studies on vegetative compatibility - incompatibility in higher plants. II. A structural study ofan incompatible heterograft between *Sedumtelephioides* (Crassulaceae) and *Solanum pennelli* (Solanaceae). *Am J Bot* **68**: 831-842.

Moore R and Walker DB, 1981a. Studies on vegetative compatibility-incompatibility in higher plants. I. A structural study of a compatible autograph in *Sedum telephoides* (Crassulaceae). *Am J Bot* **68**: 820-830.

Moreno P, Ambros S, Albiach-Marti MR *et al.,* 2008. Citrus tristeza virus: a pathogen that changed the course of the citrus industry. *Mol Plant Pathol.* **9**: 251–268.

Mosse B, 1962. Graft incompatibility in fruit trees. Tech. Comm. 28. *Comm. Bur. Hort. Plant Crops*, East Malling, England.

Mudge K, Janick J, Scofield S and Goldschmidt EE, 2009. A history of grafting. *Hort Rev.* **35**: 437–493.

Musacchi S, Pagliuca G, Kindt M *et al.,* 2000. Flavonoids asmarkers for pear-quince graft incompatibility. *J Appl Bot.* **74**: 206-211.

Nanda AK and Melnyk CW, 2018. The role of plant hormones during grafting. *J Pl Res* **131**: 49–58.

Nieminen K, Immanen J, Laxell M *et al.,* 2008. Cytokinin signaling regulates cambial development in poplar. *PNAS.* **105**: 20032–7.

Nocito FF, Espen L, Fedeli C, Lancilli C, Musacchi S, Serra S, Sansavini S, Cocucci M and Sacchi GA, 2010. Oxidative stress and senescence-like status of pear calli co-cultured on suspensions of incompatible quince microcalli. *Tree Physiol* **30**: 450–458.

Nuñez-Elisea R, Schaffer B., Fisher J *et al.,* 1999. Influence of flooding on net CO_2 assimilation, growth, and stem anatomy of *Annona* species. *Ann Bot* **74**: 771–780.

Parkinson M, CE Jeffree and Yeoman MM, 1987. Incompatibility in cultured explantgrafts between members of the Solanaceae. *New Phytol* **107**: 489-498.

Pedersen BH, 2006. Determination of graft compatibility in sweet cherry by a co-culture method. *J Hort Sci Biotech* **81**: 759–764.

Pereira IS, Pina A, Antunes LEC *et al.,* 2017. Genotypic differences in cyanogenic glycosides levels of compatible *Prunus persica P. persica* and incompatible *P. persica P. mume* combinations. *Bragantia.* **77**: 1–12.

Petkou D, Diamantidis G and Vassilakakis M, 2004. Anionic peroxidase isoform profiles from calli and barks of pear cultivars and of the quince rootstock. *Em J Biol Res* **2**: 51–55.

Pina A, Cookson SJ, Calatayud A *et al.,* 2017. Physiological and molecular mechanisms underlying graft compatibility. In: *Vegetable Grafting: Principles and Practices.* Colla, G., Perez-Alfocea, F. and Schwarz, D. eds. Croydon, London (UK): CABI, Pp; 132–154.

Pina A, Errea P, Schulz A and Martens HJ, 2009. Cell-to-cell transport through plasmodesmata in tree callus cultures. *Tree Physiol* **29**: 809–818.

Pina A and Errea P, 2005. A review of new advances in mechanism of graft compatibility–incompatibility. *Sci Hort* **106**: 1–11.

Pina A and Errea P, 2008a. Differential induction of phenylalanine ammonia-lyase gene expression in response to in vitro callus unions of *Prunus* spp. *J Pl Physiol* **165**: 705–714.

Pina A and Errea P, 2008b. Influence of graft incompatibility on gene expression and enzymatic activity of UDP-glucose pyro phosphorylase. *Pl Sci* **174**: 502–509.

Pina A, Cookson SJ, Calatayud A, Trinchera A and Errea P, 2017. Physiological and molecular mechanisms underlying graft compatibility. Vegetable Grafting: Principles and Practices. CABI, Wallingford, pp. 132–154. https: //doi. org/10.1079/9781780648972.0132.

Pina A, Errea P and Martens HJ, 2012. Graft union formation and cell-to-cell communication via plasmodesma in compatible and incompatible stem unions of *Prunus* spp. *Sci Hort* **143**: 144-150,

Pizzi A and Cameron FA, 1986. Flavonoid tannins – structural wood components for drought-resistance mechanisms of plants. *Wood Sci Technol* **20**: 119–124.

Pravdin LF and Nekrasov VI, 1953. Samoprivivki hvojnih drevesnih porod (Natural graft of woody plants). *Bot. Zhyrnal* **38**(6): 874-878.

Prodhomme D, Fonayet JV, Hevin C *et al.*, 2019. Metabolite profiling during graft union formation reveals the reprogramming of primary metabolism and the induction of stilbene synthesis at the graft interface in grapevine. *BMC Plant Biol* **19**: 1–12.

Reig G, Salazara A, Zarrouka O *et al.*, 2019. Long-term graft compatibility study of peach-almond hybrid and plum based rootstocks budded with European and Japanese plums. *Sci Hort* **243**: 392–400.

Reig G, Zarrouk O, Font i Forcada C and Moreno MA, 2018. Anatomical graft compatibility study between apricot cultivars and different plum based rootstocks. *Sci Hort.* **237**: 67–73.

Richardson FVM, tSaoir SM and Harvey BMR, 1996. A study of the graft union in *in vitro* micrografted apple. *Pl Grow Regul* **20**: 7–23.

Rodrigues AC, Machado LB, Diniz AC, Fachinello JC and de Luces Fortes GR, 2001. Evaluation of the graft compatibility in *Prunus* sp. *Rev. Bras. Frutic* **23** (2): 455-464.

Rodrigues CA, Bender-Machado L, Campos Diniz A, Fachinello JC and De Luces-Fortes GR, 2002. Peroxidases e fenóis totais em tecidos de porta-exertos de *Prunus* sp. nos períodos de crecimiento vegetativo e de donnência. *Ciencia Rural* **4**: 559–564.

Rogers WS and Beakbane AB, 1957. Stock and scion relations. *Annu Rev Pl Physiol* **8**: 217-236.

Sachs T, 1981. The control of the patterned differentiation of vascular tissues. *Adv Bot Res* **9**: 151-262.

Salesses G and Al KaL N, 1985. Simply inherited grafting incompatibility in peach. *Acta Hort* **173**: 57-62.

Santamour FS, 1988. Graft compatibility in woody plants: an expanded perspective. *J Environ Hort* **6**: 27-32.

Santamour Jr FS, McArdle AJ and Jaynes RA, 1986. Cambial iso-peroxidase patterns in Castanea. *J. Environ Hort* **4**: 14–16.

Santamour FS, 1992. Predicting graft incompatibility in woody plants. *Combined Proceedings of the International Plant Propagators Society* **42:** 131–134.

Sarooshi R, Bevington K and Coote B, 1982. Performance and compatibility of 'Muscat Gordo Blanco' grape on eight rootstocks. *Sci Hort.* **16**: 367–74.

Scaramuzzi F, 1955. Alcuni aspetti della disaffinita nel sovrinnesto degli alberi da frutto. *Riv. Ortoflorofruttic. Ital.* **39**: 523-533.

Scaramuzzi F, 1959. Alcuni aspetti dei rapporti fra virosi e disaffinita d'innesto in frutticoltura. Notiz. *Mal Piante* **47/48**: 75-102.

Schmid PPS, Schmitt ER and Zorn W, 1988. Water relations and some organic compounds in cherry leaves of grafting's with delayed incompatibility. *Acta Hort.* **227**: 90-92.

Schmid PPS and Feucht W, 1985. Compatibility in Prunus avium/Prunus cerasus grafting's during the initial phase. III. Isoelectrofocusing of proteins, peroxidases and acid phosphatases during union formation. *J Hort Sci* **60**: 311-318.

Sharma A and Zheng B, 2019. Molecular responses during plant grafting and its regulation by auxins, cytokinin's, and gibberellins. *Biomol Ther* **9**: 397.

Sharon N and Lis H, 1989. Lectins as cell recognition molecules. *Sci* **246**: 227-234.

Shimomura T and Fujihara K, 1977. Physiological study of graft union formation in Cactus. II. Role of auxin on vascular connection between stock and scion. *Int Jap Soc Hort Sci* **45**: 397-406.

Slesak I, Slesak H, Libik M and Miszalski Z, 2008. Antioxidant response system in the short-term post-wounding effect in *Mesembryanthemum crystallinum* leaves. *J Pl Physiol* **165**: 127–137.

Spiegelman Z, Ham BK, Zhang Z, Toal TW, Brady SM, Zheng Y, *et al.,* 2015. A tomato phloemÀÛÜmobile protein regulates the shootÀÛÜtoÀÛÜroot ratio by mediating the auxin response in distant organs. *Plant J.* **83**: 853–863. doi: 10.1111/tpj.12932

Sterjiades R, Dean LFD, Gamble G, Himmelsbach DS and Eriksson KEL, 1993. Extracellular laccases and peroxidases from sycamore maple (*Acer pseudoplatanus*) cell suspension cultures. Reactions with monolignols and lignin model compounds. *Planta* **190**: 75–87.

Stouffer RF and Uyemoto JK, 1976. Association of tomato ringspot virus with apple union necrosis and decline. In: *X International ymposium on Fruit Tree Virus Diseases 67*. L. Kunze ed. Heidelberg, Germany: ISHS, Pp; 203–208.

Strasburger E, 1885. Ueber verwachsungen und deren folgen. *Bot. Ges.* **3**: 34-404.

Tedesco S, Erban A, Gupta S, Kopka J, Fevereiro P, Kragler F and Pina, 2021. The impact of metabolic scion–rootstock interactions in different grapevine tissues and phloem exudates. *Metabolites* **11**: 349. https: //doi.org/10.3390/ metabo11060349.

Tiedemann R, 1989. Graft union development and symplastic phloem contact in the heterograft Cucumis sativus on *Cucurbita ficifolia. J Pl Physiol* **134**: 427–440.

Treutter D, 2006. Significance of flavonoids in plant resistance and enhancement of their biosynthesis. *Plant Biol* **7**: 581–591.

Treutter D and Feucht W, 1991. Accumulation of phenolic compounds above the graft union of cherry trees. *Gartenbauwissenschaft* **56**: 134-137.

Treutter D, Feucht W and Schmid PPS, 1986. Polyphenols of the phloem in relation to incompatibility of interspecific *Prunus* graftings (*Prunus avium* L., *Prunus cerasus* L.). 1. Flavanones and flavanols above the graft union. *Gartenbauwissenschaftg* **51**: 77–83.

Treutter DR, Galensa W Feucht and Schmid PPS, 1985. Flavonone glucosides in callus and phloem of *Prunus avium* : identification and stimulation of their synthesis. *Physiol Plant* **65**: 95-101.

Trinchera A, Pandozy G, Rinaldi S, Crinò P, Temperini O and Rea E, 2013. Graft union formation in artichoke grafting onto wild and cultivated cardoon: an anatomical study. *J Pl Physiol* **170**: 1569–1578.

Tsutsui H and Notaguchi M, 2017. The use of grafting to study systemic signaling in plants. *Plant Cell Physiol.* **58**: 1291–301.

Usenik V, Kr¡ska B, Vi¡can M and ¡Stampar F, 2006. Early detection of graft incompatibility in apricot (*Prunus armeniaca* L.) using Phenol analysis. *Sci Hort* **109**: 332–338.

Vermerris W and Nicholson R, 2006. Isolation and identification of phenolic compounds: a practical guide. In: *Phenolic Compound Biochemistry*. Springer, Dordrecht, The Netherlands, Pp: 151–196.

Wang J, Jiang L, Wu R, 2017. Plant grafting: how genetic exchange promotes vascular reconnection. *New Phytol* **214**: 56–65.

Warmund MR, Barritt BH, Brown JM Schaffer KL and Jeong BR, 1993. Detection of vascular discontinuity in bud unions of 'Jonagold' apple on 'Mark' rootstock with magnetic resonance imaging. *J Am Soc Hortic Sci.* **118**: 92–96.

Yakovleva LV, 1974. Issledovanie privivok hvoynyh porod. *Trudy botanicheskogo sada* **63**: 93-97 (in Russian).

Yeoman MM, 1984. Cellular recognition systems in grafting. In: Linkskens, H.F., Heslop-Harrison, I. (Eds.), Cellular Interaction, encyclopedia of Plant Physiology, New Series, vol. 17. Springer-Verlag, Berlin, pp. 453–472.

Yeoman MM and Brown R, 1976. Implications of the Formation of the Graft Union for Organisation in the Intact Plant. *Ann Bot* **40**: 1265–1276.

Yin H B, Yan J, Sun P, Jia Z, Zhang X *et al.,* 2012. Graft-union development: a delicate process that involves cell– cell communication between scion and stock for local auxin accumulation. *J Exp Bot* **63**: 4219-4232.

Yoda H, Hiroi Y and Sano H, 2006. Polyamine oxidase is one of the key elements for oxidative burst to induce programmed cell death in tobacco cultured cells. *Plant Physiology* **142**: 193–206.

Zarrouk O, Testillano PS, Risueño MC, Moreno MÁ and Gogorcena Y, 2010. Changes in cell/tissue organization and peroxidase activity as markers for early detection of graft incompatibility in peach/plum combinations. *J Am Soc Hort Sci* **135**: 9–17.

Zhai L, Wang X, Tang D, Qi Q, Yer H, Jiang X, Han Z, McAvoy R, Li W and Li Y, 2021. Molecular and physiological characterization of the effects of auxin-enriched rootstock on grafting. *Hort Res* **8**: 1-13.

Chapter 2

High Density Planting in Fruit Crops: Trends and Prospects

Shikha Jain, Anshuman Singh and Jai Prakash[1]*

Division of Fruits and Horticultural Technology, ICAR–Indian Agricultural Research Institute, New Delhi–110012
**Division of Crop Improvement and Biotechnology, ICAR–Central Institute for Subtropical Horticulture, Rehmankhera, Lucknow–226101, Uttar Pradesh*

India is the second largest producer of fruits in the world. The average productivity of fruits in India is, however, low as compared to many developed countries. The main reasons for low productivity are small orchards, low yielding traditional varieties, poor orchard efficiency and inadequate technological upgradation and adoption by the growers. The high-density planting technique represents a contemporary approach to fruit cultivation that involves planting trees densely, resulting in smaller or dwarf trees with modified canopies. High-density planting (HDP) involves planting fruit trees at closer spacing than the recommended standard, employing specific techniques to maximize crop yield per unit area without compromising quality. This approach has gained significant recognition over the past four decades as a key strategy in high-tech horticulture, ensuring efficient utilization of land, water, nutrients, and solar radiation while boosting production. This method aims to optimize light interception and distribution within the orchard. Additionally, various operations such as pest and disease control, weed management, and pruning are mechanized, streamlining agricultural practices. Moreover, irrigation and fertigation are automated, ensuring precise control over water and nutrient supply. The implementation of high-density planting systems yields several benefits, including precocious cropping, high and consistent fruit yields, and superior fruit quality. Furthermore, this approach reduces labor requirements, addressing the challenges posed by escalating production costs. Achieving the desired tree size and shape is facilitated through regular training, pruning, and the use of bioregulators. These technologies and techniques represent a paradigm shift in fruit production and offer significant opportunities for enhancing productivity and profitability in orchard management. Research conducted across various regions has demonstrated the efficacy of HDP in different fruit crops. For instance, in mango, planting at a density of 1600 plants/ha has shown benefits, while in guava, densities of 2222 plants/ha have been advantageous. Similarly, citrus orchards with up to 3000 plants/ha, papaya with 6400 plants/ha, banana with 4500 to 7000 plants/ha, and ultra-high density guava with 5000 plants/ha have all been found to yield higher production and returns. To maximize the benefits of high-density planting, widespread adoption of these practices is crucial. Progressive farmers play a pivotal role in this process by serving as training centers for modern fruit production techniques. By sharing their knowledge and experiences, these farmers can catalyze the adoption of high-density planting on a larger scale, driving innovation and sustainability in the fruit cultivation industry.

High-Density Planting: An Overview

In addition to diverse agro-ecological conditions that permit the cultivation of a wide range of horticultural crops, development of high-yielding cultivars and improved production technologies, and a supportive policy environment have all contributed to making India one of the significant players in global horticultural industry (Jha *et al.,* 2019; Ramappa *et al.,* 2015; Ravindran *et al.,* 2007; Reddy *et al.,* 2017; Singh *et al.,* 2016; Sankaran and Dinesh, 2020). Given its tremendous potential in improving the farm incomes and rural livelihoods, creating employment, and earning the foreign-exchange through exports, horticulture sector is increasingly being recognized a sunrise industry in India (Jha *et al.,* 2019; Mitra, 2014). Even though India continues to occupy a dominant position in global fruit production- it is the world's second-largest producer of fruits (Sharma, 2021)- poor productivity, particularly when compared to several other fruit-producing countries, remains a cause for concern (Goyal *et al.,* 2020). Numerous factors contribute to poor fruit yields in India, such as the prevalence of low-density, senile orchards, small orchard holdings, cultivation of traditional varieties that produce low yields, poor orchard management, a lack of technical knowledge among the fruit growers, and limited access to improved cultivars and production technologies (Chaudhari and Singh, 2019; Shah *et al.,* 2022; Singh *et al.,* 2014; Upadhyay *et al.,* 2018).

While the adoption of improved cultivars and crop management practices has undoubtedly increased over the past few decades (Jaiswal *et al.,* 2015; Mehta *et al.,* 2016; Yadav *et al.,* 2013), concerted efforts are still needed in order to make Indian fruit industry globally competitive in terms of fruit yield and quality, processing and value-addition, and export promotion (Raman *et al.,* 2023; Singh, 2019) while preserving the agro-biodiversity and natural resources (Chaudhari and Singh, 2019). A holistic approach is required to create fruit-based value chains that are efficient and sustainable, rather than only concentrating on production (Gulati *et al.,* 2022). Since commercialization can help India realize its full potential as a major player in the global fruit market, policy changes are necessary to support public-private partnerships, develop the infrastructure needed, and secure financial and technical resources to build on the initial successes (Mysore, 2015). In addition to augmenting fruit productivity and quality, this transition may also contribute to socio-economic development, and improve country's position on the global agricultural landscape (Li *et al.,* 2020; Singh, 2013).

High-density planting (HDP) of horticultural crops has gained traction as a crucial strategy in addressing a range of problems that frequently cause appreciable reductions in crop yields and farmers' incomes (Menzel and Lagadec, 2014). Factors such as the declining availability of arable land, escalating energy and production costs, and the growing demand for horticultural produce have fueled the adoption of HDP worldwide (Sharma *et al.,* 2009). The main goal of HDP is to maximize plant density in a given space without jeopardizing the soil health and crop productivity (Kumar, 2019). HDP seeks to minimize the gestation period, boost the yield per unit volume of trees, and improve fruit quality by optimizing the plant density (Mishra and Goswami, 2016). The early regulation of excessive vegetative growth, for instance, by using the dwarfing rootstocks, is one of the major principles underlying HDP (Mehta *et al.,* 2016). Overcrowding may occur when the fruit trees grow excessively large, causing problems such as excessive shading, reduced photosynthesis and poor fruit yields (Kumar, 2019). In order to overcome these difficulties, various approaches of high density orcharding including the techniques for ultra HDP (*e.g.*, meadow orcharding in guava) have been developed (Lal *et al.,* 2007). By using dwarf scion and rootstock cultivars in conjunction with regular topping, trimming, and hedging, such planting methods frequently significantly increase light penetration and distribution within the canopy, resulting in higher yields per unit area (Menzel and Lagadec, 2014).

The concept of HDP originated in Europe in the early 1960s, and was initially applied to apple cultivation (Stephan *et al.,* 2008). Since then, it has been widely accepted as a fundamental component of modern horticulture. The HDP seeks to maximize the input and resource use efficiencies by making the effective use of horizontal and vertical spaces (Yadav *et al.,* 2013). The role that HDP plays in raising the fruit yield has gained considerable traction over the past few decades. It is currently one of the most effective components in high-tech horticulture, helping to maximize yields by enabling the effective use of land, water, nutrients, and solar radiation (Jaiswal *et al.,* 2015). Additionally, HDP is efficient at reducing gestation periods, ensuring early returns to the fruit growers (Lauri *et al.,* 2006). Currently, HDP is one of the basic elements in the intensive production of apples and other temperate fruits in countries and regions such as Europe, North America, China,

Japan and Australia (Tripathi *et al.*, 2020). In India, high-density planting has been successfully demonstrated in papaya (Ram, 1996), guava (Lal *et al.*, 2007), litchi (Mishra *et al.*, 2014), and mango (Gunjate *et al.*, 2009).

Principles of High Density Planting

The basic underlying principles of high-density planting (HDP), which aim to maximize the fruit yield per unit area while maintaining soil fertility and ensuring sustainable resource use, are as follows:

Maximizing Plant Density

HDP seeks to maximize plant density per unit area while maintaining soil fertility and plant health. Making optimal use of both vertical and horizontal space per unit of time is the fundamental tenet of HDP. Productivity increases as a result, and resource use is optimized. Planting of more number of plants per unit area than in conventional orchards is intended to ensure the optimal use of soils, solar radiation, and crop inputs (Ladaniya *et al.*, 2020; Majid *et al.*, 2018). Standard apple plants on seedling rootstocks are planted at spacing of 6 x 6 m or 7 x 7 m, accommodating between 204 and 278 trees per hectare. Comparably, spur types on seedling rootstocks are planted at a reduced spacing (5.0 × 5.0 m), resulting in relatively higher plant density (400 plants/ha). On the other hand, when apples on dwarfing or semi-dwarfing clonal rootstocks are planted using high-density systems, planting density increases to about 3333 per hectare. Low-density apple orchards (5 x 5 m, 6 x 6 m or 7 x 7 m) produce about 10 to 12 tones fruits per hectare, which is far less than the 60–80 tones/ha produced in high-density orchards (Mir *et al.*, 2022).

Controlling Vegetative Growth

Controlling excessive vegetative growth is essential to avoiding overcrowding and maintaining the ideal canopy structure. Pruning and dwarfing rootstocks are two methods used to regulate tree size and increase fruit yield in HDP orchards (Costes and García-villanueva, 2007). According to the results of a 21-year trial, using nucellus 'Vellaikolmban' rootstock produced the highest fruit yields per unit of canopy volume and per unit land area in Alphonso scions, along with a noticeable dwarfing impact (Reddy *et al.*, 2003). When compared to traditional rootstocks such as Rangpur lime and Sunki mandarin, hybrid rootstocks (*e.g.*, TSKC × CTSW-041, LCR × TR-001, HTR-051, HTR-053 and HTR-069) resulted in greater planting densities, higher production efficiency, and higher or comparable fruit quality in Valencia orange (Yuri *et al.*, 2015). While adoption of dwarfing rootstocks has played a crucial role in the success of HDP in temperate fruit crops, such rootstocks are either mostly unavailable or not commercially used in the majority of tropical fruit crops (Mitra, 2019).

Efficient Use of Resources

Improved techniques of canopy management, irrigation and nutrient management are applied in order to ensure the efficient use of resources such as land, water, nutrients, and solar radiation (Tripathi *et al.,* 2020). Higher planting densities frequently result in more efficient use of irrigation water, chemical fertilizers, and other inputs. Olive orchards are currently planted at densities ranging from high (350-700 trees/ha) to ultra-high (1200-2500 trees/ha). Trees in such high-density orchards are trained, fertilized, and irrigated in ways that increase productivity and profitability while ensuring ease in pruning and mechanical harvesting. Generally speaking, high-density olive orchards exhibit increased root biomass and density, reduced evaporation and drainage losses, and higher yields, which in turn imply higher water use efficiency (Mairech *et al.,* 2020).

Optimal Spacing and Arrangement

Due care in layout and plant spacing is absolutely essential to ensure adequate light penetration, air circulation, and optimum resource use. Plant spacing may vary depending on the crop and its growth habit (Usha *et al.,* 2015). Efficient orchard designs increase the light interception, make better use of available space and other resources, and reduce the competition between trees for water and nutrients. Tree density, rootstock, training system, row direction and orchard layout are some of the critical factors that must be considered while designing new orchards (Javaid *et al.,* 2017). For instance, in hedgerow olive orchards, optimal combinations of row height, row width, canopy slope, and alley width that maximize both the interception of incident solar radiation and its distribution over the canopy surface are essential to facilitating the use of machines and improving the yields (Connor, 2006).

Regular Maintenance and Management

Continuous monitoring, trimming, and management techniques help to maintain the desired canopy structure, avoid crowding, and boost fruit production. Tree size and shape can be managed by using methods such as topping, hedging, and dehorning (Mishra and Goswami, 2016). Maintaining the size and shape of the trees in high-density mango orchards requires frequent pruning and the application of growth regulators (Javid *et al.,* 2017). A comparative study in Dharwad district of Karnataka, India revealed that while total annual maintenance cost was considerably higher for high-density (INR 48,132/ha) than for conventional low-density (INR 21,783/ha) mango orchards, the former were still attractive because of their easier management and better profits (Kerutagi and Deshetti, 2018).

Types of Planting Density

Low-Density Planting

This is the conventional, non-intensive system with widely spaced trees, accommodating between 100 and 250 trees per hectare. In India, low-density planting are quite common in crops such as mango. Because canopy management is seldom paid any attention and the management practices mostly aim to encourage the vegetative growth, producers with limited resources find it appealing (Ram *et al.*, 1988; 1991).

Medium-Density Planting

Planting density is moderately decreased in this system of planting to accommodate relatively more number of trees per unit area, ranging between 250 and 500 plants per hectare. Pruning is applied to achieve the desired tree shape and higher yields. This system, based on intensive care and input use, is commonly adopted in fruit crops like pomegranate, citrus, guava, mango, papaya, and banana (Ram, 1996; Lal *et al.*, 2007; Gunjate *et al.*, 2009).

High-Density Planting

Depending on fruit crop, this planting strategy consists of extremely dense spacing with 500–10,000 plants/ha. Its success is largely dependent on intensive training and pruning, use of dwarfing rootstocks, and the application of growth-controlling agro-chemicals (Tripathi *et al.*, 2020). Fruit yields are higher, but production costs per unit area are likewise substantially greater than conventional low-density orchards. Recently, super high density planting systems have also been demonstrated in apple with plant density as high as 20,000 trees per hectare (Usha *et al.*, 2015).

Status and Prospects of High-Density Planting

There has been considerable discussion regarding the potential advantages of high-density orcharding in fruit crops, although evidence of its long-term economic benefits remains limited (Ram *et al.*, 1988; 1991; 2001). Some studies have reported declines in yield after several years due to tree overcrowding. However, high-density orchards have shown the potential to provide significantly higher yields compared to traditional densities. For instance, Ram *et al.* (2001) demonstrated that high-density orchards could yield 8-9 times more than traditional densities, as seen in the case of alternate bearing Dashehari mango in North India. In this study, Dashehari mango trees were planted at a spacing of 2.5 m x 3 m (1,333 plants/ha) and managed under high-density practices, including pruning, dehorning, and paclobutrazol application after harvesting, resulting in consistent annual yields. Additionally, experiments conducted at the Central Institute for Subtropical Horticulture (CISH) in Lucknow standardized an ultra-high-density system for guava cultivation. This system accommodated 5000 plants/ha at a spacing of 2.0 m x 1.0 m and was managed through regular topping and hedging during the initial stages to control tree size and enhance yield. Similarly, Pusa Nanha papaya was found to thrive in high-density orchards, with recommended planting distances of 1.25 m x 1.25 m (6,400 plants/ha). Similar observations of increased productivity in high-density orchards have been reported by various researchers in citrus, litchi, banana, and pineapple cultivation.

Advantages of High Density Planting

High-density orcharding offers several advantages when compared to the conventional low-density orchards. First and foremost, HDP induces precocity (*i.e.*, early fruiting and maturation), resulting in quicker returns on investment compared to traditional planting methods (Musacchi *et al.*, 2011). Growers may find it financially prohibitive to plant a larger number of trees per unit area, but the greater precocity and early yields which these systems provide could quickly offset these expenditures (Anthony and Minas, 2021). HDP dramatically increases the fruit yield by optimizing the plant density per unit area. In addition to reducing the gestation period for early returns and enhancing the productivity, HDP improves the fruit quality without any adverse effects on soil and tree health (Subedi *et al.*, 2020). Furthermore, relatively controlled environmental conditions and efficient resource management in high density orchards frequently lead to better fruit quality, including increased fruit size, color, flavor, and nutritional value. With early cropping and increased yields alongside better fruit quality and reduced labor costs, HDP increases income per unit area (Tripathi *et al.*, 2020). Pruning the shoots to a length of 20 cm from the shoot tip and soil drenching of paclobutrazol in June resulted in the highest fruit yield in mango cultivar Ratna (Manohar *et al.*, 2023). HDP can result in lower expenses per unit of output in the long run. While high-density systems may demand greater focus and expertise, they are ideally suited for automation, which lowers the labor and production costs (Anthony and Minas, 2021).

Problems in Management

High-density planting offers several advantages; it also comes with challenges that require careful management and consideration to ensure sustainable productivity and fruit quality in the long run. It includes:

Competition for Resources

Over time, high-density orchards may experience heavy competition for space, nutrients, and water among densely planted trees. This can lead to decreased overall productivity and vigor of the orchard. Poor maintenance and inadequate or incorrect pruning can cause overgrowth at the top of the trees (Reginato *et al.*, 2008). Careful pruning is a crucial in high-density mango orchards given the tree's propensity for recurrent flushes, especially in humid tropical climates. A swift recovery of orchard productivity can be achieved by severely cutting large, unproductive trees and then trimming their tips (Davenport, 2006). An ideal orchard design would prioritize maximum yields while managing the orchard in ways that harvester equipment and other vehicles move through with ease. It is still unclear, nevertheless, if densities exceeding a particular level can cause competition among mature trees, jeopardizing their growth and yield and ultimately shortening the orchard's lifespan (Díez *et al.*, 2016).

Insect-Pest and Disease Management

Notwithstanding a growing worldwide interest in high-density plantings, little is known about pest and disease dynamics in high-density production systems as compared to conventional systems. Considering the notable structural variations between high-density plantings and conventional orchards, it is probable that insect abundance and disease prevalence will vary dramatically between these two systems (Lindell *et al.,* 2023). The prevalence of insect-pests and their natural enemies is expected to differ between conventional and high-density orchards on account of differences in their micro-climate (Khursheed *et al.,* 2021). The overcrowded growth of canopies in high-density orchards can create congenial conditions for pests and diseases. Reduced air circulation and increased humidity within the orchard may lead to higher incidences of pest and disease outbreaks. Tree top in overgrown trees may be beyond the reach of sprayers, such that poorly sprayed tops, leaves and fruits may act as a refuge for insect-pests, and pathogens. The overgrown trees in such orchards may also be prone to increased infestation of particular pests.

Declines in Fruit Yield and Quality

Although there is a general positive correlation between tree planting density and yield, some findings indicate that yield does not rise in direct proportion to tree density. Moreover, the incremental yield value decreases with each new tree planted (Lordan *et al.,* 2019). As trees in close plantings begin to crowd each other, yields may decline over time due to resource competition and limited access to sunlight and nutrients for each tree. This has been shown in case of mango: productivity declines as soon as trees grown in close proximity start to crowd and shade one another (Menzel and Le Lagadec, 2017). Excessive vegetative vigor reduces flowering and fruiting and makes it difficult to keep the tree contained in its allotted space. In high-density apple orchards, containment pruning of the canopy is necessary to keep trees in their designated spaces and to ensure that the lower branches of the trees receive enough light exposure (Robinson, 2007).

Challenges with Dwarfing Rootstocks

Efforts to breed dwarfing rootstocks that effectively control the growth of mature trees have been largely unsuccessful, limiting the potential of HDP to manage tree size and maintain productivity.

Canopy management challenges: Effective canopy management is essential for the success of high-density orchards. However, achieving optimal canopy structure and balance can be challenging, requiring careful and ongoing management practices.

Use of growth regulators: Growth retardants are often used to restrict growth in high-density orchards, but there are concerns about their persistence in harvested fruit and soil, potentially impacting fruit quality and environmental safety (Menzel and Lagadec, 2014).

Components of High Density Planting

High density planting (HDP) represents a modern approach to orchard management that aims to maximize yield and quality through strategic utilization of various components. These components include dwarf scion varieties, dwarfing rootstocks and inter-stocks, training and pruning techniques, the use of growth regulators, and suitable crop management practices (Subedi *et al.*, 2019).

Use of Genetically Dwarf Scion Varieties

Establishing high-density orchards become significantly easier when dealing with naturally small trees or utilizing genetically dwarf cultivars. These cultivars offer immense potential for close plantings, which can lead to higher yields and returns compared to traditional orchard layouts (Mishra *et al.*, 2003). However, despite the clear benefits, the availability of dwarfing scion cultivars remains limited as indicated below in Table 2.1. This scarcity poses a significant hurdle for growers looking to implement high-density planting systems. Without access to suitable dwarfing cultivars, achieving the desired tree size and canopy management becomes challenging. By expanding the range of options available to growers, the horticultural industry can unlock the full potential of high-density planting systems to enhance orchard productivity and profitability.

Table 2.1: Genetically Dwarf Scion Cultivars in Fruit Crops

Fruit Crop	*Varieties*	*Desirable Features*
Mango	Amrapali, Arunika, Mallika	Precocious and regular bearer
Papaya	Pusa Nanha, Pusa Dwarf	Dwarf stature
Persimmon	Jiro, Ishkei	Semi-dwarf
Guava	Lalit, Pant Prabhat	Less spreading, high yielder
Litchi	Calcuttia, China	Upright tree growth habit
	Gandaki Yogita	Dwarf and compact
Sapota	PKM-1, PKM-3	Dwarf stature
Banana	Dwarf Cavendish	High yield with dwarf stature
Peach	Red Heaven, Candor	High yield with dwarf stature
Apple	Spur varieties like Red Chief, Oregon spur, Red spur, Stark spur gold, Well spur, Starkrimson delicious	Grow to 60-70 per cent of standard cultivars in vigour and bear more spurs and yields more
Ber	Goma Yashi	Semi dwarf and suitable for dryland areas
Cherry	Compact Lambert, Meteor, North Star	High yielding, self-fruitful, dwarf

Use of Dwarfing Rootstocks and Inter-stocks

The advent and widespread adoption of size-controlling rootstock genotypes have been critical to the implementation of compact orchard systems and precision canopy management strategies that are now common in the cultivation of certain tree fruit crops. These rootstocks enable growers to maintain smaller, more manageable trees, which are essential for high-density planting. In the case of pear and plum trees, options for size-controlling rootstocks are relatively limited, posing challenges for growers looking to adopt high-density planting systems. Conversely, peach growers benefit from several promising size-controlling rootstock options. These rootstocks help to control tree vigor, allowing for the establishment of compact orchards that are conducive to intensive cultivation practices (Layne, 1987; Reighard, 2002). Rootstocks play a pivotal role in shaping various aspects of tree growth and fruit production, including vigor, precocity, productivity, fruit quality, and longevity of the grafted varieties, thereby influencing the size and shape of the canopy of the scion grafted onto them (Rom and Carlson, 1987; Lauri *et al.*, 2006; Costes and Garcia, 2007). By selecting appropriate rootstocks, growers can maintain a shallow canopy or thin hedgerow structure that allows for adequate illumination throughout the canopy's volume. This is particularly important for maximizing photosynthesis and fruit quality. In temperate fruit cultivation, a wide range of vigor-controlling rootstocks are available. These rootstocks enable growers to accommodate a large number of plants per unit area, leading to increased interception of solar energy and ultimately resulting in heavier yields and better fruit quality. By selecting rootstocks that match the desired canopy vigor and growth habit of the scion variety, growers can optimize orchard productivity and efficiency.

Efforts have been made to standardize dwarfing rootstocks, particularly in fruit crops such as ber, citrus, and grape. However, the range of available options is limited for many tropical and subtropical fruit crops compared to temperate fruits. In some cases, the use of graft-incompatible scion and rootstock combinations can induce dwarfness in the composite plant. For example, in the case of ber, grafting scion cultivars onto rootstocks such as *Zizyphus rotundifolia* or *Z. nummularia* may lead to dwarfness due to graft incompatibility. This phenomenon underscores the importance of selecting compatible graft combinations to achieve desired growth characteristics and optimize orchard performance. When apple cultivars are grafted on dwarfing rootstocks M 9 and M 27 flowered more frequently than those on M 793 (Foster *et al.*, 2016). Mango cv. Pusa Surya grafted on Olour and Kurakkan rootstocks produce dwarf trees. Rootstock K-5 and K-2 induced dwarfness in scion of mango cvs. Pusa Surya and Pusa Arunima (Dubey *et al.*, 2021). Due to its high yield efficiency and fruit quality, the hybrid HTR-051 rootstock promotes dwarfism in citrus species and appears to offer good prospects for high density orchards (de Carvalho *et al.*, 2019). Despite the challenges associated with limited options, ongoing research and breeding efforts are focused on developing new dwarfing rootstocks tailored to the specific requirements of tropical and subtropical fruit crops (Mishra *et al.*, 2003). By expanding the range of available rootstock options and enhancing our understanding of graft compatibility, growers can harness the potential of dwarfing rootstocks to optimize orchard productivity and sustainability in these regions. Collaborative efforts among researchers, nurseries, and growers are essential for evaluating and disseminating promising rootstock varieties and promoting their adoption in commercial orchards.

Table 2.2: Dwarfing Rootstocks and Inter-Stocks in Fruit Crops

Fruit Crop	*Varieties*	*Special Features*	*Reference*
Mango	Vellaikolamban	Suitable for Alphanso and Dashehari	Mukherjee and Das, 1980
	Olour	For Langra and Himsagar	Singh *et al.*, 1986
	K-2 and K-5	For Pusa Surya and Pusa Arunima	Dubey *et al.*, 2021
Guava	Pusa Srijan	Aneuploid rootstock, Tetrasomic (2n=24), Dwarfing effect on Allahabad Safeda	Sharma *et al.*, 1992
	P. friedrischathalianum, P. pumilum	Dwarfing	Edward and Shankar, 1967
Citrus	Flying Dragon	Ultra-dwarfing rootstock	Ashkenazi *et al.*, 1992
	Troyer Citrange	Dwarfing	Castle *et al.*, 1987
Apricot	Hybrid P-2308, *Prunus beyesiyi*	Dwarfing	Singh and Guptha, 1971
Ber	*Z. nummularia*	Prone to inverted bottleneck disorder	Vashishtha,1987
Pear	Quince C, Oregon-211,249	Dwarfing	-
Plum	Pixy, St. Julien 2,3	Dwarfing	Singh and Guptha, 1971
Apple	M-9, B-9	Dwarfing	Rana, 1979
	M-27	Ultra-dwarf	Pathak *et al.*, 1976
	M. baccata var. himalaica	Dwarf growth habit	
Cherry	Gisela 10, Colt, Charger, Rubira	Dwarfing	-

Use of Pruning to Control Tree Growth and Vigor

Overcrowding within orchards presents a significant challenge, impacting access for management practices and reducing light interception crucial for optimal photosynthesis, flowering, and fruit set. Various researchers, including Sharma and Singh (2006), Mishra and Lal (2013), and Nath *et al.* (2007), have explored the use of pruning as a means of canopy management in high-density orchards. Tree size control through pruning is limited to grape, apple and some temperate crops. In apple cv. Topaz grafted on M9 rootstock, trees trained as modified slender spindle had lower canopy volume compared to slender spindle and demonstrated better tree size control (Meszaros *et al.*, 2017).

Table 2.3: Influence of Rootstock on the Vegetative Growth, Fruit Quality and Mineral Uptake of Grafted Plants

Fruit Crop	*Rootstocks*	*Scion cultivar*	*Traits*	*References*
Apple	M.9, M.26, Chistock-1 and Baleng	'Red Fuji'	Dwarfing traits, scion morphology, and mineral concentration	Hayat *et al.*, 2019
Plums	Marianna GF8-1 and Marianna GF10-2 (*Prunus cerasifera* × *munsoniana*)	Rainha Claudia Verde	Tree vigour, fruit quality and fruit mineral concentration	Luo *et al.*, 2014
Citrus	Rough lemon, Rangpur lime, Troyer citrange, Jatti Khatti and sour orange	Kinnow mandarin	Root morphology, mineral uptake and nutrient absorption capacities	Ghanem *et al.*, 2011

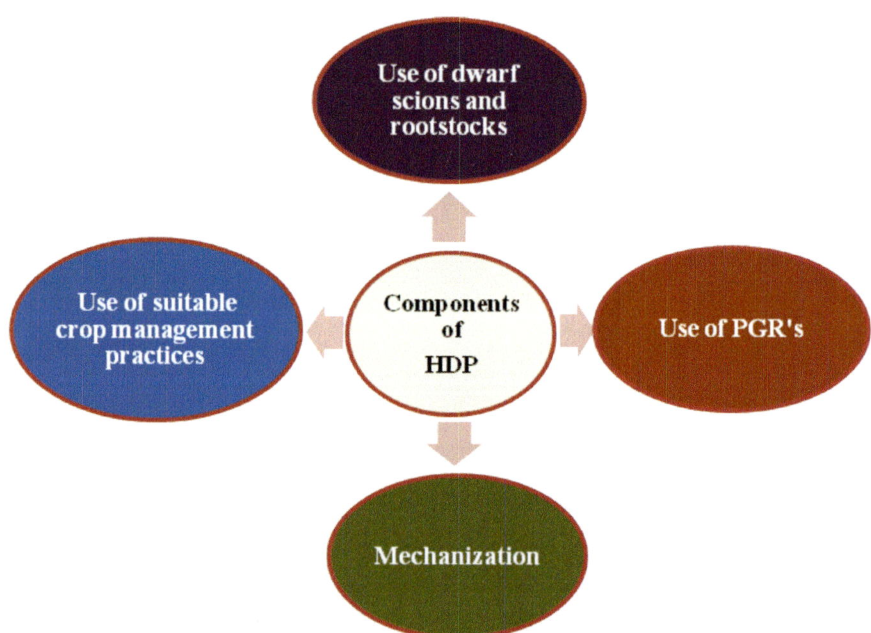

Figure 2.1: Different Components of High-Density Orchards.

Trees of cherry cv. 0900 Ziraat grafted on Gisela 5 rootstock and trained as spindle bush had lower tree height (Aglar *et al.,* 2016). Studies, such as those conducted on Dushehari mango at Pant Nagar, have highlighted the importance of pruning in maintaining vigor and promoting new shoot production, which is essential for panicle emergence. Pruning serves to increase the distribution of light throughout the canopy, thus enhancing overall photosynthetic efficiency and promoting uniform fruit development. However, it's important to note that while pruning can improve light penetration, productivity may not always be immediately restored within the same season. Crane *et al.* (1992) observed that heavy pruning of 34-year-old avocado trees led to a temporary inhibition of production for the subsequent two to three years. This underscores the need for careful consideration of pruning strategies, taking into account both short-term and long-term impacts on orchard productivity.

Despite potential setbacks, pruning remains a valuable tool for managing canopy density and promoting orchard health and productivity in high-density planting systems. Continued research and experimentation are essential for refining pruning techniques and optimizing their effectiveness within the context of specific fruit crops and orchard environments. Additionally, growers must consider factors such as tree age, growth habit, and seasonal variations in order to implement pruning practices that maximize both short-term and long-term benefits. Mango trees have a natural tendency to grow into large specimens, often reaching heights of 10 meters or more. When planted closely together, these trees tend to intertwine and shade large sections of the lower canopy, leading to a decline in productivity, typically around ten years after establishment (Sharma *et al.,* 2006). To address this challenge and facilitate the development of high-density plantings in mango orchards, effective strategies for controlling tree growth are essential. Several studies, including those by Oosthuyse (1995) and Yeshitela *et al.* (2005), have investigated the impact of pruning on mango tree physiology, growth, and yield. However, only a limited number of these studies directly address the sustainability of high-density plantings in mango orchards.

To ensure the sustainability of high-density mango plantings, pruning strategies must be carefully tailored to manage canopy density and promote light penetration while maintaining tree health and productivity. This may involve selective thinning of branches to open up the canopy, as well as strategic pruning to encourage the development of new growth and optimize fruiting potential (Sharma and Singh, 2006). Pruning intensity varies depending on the specific fruit crop and its growth habits. In some fruits like guava, Kinnow and litchi, light pruning is typically practiced. This involves minimal removal of branches to maintain tree shape, promote airflow, and encourage new growth without significantly impacting fruit production. In contrast, fruits like phalsa and ber require more severe pruning practices. Ber trees, for example, are pruned extensively during the summer months after the harvest when the trees shed their leaves. This severe pruning helps control tree size, rejuvenate growth, and promote the development of new flowering and fruiting wood for the following season. By tailoring pruning intensity to the specific needs of each fruit crop, growers can effectively manage canopy density, optimize light interception, and promote healthy growth and fruit production. However, it's essential to consider factors such as tree age, vigor, and growth habit when determining the appropriate pruning approach to ensure the long-term sustainability and productivity of orchards.

Use of Plant Growth Regulators to Control Tree Size and Vigor

Pruning often stimulates vigorous re-growth of shoots in mango and other fruit crops, which can lead to excessive vegetative growth and potentially hinder fruiting. To manage this issue and promote balanced growth, plant growth regulators (PGRs) such as Paclobutrazol, Alar, Uniconazole, and prohexadione-calcium have been utilized. Historically, daminozide has been most successfully applied in apples for growth control. However, relatively high levels of residues with potentially critical toxicological features have led to a general ban of this compound for use in edible crops. In mango orchards at Pant Nagar, Paclobutrazol treatments have shown promising results in inducing flowering and fruiting in new shoots produced after pruning, particularly when applied between September and November. These treatments not only promoted flowering and fruiting but also effectively reduced vegetative growth by 30-35 per cent, contributing to the suitability of high-density orcharding practices. Moreover, Paclobutrazol treatments did not compromise fruit quality, indicating their potential as a management tool for mango orchards. In contrast, Uniconazole was found to be more effective than Paclobutrazol in restricting shoot growth in avocado orchards in South Africa, as reported by Kohne and Kremer-kohne (1989). Similarly, prohexadione-calcium sprays on five-year-old Hass avocado trees demonstrated the ability to reduce the growth of spring shoots without adversely affecting yield, as observed by Menzel and Lagadec (2014).

Suitable Crop Management Practices

In the context of high density planting (HDP), the adoption of suitable crop management practices plays a pivotal role in nurturing and sustaining plant health. Mulching, for instance, serves as a protective layer over the soil, conserving moisture and suppressing weed growth, thereby promoting optimal plant growth and development. Fertigation, another essential practice, involves the precise application of fertilizers through irrigation systems, ensuring that plants receive adequate nutrients for robust growth. Organic farming techniques further enhance soil fertility and ecosystem health by minimizing the use of synthetic chemicals and promoting natural nutrient cycles. Integrated Nutrient Management (INM) integrates various sources of nutrients, including organic manures, biofertilizers, and chemical fertilizers, to optimize nutrient availability to plants while minimizing environmental impacts. This approach ensures a balanced supply of essential nutrients for plant growth, enhancing productivity sustainably. Integrated Pest Management (IPM) is equally crucial in HDP, focusing on the holistic management of pests by employing a combination of techniques such as biological control, cultural practices, and judicious use of pesticides. By minimizing reliance on chemical pesticides and emphasizing natural pest control mechanisms, IPM helps maintain ecological balance while effectively managing pest populations. The adoption of these crop management practices–mulching, fertigation, organic farming, INM, and IPM–within HDP is imperative for ensuring the care and maintenance of plants, optimizing productivity, and promoting sustainable agricultural practices (Mishra *et al.,* 2003).

Mechanization

Mechanization plays a pivotal role in the success of high-density planting in fruit crops, offering numerous advantages that enhance productivity and efficiency. The adoption of specialized machinery for planting, pruning, harvesting, and maintenance operations is essential to manage the closer spacing and increased plant density effectively. For planting, mechanized transplanters can be used to ensure uniform spacing and depth, which is crucial for the establishment of high-density orchards. These machines can significantly reduce labor costs and time, ensuring precise placement of each tree. Pruning, a vital practice in high-density orchards to manage tree size and optimize light penetration, can be efficiently performed using mechanical pruners or hedgers. The mechanization of pruning began in California in the early 1960s through an effort to mechanically top lemon trees (Jutras and Kretchman, 1962). Mechanical toppers were used to eliminate hand pruning of vigorous shoots at the top of the tree. The machine consisted of a modified sickle-bar mower blade mounted on towers, which were adjustable for height. In later years, topping machines used a series of circular saws mounted on a horizontal boom (Sansavini, 1978).

Mechanical pruning is based on a predetermined cutting plan: horizontal top cutting (topping) and vertical walls, or oblique hedging (house top). Consequently, hedging is normally surface pruning. However, for certain crops such as citrus, deep cuttings on alternating sides of the tree on an every-other-year basis is often employed. Robotic pruning, which is a selective pruning with accurate cuts, typically cuts the branch by an end-effector consisting of a cutting blade and anvil that uses a scissors motion. During the harvesting stage, mechanical harvesters are indispensable in high-density orchards. Maintenance operations, such as spraying and weeding, also benefit from mechanization. Specialized sprayers ensure even and targeted application of pesticides and nutrients, while mechanical weeders help manage weeds effectively without damaging the closely planted trees (Saroj and Singh, 2018). The high labor requirements of traditional tree fruit production systems are driving significant innovation in orchard management, with an urgent focus on harvesting, which often constitutes more than 50 per cent of annual production costs. In many regions, the cost of skilled harvest labor is rising while its availability is decreasing. This creates an imminent need to improve the harvest efficiency of orchard crops. Consequently, future orchard systems must integrate current and potential harvest technologies. Sweet cherry production, in particular, faces substantial challenges due to the crop's labor-intensive nature, driven by large tree sizes and high fruit counts per tree. Recent advancements have been made towards mechanizing the harvest of sweet cherries, which is critical for addressing labor shortages and reducing costs in this demanding agricultural sector. (Peterson *et al.,* 2003; Larbi *et al.,* 2015).

Utilization of New Planting Systems

The utilization of new planting systems holds immense potential for enhancing the productivity and quality of tropical and subtropical fruit crops. Despite the lack of systematic research in this area, the importance of early production of high-quality fruit for long-term profitability in orchards cannot be overstated. In the absence of such research, the standardization of planting systems becomes imperative, particularly following high-density principles to accommodate a greater number of plants per unit area, thereby maximizing productivity without compromising fruit quality. High-density planting systems have demonstrated remarkable success for earliness, improved yield, smooth handling and cultural practices using double hedge row system of planting in litchi (Mishra *et al.,* 2014), mango (Singh *et al.,* 2012, Nath *et al.,* 2007), aonla (Singh *et al.,* 2011) and guava (Lal *et al.,* 2007, Singh *et al.,* 2007). Moreover, by manipulating plant spacing using different planting systems like hedge row, double hedge row, paired planting and cluster planting proved to be an important tool to achieve high quality produce and productivity.

Opportunities for Robotics and Automation

The motivation to adopt mechanization and automation technologies in fruit production is driven primarily by the need to enhance labor productivity and address issues related to labor cost and availability. Additional factors such as improvements in cultivars and varieties, fruit quality and safety, disease and pest pressures, environmental concerns and regulations, and global market pressures also play significant roles. Over the past several decades, many tree fruit producers have actively promoted the development of mechanized and automated solutions for various production tasks, including harvesting. However, successful mechanized harvesting has largely been confined to processed fruit applications, where minor fruit damage is acceptable since the fruit is typically processed within 24 hours of harvest. The development of compact fruiting wall architectures has been particularly advantageous for mechanization, facilitating the use of platforms for efficient harvest. A Y-trellised configuration, in particular, is crucial for achieving economical mechanical harvesting with minimal fruit damage. In apple orchards, recent advancements include both the massive harvest approach, such as the shake-and-catch method developed by DeKleine and Karkee (2015), as well as the development of robotic harvest systems. While the shake-and-catch method can be used in three-dimensional (3D) orchard architectures, early trials of robotic harvest systems have highlighted the need for compact plant architectures to ensure high visibility of the fruit, as demonstrated by Silwal *et al.* (2016).

Over the last decade, numerous studies have aimed to develop advanced crop status monitoring techniques using a variety of technologies, including machine vision in the visible and near-infrared (NIR) spectral regions, ladar, ultrasonics, and, more recently, biosensors. These sensor technologies are employed to monitor critical crop factors such as yield estimation, canopy size, canopy volume, leaf density, diseases and pests pressures. For instance, Campoy *et al.* (2010) developed a tree canopy modeling and precision spraying system utilizing laser sensors and 3D maps. Chen *et al.* (2011) developed a lidar-guided sprayer designed to synchronize spray outputs with canopy structures, enhancing precision and efficiency in pesticide application. Similarly, Swanson *et al.* (2010) created a multi-modal system utilizing ladar (laser detection and ranging) for yield prediction in citrus trees, showcasing the potential of advanced technologies in agricultural management. Continuous modifications and improvements in cultural practices for mechanization are being made through ongoing research and practical experience, as highlighted by Sims (1969). For automated and mechanized systems to be successful, cultural practices must be tailored to both the machinery and the crop variety, as noted by Davis (1969). An example of this is the work of Peterson *et al.* (1999), who developed a robotic bulk harvesting system for apples. They trained the apple trees using a Y-trellis system, finding it to be compatible with mechanical robotic harvesting, thus demonstrating the importance of integrating cultural practices with technological advancements to achieve efficient and effective mechanization in orchards.

Table 2.4: Applications of Automation and Sensors in different Fruit Crops

Application	*Sensors*	*Reference*
Mechanical harvesting	3D camera (ToF *) Kinect v2	Amatya *et al.*, 2017
Robotic grapevine pruning	3 Grey Grasshopper2 colour cameras	Botterill *et al.*, 2017
	RGB camera	Naugle *et al.*, 1989
	3D camera (Cam Cube 3.0 **)	Karkee *et al.*, 2014
Robotic fruit tree pruning	Kinect v2	Chattopadhyay *et al.*, 2016

Effect of HDP and Canopy Architecture on Fruit Yield

Traditionally, tree fruit orchards such as those for apples, citrus, cherries, peaches, and pears have been characterized by low-density plantings using vigorous or seedling rootstocks. These trees were typically trained to open-center or globular architectures, resulting in tree densities of fewer than 200 trees per hectare. Minimal pruning or canopy management practices were adopted, leading to the development of large, unwieldy canopies. These expansive canopies made harvesting difficult, often necessitating the use of tall ladders, and were not conducive to mechanization of orchard management. Although some mechanization, such as mechanical pruning to restrict tree height and mechanical harvesting for processed crops like tart cherries and citrus, has been adopted, the complex canopies hinder precise management of growth and fruiting. Harvesting and pruning in these traditional systems were particularly labor-intensive tasks. The advent of size-controlling rootstocks marked a pivotal development, enabling the transition to higher efficiency orchard systems (Robinson, 2008). Close planting and the development of dense canopies in HDP systems can alter the micro-climate around the tree canopy. One of the primary advantages of HDP is the improved utilization of incident solar radiation and increased bearing surface per unit land area. However, the architecture of the tree canopy plays a crucial role in adapting trees to HDP. While traditional orchard systems can easily accommodate trees trained in open-center, central leader, or modified central leader systems, modern orcharding systems with high-density plantations require architectural engineering of plants. Various systems such as the Espalier system, V-trellis, Y-trellis, vertical axis, biaxial cordon, solaxe, spindle bush, slender spindle, tall spindle, head and spread, *etc.*, have been developed to accommodate more plants per unit area and ensure sufficient light penetration and diffusion for the production of quality apples. These systems optimize orchard space and facilitate efficient management practices, ultimately enhancing fruit quality and productivity in high-density planting systems. Sweet cherry performed well on central leader, multiple leader bush, palmette, and Y-trellis system of training (Whiting *et al.,* 2005).

Training systems that incurred the least amount of pruning (*e.g.*, a trellised palmette system, some central leader/spindle-type systems,) generally have had the highest yields but the smallest fruit size (Lang and Ophardt, 2000). Apricot has undergone a great renewal in the orchard training system and planting density has been substantially increased. The principal training systems used are vase, palmette, spindle and V-system. The V-training system is suitable for orchards with over 1200 trees ha^{-1} (De Salvador *et al.,* 2000) while, Stanica and Eremia (2014) suggested Parallel Trident System (4 x 2 m) for high density planting. Commercially grown kiwifruit plants are trained to a permanent framework, either a T-bar (like a clothesline with 4-5 wires) or pergola (solid overhead canopy with wire going in both directions). However, a T-bar system has yielded well, is easier to prune and may be better suited for small-scale production. For a T-bar trellis, vines commonly are planted 4.5m apart in rows 4.5 m apart, while, for pergola plants spaced at 4.5 to 6.0 m apart. In peach, new training system two-scaffold perpendicular V system have been suggested by many researchers. Research trials indicated that the V system is easier to prune and train than traditional systems if simple pruning and training rules are followed. The primary advantages of this system are early and sustained fruit cropping, high fruit quality and ease to management. In peaches among different training system light interception was 74 per cent in Y-shape, 71 per cent in open vase and 89 per cent in slender spindle. However, sunlight distribution in Y shape canopy was 35 per cent higher than the slender spindle canopy (DeSalvador and Dejong, 1989). At CITH, Srinagar a productivity 50 tonnes per ha was obtained in peach from V shape Tatura trellis system while the traditional vase trained trees planted at low densities of less than 400 trees per ha yielded only 15-20 t/ha.

Sensing Canopy Light Interception and Productivity

Solar radiation at the Earth's surface can be measured directly by sensors, with most Photosynthetically Active Radiation (PAR) measurements traditionally performed using hand-held devices, a method that is very time-intensive (Grossman and DeJong, 1998). This poses a significant limitation when large fields need to be sampled within a small time window (Zhang *et al.,* 2015). To address this, Giuliani *et al.* (2000) developed a photodiode-based scanner to measure whole-canopy light interception, digitizing the shadows of a single tree at different times of the day to extract tree-crown information. Building on this, Lampinen *et al.* (2012) created a mobile lightbar system to obtain PAR interception data from entire rows of trees. This system allows for the estimation of crop yield over large plots rather than just individual trees or small blocks, as demonstrated by Zarate-Valdez *et al.* (2012). Zhang *et al.* (2015) developed a mobile platform to collect PAR data under tree canopies, ensuring data collection at a constant spatial resolution by adjusting the sampling rate for speed changes. The productivity of orchards is closely linked to total light interception, a relationship that has been repeatedly demonstrated, particularly for hedgerow systems and apple orchards (Wünsche and Lakso, 2000; Grappadelli and Lakso, 2007). Canopy architecture and leaf area primarily influence the interception of incident irradiation. The need for alleyways between rows for machinery typically limits maximum light interception to approximately 70 per cent. Traditional hedgerow architectures that increase light interception often result in reduced fruit quality due to excessive intra-canopy shading. Modern orchard systems address this issue by maintaining compact fruiting walls and, in many cases, extending the canopy over the row using Y- or V-trellised architectures. Barritt (1989) reported a strong positive correlation between orchard light interception, leaf area, and fruit production. Palmer (1981) similarly found a positive correlation between light interception and yield in apple orchards, noting that trees arranged in a more rectangular pattern (2.75 m × 0.8 m) intercepted less light than those in a less dense arrangement (2.0 m × 1.25 m). Wünsche *et al.* (1996) highlighted the role of light interception by different shoot types in apple trees, emphasizing the importance of light interception by spur leaves. Their research indicated that Y-trellised systems, which intercepted 20 to 30 per cent more spur canopy light, yielded approximately 60 tons per hectare compared to 40 tons per hectare for pyramidal-shaped trees, although modern Y-trellised systems can yield 80–100 tons per hectare. These findings underscore the critical importance of both light interception and light distribution in producing high-quality fruits.

Comparison between Traditional System and High Density Planting

In the traditional system, the number of trees per hectare is limited, consisting of a few large trees, while high-density planting involves a greater density of smaller trees per hectare. This disparity impacts bearing patterns, with traditional systems typically exhibiting late bearing, whereas high-density planting facilitates early bearing, leading to quicker returns on investment. Productivity levels also vary significantly between the two systems, with traditional systems generally yielding lower outputs compared to high-density planting, which is associated with higher productivity rates. Moreover, in terms of management, traditional systems often pose challenges due to the larger tree sizes and spacing, making them more difficult to manage compared to the compact and organized layout of high-density planting, which streamlines management practices. Labor requirements follow a similar pattern, with traditional systems demanding more labor-intensive efforts for maintenance and upkeep, while high-density planting systems require fewer labor inputs due to their efficient design and management practices. Additionally, harvesting processes are simplified in high-density planting setups compared to traditional systems, where the scattered arrangement and larger tree sizes can make harvesting more laborious and time-consuming (Kumar, 2019).

Multi-Species HDP: A Viable Option for Small Farmers

In the context of small and marginal farmers in developing countries like India, mono-species high-density orcharding may not always be a viable option due to its high cost and technology intensity. For these farmers, sustainability and income generation are of paramount concern, and traditional mono-cropping may not provide sufficient income or employment opportunities. In such scenarios, a multi-species approach to high-density planting (HDP) can be more practical and beneficial. A rational strategy for multi-species HDP involves cultivating a few different crops in such a way that cultivation costs and labor requirements are spread evenly throughout the year. It's essential to prioritize staple food crops for the family while also considering economic viability. Appreciating the benefits of growing annuals and perennials in different combinations has become increasingly important for sustainable farming practices. While progress has been made in developing multiple cropping systems, particularly in plantation crops like coconut and arecanut with pepper, tree spices, tuber crops, banana, and pineapple in the interspaces, more research is needed to provide clear recommendations for different agro-climatic conditions (Nelliat and Bhat, 1979). Additionally, increasing cropping intensity can be achieved by introducing short-duration intercrops during periods when the canopy coverage of fruit trees is reduced through regular pruning, even in mature orchards. Adopting a multi-species approach to HDP not only addresses the economic constraints of small and marginal farmers but also promotes sustainable farming practices by maximizing land use efficiency and diversifying income sources. Continued research and extension efforts are essential to refine and promote these strategies for broader adoption and impact.

Status of High Density Planting in Fruit Crops

Worldwide, mango yields have typically been disappointing, ranging from 4 to 9 tonnes per hectare. One of the contributing factors to this low productivity is the wide spacing between mango trees, which is traditionally based on anticipated eventual tree size. As a result, it often takes over a decade for the canopies of these mango trees to fill the allocated space in the orchard row. Additionally, little attention is usually given to maintaining canopy size once overcrowding eventually occurs. The concept of high-density orcharding in mango gained practical traction following the release of the Amrapali variety. Unlike most commercial varieties, which are irregular bearers and traditionally planted at distances of 10-12 meters, accommodating about 70 to 100 plants per hectare, the Amrapali variety, being notably dwarf, was recommended by the Indian Agricultural Research Institute (IARI) to be grown at a spacing of 2.5 x 2.5 meters (triangular method), accommodating 1,600 plants per hectare, resulting in yields of up to 22 tons per hectare. This high-density orchard system has demonstrated significantly higher yields–up to 8 to 9 times more than traditional densities–as evidenced by research conducted by Gunjate *et al.* (2009). Furthermore, the Dushehari mango, planted at a closer spacing of 2.5 x 3 meters (equivalent to 1.333 plants per hectare), can also be successfully cultivated under high-density planting (HDP) with appropriate pruning and dehorning practices post-harvest. Additionally, successful HDP plantations (666 plants per hectare) of various commercial mango varieties, including Kesar, Alphonso, and Tommy Atkins, have been demonstrated by Reliance Agro Ltd. in Jamnagar, Gujarat, under arid agro-climatic conditions, as documented by Biswas and Kumar (2011) and Bhosale (2012). These developments underscore the potential of high-density orcharding as a means to significantly enhance mango yields and overall orchard productivity. Guava holds significant importance as a fruit crop in tropical and subtropical regions, and there's a global shift towards planting fruit trees at higher densities or adopting meadow orcharding techniques. In guava cultivation, achieving higher yields and better-quality production is realized through densely planted orchards coupled with judicious canopy management and the adoption of suitable tree training systems.

An experiment conducted at the Central Institute for Subtropical Horticulture in Lucknow revealed the suitability of meadow orcharding for ultra-high-density planting systems in guava. Under this system, guava orchards can accommodate up to 5000 plants per hectare, with a spacing of 2.0 x 1.0 meters. This arrangement is managed through regular topping and hedging during the initial stages, which helps control tree size and extends fruit availability throughout the season (Singh, 2008). In meadow orcharding, the ideal tree architecture entails a single trunk tree with no interfering branches up to 30 to 40 centimeters from the ground level, promoting a dwarf tree structure. Pruning plays a crucial role in maintaining this stature, with continued pruning operations conducted to ensure that the plants remain dwarf in stature. Pruning activities are typically carried out at specific intervals, particularly in May-June, September-October, and January-February, to manage canopy growth effectively and optimize fruit production.

Many studies have been carried out to increase the pineapple fruit yield (Hung *et al.*, 2011; Agbangba *et al.*, 2015; Sossa *et al.*, 2017). Actual crop yield is determined by many factors including planting material (genetic quality and planting material health and vigor), environmental conditions, agronomic practices (planting density and fertilizer management), pests and diseases, and their interactions (Momoh and Zhou, 2001; Tollenaar and Lee, 2002). Traditionally, pineapple cultivation has been characterized by low planting densities, typically ranging from 15,000 to 20,000 plants per hectare, resulting in sub-optimal productivity and increased production costs. However, recent advancements have identified the Kew and Queen cultivars as highly suitable for high density planting (HDP) using the double row method. This method involves planting suckers or slips spaced at 25 to 30 centimeters within the rows, with rows themselves positioned at a distance of 45 to 60 centimeters, and beds spaced at 90 to 105 centimeters apart, ultimately accommodating approximately 63,000 plants per hectare. Expanding the area under HDP presents significant opportunities for pineapple growers to boost productivity. While the average productivity of pineapple stands at 15.4 tons per hectare, adopting HDP and modifying planting systems can lead to substantially higher yields. For instance, the Kew cultivar has demonstrated the potential to achieve the highest fruit yield, reaching up to 61 tons per hectare, with a population of 63,700 plants per hectare and a spacing configuration of 22.5 x 60 x 75cm (Plant x Row x Trench) (Neri *et al.,* 2021). By embracing HDP techniques and adjusting planting systems accordingly, pineapple growers stand to significantly increase their yields, thereby enhancing profitability and competitiveness in the market.

A novel approach to High-Density Planting (HDP) in Kinnow mandarin cultivation has been developed, involving the use of Troyer Citrange rootstock and a planting distance of 2x2 meters, accommodating 2500 plants per hectare in a square system of planting. This innovative technique allows multiple shoots to emerge above the bud union. Kinnow mandarin plants grown on this rootstock exhibit precociousness, initiating fruiting as early as three years after planting in the field, with each plant yielding an average of 60 to 80 fruits. The management of such orchards proves to be highly economical due to several factors. Firstly, the use of Troyer Citrange rootstock promotes better conservation of soil moisture, which is crucial for sustaining plant health and promoting optimal fruit development. Additionally, judicious application of nutrients ensures that plants receive the necessary inputs for robust growth and fruit production, leading to improved yields. Moreover, the minimum cost of intercultural operations further contributes to the economic viability of managing these orchards. Similarly, Cleopatra mandarin rootstock has been identified as ideal for Mosambi sweet orange cultivation. For Khasi mandarin, a spacing of 5m x 5m, accommodating 400 plants per hectare, was found to be optimal (Bhosale, 2012). These findings highlight the importance of rootstock selection and planting spacing in maximizing orchard productivity and economic efficiency, ultimately benefiting fruit growers and enhancing the sustainability of citrus cultivation practices.

The availability of dwarf varieties in banana cultivation has spurred the adoption of High-Density Planting techniques, leading to increased popularity among growers. Varieties such as Dwarf Cavendish and Robusta have been identified as suitable cultivars for HDP, with planting distances of 1.2 meters x 1.2 meters, accommodating approximately 6,944 plants per hectare in a rectangular planting system. Under proper nutrient management practices, these dwarf banana varieties have demonstrated remarkable yield potentials. For instance, yield realizations range from 86 tons per hectare in Basarai planted at a spacing of 1.5 meters x 1.5 meters, to as high as 174 tons per hectare in Robusta planted at the closer spacing of 1.2 meters by 1.2 meters. These findings, documented by Athani *et al.* (2009) and Prabhu (2012), underscore the significant impact of HDP on banana cultivation, enabling growers to achieve higher yields and optimize land utilization while ensuring efficient nutrient management practices. With the development of dwarf cultivars, *viz.* Pusa Nanha and Ranchi Dwarf, it is now possible to grow papaya under the dense planting concept. Normally, papaya varieties are planted at a distance of 2.5 m x 3 m or 2.5m x 2.5m which accommodates 1333 to 1,600 plants/ha, while Pusa Nanha may be planted at a distance of 1.25 m x 1.25 m (6,400 plants/ha) as reported by Biswas *et al.* (1989). Such orchards may give 3 to 4 times higher yields (60 to 65 tonnes/ha) in comparison to the yield of traditional orchards (15 to 20 t/ha). Other cultivars like Co-1, Co-2, Pusa Dwarf and Honey Dew-1 are also suitable for high density planting. Sapota is normally planted at wide spacing of 8 m x 8 m and grows very slow and that it takes 10 to 12 years to occupy the allotted full space depending on the level of management, climate and edaphic conditions. Results under AICRP revealed that plant density of 312 plants per ha (8m x 4m) recorded highest yield (15.35 t/ha) for 15-year old PKM-1 sapota under Periyakulam conditions. At Kovvur, maximum yield (6.56 t/ha) was recorded in 7.5 m x 5.0 m spacing for seven-year old Kalipatti trees. In order to obtain higher fruit yield from unit area during the formative years, planting at 5 m x 5 m spacing is beneficial up to thirteen years. Further those trees were to be pruned (10 x 5 x 5 m canopy) and applied Paclobutrazol at 8-12 ml/tree to sustain the productivity (Anonymous, 2004).

Optimizing profitability is a challenge that strawberry farmers must face in order to remain competitive. Within this framework, plant density can play a central role (De Camacaro *et al.,* 2004). Soppelsa *et al.* (2023) conducted a study and recommended not to exceed the density of 100,000 plants ha^{-1} (economically disadvantageous) and to adopt medium or low planting densities to have strong plants (*e.g.,* for a third year of continuation) in Alpine Mountain Strawberry cultivation in Martell Valley, Italy. In recent times, high-density planting has emerged as an innovative approach in custard apple cultivation in Maharashtra. Traditionally, custard apple orchards were planted with wide spacing, typically 6 x 6 meters or 5 x 5 meters. However, farmers in the Vidarbha region are now adopting high-density planting techniques, utilizing closer spacings of 4 x 4 meters, 4 x 3 meters, 3 x 3 meters, and 4 x 2.5 meters. This method has shown promising results, with notable improvements in both growth and fruiting of the custard apple trees. The implementation of summer pruning and drip irrigation further enhances the effectiveness of high-density planting. As a result, the fruit yield per unit area has significantly increased, offering a more efficient and productive approach to custard apple cultivation (Dahapute *et al.,* 2018).

Conclusions and Future Perspectives

While there are numerous approaches to achieving High-Density Planting (HDP), successful implementations have primarily been observed in herbaceous perennials and temperate fruit crops. However, broader adoption across tropical, sub-tropical, and woody perennials remains limited. Extensive research is essential to develop suitable methods for controlling excessive canopy vigor and designing optimal planting systems tailored to the unique requirements of these crops. In many cases, an integrated approach involving multiple methods may be necessary to realize successful HDP. This could entail a combination of techniques such as pruning, rootstock selection, canopy management, and soil nutrient management. For modeling viable multi-species high-density cropping systems (HDMSCS), precise information on root distribution patterns and canopy characteristics, including architecture, vigor, light transmission, shade tolerance, and soil microflora interactions, is imperative. Furthermore, thorough investigations into the economic aspects and sustainability of various HDP systems are essential. By enhancing our understanding of these aspects, we can advance the adoption of HDP across a wider range of crops and improve agricultural productivity and sustainability in diverse agro-climatic regions.

Fruit production in India is undergoing a transformative phase characterized by significant changes in cultivars, rootstocks, cultural practices, handling, and marketing strategies. These changes are crucial for shifting the focus towards high-density plantings aimed at achieving early and sustained production of high-quality fruits. To successfully implement high-density plantings of tropical, sub-tropical, and temperate fruits, it is essential to formulate need-based, crop-specific strategies tailored to the unique requirements of each fruit crop. Dwarf and compact trees in high-density orchards offer numerous advantages, including higher yield and economic returns per unit area, efficient use of fertilizers and water, improved plant protection measures, weed control, and easier harvest operations. Developing crop-specific strategies for high-density plantings is particularly important in India, where a wide variety of fruit crops are grown. For indeterminate crops, the focus should be on maximizing yield per tree, while for determinate crops, increasing the number of trees per unit area is ideal for boosting productivity while minimizing reduction in yield per tree. Emphasis should also be placed on the development and evaluation of rootstocks with varying degrees of tree size control to suit closer spacing. Although some rootstocks and hybrids in mango and citrus have been identified for their dwarfing characteristics, they may not be reliable under different agro-climatic conditions. Standardization of training and pruning techniques is another critical aspect of successful high-density plantings.

REFERENCES

Agbangba CE, Dagbenonbakin GD, Djogbenou CP, Houssou P, Assea ED and Sossa EL, 2015. Influence de la fertilisation minérale sur la qualité physico-chimique et organoleptique du jus d'ananas transformé de Cayenne lisse au Bénin. *Int. J. Biol. Chem. Sci.,* **9**: 1277–1288.

Aglar E, Yildiz K and Long LE, 2016. Effects of rootstocks and training systems on the early performance of '0900 Ziraat' sweet cherry. *Notulae Botanicae Horti Agrobotanici Cluj- Napoca,* **44**(2): 573-578.

Amatya S, Karkee M, Zhang Q and Whiting MD, 2017. Automated detection of branch shaking locations for robotic cherry harvesting using machine vision. *Robotics,* **6**: 31.

Anonymous, 2004. *Research Report of All India Coordinated Research Project and ICAR Ad-hoc schemes on Tropical Fruits* (Eds., Reddy, B.M.C., Patil, P. and Thomas, R.R.). p. 105-122.

Anthony BM and Minas IS, 2021. Optimizing peach tree canopy architecture for efficient light use, increased productivity and improved fruit quality. *Agronomy,* **11**(10): 1961.

Ashkenazi S, Asor Z, Rasis A and Rosenberg O, 1992. Flying dragon trifoliate as a dwarfing interstocks for citrus trees. In: International Citrus Congress, 7., 1992, Acireale. Proceedings. Hartley Grove: South Australian R and D Institute, Abstract, 201

Athani SI, Revanappa and Dharmatti PR, 2009. Effect of plant density on growth and yield in banana. *Karnataka J. Agric. Sci.,* **22**(1): 143-146.

Barritt BH, 1989. Influence of orchard system on canopy development, light interception and production of third-year Granny Smith apple trees. *Acta Horticulturae,* **243**: 121–130.

Bhosale J, 2012. High-density cultivation: the new mantra for low cost, better yield. The Economic Times, July 12, 2012.

Biswas B, Sen SK and Maiti SC, 1989. Effect of plant density on growth, yield and chemical composition in papaya fruits varieties at Ranchi. *Prog. Hort.,* **21**: 280-84.

Biswas BC and Kumar L, 2011. Revolution in Mango Production: Success Stories of Some Farmers. Fertilizer Marketing News, pp-1-8.

Botterill T, Paulin S, Green R, Williams S, Lin J, Saxton V, Mills S, Chen X and Corbett-Davies SA, 2017. robot system for pruning grape vines. *J. Field Robot.,* **34**: 1100–1122.

Campoy J, Gonzalez-Mora J and Dima C, 2010. Advanced sensing for tree canopy modeling and precision spraying. ASABE Annual International Meeting, ASABE Paper No. 1009470. *American Society of Agricultural and Biological Engineers,* St Joseph, Michigan.

Castle WS, Youtsey CO and Hutchison DJ, 1987. Rangpur x Troyer citrange, a hybrid citrus rootstock for closely spaced trees. *Proc. Fla. Sta. Hort. Soc.,* **99**: 33-35.

Chattopadhyay S, Akbar SA, Elfiky NM, Medeiros H and Kak A, 2016. Measuring and modeling apple trees using time-of-flight data for automation of dormant pruning applications. In Proceedings of the Applications of Computer Vision 2016 IEEE Winter Conference, Lake Placid, NY, USA, 7–10 March 2016; Institute of Electrical and Electronics Engineers: Piscataway, NJ, USA, pp. 1–9.

Chaudhari SK and Singh A, 2019. NRM Interventions for enhanced mango productivity and quality. *Advanced Agricultural Research and Technology Journal*, **3**: 9-29.

Chen Y, Zhu H, Ozkan HE, Derksen RC and Krause CR, 2011. An experimental variable-rate sprayer for nursery and orchard applications. ASABE Annual International Meeting, Paper No. 1110497. *American Society of Agricultural and Biological Engineers*, St Joseph, Michigan.

Connor DJ, 2006. Towards optimal designs for hedgerow olive orchards. *Australian Journal of Agricultural Research*, **57**(10): 1067-1072.

Costes E and García-villanueva E, 2007. Clarifying the effects of dwarfing rootstock on vegetative and reproductive growth during tree development: A study on apple trees. *Annals of Botany*, **100**(2): 347-357.

Crane JH, Schaffer B, Davenport TL and Balerdi C, 1992. Rejuvenation of mature, non productive Lula and Booth 8 avocado grove by topping and tree removal. *Proc.Fla.Hortic. Soc.*, **105**: 282-285.

Dahapute VM, Joshi PS, Tayde SA and Nagre PK, 2018. Effect of severity pruning on growth, yield and quality of custard apple. *International Journal of Chemical Studies*, **6**(2): 1606-1609.

Davenport TL, 2006. Pruning strategies to maximize tropical mango production from the time of planting to restoration of old orchards. *Hort Science*, **41**(3): 544-548.

Davis V, 1969. Mechanization of Fruit and Pod Vegetables Introduction. Report No. 16, Rural Manpower Center. Michigan State University, East Lansing, Michigan, pp. 205–213.

De Camacaro MEP, Camacaro GJ, Hadley P, Dennett MD, Battey NH and Carew JG, 2004. Effect of Plant Density and Initial Crown Size on Growth, Development and Yield in Strawberry Cultivars Elsanta and Bolero. *J. Hortic. Sci. Biotechnol.*, **79**: 739–746.

de Carvalho LM, de Carvalho HW, de Barros I, Martins CR, Soares Filho WDS, Girardi EA and Passos OS, 2019. New scion-rootstock combinations for diversification of sweet orange orchards in tropical hardsetting soils. *Scientia Horticulturae*, **243**: 169-176.

De Kleine ME and Karkee M, 2015. A semi-automated harvesting prototype for shaking fruit tree limbs. *Transactions of the ASABE*, **58**(6): 1461–1470.

De Salvador FR and Dejong TM, 1989. Observation of sunlight interception and penetration into the canopies of peach trees in different planting densities and pruning configuration. *Acta Horticulturae,* **254**: Ig 341-346.

De Salvador FR, Nuzzo V and Mattatelli B, 2000. Gestionedell'albero e sistemid' impiantodell' albicocco. *Rivista di Frutticoltura e Ortofloricoltura J.,* **4**: 19-26.

Díez C M, Moral J, Cabello D, Morello P, Rallo L and Barranco D, 2016. Cultivar and tree density as key factors in the long-term performance of super high-density olive orchards. *Frontiers in Plant Science,* **7**: 1226.

Dubey AK, Sharma RM and Kumar A, 2021. Long term performance of mango varieties on five polyembryonic rootstocks under subtropical conditions: Effect on vigour, yield, fruit quality and nutrient acquisition. *Scientia Horticulturae,* **280**: 109944.

Edward JC and Shanker G, 1964. Rootstock trial for guava. *Allahabad Farmer,* **38**: 249-50.

Foster TM, van Hooijdonk BM, Friend AP, Seleznyova AN and McLachlan ARG, 2016. Apple rootstock-induced dwarfing is strongly influenced by growing environment. *Journal of Horticulture,* **3**: 180.

Ghanem ME, Albacete A, Smigocki AC, Frébort I and Pospíšilová H, 2011. Root-synthesized cytokinins improve shoot growth and fruit yield in salinized tomato (*Solanum lycopersicum* L.) plants. *Journal of Experimental Botany,* **62**: 125–40.

Giuliani R, Magnanini E, Fragassa C and Nerozzi F, 2000. Ground monitoring the light-shadow windows of a tree canopy to yield canopy light interception and morphological traits. *Plant, Cell and Environment,* **23**: 783–796.

Goyal SK, Goyal N, Agarwal S and Goyal M, 2020. Growth and instability in production of major fruits grown in India. *International Journal of Education and Management Studies,* **10**(1): 50-52.

Grappadelli LC and Lakso AN, 2007. Is maximizing orchard light interception always the best choice? *Acta Horticulturae,* **732**: 507–518.

Grossman YL and DeJong TM, 1998. Training and pruning system effects on vegetative growth potential, light interception, and cropping efficiency in peach trees. *Journal of the American Society for Horticultural Science,* **123**: 1058–1064.

Gulati A, Ganguly K and Wardhan H, 2022. *Agricultural value chains in India: Ensuring competitiveness, inclusiveness, sustainability, scalability, and improved finance* (p. 298). Springer Nature.

Gunjate RT, Kumbhar AR, Thiamaiah IM and Amin SM, 2004. Performance of some Indian and exotic mango cultivars under high density plantation in arid conditions of Gujarat (India). *Acta Hortic.,* **645**: 347–351.

Hayat F, Qiu C, Xu X, Wang Y and Wu T, 2019. Rootstocks Influence Morphological and Biochemical Changes in Young 'Red Fuji'Apple Plants. *International journal of agriculture and biology,* **21**: 1097–105.

Hung N, Thoa D and Huong N, 2011. Effect of planting density on growth, development and yield of irrigated pineapple in Nghe An Province. *Acta Hort.,* **902**: 307–311.

Jaiswal M, Gupta S and Verma S, 2015. Adoption Behavior of Banana Growers in Burhanpur District. *Journal of Community Mobilization and Sustainable Development,* **10**(1), 112-116.

Javaid K, Qureshi SN, Masoodi L, Sharma P, Fatima N and Saleem I, 2017. Orchard designing in fruit crops. *Journal of Pharmacognosy and Phytochemistry,* **6**(5): 1081-1091.

Jha GK, Suresh A, Punera B and Supriya P, 2019. Growth of horticulture sector in India: Trends and prospects. *Indian Journal of Agricultural Sciences,* **89**(2): 314-321.

Jutras PJ and Kretchman DW, 1962. A topping machine for Florida citrus growers. *Proceedings of the Florida State Horticultural Society,* **75**: 32-35.

Karkee M, Adhikari B, Amatya S and Zhang Q, 2014. Identification of pruning branches in tall spindle apple trees for automated pruning. *Comput. Electron. Agric.,* **103**: 127–135.

Kerutagi MGM and Deshetti B, 2018. Comparative Economics of Traditional *viz.* High Density Mango Cultivation in Karnataka. *Indian Journal of Hill Farming, Indian Journal of Hill Farming,* 76-86.

Khursheed S, Bhat ZA, Rather GH, Itoo H, Malik A R and Pandit BA, 2021. Occurrence of insect and mite pests and their natural enemies under high density apple agro-ecosystems in Kashmir. *Journal of Entomology and Zoology Studies,* **9**(1): 993-998.

Kohne J and Kremer-Kohne.,1989. Comparison of growth regulators paclobutrazol and uniconazole on avocado. *S. Afr. Avocado Growers Assoc. Yearb.,* **12**: 38-39.

Kumar N, 2019. High Density Planting in Mango-Prospects and Problems. *Adv Agri Res Technol J.,* **3**: 47-53.

Lal S, Tiwari JP and Mahajan AR, 2007. Studies on planting systems in guava (*Psidium guajava* L.) cv. Sardar. *Acta Hortic.,* **735**: 263-266.

Lampinen BD, Udompetaikul V, Browne GT, Metcalf SG and Stewart WL, 2012. A mobile platform for measuring canopy photosynthetically active radiation interception in orchard systems. *HortTechnology,* **22**(2): 237–244.

Lang GA and Ophardt DR, 2000. Intensive crop regulation strategies in sweet cherries. *Acta Horticulturae,* **514**: 227-234.

Larbi PA, Karkee M, Amatya S, Zhang, Q and Whiting M, 2015. Modification and field evaluation of an experimental mechanical sweet cherry harvester. *Applied Engineering in Agriculture,* **31**: 387–397.

Lauri PÉ, Maguylo K and Trottier C, 2006. Architecture and size relations: An essay on the apple (*Malus× domestica*, Rosaceae) tree. *American Journal of Botany,* **93**(3): 357-368.

Layne REC, 1987. Peach Rootstocks. In: Rom, R.C. and Carlson, R.F. (eds) *Rootstocks for Fruit Crops.* Wiley, New York, pp. 185–216.

Li X, Yadav R and Siddique KH, 2020. Neglected and underutilized crop species: the key to improving dietary diversity and fighting hunger and malnutrition in Asia and the Pacific. *Frontiers in Nutrition,* **7**: 593711.

Lindell CA, Irish-Brown A, Rothwell NL and Wallis AE, 2023. Pest and disease risk and management in high-density perennial crops: Current knowledge and areas of future research. *Crop Protection,* **165**: 106150.

Lordan J, Gomez M, Francescatto P and Robinson T L, 2019. Long-term effects of tree density and tree shape on apple orchard performance, a 20 year study–part 2, economic analysis. *Scientia Horticulturae,* **244**: 435-444.

Luo X, Chen Z, Gao J and Gong Z, 2014. Abscisic acid inhibits root growth in Arabidopsis through ethylene biosynthesis. *The Plant Journal,* **79**: 44–55.

Mairech H, López-Bernal Á, Moriondo M, Dibari C, Regni L, Proietti P, Villalobos FJ and Testi L, 2020. Is new olive farming sustainable? A spatial comparison of productive and environmental performances between traditional and new olive orchards with the model OliveCan. *Agricultural Systems,* **181**: 102816.

Manohar A, Bhaskar J and Suma A, 2023. Shoot tip pruning and paclobutrazol soil drench on the yield of mango hybrid Ratna under HDP system. *Journal of Tropical Agriculture,* **61**(1): 140-142.

Mehta BM, Sonawane M and Thakor RF, 2016. Role of Co-Operative Societies in Adoption of Improved Production Technology by Sapota Growers. *Indian Research Journal of Extension Education,* **13**(3): 59-63.

Menzel CM and Le Lagadec MD, 2017. Can the productivity of mango orchards be increased by using high-density plantings?. *Scientia Horticulturae,* **219**: 222-263.

Menzel CM and Lagadec MDL, 2014. Increasing the productivity of avocado orchards using high density plantings: A review. *Sci Hort.,* **177**: 21-36.

Mészáros M, Laòar L, Sus J and Námìstek J, 2017. Comparison of two training methods applied to apple trees trained to slender spindle during the first years after planting. *Acta Universitatis Agriculturae et Silviculturae Mendelianae Brunensis,* **65**(4): 1241-1245.

Mishra DS and Goswami AK, 2016. High density planting in fruit crops. *HortFlora Res. Spectrum*, **5**(3): 261-264.

Mishra DS and Lal RL, 2013. Canopy management in overgrown litchi orchards for improving yield and quality. In *Canopy Management and High Density Planting in Subtropical Fruit Crops* (Eds.) Singh, V.K. and Ravishankar, H., Army Printing Press Lucknow pp. 225-228.

Mishra DS, Lal RL and Chand S, 2014. Precision litchi culture technologies in Uttarakhand. *Recent Plasticulture Approaches Towards Precision Horticulture* (Eds.), pp. 73-77.

Mishra DS, Misra KK and Tiwari JP, 2003. High density orcharding in tropical and subtropical fruits – a review. In: Singh, S.P.(ed.) *Adv. in Hort. and Forestry*, 9. Scientfic Publishers (India), Jodhpur, pp. 179-195.

Mitra S, 2014. Tropical and subtropical fruits in India. *Journal of Tropical Crop Science;*https: //agris.fao.org/search/en/providers/122323/ records/647760dc5eb437ddff77baf3.

Mitra SK, 2019. High density orcharding and canopy management of some tropical and subtropical fruits. *Acta Hortic.*, **1255**: 127-138.

Momoh E and Zhou W, 2001. Growth and yield responses to plant density and stage of transplanting in winter oilseed rape (*Brassica napus* L.). *J. Agron. Crop Sci.*, **186**: 253–259.

Mukherjee SK and Das D, 1980. Anatomical screening of mango seedling for use as dwarfing rootstock. *Sci. Cult.*, **46**: 333-36.

Musacchi S, Serra S and Ancarani V, 2011. Comparison among pear training systems and rootstocks for high density planting (HDP) of the cultivar 'Abbé Fétel'. *Acta Hortic.*, **909**: 251-258.

Mysore S, 2015. Technology Commercialization through Licensing: Experiences and Lessons-A Case Study from Indian Horticulture Sector. *Journal of Intellectual Property Rights*, **20**(6): 363.

Nath V, Das B and Rai M, 2007. Standardization of high density planting in mango (*Mangifera indica*) under sub-humid Alfisols of Eastern India. *Indian J. Agri. Sci.*, **77**: 3-7.

Naugle JA, Rehkugler GE and Throop JA, 1989. Grapevine cordon following using digital image processing. *Trans. ASAE*, **32**: 309–0315.

Nelliat EV and Bhat KS (Eds.), 1979. Multiple cropping in coconut and arecanut garden (eds) Technical Bulletin 3. C.P.C.RI., Kasargod, 54 p.

Neri J, Mori M, Valqui V, Huaman EH, Silva RC and Oliva M, 2021. Effect of Planting Density on the Agronomic Performance and Fruit Quality of Three Pineapple Cultivars (*Ananas comosus* L. Merr.). *International Journal of Agronomy*, **4**: 5559564.

Oosthuyse S, 1995. Pruning of mango trees: an update. *South Afr. Mango Grow. Assoc. Yearb.,* **15**: 1–9.

Palmer JW, 1981. Computed effects of spacing on light interception and distribution within hedgerow trees in relation to productivity. *Acta Horticulturae,* **114**: 80–88.

Pathak RK and Awasthi DN, 1976. Standardization of propagation techniques in temperate fruits in India. ICAR Fruit Research Workshop, Ranikhet, UP, 17-20 November 1976.

Peterson D, Whiting MD and Wolford S, 2003. Fresh market quality sweet cherry harvester. *Applied Engineering Agriculture,* **19**: 539–543.

Peterson DL, Bennedsen BS, Anger WC and Wolford SD, 1999. A systems approach to robotic bulk harvesting of apples. *Transactions of the ASAE,* **42**(4): 871–876.

Prabhu MJ, 2012. High-density planting increases banana yield and brings hope. The Hindu, August 23, 2012.

Ram S, 1996. High density orcharding in mango. *Res. Bull.* 122, DES, GBPUAT, Pantnagar, pp. 1-24.

Ram S and Sirohi SC, 1988. Studies of high density orcharding in mango cv.Dashehari. *Acta Hortic.,* **231**: 339–344.

Ram S and Sirohi SC, 1991. Feasibility of high density orcharding in Dashehari mango. *Acta Hortic.,* **291**: 207–211.

Ram S, Singh CP and Shukla P, 2001. Effect of different planting densities on growth and yield of mango. *Indian J. Hortic.,* **58**: 191–195.

Raman MS, Pant DK, Singh A and Kumar R, 2023. Competitiveness of fruits' and vegetables' exports from India. *Economic Affairs,* **68**: 1379-1386.

Ramappa KB, Upadhyay J and Nagaraju Y, 2015. Growth of horticulture sector in Karnataka-post reform period. *Indian Journal of Economics and Development,* **11**(3): 661-672.

Rana SS, 1979. Anatomical studies on the clones of apple rootstocks with reference to their growth habit. M.Sc. thesis, HPKV, College of Agriculture, Solan.

Ravindran C, Kohli A, Murthy BS and Srinivasa A, 2007. Fruit production in India. *Chronica Horticuturae,* **47**: 21-26.

Reddy AGK, Kumar JS, Maruthi V, Venkatasubbaiah K and Rao CS, 2017. Fruit production under climate changing scenario in India: a review. *Environment and Ecology,* **35**: 1010-1017.

Reddy YTN, Kurian RM, Ramachander PR, Singh G and Kohli RR, 2003. Long-term effects of rootstocks on growth and fruit yielding patterns of 'Alphonso' mango (*Mangifera indica* L.). *Scientia Horticulturae,* **97**(2): 95-108.

Reginato GH, Carrasco Ó and García de Cortázar V, 2008. Evolution of planting and training systems in apple and peach orchards over 25 years of intensive fruit industry development in Chile. *Acta Hortic.,* **772**: 431-440.

Reighard GL, 2002. Current directions of peach rootstock programs worldwide. *Acta Horticulturae,* **592**: 421–427.

Robinson TL, 2007. Common mistakes in planting and establishing high-density apple orchards. *New York Fruit Quarterly, Geneva,* **15**(4): 1-7.

Robinson TL, 2008. The evolution towards more competitive apple orchard systems in the USA. *Acta Horticulturae,* **772**: 491–500.

Rom RC and Carlson RF (eds), 1987. *Rootstocks for Fruit Crops.* Wiley, New York.

Sankaran M and Dinesh M R, 2020. Biodiversity of tropical fruits and their conservation in India. *Journal of Horticultural Sciences,* **15**(2): 107-126.

Sansavini S, 1978. Mechanical pruning of fruit tree. *Acta Horticulturae,* **65**: 183-197.

Saroj and Singh KK, 2018. Density Concept of Orcharding. *International Journal of Advanced Scientific Research and Management,* Special Issue I, 56-61.

Shah ZA, Dar M.A, Dar EA, Obianefo CA, Bhat AH, Ali MT, El-Sharnouby M., Shukry M, Kesba H and Sayed S, 2022. Sustainable fruit growing: An analysis of differences in apple productivity in the Indian state of Jammu and Kashmir. *Sustainability,* **14**(21): 14544.

Sharma A, 2021. Trends in production and export of major fruits in India. *Indian Journal of Economics and Development,* **17**(2): 462-467.

Sharma RR and Singh R, 2006. Pruning intensity modifies canopy microclimate, and influences sex ratio, malformation incidence and development of fruited panicles in 'Amrapali' mango (*Mangifera indica* L.). *Sci. Hort.,* **109**: 118-122.

Sharma RR, Singh R and Singh DB, 2006. Influence of pruning intensity on light penetration and leaf physiology in high-density orchards of mango trees. *Fruits,* **61**: 117–123.

Sharma S, Rehalia AS and Sharma SD, 2009. Vegetative growth restriction in pome and stone fruits- A review. *Agri Rev.,* **30**: 13.

Sharma YK, Goswami AM and Sharma RR, 1992. Effect of dwarfing aneuploid guava rootstock in high density orcharding. *Indian J. Hort.,* **49**: 31-36.

Silwal A, Karkee M and Zhang Q, 2016. A hierarchical approach to apple identification for robotic harvesting. *Transactions of the ASABE,* **59**(5): 1079–1086.

Sims WL, 1969. Cultural Practices for Fruit Vegetables. Report No.16, Rural Manpower Center. Michigan State University, East Lansing, Michigan, pp. 225–237.

Singh AK, Singh CP and Pandey G, 2012. Effect of planting systems on growth, yield and quality of mango (*Mangifera indica* L.) cv. Dashehari. *Ann. Hort.,* **5**(2): 173-178.

Singh AK, Singh S, Appa Rao VV, Bagle BG and More TA, 2011. Effect of high density planting systems on the productivity of NA-7 aonla under rainfed conditions. *Indian J. Hort.,* **68**(4): 461-465.

Singh G, 2008. High density and meadow orcharding of guava, Extn. Bull. 35, (1st Ed) CISH, Rehmankhera.

Singh G, Singh AK and Mishra D, 2007. High density planting in guava. *Acta Hort.,* **735**: 235-241.

Singh K, Verma SR, Sharma FL and Punjabi NK, 2014. Constraints and obstacles perceived by mango growers in adoption of scientific mango production technology. *Ind. J. Extn. Educ. and RD,* **22**: 124-133.

Singh NP, Srivastava RP and Chadha KL, 1986. Screening of dwarfing mango rootstocks at nursery stage on the basis of anatomical characters. *Indian J. Hort.,* **43**: 18-22.

Singh, R. N. and Guptha, P. N. 1971. Rootstock problem in stone fruits and potentialities of wild species found in India. *Punjab Hort. J.,* **11**: 157-175.

Singh RP, 2013. Horticultural (high value agricultural) crops diversification in eastern India: I–Geographical-agro-socio-ecological and economical aspects and constraints analysis. *International Journal of Innovative Horticulture,* **2**(1): 12-27.

Singh S, 2019. Examining global competitiveness of Indian agribusiness in the twenty-first-century Asian context: Opportunities and challenges. *Millennial Asia,* **10**(3): 299-321.

Singh S, Waman AA, Bohra P, Gautam RK and Roy SD, 2016. Conservation and sustainable utilization of horticultural biodiversity in tropical Andaman and Nicobar Islands, India. *Genetic Resources and Crop Evolution,* **63**(8): 1431-1445.

Soppelsa S, Gasser M and Zago M, 2023. Optimizing Planting Density in Alpine Mountain Strawberry Cultivation in Martell Valley, Italy. *Agronomy,* **13**(5): 1422.

Sossa EL, Agbangba CE, Accalogoun SGGS, Amadji GL, Agbossou KE and Hounhouigan DJ, 2017. Residues management practices and nitrogen-potassium fertilization influence on the quality of pineapple (*Ananas comosus* [L.] Merrill) sugarloaf fruit for exportation and local consumption. *Agronomy,* **7**: 26.

Stanica F and Eremia A, 2014. Behaviour of some new apricot cultivars under the parallel trident planting system. *Acta Horticulturae,* **1058**: 129-136.

Stephan J, Sinoquet H, Donès N, Haddad N, Talhouk S and Lauri PÉ, 2008. Light interception and partitioning between shoots in apple cultivars influenced by training. *Tree Physiol.,* **28**: 331–342.

Subedi GD, Gurung CR, Poudel KK, Giri RK and Gurung YR, 2019. An experience of high density planting of apple in Nepal. Proc. of the 10th National Horticulture Seminar organized by Nepal Horticulture Society, 1-9 February 2019, Kathmandu.

Subedi GD, Atreya PN, Gurung CR, Giri RK and Gurung YR, 2020. High density cultivation of major fruit crops: Opportunities and challenges in Nepal. In *Proceeding of National Horticulture Seminar* (pp. 94-107).

Swanson M, Dima C and Stentz A, 2010. A multi-modal system for yield prediction in citrus trees. ASABE Annual International Meeting, Paper No. 1009474. *American Society of Agricultural and Biological Engineers*, St Joseph, Michigan.

Tollenaar M and Lee E, 2002. Yield potential, yield stability and stress tolerance in maize. *Field Crops Res.,* **75**: 161–169.

Tripathi VK, Kumar S, Dubey V and Nayyer MA, 2020. High-Density Planting in Fruit Crops for Enhancing Fruit Productivity. In *Sustainable Agriculture* (pp. 253-267). Apple Academic Press.

Upadhyay AP, Papnai G and Singh P, 2018. Problems and Prospects of Guava Producers in Allahabad District of Uttar Pradesh, India. *IOSR J. Humanities and Social Sci.,* **23**(6): 1-7.

Usha K, Thakre M, Goswami AK and Nayan DG, 2015. Fundamental of Fruit production, Division of Fruits and Horticultural Technology, Indian Agricultural Research Institute, New Delhi.

Vashishtha BB, 1987. Hortcultural qualities. In *Bordi* (*Zizyphus nummularia*) (Mann, H.S. and Saxena, S.K., Eds.), pp. 31-33.

Whiting MD, Lang G and Ophardt D, 2005. Rootstock and training system affect sweet cherry growth, yield and fruit quality. *Hort. Sci.,* **40**(3): 582-586.

Wünsche JN and Lakso AN, 2000. The relationship between leaf area and light interception by spur and extension shoot leaves and apple orchard productivity. *HortScience,* **35**: 1202–1206.

Wünsche JN, Lakso AN, Robinson T, Lenz F and Denning S, 1996. The bases of productivity in apple production systems: the role of light interception by different shoot types. *Journal of the American Society for Horticultural Science,* **121**: 886–893.

Yadav BC, Choudhary R and Saran PL, 2013. Adoption of improved production technology of Mandarin in Rajasthan, India: A review. *African Journal of Agricultural Research,* **8**(49): 6590-6600.

Yeshitela T, Robbertse PJ and Stassen PJC, 2005. Effects of pruning on flowering, yield and fruit quality in mango (*Mangifera indica*) cultivars. *Aust. J. Exp. Agric.,* **45**: 1325–1330.

Yuri CR, Eduardo SS, Eduardo AG, Helton C de L, Abelmon da SG., Orlando SP and Walter dos SS,. 2015. Dwarfing rootstocks for 'valencia' sweet orange. *Acta Hortic.,* **1065**: 351-354.

Zarate-Valdez J, Whiting M, Lampinen B, Metcalf S, Ustin S and Brown P, 2012. Prediction of leaf area index in almonds by vegetation indexes. *Computers and Electronics in Agriculture,* **85**: 24–32.

Zhang J, Whiting M and Zhang,. 2015. Diurnal pattern in canopy light interception for tree fruit orchard trained to an upright fruiting offshoots (UFO) architecture. *Biosystems Engineering,* **129**: 1–10.

Chapter 3

Intercropping Management in Orchards

Amit Kumar Goswami, Chavlesh Kumar, B.R. Vinod, H.R. Raghavendra and Kritidipta Pramanik

Division of Fruits and Horticultural Technology, ICAR–Indian Agricultural Research Institute, New Delhi–110012

Intercropping in orchard systems offer significant environmental, economic, and ecological benefits, enhancing sustainability and resilience. By integrating crops such as legumes, grasses, and aromatic plants into orchards, farmers can improve soil health, promote biodiversity, and reduce pest and disease pressures. Intercropping enhances natural soil fertility through nitrogen fixation, increases microbial activity, and boosts carbon sequestration. It also provides habitats for beneficial organisms, supporting natural pest control and reducing reliance on chemical inputs. Economically, intercropping diversifies income streams and improves profitability, as shown by increased yields and returns in various systems of orchards. Moreover, intercropping mitigates the environmental footprint by conserving soil moisture, reducing erosion, and lowering greenhouse gas emissions. Studies demonstrate that integrating crops in orchards can enhance water retention, nutrient availability, and microclimate conditions, making these systems more adaptable to climate change. Overall, intercropping serves as a powerful strategy to optimize orchard productivity, enhance ecosystem services, and support sustainable farming practices.

Introduction

A cropping system refers to the arrangement of crops cultivated on a specific area over the course of a year. The variations in horticultural practices and cropping systems worldwide are largely influenced by local factors such as climate, soil type, economic conditions, and social structures. The primary determinants of crop growth and the viability of a cropping system are water balance, radiation, temperature, and soil conditions. Intercropping, a form of mixed cropping, is defined as the practice of growing two or more crops simultaneously in the same area (Andrew and Kassam, 1976). Intercropping is an ancient agricultural practice that has been revitalized in recent years as a means to enhance sustainability in modern farming. It involves the simultaneous cultivation of multiple crop species within the same field, fostering biodiversity and optimizing the use of natural resources with minimal inputs. By promoting ecological balance and improving soil health, intercropping offers a powerful tool for enhancing both productivity and environmental stewardship. This approach not only contributes to food production in healthier environmental conditions but also enhances the resilience of agroecosystems by increasing their overall stability (Burgess *et al.*, 2022).

Concept of Intercropping

Intercropping is an agricultural practice that involves growing multiple crops together in the same area, founded on the principle of diversity. This approach mimics nature by integrating plants and animals into a varied landscape, which in turn enhances ecosystem stability. The increased diversity not only boosts productivity compared to monoculture but also aids in pest management by disrupting pest populations. Furthermore, intercropping promotes environmental sustainability by mitigating the negative effects of pollution on agricultural productivity (Mousavi and Eskandari, 2011). It also provides protection against severe weather conditions, such as wind, frost, and drought, making crops more resilient to environmental stressors. This system takes various forms, such as mixed intercropping, row intercropping, strip intercropping, relay intercropping and multistoried intercropping or multi-tier intercropping. These configurations can be tailored to fit specific agroecological contexts, allowing for flexibility and adaptability in farm management. For example, in mixed intercropping, crops are grown together without distinct row arrangements, while row and strip intercropping involve planting crops in alternating rows or strips. Relay intercropping extends this diversity further by overlapping growth periods of different crops, allowing for continuous production cycles (Maitra *et al.*, 2021). Multi-storied intercropping is a prevailing practice in plantation systems like coconut and areca nut farming. This method involves growing crops of different heights, rooting depths, and growth durations together in the same space. The goal is to make the most efficient use of vertical growing space.

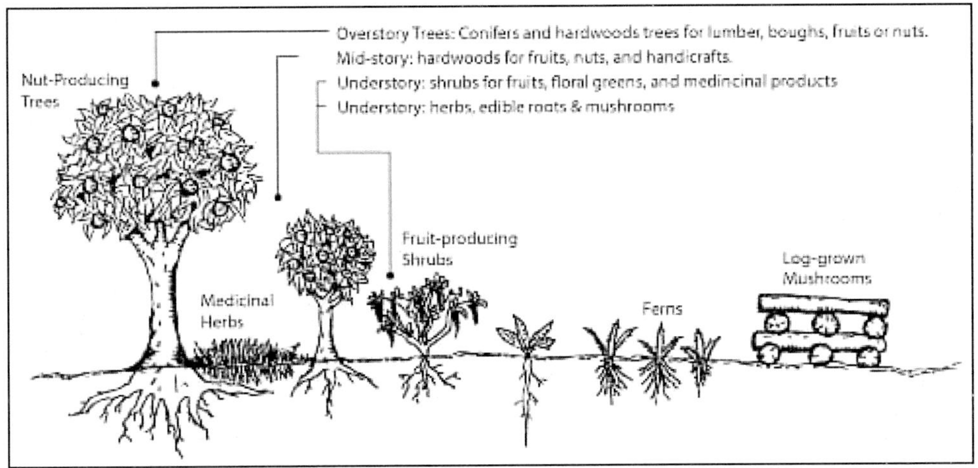

Figure 3.1: Multi-Storied Intercropping.

A key advantage of intercropping is its ability to efficiently use resources through complementary interactions between crops. For instance, crops with different root depths can extract nutrients and water from varying soil layers, reducing competition and improving nutrient cycling. This practice also enhances soil organic matter, improves fertility, and often results in higher yields per unit area compared to monocultures (Zhi *et al.,* 2007). Beyond productivity, intercropping offers multiple ecosystem services. These include improved soil structure, reduced erosion, enhanced water quality, better pest and disease management, and support for biodiversity, including pollinators and wildlife. Moreover, the practice has the potential to mitigate climate change impacts by sequestering carbon and creating favourable microclimates that protect crops from extreme temperatures (Bybee-Finley and Ryan, 2018). Intercropping plays a vital role in contributing to sustainable development goals (SDGs), including zero hunger (SDG 2), climate action (SDG 13), and land restoration (SDG 15). Resource partitioning and facilitation are the primary mechanisms that drive intercropping's success. In resource partitioning, crops make more efficient use of available nutrients, water, and light, while in facilitation, one crop enhances the growing conditions for another (Loreau and Hector, 2001; Singh *et al.,* 2020). As agriculture continues to evolve, intercropping in orchards offers a promising strategy for sustainable intensification, balancing productivity with environmental conservation. Through careful management of spatial and temporal relationships between crops, farmers can optimize resource use, reduce inputs, and support long-term farm resilience. This chapter will explore the principles, benefits, and practical strategies for implementing intercropping in orchard systems, with a focus on improving productivity and sustainability.

History

Intercropping has been a long-standing practice in traditional agriculture worldwide. For centuries, farmers have cultivated various crops together, and these mixed cropping systems likely represent some of the earliest forms of agriculture (Plucknett and Smith, 1986). Ancient Greece, around 300 B.C., was familiar with different types of intercropping. Theophrastus, a renowned Greek philosopher and scientist, observed that wheat, barley, and certain legumes could be planted at various times during the growing season, often in combination with vines and olives, suggesting a knowledge of intercropping practices (Papanastasis *et al.,* 2004). Studies promote this ancient cropping system as a substitute for more sustainable agriculture with low input and stabilized yield (Ren *et al.,* 2017). Intercropping is a widely practiced agricultural technique in many tropical regions, particularly among small-scale traditional farmers. Traditional multiple cropping systems are estimated to account for 15-20 per cent of the world's food supply (Altieri, 1999). In Latin America, farmers frequently intercrop beans with maize, potatoes, and other crops, with 70-90 per cent of bean production and 60 per cent of maize production occurring in intercropped systems (Francis, 1986). Quantitative studies indicate that intercropping is prevalent in other regions as well, with 89 per cent of cowpeas in Africa, 90 per cent of beans in Colombia, and varying percentages of cropped land devoted to intercropping, ranging from 17 per cent in India to 94 per cent in Malawi (Vandermeer, 1992).

However, intercropping is less prevalent in mechanized agricultural sectors of Europe, North America, and parts of Asia, due to a shift toward market-oriented agriculture and intensive monocropping systems (Horwith, 1985). Despite its advantages, agricultural intensification over the past 50 years, characterized by plant breeding, mechanization, and increased fertilizer and pesticide use, has led to the decline of intercropping in many farming systems. However, intercropping has been shown to produce higher and more stable yields while minimizing input use, emphasizing the production of healthy, safe, and high-quality food in an environmentally sustainable manner. In the organic sector, intercropping is considered an effective means of self-regulation and resilience for organic agroecosystems, particularly in the absence of agrochemicals allowed under organic agriculture principles and European Union regulations (Lammerts van Bueren *et al.,* 2002).

Traditional vs. Modern Intercropping Approaches

Traditional systems emphasize biodiversity, sustainability, and low-input practices, making them resilient but often less productive on a commercial scale. On the other hand, modern intercropping incorporates advanced technologies, precision farming, and high-input strategies to enhance productivity and profitability but can sometimes reduce biodiversity and increase environmental impact. Finding a balance between the two approaches, particularly in the context of sustainable agriculture, will be key to addressing future food security and environmental challenges. This discussion will highlight the traditional as well as contemporary methods of intercropping.

Traditional Intercropping Approaches

1. **Poly-cropping for subsistence:** Traditional intercropping systems, particularly in developing countries, have been largely subsistence-oriented. These systems are characterized by small-scale, low-input farming practices where farmers use a variety of crops to ensure food security. These combinations benefit from natural synergies like nitrogen fixation by legumes, which enriches soil fertility for companion crops like maize.
2. **Low-input and sustainability-focused:** Traditional methods are often rooted in local knowledge and focus on the sustainability of farming systems. Limited external inputs like chemical fertilizers or pesticides are used. Instead, natural practices such as mulching, organic manuring, and hand-weeding are emphasized. The key goal is the conservation of biodiversity and resilience to climatic variability (Altieri, 1999).
3. **Complementary growth patterns:** Farmers in traditional systems select crop combinations with complementary growth habits to optimize the use of spatial and temporal resources. For example, tall-growing crops like maize are combined with creeping crops like beans. This allows for efficient use of vertical space and better light utilization (Vandermeer, 1992).
4. **Local crop varieties:** Traditional intercropping often incorporates locally adapted crop varieties, which are better suited to the specific environmental conditions of the region. This enhances the system's resilience to pests, diseases, and erratic weather patterns.

Modern Intercropping Approaches

1. **Precision and mechanized intercropping:** Modern intercropping systems have incorporated advancements in precision agriculture. Technologies like GPS-guided machinery and drones are used for efficient crop management and resource application. These innovations allow farmers to optimize plant spacing and improve irrigation efficiency. Moreover, precision technologies help monitor crop health and manage pests more effectively (Zhang *et al.,* 2019).
2. **Agroecological and climate-smart approaches:** Modern intercropping often involves incorporating agroecological principles, emphasizing both high productivity and sustainability. Climate-smart practices like conservation tillage, cover cropping, and integrating agroforestry are commonly integrated into modern systems. These methods reduce greenhouse gas emissions, improve carbon sequestration, and build resilience to extreme climate events (Kassam *et al.,* 2011).
3. **High-input, commercial focus:** Unlike traditional subsistence farming, modern intercropping systems are primarily geared toward commercial

production goals. These systems often depend on the use of synthetic fertilizers, herbicides, and pesticides to enhance yields. Modern intercropping is tailored to optimize market-oriented outcomes, including improved crop quality, yield stability, and profitability (Lithourgidis *et al.,* 2011).

4. **Research-driven crop combinations:** In contrast to traditional methods rooted in local knowledge, modern intercropping systems are frequently guided by scientific research to identify the best crop combinations. Researchers have explored how different plant species interact regarding root systems, nutrient uptake, and pest resistance, contributing to the development of more efficient intercropping patterns (Brooker *et al.,* 2015).

Types of Interactions in Intercropping in Orchards

In orchards, intercropping involves growing different plant species together, usually combining trees with crops. The nature of these interactions can significantly influence the overall productivity and sustainability of the system. There are three main types of interactions observed in orchard intercropping systems:

Complementary Interactions

This type of interaction occurs when both the tree and the intercrop benefit from being grown together, leading to an increase in production for both. Complementary relationships often arise due to factors like enhanced resource use efficiency (water, light, or nutrients), which benefits both species without direct competition. For example, crops planted between trees may benefit from partial shading and enhanced soil fertility, while trees gain from improved ground cover and soil moisture retention provided by the crops (Yin *et al,* 2020).

Figure 3.2: Olive Orchard Intercropped with Winter Wheat.

This synergy ensures that both components contribute to a greater yield than if they were grown separately. An example of a complementary interaction is seen in an olive orchard intercropped with winter wheat. The growth of the olive tree is unaffected by the wheat, as the tree grows slowly during winter and early spring while the wheat grows quickly. For this system to work effectively, it is important to prevent the wheat's maturation from coinciding with the olive tree's vegetative and reproductive growth stages (Wang, 2019).

Supplementary Interactions

In supplementary interactions, the yield of one component, typically the tree, increases while the yield of the other component (crop) remains unaffected. This can occur when the tree benefits from factors such as better soil health or microclimate conditions provided by the intercrop, but the intercrop does not experience any noticeable changes in its growth (Glaze *et al*, 2020). For instance, leguminous intercrops can fix nitrogen in the soil, enhancing tree growth without altering the crop's productivity. This type of interaction supports orchard health while maintaining stable yields for the intercrop.

Figure 3.3: Citrus Orchard Intercropped with Cowpea.

In citrus orchards, intercropping with legumes such as cowpea or beans is a common practice. These legumes form a symbiotic relationship with nitrogen-fixing bacteria, enriching the soil with nitrogen (Martins *et al.,* 2021). As a result, the citrus trees experience improved soil fertility, leading to better growth and higher yields, while the legume crops maintain stable production. Similarly, in apple orchards, herbs like mint or basil are often intercropped between the trees. These aromatic herbs help deter pests and create a more favourable microclimate by providing ground cover, which reduces pest pressure and retains soil moisture. The apple trees thrive with these added benefits, and the herbs grow well without significant impact on their yield (Dikr, 2022). Pear orchards, on the other hand, are frequently intercropped with cover crops such as clover or alfalfa. These cover crops enhance soil structure and fertility, promoting the growth of pear trees. Additionally, they suppress weeds and help control soil erosion, all while remaining unaffected by the pear trees' presence (Das *et al.,* 2021).

Competitive Interaction

Competitive interactions take place when the yield of one component (usually the tree) increases at the expense of the other (the crop). This happens when both components compete for limited resources like water, nutrients, or light. In many cases, trees have deeper root systems and a more dominant canopy, which may outcompete shallow-rooted crops for resources. While trees benefit from this competition by gaining access to more resources, the intercrop may suffer from reduced yields. Competitive interactions are common when plant spacing or species selection is not optimized, leading to resource imbalances within the system.

Components of Fruit Based Intercropping System

Fruit-based intercropping systems are designed to optimize the use of resources by combining two key components: the main crop, which typically consists of perennial fruit trees, and the intercrop, which comprises annual or short-duration crops planted alongside the main crop. This integration allows for enhanced land utilization, improved microclimate conditions, and maximized income during the orchard's juvenile phase.

Main Crop

The main crop refers to fruit species with larger canopies and a prolonged productive lifecycle. These crops gradually dominate the orchard space over time, typically occupying the entire land area after 20–25 years. However, during the initial 10 years, only 25–30 per cent of the land is effectively utilized by the main crop, providing space for intercrops in the early stages. Main crops are usually planted at wider spacing to accommodate their growth and to facilitate the intercropping system.

Criteria for Selecting Main Crops

When selecting main crops, several important criteria should be considered to ensure their suitability and success. The crops should exhibit xeric characteristics such as a deep taproot system, high water retention in tissues, sunken stomata, waxy coatings, thick cuticles, and pubescence, making them resilient to drought and heat stress. They should also be adapted to moisture availability, able to complete their vegetative and reproductive phases during periods when moisture is most abundant. A deep root system is essential, allowing the crops to access nutrients and water from deeper soil layers, thus minimizing competition with understorey crops.

The branching habits of the trees should permit sufficient light penetration for crops growing beneath them, ensuring a balanced ecosystem. Additionally, the species must have good coppicing ability, responding well to pruning to manage canopy size and promote healthy growth. A high contribution of leaf litter, especially during the understorey crop's growth phase, along with rapid decomposition, helps enhance soil fertility. Lastly, the main crops should possess natural resistance to pests and diseases, reducing the need for extensive management and supporting sustainable cultivation.

Intercrop

The intercrop refers to crops grown during the early years of orchard establishment when the main crop has not yet fully occupied the available space. Intercrops are usually annual, location-specific crops, or short-duration fruit crops known as "fillers". These crops are selected based on their ability to generate additional income and maximize land utilization during the juvenile phase of the main crop. As the orchard matures, shade-tolerant crops are preferred to minimize competition with the growing canopy of the main crop. Filler plants, often with short growth cycles, are planted to exploit the unused land effectively while ensuring they do not interfere with the long-term productivity of the main crop.

Criteria for Selecting Intercrops

When selecting intercrops, several key factors should be considered for optimal integration with the main crop. Intercrops should be early-maturing, with a short growth cycle that allows them to be harvested before the main crop fully occupies the space. Shade tolerance is essential, as these crops must thrive under reduced light caused by the shading of the main crop. To ensure harmony, intercrops should offer minimal competition for resources like water, nutrients, and light, allowing both crops to grow effectively side by side.

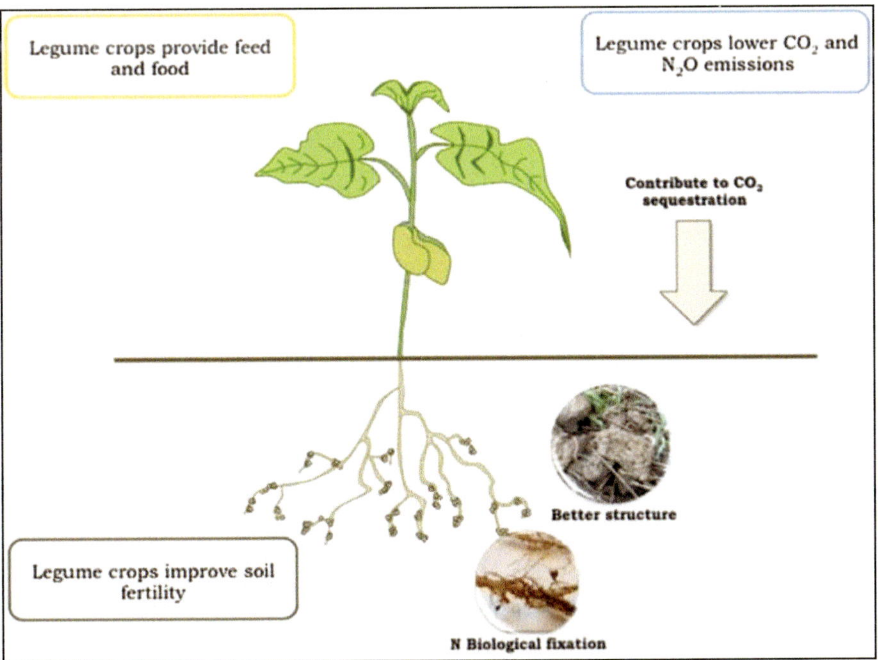

Figure 3.4: Importance of Legumes as Intercrop.

It's also important that intercrops have a low susceptibility to common diseases shared with the main crop, reducing the risk of spreading infections. Harvesting should be non-mechanized to prevent any disturbance to the main crop or damage to the soil structure. The economic lifespan of the intercrops should be shorter than that of the main crop, avoiding overlap with its productive phase. Additionally, intercrops should be well-suited to local agro-climatic conditions, labour availability, and market access to ensure they thrive in the region. Including leguminous crops like peas, soybeans, or cowpeas is often beneficial, as they contribute to soil fertility through nitrogen fixation.

Leguminous crops have been shown to be the most profitable intercrops in fruit-based systems, particularly in the first 7–10 years. Beyond their nitrogen-fixing properties, these crops contribute to the orchard ecosystem by improving soil health. Commonly recommended leguminous intercrops include peas, cowpeas, chickpeas, beans, lentils, and other pulse crops. The introduction of oilseeds like soybean, sesame, sunflower, and Brassica into intercropping systems has transformed traditional agronomic practices. In arid regions, stress-resistant grasses such as *Cenchrus ciliaris* have also demonstrated strong performance. Natural grasses, clover, rye grass, *Vulpia myuros,* and alfalfa are widely used for orchard intercropping (Sun *et al.,* 2022).

Additionally, many medicinal and aromatic plants thrive in partial shade and moist environments, making them well-suited for intercropping with fruit trees. Vegetables in intercropping systems have also proven beneficial in terms of economic returns. Filler fruit crops like guava, papaya, banana, pomegranate, and phalsa are ideal due to their short stature, prolific bearing, and potential to generate additional income during the early years of orchard establishment. These crops provide a steady income until the primary fruit trees reach full productivity.

Key Considerations for Selecting Optimal Crop Combinations

When planning intercrops in an orchard, choosing the right crop combinations is crucial to ensure both the health of the orchard and the productivity of the crops. Various factors need to be considered when selecting which crops to grow together. These considerations include the type of produce, growth patterns, input requirements, rooting structures, and the potential for improving soil fertility. Additionally, several external factors, such as ecological conditions and the farmer's financial stability, also significantly impact the cropping pattern.

Produce

The type of produce plays a central role in selecting the right intercrops. Farmers often grow crops for multiple purposes–food, fodder, firewood, or even medicinal uses. Some plantation crops, like tea and coffee, are grown exclusively for cash, while fruits and vegetables can serve both household consumption and as sellable products. When choosing intercrops, it's important to ensure there's a market for surplus produce. For example, growing food crops like cereals and legumes as intercrops in an orchard allows farmers to use them for personal consumption while also profiting from the excess.

Growth Pattern

Understanding the growth patterns of different crops is essential when deciding on intercropping combinations. Factors such as the amount of rainfall, soil moisture, and the seasonality of the crops play a significant role in determining what can grow well together. Certain crops thrive in particular seasons, while others may struggle in the same conditions. By aligning intercrops with the growth cycles of the primary orchard crop, farmers can optimize both crop yields and resource use.

Input Requirements

Availability of inputs such as seeds, fertilizers, and pest control solutions can heavily influence crop selection. Some crops may demand higher levels of fertilizers or insecticides, while others, such as legumes, might naturally enrich the soil by fixing nitrogen. Ensuring that the necessary inputs are available and affordable is vital for successful intercropping.

Rooting Pattern

The rooting pattern of crops also affects the suitability of intercrop combinations. Tall fruit trees have deep roots that can penetrate up to 6 meters into the soil. This deep rooting helps improve soil structure and porosity, making them a good choice for compacted soils. Intercropping these deep-rooted plants with shallow-rooted species can create a balanced ecosystem in the orchard soil.

Soil Fertility Improvement

Legumes are known for their ability to fix nitrogen from the air, enriching the soil for the benefit of surrounding crops. When intercropped with fruit and plantation crops not only improve soil fertility but also reduce the need for additional fertilizers. This synergy makes legumes a valuable addition to orchard systems, promoting sustainable farming practices.

Soil Coverage

Certain crops, like tall cereals, do not provide adequate soil cover due to their upright leaves and wide spacing. On the other hand, legumes such as groundnut, cowpea, and beans, or short grasses, cover the ground quickly after planting. These crops are referred to as "cover crops" when their primary function is to protect the soil, preventing erosion and moisture loss. Choosing crops that complement each other in terms of ground coverage can lead to better soil health and reduced weed growth.

Feasibility of Space and Time Utilization in Orchard Intercropping

Horticultural crops like mango, coconut, areca nut, and cashew are typically perennial with long pre-bearing periods. These tall crops, spaced widely (*e.g.*, coconut at 9 x 9 meters), often leave up to 70 per cent of interspace underutilized during early growth. Efficient planning and intercropping can optimize this space (Hegde *et al.,* 2015). Crops with complementary growth habits, such as banana, cocoa, turmeric, ginger, and pineapple, thrive under the shade of taller trees and adapt well to the orchard's microclimate. Intercropping not only improves space and time management but also ensures a staggered harvest for consistent income. Additionally, by-products like biomass from these crops enrich the soil, supporting sustainable farming practices and enhancing orchard health (Sau *et al.,* 2017).

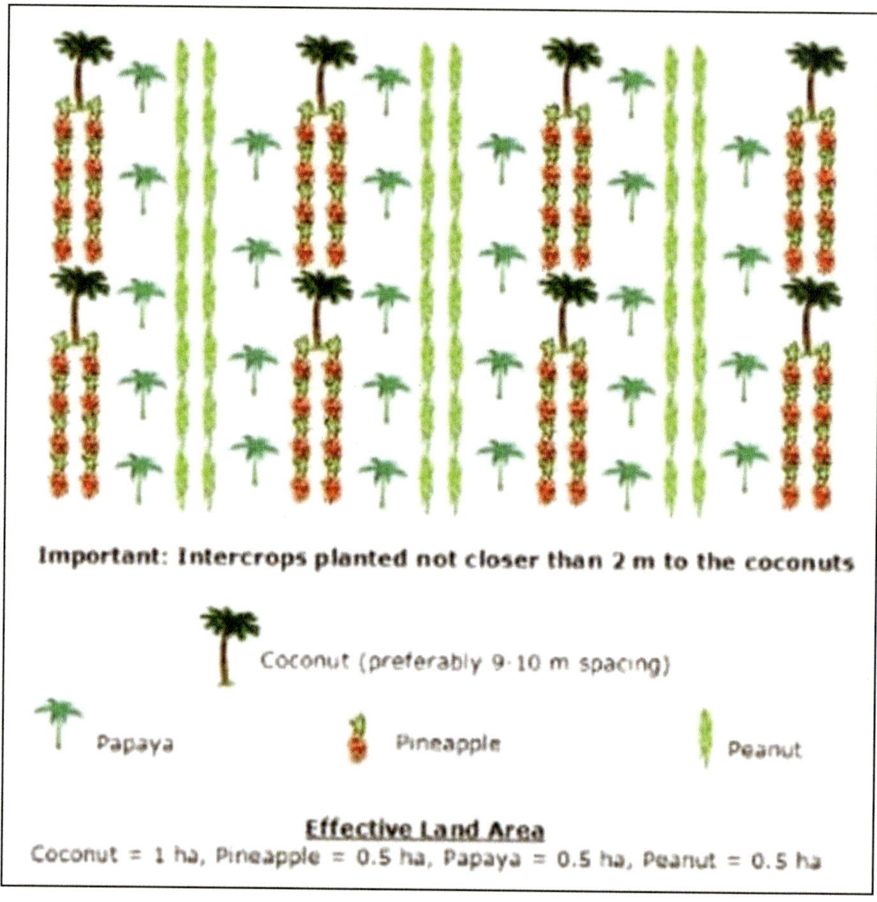

Figure 3.5: Coconut Based Cropping System.

Design and Layout of Intercropping Systems

Designing an effective intercropping system within orchards involves a complex interplay of ecological, agronomic, and environmental considerations. One of the primary questions in the design process is which traits should be prioritized to maximize the system's efficiency and productivity (Moore *et al.,* 2022). According to ecological theory, understanding interspecies interactions and their effect on productivity is critical when developing intercropping systems (Brooker *et al.,* 2015; Litrico and Violle, 2015). These interactions often involve competition for resources, such as light, water, and nutrients, which can reduce the productivity of one or both components.

Competition and Resource Allocation

Intercropping systems typically encounter two types of competition: exploitation competition, where species compete for the same resources, and interference competition, where one species alters the resource acquisition behaviour of another (Case and Gilpin, 1974; Schoener, 1983). Asymmetric exploitation competition is a frequent challenge in intercropping systems, where one species has a competitive advantage, leading to reduced productivity and a loss of potential benefits (Bybee-Finley *et al.,* 2016). This competitive advantage is often influenced by growing conditions–low nitrogen levels may favour legumes, while arid conditions could favour drought-tolerant species. The role of genotype × environment (G×E) interactions and genotype × environment × management (G×E×M) interactions become particularly important in addressing these challenges. These interactions inform breeding programs designed for specific regional conditions and management approaches. Over time, the competition between intercropped species can change depending on their phenology, stress tolerance, and persistence (Ginakes *et al.,* 2020). Understanding these temporal dynamics is crucial for selecting crop pairings and determining optimal planting times.

Temporal and Spatial Niche Differentiation

Niche differentiation is crucial for minimizing competition and enhancing system efficiency in intercropping. By allowing species to exploit different resources or use the same resources at different times, it reduces competitive interactions (Litrico and Violle, 2015). For instance, species with varying rooting depths or plant structures can access different soil layers, and differences in growth periods can lessen overlap. Choosing species with high phenotypic plasticity–adjusting well to varying conditions–further supports niche differentiation, improving resource sharing and reducing competition (Zhu *et al.,* 2015).

Facilitation

Facilitation is vital for successful intercropping, involving positive interactions where one species enhances the growth or survival of another. This can be through improved nutrient availability, such as nitrogen fixation, or by mitigating environmental stresses (Li *et al.,* 2014). Some crops may also suppress pests or diseases, benefiting their intercropped partners. Additionally, facilitation can involve allelopathic effects, where one plant releases chemicals that inhibit weeds. Choosing crops with low allelochemical production or resistance to these chemicals supports productivity. Facilitation is more common in perennial systems with longer growth cycles, offering more time for beneficial interactions compared to annual systems (Bybee-Finley and Ryan, 2018).

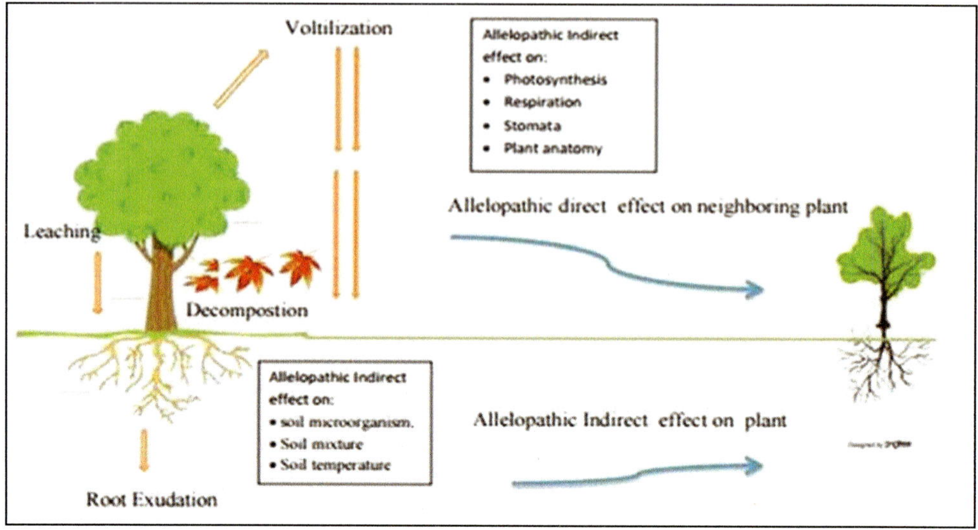

Figure 3.6: Allelopathic Effect.

Balancing Productivity and Ecosystem Services

Designing intercropping systems requires aligning goals with crop functions. If one crop is planted for ecosystem services like soil health or biodiversity, reducing the yield of a cash crop partner may be acceptable. However, if both crops have comparable value, some yield reduction might be acceptable if the system's overall benefits, such as improved land use efficiency, outweigh the losses.

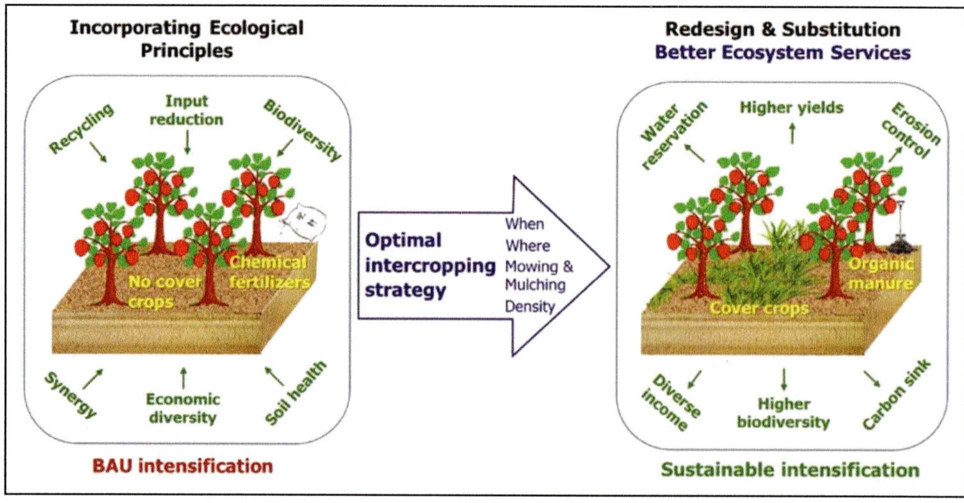

Figure 3.7: Optimal Intercropping Strategy in Apple Orchard (Gao *et al.*, 2022).

The key to successful intercropping is enhancing complementarity through niche differentiation and facilitation, which improves resource use efficiency and productivity while minimizing competition (Loreau and Hector, 2001). Careful species selection, spatial and temporal arrangements, and promoting positive interactions are essential for maximizing intercropping benefits in orchard systems.

Features and Advantages of Intercropping Systems in Orchards

Intercropping, the practice of growing multiple crops together in the same field, offers a range of benefits to orchard systems, making it a sustainable and profitable approach to farming. This system capitalizes on different plant species' complementary use of resources, including water, nutrients, and space. Table 3.1 gives the overview of intercropping systems in orchard. The following sections outline the key features and advantages of intercropping in orchard management, based on insights from several studies and research findings.

Enhanced Resource Utilization

Intercropping ensures more efficient utilization of resources and leads to higher productivity compared to growing crops individually. This system enhances the efficient use of natural resources like water, sunlight, and soil nutrients. By combining species with different root depths and nutrient needs, intercropping minimizes competition and maximizes the use of various soil layers. Common intercrops in orchards, such as legumes, vegetables, and cereals, complement the nutrient and water needs of the main tree crops (Maitra *et al.*, 2019). For instance, research comparing walnut and maize grown in monoculture systems versus an intercropped walnut-maize system demonstrated significant resource efficiency. The land equivalent ratio revealed that 32 per cent more land would be required to achieve similar yields if these crops were grown separately. Additionally, the water equivalent ratio showed that 31 per cent more water would be needed in monoculture systems to achieve the same output (Zalac *et al.*, 2023). This highlights the clear advantage of intercropping in terms of both land and water use efficiency.

Improved Soil Health and Biodiversity

Monocropping can cause soil degradation, erosion, and loss of biodiversity due to intensive tillage and exposed soil. Intercropping counters these issues by adding ground cover that reduces erosion, improves water retention, and boosts soil fertility with organic matter and root exudates (Morugan-Coronado *et al.*, 2020). Leguminous crops enhance soil quality by fixing atmospheric nitrogen (Sastre *et al.*, 2018). In Arecanut gardens, cover crops like *Pueraria javanica* meet a significant portion of nutrient needs. The Arecanut + cocoa system benefits from biomass recycling, boosting soil health (Bhat *et al.*, 2014). Intercropping also increases microbial activity, soil organic carbon, and carbon sequestration, vital for long-term soil health (Chamkhi *et al.*, 2022). Aromatic plants and liquorice improve microbial diversity and soil nitrogen in orchards (Zhang *et al.*, 2021; Li *et al.*, 2022). Additionally, intercropping with Chinese fir and Kura clover enhances soil nitrogen, microbial diversity, and reduces erosion (Zhang *et al.*, 2020; Kremer and Kussman, 2011).

Increased Ecosystem Services

Incorporating legumes, grasses, or aromatic plants into orchard alleys boosts soil health and provides essential ecosystem services. These intercrops suppress weeds, attract pollinators, and offer habitats for beneficial organisms like predatory arthropods, crucial for natural pest control (Almagro *et al.*, 2016). Aromatic plants such as lavender and rosemary enhance ecological diversity and resilience to climate challenges (Lee *et al.*, 2019). In insecticide-free apple orchards, intercropping flowering plants increases predator populations, improving biological pest control (Zhang *et al.*, 2022). Additionally, the main fruit trees modify the microclimate by reducing temperature fluctuations, mitigating wind, and increasing humidity, which optimizes water use and stabilizes growing conditions (Gosme *et al.*, 2016).

Pest and Disease Management

Intercropping enhances biological pest control by increasing habitat diversity for natural enemies like predators and parasitosis. Flowering plants in orchards provide nectar, pollen, and shelter for beneficial insects, promoting their presence. For instance, intercropping with *Cnidium monnieri* in apple orchards attracted predators such as ladybeetles and lacewings, reducing aphid populations (Zhang *et al.*, 2020). Aromatic plants like rosemary and basil also lower pest densities without harming natural enemies (Li *et al.*, 2021; Liu *et al.*, 2023). In citrus orchards, intercropping with guava reduced infestations of citrus pests (Gottwald *et al.*, 2010; Gottwald, 2014). This approach reduces the need for insecticides and supports orchard resilience.

Yield Stability and Resilience

While some farmers fear that intercropping might reduce yields due to resource competition, research indicates otherwise. Intercropping can enhance soil health, improving water retention and nutrient availability, which can maintain or even increase crop yields (Daryanto *et al.*, 2018). Annual crops in orchard alleys offer management flexibility, as they can be removed during critical growth stages to avoid impacting the main crop's productivity (Palese *et al.*, 2014). For example, intercropping cluster bean and wheat improved Kinnow fruit yield, with increases in plant height and fruit numbers (Gill *et al.*, 2018). Additionally, ground storey crops like Indian aloe and mustard in ber-based farming systems provided substantial yields alongside main crops (Verma and Saroj, 2018). Intercropping with crops such as pepper, cocoa, and banana can boost productivity (Bhat *et al.*, 2014), and cultivating brinjal in mango orchards even increased mango yields by 8.6 per cent (Singh *et al.*, 2015). These studies underscore the productivity benefits of intercropping in orchards.

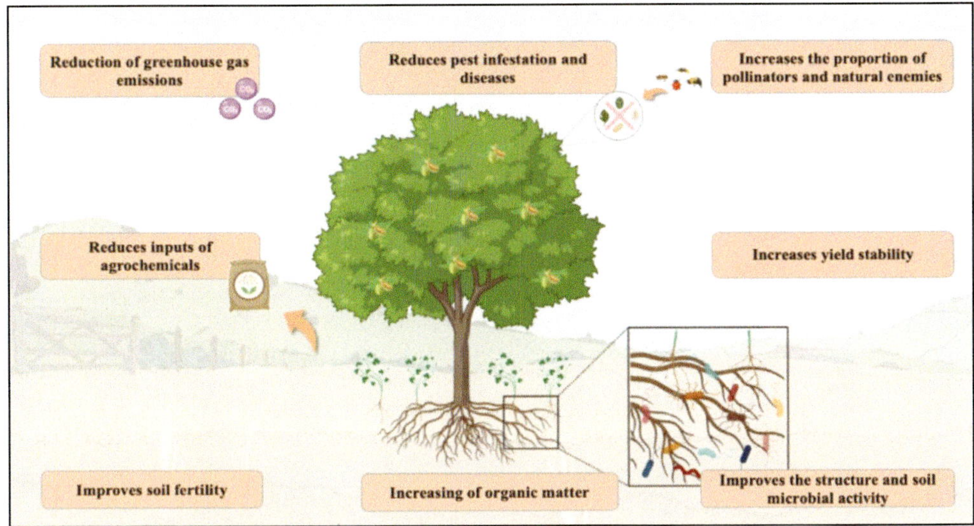

Figure 3.8: Advantages of Intercropping System.

Economic Benefits

Intercropping can significantly enhance the profitability of orchard systems by diversifying production and providing additional income streams. Integrating crops like fava beans, purslane, or barley with fruit trees can generate extra products, reducing market risks and boosting overall farm revenue (Maitra *et al.,* 2021). It also lowers production costs by improving soil fertility and reducing the need for synthetic fertilizers and pesticides (Chamkhi *et al.,* 2022). Legumes fix nitrogen, promoting sustainability, while medicinal and aromatic plants can increase productivity by 11-53 per cent and offer substantial returns (Sujatha *et al.,* 2016). Though initial costs may rise, selecting appropriate intercrops can mitigate market risks and enhance economic resilience, as shown by net returns of ₹60,000-1,10,000 per hectare in ber-based systems and ₹4.5 for every rupee invested in Arecanut plantations (Verma and Saroj, 2018; Bhat *et al.,* 2014).

Environmental Sustainability and Climate Resilience

Intercropping reduces the environmental footprint of orchard farming by enhancing resource conservation and lowering greenhouse gas emissions. A study in mandarin orchards showed that integrating crops like fava bean, purslane, and barley/vetch mix had no negative environmental impacts, supporting both environmental and economic sustainability (Martin-Gorriz *et al.,* 2022). Intercropping also helps conserve soil moisture and mitigate drought effects, particularly in arid regions, by reducing evapotranspiration and soil erosion (Morugan-Coronado *et al.,* 2020). Trees in orchards modify microclimate conditions, such as lowering temperatures and increasing humidity (Zalac *et al.,* 2023), and their deep roots enhance soil carbon storage and provide bio irrigation or biofertilization benefits (Burgess *et al.,* 2022). Overall, intercropping improves ecosystem resilience to climate change and extreme weather.

Table 3.1: Intercropping Systems in Orchards

Main Crop	*Intercrop*	*Advantages/Positive Effect*	*Reference*
Areca nut	Pandan (*Pandanus amaryllifolius*)	Intercropping had a beneficial effect on soil health, promoting greater microbial diversity and maintaining a stable ecological balance.	Zhong *et al.,* 2022
Pear	*Centaurea cyanus, Saturela hortensis, Nepeta cataria, Ageratum houstonianum,* and *Ocimum basilicum.*	Aromatic plants were particularly effective in reducing pest populations. Additionally, they had a positive impact on increasing the biodiversity of arthropod species within the orchard.	Song *et al.,* 2010
Apple	licorice	Intercropping licorice significantly enhances soil biodiversity by increasing microbial species, as well as boosting microbial nitrogen and carbon levels. This practice also promotes better apple tree growth and strengthens disease resistance.	Li *et al.,* 2022
Pear	*Ocimum basilicum, Satureja hortensis* and *Ageratum houstonianum*	Root exudates from aromatic plants play a crucial role in shaping soil microbial diversity, structure, and functional groups involved in the nitrogen cycle during the mineralization of organic matter. Intercropping with aromatic plants, particularly basil and summer savory, has been shown to enhance nitrogen release in orchard soils.	Zhang *et al.,* 2021
Coconut	Cocoa, Banana, Moringa and Pineapple	This multi-tier cropping system, which utilized 75 per cent NPK combined with organic recycling through vermicompost, produced the highest nut yield of 182 nuts per palm. It also generated the highest net income of ₹3.80 lakhs per hectare and achieved a B:C ratio of 2.71.	Nimbolkar *et al.,* 2016

Main Crop	*Intercrop*	*Advantages/Positive Effect*	*Reference*
Apple	Marigold	Intercropping with marigold improved soil moisture, organic matter content, nitrogen levels, and enzyme activity. It also boosted the presence of beneficial plant growth-promoting bacteria in the rhizosphere, including species from the Rhizobiales, Pseudomonadales, and Bacillales orders.	Xue *et al.*, 2023
Macadamia	Coffee	Macadamia nut production in an intercropping system with coffee saw a remarkable 251 per cent increase compared to rainfed monocropping. This system led to a 32 per cent increase in the number of fruits and a 30 per cent rise in nut-in-shell yield per tree compared to monoculture, ultimately enhancing overall macadamia production.	Perdoná *et al.*, 2015
Apple	*B. napus, V. villosa, C. monnieri,* and *S. tenuifolia*	Intercropping flowering plants in an apple orchard, increased predator populations and enhanced the biological control of the aphid *Aphis spiraecola.*	Zhang *et al.*, 2022
Mango	Cowpea and Indian mustard	The intercropping system consistently achieved the highest values for system productivity, gross returns, net returns, and B:C ratio in both crop seasons.	Mirjha and Rana, 2016
Cashew	Mango ginger, elephant foot yam, turmeric, east Indian arrowroot, and taro	The intercropping system of cashew and turmeric achieved a notably higher cashew equivalent yield, reaching 3,521.58 kg/ha, primarily due to turmeric's higher output. Meanwhile, the cashew-taro intercropping system demonstrated superior land use efficiency, with a land equivalence ratio of 1.67 and a production efficiency of 54.80 kg/ha/day, outperforming other intercropping combinations	Ramteke *et al.*, 2022

Main Crop	*Intercrop*	*Advantages/Positive Effect*	*Reference*
Jackfruit	Cabbage, cauliflower, broccoli, tomato, brinjal, palak, peas, radish, cucumber, amaranthus	The jackfruit and tomato intercropping system shows a higher B:C ratio of 2.86. Intercropping not only promotes the growth of young jackfruit trees but also enhances soil fertility in the orchard.	Santhosh-kumar *et al.,* 2022
Mandarin	Purslane and fava bean	Intercropping can be a profitable strategy that helps farmers manage the risks associated with fluctuations in crop prices.	Martin-Gorriz *et al.,* 2022
Olive	Barley + pea, Barley + vetch, Triticale + pea, Triticale + vetch	The highest biomass yield was achieved with a triticale and pea intercrop. Using intercrops can boost biomass production for animal feed and enhance soil quality in olive tree plantations.	Ozturkmen *et al.,* 2020
Kinnow mandarin	Cluster been + wheat Cluster been + barley Cluster been + fenugreek Green gram + wheat Green gram + barley Green gram + fenugreek	The cluster bean and wheat intercropping system demonstrated superior performance, as indicated by several key metrics. This system resulted in an average plant height of 3.31 cm, a gross income of ₹41,180, a fruit count of 175, and a TSS percentage of 10.8 per cent for Kinnow fruit.	Gill *et al.,* 2018
Jujube	Cotton	Cotton proved far more competitive than jujube, especially under water scarcity conditions. Cotton roots encroached on the space occupied by jujube roots, significantly impeding the lateral growth of the jujube roots.	Ai *et al.,* 2021
Pomegranate	Sweet basil and Rosemary	The combined benefits in terms of land equivalent ratio, area time equivalent ratio, and land utilization efficiency were highest when pomegranate was intercropped with certain aromatic plants.	Dikr, 2022

Case Studies and Success Stories

Case Study 1: Guava-Based Intercropping for Sustainable Farming

In the Haroti region of Rajasthan, traditional farming methods and crops like soybean, wheat, and coriander have long dominated the landscape. However, these crops often face challenges such as erratic rainfall, pest infestations, and market gluts that result in low prices and reduced profitability. With these issues in mind, one farmer, Sh. Ghanshyam, sought innovative solutions to integrate guava-based intercropping into his farm.

Implementing Guava-Based Intercropping

Ghanshyam planted guava trees in a 3-hectare orchard, spaced 5 × 5 meters apart, and began experimenting with various intercropping combinations. These included guava with onion, tomato, garlic, and potato, along with other vegetables like brinjal, pumpkin, and marigold. He also incorporated relay cropping techniques, such as garlic and bottle gourd, to maximize the utilization of space between the guava trees.

Figure 3.9: Guava + Onion.

Figure 3.10: Guava + Garlic.

Economic and Environmental Impact

The intercropped system not only enhanced the productivity of the land but also provided higher economic returns compared to traditional sole cropping systems. For example:

- Sole vegetable crops like onion, garlic, and tomato gave an average net return of ₹281,360/ha with a Benefit-Cost (B:C) ratio of 2.55.
- Guava as a sole crop yielded a higher net return of ₹385,260/ha with a B:C ratio of 3.25.
- However, the guava-based intercropping system generated a significantly higher net return of ₹611,983/ha, with a B:C ratio of 4.13, nearly doubling the profitability compared to sole cropping.

In addition to the economic benefits, this system improved soil fertility, reduced weed growth, and enhanced water-use efficiency. The integration of various vegetable crops also mitigated the risk of total crop failure due to pests or poor market conditions, providing Ghanshyam with more consistent income throughout the year.

Case Study 2: Maximizing Mango Orchard Yields with Intercropping Systems

At the Central Institute for Subtropical Horticulture in Lucknow, a groundbreaking experiment was carried out to explore the effects of various intercrops on the yield of the popular mango cultivar 'Dashehari.' The goal was to identify how intercropping could boost mango production, enhance economic returns, and support sustainable farming practices (Singh *et al.,* 2015).

The research focused on intercropping vegetables like Brinjal ('Rajni'), Bottlegourd ('Varad'), Cauliflower ('Girja') and Cabbage ('Indam 296').

Effect on Yield

- ✰ Brinjal emerged as the top performer, with a yield of 19.38 tons per hectare, significantly outperforming other intercrops. Bottlegourd followed with a yield of 13.54 tons per hectare. Whereas, Cabbage yielded the least, at 8.50 tons per hectare.
- ✰ Not only did brinjal outperform the other crops in terms of yield, but it also positively impacted mango production. The intercropping of brinjal led to an 8.6 per cent increase in mango yield, translating to an additional 4.55 tons per hectare.

Economic Impact

1. The economic benefits of intercropping were substantial. The highest monetary return was achieved with brinjal intercropping in the mango orchard, amounting to ₹160,300 per hectare. In contrast, the sole mango crop yielded only ₹49,920 per hectare.
2. Intercropping not only boosted income but also provided additional employment opportunities, particularly during the off-season and pre-production phases. This was especially beneficial for small and marginal farmers who face economic challenges during periods when mango production alone might not be sufficient.

Case Study 3: Securing Livelihood through Ber-Based Cropping System

In the hot arid regions of India, farmers face numerous challenges, including scarce and erratic rainfall, extreme temperatures, recurrent droughts, and poor soil conditions. These environmental constraints make traditional monocropping risky, with low yields and unstable incomes that are often insufficient to sustain families. To address this, the adoption of a ber-based cropping system has emerged as a sustainable solution, particularly for small and marginal farmers.

The Ber-Based Cropping System

To enhance productivity and sustainability, the Central Institute for Arid Horticulture (ICAR–CIAH) initiated research on the ber-based cropping system under the limited irrigated conditions of western Rajasthan. The system involved integrating the ber cultivar 'Gola' with groundstorey crops such as groundnut, wheat, cluster bean, mustard, and Indian aloe.

Key Findings

Resource Efficiency

There was no competition between ber and the groundstorey crops during the initial establishment phase. In fact, the inputs applied to the ground crops, such as water and nutrients, boosted the growth and Vigor of the ber plants, particularly when paired with Indian aloe.

Enhanced Yield

The system produced multiple outputs, including ber fruits, green pods from cluster beans, and dry pods from groundnuts. The combination of ber and Indian aloe yielded the highest fruit production compared to other crop pairings. This diversification provided farmers with a steady and reliable income throughout the year.

Figure 3.11: Ber + Cluster Bean.

Water and Nutrient Management

Given the limited water availability in arid lands, the system incorporated water harvesting techniques and judicious irrigation methods, such as drip irrigation, to conserve water. The fertility of the soil was maintained through the residues of intercrops and targeted nutrient management, especially in micronutrient-deficient soils.

Economic Benefits

The ber-based cropping system proved economically viable. Farmers could achieve a net return of ₹60,000 to ₹110,000 per hectare annually by intercropping cluster beans during the kharif season and mustard during the rabi season, alongside ber fruit production. This system provided a much-needed boost to farmers' livelihoods in an otherwise challenging environment.

Case Study 4: Enhancing Productivity and Income through Arecanut-Based Cropping System

Arecanut (*Areca catechu*), commonly grown in the West Coast and North East regions of India, is a significant plantation crop known for its economic value. However, the profitability of arecanut farming faces challenges due to price fluctuations, pest and disease outbreaks, waterlogging, and drought. For small and marginal farmers, relying solely on arecanut is risky, leading to unstable income. To overcome these issues, intercropping and mixed cropping systems have been developed, offering a solution for increasing productivity and generating steady income.

The Arecanut Advantage

Arecanut's compact crown structure, which stands 10–15 meters above the ground, allows ample sunlight to reach the understory crops, creating an ideal environment for shade-loving plants.

Intercropping Systems and Benefits

A wide variety of crops are compatible with arecanut plantations. In the early stages, annuals like banana, ginger, turmeric, and vegetables can be cultivated as intercrops. As the plantation matures, shade-loving crops such as pepper, cocoa, betel vine, and medicinal plants thrive under the arecanut canopy.

1. **Arecanut + banana:** Banana acts as a nurse crop, providing shade in the early stages of arecanut growth. Varieties such as Red Banana and Robusta are ideal for intercropping and generate interim income for farmers. This system also increases soil moisture content and protects crops from sunscald.
2. **Arecanut + cocoa + pepper:** This high-density multispecies cropping system combines cocoa, black pepper, and banana. It helps in recycling organic matter, reducing fertilizer needs by 50-66 per cent, and improving soil fertility. Intercropping cocoa with arecanut enhances productivity by 650-900 kg/ha, and pepper can yield around 1300 kg/ha, offering farmers a steady income source.
3. **Medicinal and aromatic plants:** Growing medicinal plants like vetiver, long pepper, and basil in arecanut plantations provides high returns, with profits ranging from ₹1.95 to ₹4.25 per rupee invested. These crops can

increase productivity by 11-53 per cent and boost soil fertility through nitrogen fixation.

Figure 3.12: Arecanut + Banana.

Figure 3.13: Arecanut + Cocoa + Pepper.

Economic and Environmental Gains

Arecanut-based cropping systems not only improve productivity but also enhance farmers' income. The net returns from intercropping systems can reach up to ₹4.5 per rupee invested, highlighting the economic viability of these models. Additionally, the system promotes better soil health by reducing erosion, increasing organic matter, and improving nutrient availability, especially for nitrogen and potassium.

Limitations of Intercropping

Intercropping offers several limitations that impact its effectiveness in agricultural systems (Mir *et al.,* 2022). Mechanization challenges may arise, as the practice can limit the use of large-scale machinery, making operations more labour-intensive. Harvesting difficulties can occur when multiple crops are involved, complicating the process. The system demands increased management, requiring more attention, labour, and knowledge to balance the needs of different crops. In some cases, intercropping can lead to reduced yields of the primary crop if the intercrops are poorly matched and compete for resources. While leguminous plants may help with nitrogen fixation, limited nitrogen fixation can still be a concern if non-legume species dominate the intercrop mix. Additionally, herbicide use limitations may arise, as certain chemicals may harm intercrops. The potential for allelopathy, where certain plants release chemicals that inhibit the growth of others, also poses a challenge. Finally, intercropping can increase drudgery, making it labour-intensive, and intercultural operation challenges can arise when managing different crops simultaneously. These factors highlight the need for careful planning and management to implement intercropping systems successfully.

Intercropping plants of varying heights can lead to mutual shading, where taller plants cast shadows over shorter ones (Liu and Song, 2012). This can reduce photosynthesis and crop yield for the smaller species (Peng *et al.,* 2009). However, intercropping can also increase air humidity, which can promote plant growth by boosting photosynthesis and leading to higher crop yields (Li *et al.,* 2020). Resource scarcity can intensify the challenges of intercropping. While much research has focused on the physical characteristics of successful intercropped species, our understanding of how plants physiologically respond to water limitations within these systems remains limited (Ai *et al.,* 2021).

Financial Constraint

Multiple cropping practices face challenges due to inefficient markets, limited access to knowledge and financial resources, and unclear land tenure rights. Successful adoption of these practices requires supportive policies, strong institutions, and market demand to influence farmer and consumer behaviour (Alam *et al.,* 2014; Dhandapani *et al.,* 2020). Agroforestry, despite its advantages, is hindered by legal barriers, unfavourable policies, and poor coordination among government agencies, often due to the separation of forest and agricultural policies, missing broader landscape-level benefits (Mbow *et al.,* 2014).

Agronomic Constraints

The selection and spatial arrangement of plants in multiple cropping systems are vital in determining competition and benefits. Crafting the optimal combination is complex, requiring more skills and knowledge than monoculture (Burgess *et al.,* 2022). This includes precise timing of field operations, adjusting cultivation practices, and careful use of inputs like fertilizers and mulches. Crop selection depends on region, farmer expertise, intercropping, irrigation, marketing, and soil conditions. Drought-resistant crops perform well in arid areas, while root crops and certain vegetables suit low-fertility soils. Coastal laterite soils favour root crops, tobacco, and vegetables, with crops like pineapple, banana, ginger, and spices intercropped with oil palm (Awasthi *et al.,* 2018).

Varietal Constraints

Although multiple-cropping systems offer numerous potential benefits, there has been little focus on breeding and genetic improvement for these systems. Currently, no crop varieties are specifically developed for multiple cropping, and those used in intercropping or agroforestry were originally bred for monoculture. As a result, their performance in these alternative systems often remains untested (Brooker *et al.,* 2015). This lack of adaptability limits the advantages of such systems (Saxena *et al.,* 2018). Simulation studies suggest that breeding programs for intercropping that incorporate genomic selection can achieve faster genetic progress than those relying solely on phenotypic data (Bancic *et al.,* 2021). This approach can help shorten the time needed to develop new varieties while improving the intensity and precision of selection. However, breeding for multi-species systems requires more complex and integrated goals for breeders (Duc *et al.,* 2015).

Allelochemicals

Allelochemicals are present in nearly all plant tissues–leaves, fruits, stems, and roots–and are released through volatilization, root exudation, and decomposition. Leaves are the most consistent source, while roots typically contain fewer toxins. According to Rice (1984) and Putnam (1985), there are four primary ways allelochemicals are released: volatilization, leaching, root exudation, and decomposition of plant residues. They influence various biological processes, including mineral uptake, hormone balance, and photosynthesis. Allelochemicals, which include simple hydrocarbons, organic acids, polycyclic structures, phenols, flavonoids, and amino acids, play a crucial role in plant development and ecosystem interactions. These compounds affect plant interactions and ecological processes by altering physiological and biochemical pathways (Awasthi *et al.,* 2018).

Conclusion

Intercropping in orchard systems is a practical and sustainable method to tackle soil degradation, pest control, and climate adaptation. By incorporating crops like legumes, grasses, and aromatic plants, farmers can optimize resource use, enhance soil fertility, and boost water retention, particularly in water-stressed areas. This practice also reduces erosion and fosters biodiversity, creating habitats for beneficial organisms that aid in natural pest control and decrease reliance on chemical inputs. Economically, intercropping diversifies income streams, which is particularly advantageous during long pre-bearing periods of orchard trees. Additionally, it moderates microclimate conditions, improving temperature regulation and humidity, which supports better tree growth and crop yields. Overall, intercropping enhances the resilience of orchard systems to environmental stresses, ensuring long-term sustainability and productivity.

REFERENCES

Ai P, Ma Y and Hai Y, 2021. Influence of jujube/cotton intercropping on soil temperature and crop evapotranspiration in an arid area. *Agricultural Water Management,* **256**: 107118.

Alam M, Olivier A, Paquette A, Dupras J, Reveret JP, Messier C, 2014. A general framework for the quantification and valuation of ecosystem services of tree-based intercropping systems. Agrofor. Syst. **88**: 679–691.

Almagro M, de Vente J, Boix-Fayós C, García-Franco N, Melgares de Aguilar J, González D and Martínez-Mena M 2016. Sustainable land management practices as providers of several ecosystem services under rainfed Mediterranean agroecosystems. *Mitigation and adaptation strategies for global change,* **21**: 1029-1043.

Altieri MA, 1999. The ecological role of biodiversity in agroecosystems. *Agriculture, Ecosystems and Environment*, **74**(1-3): 19-31.

Andrews DJ and Kassam AH, 1976. The importance of multiple cropping to increasing world food supplies. In: Stelly, M. (Ed.), Multiple Cropping. ASA Special Publication 27, Madison, WI, pp. 1–10

Awasthi OP, 2018. Fruit based diversified cropping system: an alternative approach for nutritional and economic security. *Progressive Horticulture*, **50**(1 and 2): 41-46.

Bancic J, Werner C, Gaynor R, Gorjanc G, Odeny D, Ojulong H, Dawson I, Hoad S and Hickey J, 2021. Modeling illustrates that genomic selection provides new opportunities for intercrop breeding. *Front. Plant Sci.* **12**: 605172.

Bhat R, Sujatha S and Chowdappa P, 2014. Arecanut-based cropping system fetches more return. *Indian Horticulture*, **59**(6).

Brooker RW, Bennett AE, Cong WF, Daniell TJ, George TS, Hallett PD and White PJ, 2015. Improving intercropping: a synthesis of research in agronomy, plant physiology and ecology. *New Phytologist*, **206**(1): 107-117.

Burgess AJ, Cano MEC and Parke B, 2022. The deployment of intercropping and agroforestry as adaptation to climate change. *Crop and Environment*, **1**(2): 145-160.

Bybee-Finley KA and Ryan MR, 2018. Advancing intercropping research and practices in industrialized agricultural landscapes. *Agriculture*, **8**(6): 80.

Bybee–Finley K A, Mirsky SB and Ryan MR, 2016. Functional diversity in summer annual grass and legume intercrops in the northeastern United States. *Crop Science*, **56**(5): 2775-2790.

Case TJ and Gilpin ME, 1974. Interference competition and niche theory. *Proceedings of the National Academy of Sciences*, **71**(8): 3073-3077.

Chamkhi I, Cheto S, Geistlinger J, Zeroual Y, Kouisni L, Bargaz A and Ghoulam C, 2022. Legume-based intercropping systems promote beneficial rhizobacterial community and crop yield under stressing conditions. *Industrial Crops and Products*, **183**: 114958.

Daryanto S, Fu B, Wang L, Jacinthe PA and Zhao W, 2018. Quantitative synthesis on the ecosystem services of cover crops. *Earth-Science Reviews*, **185**, 357-373.

Das B, Kandpal BK and Devi HL, 2021. Cover crops for orchard soil management. In Cover Crops and Sustainable Agriculture (pp. 147-168). CRC Press.

Dhandapani S, Girkin N, Evers S, Ritz K, Sjogersten S, 2020. Is intercropping an € environmentally-wise alternative to established oil palm monoculture in tropical peatlands? Front. For. Glob. Chang. **3**, 70.

Dikr W, 2022. Role of intercropping some aromatic and medicinal plants with fruit vegetables crops, a Review. *Glob. Acad. J. Agric. Biosci*, **4**: 22-30.

Duc G, Agrama H, Bao S, Berger J, Bourion V, De Ron A, Gowda C, Mikic A, Millot D, Singh K, Tullu A, Vandenberg A, Vaz Patto M, Warkentin T and Zong X, 2015. Breeding annual grain legumes for sustainable agriculture: new methods to approach complex traits and target new cultivar ideotypes. CRC. Crit. Rev. Plant Sci. **34**: 381–411.

Francis CA, 1986. Distribution and importance of multiple cropping. In: Francis CA (ed) Multiple Cropping Systems. MacMillan Publishing Company, New York, USA.

Gao X, He N, Jia R, Hu P and Zhao X 2022. Redesign of dryland apple orchards by intercropping the bioenergy crop canola (*Brassica napus* L.): Achieving sustainable intensification. *GCB Bioenergy*, **14**(3): 378-392.

Gill MS, Khehra S and Gupta N, 2018. Impact of intercropping on yield, fruit quality and economics of young Kinnow mandarin plants. *Journal of Applied and Natural Science*, **10**(3): 954-957.

Ginakes P, Grossman J, Baker J and Sooksa-nguan T, 2020. Tillage intensity influences nitrogen cycling in organic kura clover living mulch. *Nutrient Cycling in Agroecosystems*, **116**: 71-82.

Glaze S, Hashemi M, Sadeghpour A, Jahanzad E, Afshar RK, Liu X and Herbert SJ, 2020. Understanding intercropping to improve agricultural resiliency and environmental sustainability. Advances in agronomy, **162**: 199-256.

Gosme M, Dufour L, Inurreta Aguirre HD and Dupraz C, 2016. Microclimatic effect of agroforestry on diurnal temperature cycle. In *Proceedings of the 3rd European Agroforestry Conference, European Agroforestry Federation, Montpellier, France* (pp. 23-25).

Gottwald TR, Hall DG, Beattie GAC, Ichinose K, Nguyen MC, Le QD and Hilf ME, 2010. Investigations of the effect of guava as a possible tool in the control/ management of huanglongbing. In *International Organization of Citrus Virologists Conference Proceedings* (*1957-2010*) (Vol. 17, No. 17).

Gottwald T R, Hall DG, Kriss AB, Salinas EJ, Parker PE, Beattie GAC and Nguyen MC, 2014. Orchard and nursery dynamics of the effect of interplanting citrus with guava for huanglongbing, vector, and disease management. *Crop Protection*, **64**: 93-103.

Hegde NK, Vijayakumar BN and Pushpa P, 2015. Importance of multistoried cropping systems in Horticulture. *Winter school on multi-storeyed cropping system and canopy management in horticultural crops from 28th Sept. to 18th Oct*, 10-13.

Horwith B, 1985. A role for intercropping in modern agriculture. *BioScience*, **35**: 286-291.

Kassam A, Friedrich T and Derpsch R, 2011. Conservation agriculture in the 21st century: A paradigm of sustainable agriculture. *European Journal of Agronomy*, **34**(2): 33-34.

Kremer RJ and Kussman RD, 2011. Soil quality in a pecan–kura clover alley cropping system in the Midwestern USA. *Agroforestry systems*, **83**: 213-223.

Lammerts Van Bueren ET, Struik PC and Jacobsen E 2002. Ecological concepts in organic farming and their consequences for an organic crop ideotype. *Nether J Agr Sci* **50**: 1-26.

Li C, Wang R, Zhou X, Cheng H, Liu W and Li Y, 2020. Effects of water and fertilizer regulation on photosynthetic characteristics, yield and water use of apple-soybean intercropping system under drip irrigation. *Journal of Soil and Water Conservation*, **34**: 299–310.

Li L, Tilman D, Lambers H and Zhang FS, 2014. Plant diversity and overyielding: insights from belowground facilitation of intercropping in agriculture. *New phytologist*, **203**(1): 63-69.

Li XW, Lu XX, Zhang Z J, Huang J, Zhang JM, Wang LK. and Lu, YB, 2021. Intercropping rosemary (*Rosmarinus officinalis*) with sweet pepper (*Capsicum annum*) reduces major pest population densities without impacting natural enemy populations. *Insects*, **12**(1): 74.

Li X, Chu Y, Jia Y, Yue H, Han Z and Wang Y, 2022. Changes to bacterial communities and soil metabolites in an apple orchard as a legacy effect of different intercropping plants and soil management practices. *Frontiers in Microbiology*, **13**: 956840.

Lithourgidis A, Dordas C, Damalas CA and Vlachostergios D, 2011. Annual intercrops: an alternative pathway for sustainable agriculture. *Australian Journal of Crop Science* **5**(4): 396-410.

Litrico I and Violle C, 2015. Diversity in plant breeding: a new conceptual framework. *Trends in Plant Science*, **20**(10): 604-613.

Liu Y, Zhang Y, Xiao T, Wu Y, Li Y, He J and Yao B, 2023. Effect of Intercropping on Fruit Yield and Financial Benefits of *Rosa roxburghii* Tratt Orchard in Southwest China. *Agronomy*, **13**(12): 2953.

Loreau M and Hector A, 2001. Partitioning selection and complementarity in biodiversity experiments. *Nature*, **412**(6842): 72-76.

Maitra S, Hossain A, Brestic M, Skalicky M, Ondrisik P, Gitari H and Sairam M, 2021. Intercropping–A low input agricultural strategy for food and environmental security. *Agronomy*, **11**(2): 343.

Maitra S, Palai JB, Manasa P and Kumar DP, 2019. Potential of intercropping system in sustaining crop productivity. *International Journal of Agriculture, Environment and Biotechnology*, **12**(1): 39-45.

Martin-Gorriz B, Zabala JA, Sánchez-Navarro V, Gallego-Elvira B, Martínez-García V, Alcon F and Maestre-Valero JF, 2022. Intercropping practices in mediterranean mandarin orchards from an environmental and economic perspective. *Agriculture*, **12**(5): 574.

Martins CR, Rodrigues GS and Barros ID, 2021. Assessment of economic and environmental performance in citrus-based intercropping systems. *Revista Brasileira de Fruticultura*, **43**: e-463.

Mbow C, Van Noordwijk M, Luedeling E, Neufeldt H, Minang P, Kowero G, 2014. Agroforestry solutions to address food security and climate change challenges in Africa. *Curr. Opin. Environ. Sustain.* **6**: 61–67.

Mir MS, Saxena A, Kanth RH, Raja W, Dar KA, Mahdi SS and Mir SA, 2022. Role of intercropping in sustainable insect-pest management: a review. *International Journal of Environment and Climate Change*, **12**(11): 3390-3403.

Mirjha PR and Rana DS, 2016. Yield and yield attributes, system productivity and economics of mango (Mangifera indica)-based intercropping systems as influenced by mango cultivars and nutrient levels. *Indian Journal of Agronomy*, **61**(3): 307-314.

Moore VM, Schlautman B, Fei SZ, Roberts LM, Wolfe M, Ryan MR and Lorenz AJ, 2022. Plant breeding for intercropping in temperate field crop systems: a review. *Frontiers in Plant Science*, **13**: 843065.

Mortensen DA and Smith RG, 2020. Confronting barriers to cropping system diversification. *Frontiers in Sustainable Food Systems*, **4**: 564197.

Morugán-Coronado A, Linares C, Gómez-López MD, Faz Á and Zornoza R, 2020. The impact of intercropping, tillage and fertilizer type on soil and crop yield in fruit orchards under Mediterranean conditions: A meta-analysis of field studies. *Agricultural Systems*, **178**: 102736.

Mousavi SR and Eskandari H, 2011. A general overview on intercropping and its advantages in sustainable agriculture. *Journal of Applied Environmental and Biological Sciences*, **1**(11): 482-486.

Nimbolkar PK, 2016. Multi storied cropping system in horticulture-a sustainable land use approach. *International Journal of Agriculture Sciences, ISSN*, 0975-3710.

Ozturkmen AR, Ramazanoglu E, Almaca A and Çakmakli M, 2020. Effect of intercropping on soil physical and chemical properties in an olive orchard. *Applied Ecology and Environmental Research*, **18**(6): 103-108.

Palese AM, Vignozzi N, Celano G, Agnelli AE, Pagliai M and Xiloyannis C, 2014. Influence of soil management on soil physical characteristics and water storage in a mature rainfed olive orchard. *Soil and Tillage Research*, 144: 96-109.

Papanastasis VP, Arianoutsou M andLyrintzis G, 2004. Management of biotic resources in ancient Greece. Proceedings of the 10th Mediterranean Ecosystems (MEDECOS) Conference, 25 April-01 May 2004, Rhodes, Greece, pp. 1-11.

Peng X, Zhang Y, Cai J, Jiang Z, Zhang S, 2009. Photosynthesis, growth and yield of soybean and maize in a tree-based agroforestry intercropping system on the Loess Plateau. *Agrofor. Syst.* **76**: 569–577.

Perdoná MJ and Soratto RP, 2015. Higher yield and economic benefits are achieved in the macadamia crop by irrigation and intercropping with coffee. *Scientia Horticulturae*, **185**: 59-67.

Plucknett DL and Smith NJH, 1986. Historical perspectives on multiple cropping. In: Francis CA (ed) Multiple Cropping Systems. MacMillan Publishing Company. New York, USA.

Ramteke V, Thakur P, Kerketta A, Paramesh V, Nirala YS, Netam RS and Adiga JD, 2022. Assessment of Cashew-Based Intercropping System in Chhattisgarh, India. *National Academy Science Letters*, **45**(6): 485-489.

RenYY, Wang XL, Zhang SQ, Palta JA and Chen YL, 2017. Influence of spatial arrangement in maize-soybean intercropping on root growth and water use efficiency. *Plant and Soil*, **415**(1): 131-144.

Rice EL, 1984. Allelopathy, Second Edition, Pub. Academic Press Inc. New York, London.

Santhoshkumar GM, Nilesh Bhowmick Choudhury A, 2022. Performance of Intercropping under Juvenile Jackfruit Orchard in Sub-Himalayan Terai Region of West Bengal. *Biological Forum – An International Journal*, **14**(2): 498-502.

Sastre B, Marques MJ, García-Díaz A and Bienes R, 2018. Three years of management with cover crops protecting sloping olive groves soils, carbon and water effects on gypsiferous soil. *Catena*, **171**: 115-124.

Sau S, Sarkar S, Das A, Saha S and Datta P, 2017. Space and time utilization in horticulture based cropping system: an income doubling approach from same piece of land. *Journal of Pharmacognosy and Phytochemistry*, **6**(6): 619-624.

Saxena K, Choudhary A, Saxena R, Varshney R, 2018. Breeding pigeonpea cultivars for intercropping: synthesis and strategies. *Breed. Sci.* **68**: 159–167.

Schoener TW, 1983. Field experiments on interspecific competition. *The American Naturalist*, **122**(2): 240-285.

Singh D, Mathimaran N, Boller T and Kahmen A, 2020. Deep-rooted pigeon pea promotes the water relations and survival of shallow-rooted finger millet during drought–Despite strong competitive interactions at ambient water availability. *PLoS One*, **15**(2): e0228993.

Singh VK, Kumar Soni M and Singh K, 2015. Increasing profitability of mango (*Mangifera indica* L.) orchard through intercropping. In *Global Conference on*

Augmenting Production and Utilization Of Mango: Biotic and Abiotic Stresses, Acta Hort. **1066**: 151-157.

Song BZ, Wu HY, Kong Y, Zhang J, Du YL, Hu JH and Yao YC, 2010. Effects of intercropping with aromatic plants on the diversity and structure of an arthropod community in a pear orchard. *BioControl*, **55**: 741-751.

Sujatha S, Bhat R and Chowdappa P, 2016. Cropping systems approach for improving resource use in arecanut (*Areca catechu*) plantation. *The Indian Journal of Agricultural Sciences*, **86**(9): 1113-20.

Sun Y, Chen L, Zhang S, Miao Y, Zhang Y, Li Z and Yao Y, 2022. Plant interaction patterns shape the soil microbial community and nutrient cycling in different intercropping scenarios of aromatic plant species. *Frontiers in Microbiology*, **13**: 888789.

Vandermeer JH, 1992. The ecology of intercropping. *Cambridge University Press.*

Verma AK and Saroj PL, 2018. Securing livelihood through ber-based cropping system. *Indian Horticulture*, **63**(5): 7-8.

Wang J, 2019. Olive (*Olea europaea*) tree growth and development and the effect of intercropping with wheat (*Triticum aestivum*) or wheat undersown with white clover (*Trifolium repens*): A thesis submitted in partial fulfilment of the requirements for the Degree of Master of Applied Science at Lincoln University (Doctoral dissertation, Lincoln University).

Xue X, Chen R, Xu C, Zhang C, Dong L, Zhao X and Wang X, 2023. Apple-marigold intercropping improves soil properties by changing soil metabolomics and bacterial community structures. *Frontiers in Microbiology*, **14**: 1195985.

Yin W, Chai Q, Zhao C, Yu A, Fan Z, Hu F and Coulter JA, 2020. Water utilization in intercropping: A review. *Agricultural Water Management*, **241**: 106335.

Žalac H, Herman G, Ergoviæ L, Joviæ J, Zebec V, Bubalo A and Iveziæ V, 2022. Ecological and Agronomic Benefits of Intercropping Maize in a Walnut Orchard–A Case Study. *Agronomy*, **13**(1), 77.

Zhang X, Huang G and Wang L, 2019. Advances in the mechanization and automation of intercropping systems. *Agronomy Journal*, **111**(1): 1-10.

Zhang X, Ouyang F, Su J, Li Z, Yuan Y, Sun Y and Ge F, 2022. Intercropping flowering plants facilitate conservation, movement and biocontrol performance of predators in insecticide-free apple orchard. *Agriculture, Ecosystems and Environment*, **340**: 108157.

Zhang Y, Han M, Song M, Tian J, Song B, Hu Y and Yao Y, 2021. Intercropping with aromatic plants increased the soil organic matter content and changed the microbial community in a pear orchard. *Frontiers in Microbiology*, **12**: 616932.

Zhang Y, Hou L, Li Z, Zhao D, Song L, Shao G and Sun Q, 2020. Leguminous supplementation increases the resilience of soil microbial community and nutrients in Chinese fir plantations. *Science of the Total Environment,* **703**: 134917.

Zhi C, Shuoshi M, Guisheng F, Yonglai Z, 2007. Effect of wind erosion resistance of soil for wheat and potato strip intercropping farmland. *Trans. Chin. Soc. Agric. Eng.* **23**(3): 51–54.

Zhong Y, Zhang A, Qin X, Yu H, Ji X, He S and Tang J, 2022. Effects of Intercropping Pandanus amaryllifolius on soil properties and microbial community composition in Areca Catechu plantations. *Forests,* **13**(11): 1814.

Zhu J, van der Werf W, Anten NP, Vos J and Evers JB, 2015. The contribution of phenotypic plasticity to complementary light capture in plant mixtures. *New Phytologist,* **207**(4): 1213-1222.

Chapter 4

Water Conservation in Orchards

Rajender Kumar and Vinay Kumar*

School of Organic Farming, PAU, Ludhiana–141004, Punjab
**Department of Agronomy, PAU, Ludhiana–141004, Punjab*

Water is one of Earth's most vital resources, essential for all forms of life and crucial for human civilization. The planet's water resources include oceans, rivers, lakes, glaciers, groundwater, and atmospheric moisture. Oceans, which make up about 97 per cent of Earth's water, are the largest reservoir but contain saline water, making it unsuitable for direct human consumption. Freshwater resources, vital for drinking, agriculture, and industry, constitute only about 3 per cent of the total water supply, with the majority locked in ice caps and glaciers. Groundwater, rivers, and lakes provide accessible sources of freshwater, though their distribution is uneven across the globe. Effective management and conservation of these water resources are critical to addressing challenges such as water scarcity, pollution, and the impacts of climate change.

Water Resource scenario

In India, the availability and demand for water present a complex challenge. It also highlights a pressing need for multifaceted approaches to address the country's water challenges comprehensively. The country receives a substantial annual precipitation of 3880 billion cubic meters (BCM), contributing to an average annual water resource of 1999.2 BCM from more than 20 major rivers and tributaries (CWC, 2019). However, the utilizable surface water is limited to 690 BCM due to physiographic and other constraints. Groundwater, another critical source, is estimated at 449 BCM in total replenishable volume, with an annual utilizable resource of 407 BCM, considering allocations for natural discharge (CGWB, 2023). Furthermore, while groundwater serves as a crucial supplement, it's overexploitation poses risks to sustainability and ecosystem health.

Despite these resources, the demand for water is significant, with a total potential utilization of 710 BCM (NITI Aayog, 2021). This demand is driven by various sectors, with agriculture requiring 557 BCM, domestic use at 56 BCM, industry at 43 BCM, and other sectors at 54 BCM. Projections indicate a worrying trend of increasing demand, expected to reach 843 BCM by 2025 and 1180 BCM by 2050. The escalating demand for water, propelled by agricultural, domestic, industrial, and other sectors, exacerbates the strain on available resources. Agriculture, a cornerstone of the Indian economy, remains the largest consumer, accounting for a significant portion of the total water utilization. Balancing agricultural needs with efficient water management practices is imperative for ensuring food security while conserving water resources. Similarly, urbanization and industrialization drive demand from domestic and industrial sectors, necessitating integrated strategies for conservation and efficient usage. Compounding the issue, per capita water availability has been steadily declining, falling from 1820 m^3 in 2001 to 1486 m^3 in 2021 (NITI Aayog, 2021). Future estimates are even more concerning, with anticipated levels of 1367 m^3 by 2031 and 1228 m^3 by 2051. These figures underscore the urgent need for comprehensive water management strategies to ensure sustainable access to this vital resource for India's growing population and economy.

Need for Sustainable Water Use in Fruit Crops

Addressing the intricate interplay between water availability, demand, and sustainability requires concerted efforts across multiple fronts. By adopting a holistic and collaborative approach, world can navigate its water challenges and ensure a secure and resilient water future for its people and ecosystems. Water conservation involves saving, preserving, and managing water resources to meet needs while ensuring sustainability. Driven by increasing demand surpassing supply, it aims to close this gap. Studies show underground water depletion exceeds recharge rates, and growing demands from agriculture, industries, and human consumption strain limited resources, threatening future sustainability. Globally, irrigation consumes the largest share of water, accounting for 70 per cent of freshwater usage. In developing countries, more than 80 per cent of freshwater withdrawals are dedicated to irrigating crops. Effective water management requires government policies, farmer involvement, and stakeholder collaboration. Despite infrastructure like reservoirs and wells to capture rainwater, modern techniques such as farm-level water harvesting, subsurface irrigation, and precision irrigation are needed. Sustainable water management is an important aspect in fruit production which varies with prevailing climatic conditions and specific zones *viz.* humid, arid and semi-arid climates. Both under and over irrigations may pose adverse effect on fruit productions. In sufficient watering thereby water stress can cause poor plant growth, yield and quality of fruits. In contrary, over irrigation leads to various problems like water logging, nutrient leaching, weed infestation, increased cost of cultivation, higher incidence of insect- pests and diseases *etc.* Hence, meeting irrigation requirements by appropriate irrigation scheduling and irrigation methods for orchards is obligatory for better plant/tree growth, higher fruit production, better fruit quality and consequently, improved water productivity.

Limitations of Conventional Irrigation Methods

Irrigation, the artificial application of water to plants, is essential when rainfall is insufficient to meet crop moisture needs. Various methods, categorized into traditional and modern techniques, affect water conservation, monitoring, scheduling, and control. Traditional surface irrigation methods, such as flood, furrow, border, check basin, wild flooding, and manual watering, rely on gravity flow and proper soil leveling for even water distribution. Despite being widely used, these methods have limited water-saving capabilities due to high evaporation and uncontrolled irrigation volumes, leading to issues like surface runoff, deep percolation, leaching, and nutrient loss. Flood irrigation, common in South Asia, is inefficient and often leads to waterlogging and reduced crop productivity. Border and basin irrigation methods involve creating leveled strips or basins to hold water, commonly used for tree crops and orchards. While traditional methods have been used for centuries, modern techniques like sprinkler and drip irrigation offer higher water efficiency, better monitoring, and control, leading to higher crop yields and more precise irrigation.

Efficient Irrigation Techniques

Microirrigation Systems

Microirrigation systems including drip and trickle irrigation, deliver water directly to the plant root zone in small, precise amounts. These systems use networks of tubes or pipes with emitters or drippers to provide water to individual plants or specific areas. Water is applied slowly and uniformly, enhancing efficiency and reducing evaporation or runoff. Suitable for fruit crops, micro irrigation minimizes water wastage and optimizes plant growth and productivity.

Mulching

Mulching refers to spreading materials such as straw, leaves, sawdust, or plastic sheets in the inter-row spaces of orchards. The choice of mulch depends on soil and weather conditions. Mulching conserves moisture, moderates soil temperature, reduces evaporation, and decreases irrigation frequency. It also helps control weeds, adds organic matter, reduces soil erosion, and increases rainwater absorption by reducing runoff.

Partial Rootzone Drying

It is a regulated irrigation technique that alternates between wetting and drying different parts of the root zone in a cyclic manner. By irrigating alternate furrows and keeping one side of the plant moist while the other remains dry, this method conserves water without harming orchard productivity. Essential elements include furrows and irrigation pipes on both sides of plants to allow switching. Typically, a 10-15day cycle is used. This technique reduces water application volume and triggers physiological changes in plants, optimizing water use and maintaining fruit yield and quality. Proper monitoring of soil moisture is crucial for success.

Deficit Irrigation

It involves planning the frequency and quantity of water for orchards to save water without compromising productivity. By providing less than full irrigation and focusing on critical growth stages, this technique maintains yield and quality while enhancing water productivity. Limited water supply during non-critical periods reduces evaporative demands through altered stomatal functioning. Successful deficit irrigation requires high skill and expertise in scheduling.

Irrigation Scheduling

An optimal irrigation schedule must specify the timing, quantity and method of water application. Planning irrigation based on factors like crop type, growth stage, soil type, weather conditions, and evapotranspiration rates to optimize water use.

Automated Irrigation Systems

Using automated systems with timers and controllers to ensure timely and efficient water application.

Fruit crop-specific water management is necessary as different fruit crops have unique water requirements and growth cycles. Major fruit crops, including citrus, apples, bananas, grapes, and mangoes, are vital to global agriculture and economies. Tailored water management ensures higher yields and maximize efficiency, reduce costs, and support long-term agricultural productivity.

Citrus

Citrus is a widely cultivated fruit tree species and makes up a significant portion of global fruit production, accounting for about 16 per cent (138 MT) of total fruit production worldwide (FAO, 2020). Citrus production heavily depends on water resources, which are increasingly affected by climate change (IPCC, 2018). An inadequate water supply can significantly reduce citrus yield and fruit quality, resulting in economic losses (Capra *et al.,* 2011). Effective irrigation management is essential to mitigate adverse effects on soil and crops. In north-west Indian states, particularly Punjab, groundwater resources are depleting rapidly, with the water table declining in most areas. The primary cause of groundwater overuse in Punjab is rice cultivation, which demands large amounts of water. Excessive irrigation through flood methods exceeds the soil's water storage capacity, leading to drainage losses. Notably, citrus cultivation also contributes significantly to water drainage losses, with 49 per cent of total water input lost beyond the root zone, second only to rice at 67 per cent (Jalota *et al.*, 2021). A recent CGWB report (2022) found nitrate levels in Punjab's groundwater ranging from 0 mg/L to 775 mg/L, with 22.4 per cent of samples exceeding the permissible limit of 45 mg/L (BIS standard). Higher nitrate concentrations are prevalent in the citrus-growing southwestern districts.

Figure 4.1L: Flood Irrigation in *Kinnow.*

Water requirements for citrus trees range from 651 to 997 mm annually, depending on regional climate, soil type, and tree growth stage. The most water-sensitive stages for citrus are flowering, fruit set, and fruit development, where insufficient soil moisture can significantly reduce yield. Reducing irrigation during non-critical growth stages, such as the initial and final fruit growth periods, can help maintain citrus production with better water efficiency in water-scarce regions. In many developing countries, soil fertilization combined with flood irrigation using canal water is common in citrus orchards, but this practice often leads to nutrient and water loss through conveyance, percolation, evaporation, and uneven distribution. Flooded conditions can also promote diseases like root rot and collar rot. Frequent, light irrigation is more beneficial, and irrigation water with salt concentrations exceeding 1000 ppm can be harmful to citrus trees.

Excessive vegetative growth from flood irrigation in citrus-bearing trees can reduce productivity. Conversely, applying water stress during specific growth stages can enhance yield and fruit quality (Cheng *et al.*, 2021). Deficit irrigation has been shown to be an effective water-saving technique, particularly in drip-irrigated Nagpur mandarin. Research by Panigrahia and Srivastava (2016) demonstrated that maintaining a 50-60 per cent soil water deficit in the effective root zone (0–0.40 m) during both the initial and final fruit growth periods can create beneficial water stress, leading to larger fruit size and improved internal qualities such as total soluble solids and acidity. Additionally, a study by Al-Qthanin *et al.* (2024) in Egypt's arid climate found that organic mulching with rice straw, combined with reduced irrigation levels (100, 85, 70 per cent ETc), increased Valencia orange yields and water productivity. The study showed that mulching improved nutrient levels, photosynthetic pigments, and fruit retention while conserving 15.4 per cent of water at 85 per cent ETc, achieving similar cost efficiency to traditional irrigation methods.

In recent years, ensuring adequate water supply for crops during critical fruit growth stages has become a major challenge in many states due to frequent canal closures for routine maintenance by the irrigation department. This also disrupts proper nutrient management in orchards. Micro irrigation systems offer a solution by not only conserving water and nutrients but also ensuring better fruit retention during key growth stages in March and April, even when water availability is not an issue. Drip irrigation has been shown to save 30 per cent of irrigation water while increasing fruit yield by 50 per cent compared to basin irrigation in citrus orchards in central India (Panigrahi *et al.,* 2012). Kumar *et al.* (2023) found that drip-irrigated and fertigated trees had a 20 per cent higher fruit set, 169 g heavier fruits, 7.7 cm larger fruit size, and a 33 per cent increase in yield compared to trees under flood irrigation with soil fertilization. Fertigation with only 75 per cent of the recommended dose of nitrogen and phosphorus resulted in similar growth and productivity of *kinnow* mandarin as the full dose, saving 25 per cent of N and P usage while improving fruit grade and quality compared to conventional flood irrigation and soil fertilization practices.

The drip irrigation system enables efficient and judicious use of water along with improvement in yield. For efficient use of water sub lateral loops of the drippers should be used around each plant or tree. Initially, 1-2 drippers per plant or tree are recommended, increasing to 4-5 as the soil type and plant age dictate. The monthly water requirements vary based on the tree's age (Table 4.1, Anonymous, 2021). The amount of irrigation water may fluctuate by 10-15 per cent depending on the weather conditions. *Kinnow* mandarin trees can occasionally be irrigated with a mix of fresh and saline water (EC around 2 mmhos/cm) through drip irrigation, serving as life-saving irrigation during canal closures or water scarcity.

New tools and strategies, such as using infrared thermometry (IRT) to measure canopy temperature, have proven to be simple and cost-effective methods for monitoring crop water status. Thermal indicators like the Crop Water Stress Index (CWSI) and Stomatal Conductance Index are derived from these measurements and offer reliable proxies for assessing water stress. Wang *et al.* (2009) demonstrated that the spectral absorption valley around 1450 nm is influenced by leaf water content, while the 1550–1750 nm range is ideal for monitoring canopy water content via spectral imaging. Dzikiti *et al.* (2011) utilized FieldSpec Pro, covering the

350–2500 nm spectral range, to detect water status in Valencia orange trees. Further comparisons of object-based physiological indices from UAV images showed that PRI Photochemical Reflectance Index formulations are more effective in tracking citrus canopy water status than chlorophyll ratio indices (Stagakis *et al.,* 2012). Additionally, Shivers *et al.* (2019) employed paired thermal and hyperspectral aerial imagery to quantify land surface temperature variability and assess crop stress in citrus orchards. UAV technology, in particular, holds promise for enhancing both the spatial and temporal resolution of vegetative status data, effectively bridging the gap between leaf-level and canopy-level understanding.

Table 4.1: Drip Irrigation Schedule (litres/day/plant) for Kinnow Plants with Standard Spacing of 6×6m (Anonymous, 2021)

Month	*Age of Plants (Years)*				
	0-2	*3-4*	*5-6*	*7-8*	*9 and more*
January	3	6	9	12	15
February	6	12	18	24	30
March	9	18	27	36	45
April	13	25	39	52	65
May	16	32	48	64	80
June	17	34	51	68	85
July	13	26	39	52	65
August	12	24	36	48	60
September	11	22	33	44	55
October	8	16	24	32	40
November	5	10	15	20	25
December	3	6	9	12	15

Mango

Although mango has relatively good drought tolerance but marketable fruit production requires sufficient irrigation. Water stress can lead to reduced vegetative growth, hinder root development, increase early-season fruit drop, and decrease fruit size (Bhriguvanshi *et al.,* 2015). Young mango plants need frequent irrigation during dry periods, while mature trees in northwest India require irrigation at every 10-12 days during summer. No irrigation is necessary during the winter for 2-3 months. Irrigation should align with late winter fertilizer application. Interestingly, water stress plays a key role in regulating flowering and fruiting by triggering floral induction. Studies indicate that deficit irrigation not only conserves water but also enhances fruit quality (Table 4.6). While deficit irrigation during mango fruit development reduces the net photosynthetic rate, transpiration rate, and stomatal conductance, it does not significantly affect intercellular CO_2 concentration or leaf temperature (Zuazo *et al.,* 2021). However, deficit irrigation at 75 or 50 per cent ETc during early blooming and setting stages can significantly reduce yield, whereas applying 50 per cent ETc during the maturity stage has minimal impact on yield and improves water use efficiency(Santos *et al.,* 2016). Mild water deficit speeds up starch breakdown and increases fructose content in mango, leading to better fruit quality (Lechaudel *et al.,* 2005).

Application of sugarcane leaf mulching (SM) in mango orchards significantly improved photosynthetic rate, stomatal conductance, transpiration rate, leaf water-use efficiency and carboxylation efficiency as compared toun-mulched (NM) trees in a dry, hot climate in China (Liu *et al.,* 2021). SM also increased fresh fruit yield, irrigation water-use efficiency, solid-to-acid ratio, sugar-to-acid ratios but decreased titratable acid content compared to NM. Similarly, application of deficient irrigation at 75 per cent ETc (DI75) boosted carotenoid content, total soluble solids, solid-to-acid ratio, and sugar-to-acid ratio, fresh fruit and reduced titratable acid as compared to full irrigation (FI). However, DI50 reduced fresh fruit yield by 22-24 per cent. Combining SM + DI50 resulted in highest mango yield, better quality, and improved water-use efficiency, offering a strong basis for scientific mango management in hot, dry areas. Further integration of drip irrigation with organic mulch results in the highest IWUE, followed by loop drip, single-line drip, and bubbler drip. Bubbler systems has been proved for the lowest efficiency. Drip irrigation with 60 per cent water and organic mulch was 3.3 times more efficient than surface irrigation.

Guava

For guava, efficient use of water and nutrients is crucial for high yields and fruit quality. Young guava plants need weekly irrigation in summer and 2-3 times during winter (Table 4.1). For bearing trees, irrigation should beapplied every 2-3 weeks in summer and monthly in winter for optimal flowering and fruit set. Heavy irrigation should be avoided during peak flowering to prevent flower drop. Improvements in guava yields were observed under deficit irrigation (DI75) compared to full irrigation (Jat *et al.,* 2022). Increasing irrigation from 50 to 75 per cent ETc under DI boosted yields up to 38 per cent. The improved yields under DI75 are likely due to moderate water deficit suppressing vegetative growth and redirecting energy toward fruiting. Guava's hardiness in semi-arid and arid conditions may also contribute to better performance under moderate water stress.

Among different mulching materials, silver-black (MSB), black (MB), organic (MOM), Jat *et al.* (2022) found that MSB yielded 14-16 per cent more fruit compared to un-mulched plants. MSB's higher yields are attributed to improved microclimate control, better soil moisture, and reduced weed competition, leading to enhanced moisture and nutrient availability.

Drip irrigation in guava can further reduce water usage by 40–70 per cent and boost productivity by 10–55 per cent, depending on soil and climate conditions

(Deshmukh and Sen, 2000). Enhanced water use efficiency directly increases biomass production relative to crop consumption. The drip irrigation schedule for guava planted at 6×5 m spacing results in 35-40 per cent saving of irrigation water besides improving fruit yield and quality (Table 4.2).

Table 4.2: Drip Irrigation Schedule (litres/day/plant) for Guava Plants at 6×5m Spacing (Anonymous, 2021)

Month	*Age of Plants (Year)*								
	1	*2*	*3*	*4*	*5*	*6*	*7*	*8*	*9*
January	0.5	0.8	2.3	2.8	5.3	5.7	7.0	8.0	8.4
February	0.6	2.3	4.5	5.7	9.6	10.2	11.4	12.5	13.2
March	1.2	3.4	7.5	9.9	15.9	17.0	18.4	19.2	19.8
April	2.4	8.5	16.0	17.9	28.8	30.5	32.5	33.6	34.5
May	3.0	14.9	19.2	23.0	37.5	39.5	41.5	43.4	44.0
June	3.3	13.1	18.6	27.0	47.8	49.2	51.4	52.6	53.0
July	2.4	10.0	11.9	19.7	28.1	30.5	32.4	33.5	33.8
August	1.9	6.9	11.6	14.9	20.6	22.5	24.3	24.8	25.0
September	1.4	7.6	9.7	14.0	18.4	20.2	21.5	22.7	23.0
October	1.3	5.4	8.5	10.8	13.2	13.8	15.0	16.2	16.8
November	0.8	3.0	5.4	4.0	8.8	9.2	10.4	11.1	12.1
December	0.4	2.1	3.3	5.0	5.2	6.0	6.8	7.2	7.8

Scheduling of drip irrigation in guava at 80 and 90 per cent of daily pan evaporation through drip irrigation showed maximum irrigation water use efficiency (IWUE) and fruit yield of high density plantation guava along with significant water saving over local control (Lakhawat *et al.,* 2024). For ultra-high-density guava planting, reducing water from 80 to 60 per cent PE had no significant impact on fruit yield per plant (Table 4.3). Drip irrigation at 80 per cent CPE along with 100 per cent recommended dose of fertilizers and mulching with silver-black plastic mulch further improves guava yield, number of fruits per plant, fruit weight, ascorbic acid, total sugars and leaf NP K and Zn contents (Singh *et al.,* 2021). Drip irrigated guava plants have been reported to start bearing at 3rdyear of planting while flood irrigated plants start bearing after fourth year of planting (Mandal *et al.,* 2007). Additionally, fruits from drip-irrigated plants had superior fruit weight, total soluble solids, total sugars, and vitamin C content compared to those from flood-irrigated plants. The spacing of plants influenced the installation cost of the drip irrigation system, with a 5 x 5 m spacing yielding higher net profit and a favorable cost-benefit ratio of 2.20.

Table 4.3: Number of Fruits, Fruit Yield and Irrigation Water Use Efficiency of Guava as Influenced by Irrigation Scheduling at Udaipur (Lakhawat *et al.*, 2024)

	Number of Fruits per Plant	*Fruit Yield (t/ha)*	*Irrigation Water Use Efficiency (t/ha-cm)*
Local control	71.7	24.8	0.169
Irrigation at 70 per cent PE	56.8	18.6	0.205
Irrigation at 80 per cent PE	90.1	37.3	0.359
Irrigation at 90 per cent PE	81.8	32.1	0.275
Irrigation at 100 per cent PE	75.0	28.3	0.218
CD at 5 per cent	1.6	3.4	-

Pomegranate

Pomegranate trees are drought-tolerant; however, irrigation is necessary during the dry summer to optimize growth, yield, and fruit quality for commercial production. In north-west India, light irrigation is required in winters during long dry spells. In summer, it need irrigation at 10-15 days interval. Deficit irrigation in pomegranate significantly reduced fruit yield per tree compared to full irrigation (Beelagi *et al.,* 2023). The yield declined by 12 per cent with I80 irrigation and by 24 per cent with I60, mainly due to fewer fruits per plant and reduced fruit weight. However, mulch application, regardless of irrigation level, boosted fruit yield by an average of 30 per cent, primarily by enhancing fruit size. Ghosh and Bera (2015) reported that rice straw mulch conserved the highest soil moisture, while unmulched conditions had the lowest. In pomegranate cv. Ruby, black polyethylene mulch was found to be more effective at moisture retention compared to white polyethylene. For pomegranate cv. Bhagwa, Yograj (2016) compared different mulching materials and observed that black polyethylene provided the highest benefit-cost ratio, followed by silver polyethylene, with unmulched plants showing the lowest ratio.

Irrigation water requirement of pomegranate varies according to growth stages and methods of irrigation (Meshram *et al.,* 2017). For cv. Bhagwa, total water consumption was 4,572 liters with subsurface drip irrigation (SSDI) and 6,230 liters with surface drip irrigation (SDI) (Table 4.4). SSDI demonstrated higher water use efficiency due to its superior water-saving capacity. Unlike the SDI system, which frequently encountered water leakage from joint openings, the SSDI system had no issues with water leakage or emitter blockage. Overall, water usage in the SSDI system was 73 per cent lower than in the SDI system.

Table 4.4: Water Requirement of Pomegranate (liters/plant) at different Phenological Stages under different Irrigation Methods (Adapted from Meshram *et al.*, 2017)

Critical Stages for Irrigation	*Months*	*ET (mm)*	*WRinliters*	
			SSDI	*SDI*
New leaf initiation (22-25 days)	November	172.8	360	480
Development (flowering to fruit setting, 70-80 days)	December	184.2	428	570
	January	190.2	484	645
	February	184.9	487	650
	March	196.8	744	992
Maturity (fruit development, 60-70 days)	April	179.9	744	1032
	May	214.6	756	1141
Harvesting (fruit removal, 45-60 days)	June	153.4	540	720
	Total	1476.6	4572	6230

Table 4.5: Recommendation for Scheduling Irrigation in Fruit Crops in North-Western India (Anonymous, 2021)

Sl.No.	*Fruit Crop*	*Recommended Irrigation Scheduling*
1	**Citrus**	Young plants up to 3-4 years old should be irrigated weekly, while older trees should be irrigated every 2-3 weeks, depending on climate, rainfall, and soil type. Irrigation is critical before sprouting in February, after fruit set in April, and during hot weather to prevent adverse effects on tree growth and excessive flower/fruit shedding.
2	**Mango**	Young plants need frequent irrigation during dry periods. Bearing trees, however, typically require irrigation every 10-12 days from April to late June. One irrigation should be provided when fertilizers are added in February. No irrigation is necessary from October to December for 2-3 months.
3	**Guava**	Young guava plants need weekly irrigation in summer and 2-3 times during winter. Bearing trees require irrigation every 2-3 weeks in summer and monthly in winter for optimal flowering and fruit set. Avoid heavy irrigation during peak flowering to prevent excessive flower drop.
4	**Pear**	Pear trees need frequent irrigation after transplanting, with intervals of 5 to 7 days during summer and 15 days in July-August. Irrigation should be reduced after fruit harvest and stopped entirely in January. Bearing trees should be flood-irrigated every 5 to 7 days in summer to increase fruit size.

Sl.No.	*Fruit Crop*	*Recommended Irrigation Scheduling*
5	**Ber**	Irrigation is crucial during fruit development from October to February, with intervals of 3 to 4 weeks depending on weather conditions. Regular irrigation during this period enhances fruit size, quality, and reduces shedding. Irrigation should be stopped in the second half of March to prevent damage and delayed ripening of fruits on low-lying branches.
6	**Litchi**	Irrigate young plants twice a week from April to June and weekly in winter. For bearing trees, irrigation is crucial during fruit development from mid-May to late June. Two irrigations per week during this period reduces fruit cracking and promotes proper size development. After the monsoon, young orchards should be irrigated weekly and fortnightly from September onwards. For bearing orchards, apply irrigation in late November or early December to protect against frost.
7	**Peach**	In peach trees, fruit development from March to June is critical for irrigation. Avoid moisture stress, particularly 25-30 days before maturity, as this period is crucial for maximum weight gain.
8	**Amla**	Young plantations should be irrigated every 10-15 days during summer. Avoid irrigating bearing plants during flowering.
9	**Plum**	Young plum plants up to 3-4 years old need irrigation every 4-5 days from April to June. As deciduous trees, plums require less frequent irrigation in winter, only during prolonged droughts. Bearing trees should be regularly irrigated from fruit set to maturity, with a 4-5 day interval in summer.
10	**Banana**	Banana requires wet soil conditions for optimal growth. Even slight moisture stress can reduce finger size and leaf color, while excess water may cause pseudostem breakage due to rotting.
11	**Sapota**	Sapota can tolerate some drought but needs assured irrigation for commercial production. Irrigate every 30 days in winter and every 10-12 days in summer. During flowering, apply frequent irrigation to maintain humid conditions for better fruit set. Young plants should be watered every 6-8 days in hot, dry summer, and as needed during rainy and winter seasons.
12	**Grapes**	One irrigation should be applied after pruning in the first fortnight of February and another in the first week of March. Following fruit set in April, irrigation should occur every 10 days until the first week of May. During the rest of May, irrigation should be weekly, and in June, it should be every 3 to 4 days. From July to October, irrigate only if rainfall is insufficient. Finally, from November to January, provide one irrigation if the soil becomes extremely dry.In light soils, alternate saline sodic water with canal water or mix with 6 kg/vine of press mud to improve yield and soil health.

Sl.No.	Fruit Crop	Recommended Irrigation Scheduling
13	**Pomegranate**	In winter, provide light irrigation during long dry spells. In summer, irrigate every 10-15 days.
14	**Loquat**	Loquat is drought-tolerant, but performs best with judicious irrigation. Three to four irrigations are recommended from fruit growth to maturity.
15	**Fig**	Fig trees are stressed by hot, dry periods due to their shallow roots. Consistent moisture is crucial during fruit development. Sudden water increases during ripening can cause fruit to split.
16	**Date palm**	Date palms are more resistant to drought and salinity compared to other fruit crops. One irrigation before spathe emergence is crucial for normal fruiting. After fruit set, regular irrigation is needed. In light soils, irrigate every 10-15 days in summer and every 30-40 days in winter.
17	**Bael**	Established plants can withstand intermittent drought, but young plants require frequent watering. During summer, when trees shed their leaves, water requirements are significantly reduced.
18	**Banana (cv. Grand Naine)**	Irrigation should be performed every 7-8 days during March-April, and every 4-6 days in May-June. From July onwards, adjust irrigation to every 7-8 days as needed during the rains. From October to February, irrigate every 10-15 days.

Table 4.6: Effects of Water Conservation Strategies across different Fruit Crops

Fruit Crop	Strategy/Technique	Impact	Reference
Citrus	Irrigation with saline reclaimed water	Leaf chlorophyll content as an effective indicator for salt and water stress	Romero-Trigueros *et al.* (2014)
Nagpur Mandarin	Continuous trenching, bunding, and staggered trenching	Enhanced yield and reduced soil erosion	Panigrahi *et al.* (2009)
Mango	Regulated deficit irrigation	Increased post-harvest fruit yield; sensitive to deficit during final fruit growth stage	Levin *et al.* (2018)
Strawberries	Precision farming using multiplatform approach	Achieved 11-33 per cent water savings	Perea *et al.* (2017)
Plum	Regulated deficit irrigation	Achieved 30 per cent water savings with improved water use efficiency	Intrigliolo and Castel (2010)

Fruit Crop	*Strategy/Technique*	*Impact*	*Reference*
Apricot	Soil management with mini-catchments and perforated topsoil	Improved water infiltration, 6-9 per cent water savings, and reduced runoff on loamy soil	Abrisqueta *et al.* (2007)
Jujube	Mulching with jujube branches, twigs, and leaves	Reduced surface runoff and sediment loss	Wang *et al.* (2015)

Strawberry

Strawberry cultivation requires consistent irrigation to ensure high yields and fruit quality due to the plant's shallow root system, extensive leaf area, and high fruit water content (Perea *et al.,* 2017). Water needs increase significantly under plastic tunnels, where plants depend entirely on irrigation. This often leads to either over- or under-irrigation. Drip irrigation, however, offers precise nutrient distribution within the root zone, boosting nutrient efficiency. Maintaining adequate water levels is crucial, especially during growth and fruit development. Strawberries are sensitive to excess soil moisture, so daily drip irrigation is recommended using lateral pipes with 2.2 liters per hour drippers spaced 30 cm apart (Anonymous, 2021). Fertigation should begin six days post-transplanting, applying a total of 44 kg N, 32 kg P_2O_5, and 40 kg K_2O_5 per acre according to the schedule (Table 4.7). Combining polyethylene mulch with fertigation enhances both vegetative growth and fruit yield.

Table 4.7: Drip Irrigation Schedule for Strawberry (drip discharge 2.2 lps)

Months					
November	*December*	*January*	*February*	*March*	*April*
Time (minutes/day)					
10	14	12	22	29	38

Conclusion and Suggestions

Effective water management and conservation in orchards are essential for ensuring sustainable fruit production and addressing the challenges posed by water scarcity. Traditional irrigation methods, such as flood and furrow irrigation, often result in significant water waste due to issues like evaporation, runoff, and nutrient leaching, leading to decreased crop productivity. In contrast, modern irrigation techniques offer more precise and efficient water usage.

Microirrigation systems, including drip and trickle irrigation, deliver water directly to the plant root zone in controlled amounts, significantly reducing water wastage and enhancing plant growth and productivity. Mulching, which involves spreading organic or inorganic materials around the plants, conserves soil moisture, regulates temperature, and reduces irrigation frequency. Techniques like partial root zone drying and deficit irrigation are designed to optimize water use by carefully managing irrigation schedules and targeting specific growth stages. These methods allow for reduced water application without compromising fruit yield or quality. Automated irrigation systems, equipped with timers and controllers, further improve efficiency by ensuring that water is applied at the right time and in the right amount.

Fruit crop-specific water management is crucial because different crops have distinct water requirements and growth cycles. Major fruit crops like citrus, mango, guava, and pomegranate play significant roles in global agriculture and economies, and tailored water management is key to optimizing their productivity. Citrus, which accounts for a large portion of global fruit production, is highly dependent on water, and effective irrigation practices like deficit and drip irrigation have proven essential. These methods not only conserve water but also improve yield and fruit quality by targeting specific growth stages and reducing unnecessary water loss. For mangoes, while they are relatively drought-tolerant, proper irrigation is still necessary to ensure marketable fruit production. Strategic deficit irrigation combined with mulching can enhance fruit quality and water use efficiency, making it a valuable practice in regions with limited water resources. Guava, known for its resilience, benefits from efficient water management through drip irrigation and mulching, which significantly reduce water usage and increase yields. Deficit irrigation in guava can also help redirect energy from vegetative growth to fruit production, leading to improved productivity. Pomegranates, despite their ability to withstand drought, require careful irrigation management, especially during critical growth stages, to maximize fruit size and quality. Subsurface drip irrigation has shown superior water efficiency, making it an ideal choice for pomegranate cultivation in water-scarce areas.

The irrigation scheduling recommendations for fruit crops in north-western India emphasize the need for tailored irrigation practices based on crop type, growth stage, and environmental conditions. Young plants typically require more frequent watering, while established trees benefit from less frequent, strategically timed irrigation to optimize fruit quality and yield. Critical periods for irrigation include pre-sprouting, fruit set, and periods of high temperature, with adjustments made for soil type and rainfall.

In conclusion, sustainable water management in orchards requires a shift from traditional, less efficient practices to modern, scientifically-backed irrigation methods. By optimizing water use through tailored irrigation strategies, adopting advanced technologies, and integrating practices like mulching, orchard managers can ensure the long-term sustainability of water resources while maximizing fruit yield and quality. This approach not only addresses the immediate challenges of water scarcity but also supports the broader goals of environmental conservation and agricultural resilience in the face of climate change.

Based on the review of studies and research work done on various aspects of water management and conservation in orchards, some suggestions have been made as follows:

- ✰ **Implement micro irrigation systems**: Install drip or trickle irrigation to deliver precise amounts of water to plant roots. Regularly maintain and inspect systems to ensure efficient operation.
- ✰ **Adopt mulching practices**: Apply organic or synthetic mulch around plants to conserve moisture and reduce evaporation. Choose mulch that suits the local climate and crop needs.
- ✰ **Partial root zone drying and deficit irrigation strategies**: Alternate between wetting and drying different parts of the root zone to improve water efficiency. Monitor soil moisture levels to adjust watering practices accordingly. Provide less water during non-critical growth stages to reduce water use without significantly affecting yield. Customize irrigation schedules based on crop type and growth stages.
- ✰ **Irrigation scheduling and automated irrigation systems**: Develop schedules based on crop requirements, soil moisture, and weather conditions. Install timers and controllers to automate irrigation processes, ensuring consistent and efficient water use. Use weather forecasts and soil moisture sensors to fine-tune irrigation timings.
- ✰ **Modern monitoring tools**: Use Infrared Thermometry (IRT) to assess canopy temperature and detect water stress. Employ UAV technology to monitor and analyze crop water status for better decision-making.

REFERENCES

Abrisqueta JM, Plana V, Mounzer OH, Mendez J and Ruiz-Sanchez MC, 2007. Effects of soil tillage on runoff generation in a Mediterranean apricot orchard. *Agricultural Water Management*, **93**(1): 11-18.

Al-Qthanin RN, AbdAlghafar MI, Mahmoud DS, Fikry AM, AlEnezi NA, Elesawi IE, AbuQamar SF, Gad MM and El-Tarabily KA, 2024. Impact of rice straw mulching on water consumption and productivity of orange trees [*Citrus sinensis* (L.) Osbeck]. *Agricultural Water Management*, **298**: 108862.

Anonymous, 2021. Package of practices for cultivation of fruits-2021. Punjab Agricultural University, Ludhiana. pp: 1-188.

Beelagi R, Singh VP, Jat R, Singh PK, Rai R, Singh A, Basile B, Mataffo A, Corrado G and Kumar P, 2023. Enhancing the fruit yield and quality in pomegranate:

Insights into drip irrigation and mulching strategies. *Plants*, **12**: 3241. https: //doi.org/10.3390/plants12183241

Bhriguvanshi SR, Adak T, Kumar K, Singh VK, Singh VK and Singh A, 2015. Drip irrigation scheduling of Mango cv. Langra under subtropical condition of Uttar Pradesh. *Journal of Soil and Water Conservation*, **14**(2): 133–140.

Capra A, Consoli S and Scicolone B, 2011. Economic analysis of citrus orchards under deficit irrigation in South Italy. *Acta Horticulturae*, **922**: 209–216. https: //doi.org/10.17660/actahortic.2011.922.27.

Central Ground Water Board (CGWB), 2023. Dynamic Ground Water Resources of India-2023. Department of Water Resources, River Development and Ganga Rejuvenation, Ministry of Jal Shakti, Government of India, New Delhi. pp: 431.

Central Water Commission (CWC), 2019. Reassessment of water availability in India using space inputs. Basin Planning and Management Organisation, Central Water Commission, New Delhi. pp: 99.

Cheng M, Wang H, Fan J, Zhang S, Liao Z, Zhang F and Wang Y, 2021. A global meta-analysis of yield and water use efficiency of crops, vegetables and fruits under full, deficit and alternate partial root-zone irrigation. *Agricultural Water Management*, **248**: 106771.

Deshmukh MM and Sen NL, 2000. Studies on irrigation scheduling and point of application through drip irrigation to guava (*Psidium guajava* L.) variety–Allahabad safeda. *Haryana Journal of Horticultural Sciences*, **29**(3/4): 156-158.

Dzikiti S, Verreynne SJ, Stuckens J, Strever A, Verstraeten WW, Swennen R, Theron KI and Coppin P, 2011. Seasonal variation in canopy reflectance and its application to determine the water status and water use by citrus trees in the Western Cape, South Africa. *Agricultural and Forest Meteorology*, **151**: 1035–1044.

Food and Agriculture Organization (FAO), 2020. FAOSTAT database. Food and Agriculture Organization, Rome. Retrieved from www.fao.org/faostat.

Ghosh SN and Bera B, 2015. Effect of mulching on soil moisture, yield and quality of pomegranate. *Indian Journal of Soil Conservation*, **43**: 92–95.

Intergovernmental Panel on Climate Change (IPCC), 2018. Global Warming of 1.5°C: An IPCC Special Report on the impacts of global warming of 1.5°C above pre-industrial levels and related global greenhouse gas emission pathways. Retrieved from https: //www.ipcc.ch/sr15/.

Intrigliolo DS and Castel JR, 2010. Response of plum trees to deficit irrigation under two crop levels: tree growth, yield, and fruit quality. *Irrigation Science*, **28**(6): 525-534.

Jalota SK, Vashisht BB and Kaur S, 2021. Field water budget of crops as influenced by soil texture, climate change, and management interventions. *Agricultural Research Journal*, **58**: 373-382.

Jat R, Singh VP, Ali Abed S, Al-Ansari N, Singh PK, Vishwakarma DK, Choudhary A, Al-Sadoon MK, Popat RC and Jat SK, 2022. Deficit irrigation scheduling with mulching and yield prediction of guava (*Psidium guajava* L.) in a subtropical humid region. *Frontiers in Environmental Science,* **10**: 1044886.

Kumar A, Singh H and Pathania S, 2023. Effect of N and P fertigation on fruit yield and quality of *kinnow. Indian Journal of Agricultural Sciences,* **93**(5): 517–522.

Lakhawat SS, Sharma V, Singh TK, Patil P, Priyadevi S and Gutam S, 2024. Effects of pan evaporation-based drip irrigation levels on the performance of guava grown in Udaipur and Rewa regions of India. *Journal of Agrometeorology,* **26**(1): 69–73.

Lechaudel M, Joas J, Caro Y, Genard M and Jannoyer M, 2005. Leaf ratio and irrigation supply affect seasonal changes in minerals, organic acids, and sugars of mango fruit. *Journal of the Science of Food and Agriculture,* **85**: 251–260.

Levin AG, Peres M, Noy M, Love C, Gal Y and Naor A, 2018. The response of field-grown mango (cv. Keitt) trees to regulated deficit irrigation at three phenological stages. *Irrigation Science,* **36**(1): 25-35.

Liu X, Zhang Y, Leng X, Yang Q, Chen H, Wang, X and Cui N, 2022. Exploring the optimisation of mulching and irrigation management practices for mango production in a dry hot environment based on the entropy weight method. *Scientia Horticulturae,* **291**: 110564.

Mandal G, Kumar S, Kumar R and Singh R, 2007. Effect of drip irrigation and plant spacing on yield, quality, and economic return of guava (*Psidium guajava* L.) grown in saline soil. *Acta Horticulturae,* **735**: 427-432.

Meshram S, Kadam V and Suryawanshi P, 2017. Microirrigation for sustainable production of pomegranates. *Indian Horticulture,* March-April 2017, 3-4. ICAR.

NITI Aayog, 2021. Water resources sector report. Development Monitoring and Evaluation Office, NITI Aayog, Government of India. pp: 196.

Panigrahi P, Srivastava AK, 2016. Effective management of irrigation water in citrus orchards under a water-scarce hot sub-humid region. *Scientia Horticulturae,* **210**: 6–13.

Panigrahi P, Srivastava AK and Huchche AD, 2009. Influence of in-situ soil and water conservation measures on the performance of Nagpur mandarin. *Journal of Agricultural Engineering (ISAE),* **46**(3): 37-40.

Panigrahi P, Srivastava AKand Huchche AD, 2012. Effects of drip irrigation regimes and basin irrigation on Nagpur mandarin agronomical and physiological performance. *Agricultural Water Management,* **104**: 79–88.

Perea RG, Garcia IF, Arroyo MM, Rodríguez Díaz JA, Poyato EC and Montesinos P, 2017. Multiplatform application for precision irrigation scheduling in strawberries. *Agricultural Water Management,* **183**: 194-201.

Preet MS, Kumar R, Singh VP, Dongariyal AN and Srivastava R, 2021. Response of guava to integrated nutrient and water management. *Indian Journal of Horticulture*, **78**: 189–197.

Romero-Trigueros C, Nortes PA, Pedrero F, Mounzer O, Alarcón JJ, Bayona JM and Nicolás E, 2014. Assessment of the sustainability of using saline reclaimed water in grapefruit in the medium to long term. *Spanish Journal of Agricultural Research*, **12**(4): 1137-1148.

Santos RMD, Donato SLR, Faria LN, Coelho EF and Cotrim JPRF, 2016. Irrigation strategies with deficit in 'Tommy Atkins' mango tree. *EngenhariaAgrícola*, **36**(6): 1096–1109.

Shivers SW, Roberts DA and McFadden JP, 2019. Using paired thermal and hyperspectral aerial imagery to quantify land surface temperature variability and assess crop stress within California orchards. *Remote Sensing of Environment*, **222**: 215–231.

Stagakis S, Gonzalez-Dugo V, Cid P, Guillen-Climent M and Zarco-Tejada PJ, 2012. Monitoring water stress and fruit quality in an orange orchard under regulated deficit irrigation using narrow-band structural and physiological remote sensing indices. *ISPRS Journal of Photogrammetry and Remote Sensing*, **71**: 47–61.

Wang J, Huang J, Wu P, Zhao X, Gao X, Dumlao M and Si BC, 2015. Effects of soil management on surface runoff and soil water content in jujube orchard under simulated rainfalls. *Catena*, **135**: 193-201.

Wang J, Xu R and Yang S, 2009. Estimation of plant water content by spectral absorption features centered at 1,450 nm and 1,940 nm regions. *Environmental Monitoring and Assessment*, **157**: 459.

Yograj S, 2016. Studies on the influence of different types of mulch on growth, yield and quality of pomegranate (*Punica granatum* L.) cv. Bhagwa. M.Sc. Thesis, Department of Fruit Science, University of Horticultural Sciences, Bagalkot, 114 p.

Zuazo VHD, Garcia-Tejero IF, Rodríguez BC, Tarifa DF, Ruiz BG and Sacristan PC, 2021. Deficit irrigation strategies for subtropical mango farming: A review. *Agronomy for Sustainable Development*, **41**: 13.

Chapter 5

Water Management in Fruit Crops

Dinesh Kumar

ICAR–Central Institute for Subtropical Horticulture, Lucknow–226101, Uttar Pradesh

Water is the most precious and common renewable natural resource for sustaining life and environment as it provides life support system for human beings, plants and animals. It is also a vital part of socio-ecological system of the nation. The declining per capita of fresh water availability especially in the developing countries is causing a serious threat. Therefore, there is need to go for proper planning, development, management and optimal utilization of water resources. It is of paramount importance for food, environmental, social and economic security of the country. This definitely pay our attention for efficient methods of water application for sustaining living organisms and natural habitat on the earth.

Increasing water productivity is an important key in improved water management for sustainable horticulture production and nutritional security. Concept of water productivity in agriculture production system is focussed on producing more food with the limited water resources (more crop per drop). This can be improved by choosing right and well adopted crop to agro-ecosystem, minimized unproductive water loss by maintaining vigorously healthy growing crops through optimized water and nutrient management practices. Water scarcity in India is looming and agriculture sector consumes about 80 per cent of total water resources. According to Central Water Commission by 2050, the total water demand will over shoot supply in the country and the share of irrigation water will decline. Hence there is urgent need for judicious and efficient use of water in agriculture to ensure higher water productivity of Indian Agriculture.

Although, India occupies only 3.29 million square kilometre geographical area, which is confined to 2.4 per cent of the world's land area, and supported by 17 per cent of world's population. There are pressure on cultivable land for agriculture continues to be high as Indian agriculture supports 4.5 per cent of water. Looking into population growth, declining land availability, water coupled with challenges of climate change has created much greater than before and has to be addressed with strategic approaches, utilizing innovation in agriculture through science and technology. In general, global warming will have impact on agricultural crops due to erratic rainfall pattern, reduction in total number of rainy days and high intense rain resulted increase in biotic and abiotic stresses. Increasing temperature will have more effect on crop reproductive biology and reduced water may affect the productivity but adoptive mechanism like time adjustment and productive use of water shall reduce the negative impact. The adequate scientific knowledge on these aspect may address all the problems and which can convert the challenges into opportunities.

Indian subtropics contribute about 35 per cent of total fruit production out of which mango and guava contribute more than 50 per cent of total fruit production. The water and nutrient play an important role in improving productivity and quality of subtropical fruits. Water productivity can be improved by adopting modern method of irrigation such as microirrigation system. In order to enhance livelihood and profitability of medium and small farmers not only interventions such as increasing productivity, enhancing quality, reducing farmer level post harvest losses and intelligent marketing but also developing fruit based cropping system will go along way. Enhancing income of mango and guava farmers through various interventions will significantly improve the economy and livelihood of rural masses.

The traditional method of irrigation *i.e.* flood, furrow and basin method of irrigation resolves round the concept of replenishing the moisture level to field capacity. The system do not permit the restricting of water to meet the requirement at the root zone, thus leading to excessive percolation and other losses which leads of water logging, soil salinity and even drought like conditions at last point. These conditions have created the low productivity levels in irrigated crops also. To meet the food, income and nutritional security of the projected population, the additional requirement of food grains, fruits and vegetables from limited land and declining water resources in the country.

Methods of Irrigation

Different irrigation methods have been used in orchards depending on water availability, topography, fruit plant species, age of tree, inter-crop, *etc.*

Surface Irrigation

It is the application of water by gravity flow to the surface of the field. It is most common method of irrigation due to easy in layout and operation. It is one of the common and economical method of irrigation when properly designed and operated.

Figure 5.1: Surface Irrigation in Mango.

Basin Irrigation

Basin irrigation is a class of surface irrigation methods in which area is divided into number of checks or basins. For good crop growth it is very important that the right quantity of water is supplied to the root zone and is wetted uniformly around the tree canopy. In this method, earthing up or raised the soil around the tree trunk to avoid the direct contact of water to main trunk to avoid fungal infection. Almost 60-70 per cent of root activity is confined to 45-60 cm top soil and spread within the canopy area of tree.

Furrow Irrigation

Furrows are small channels which carry water down the land slop between crop rows. The water infiltrates into the soil as it moves along the slope. The irrigation water flow from the field channel into the furrows by opening up the bank of the channel. The size and shape of the furrows depends on the crop grown, slope type, equipment to be used and spacing between crop rows. The size and shape of the furrows can be modified widely to suit the crop geometry, irrigation system and management. The planting on the top of the ridge is preferred in the high rainfall regions and planting at the bottom of the furrow is preferred in low rainfall regions.

Pitcher Irrigation

It is a very useful technique to establish young plants in sandy wasteland area having little or no irrigation facility. Small pitcher of about 3-5 litre capacity having a hole at bottom fitted with a wick is buried alongside a newly planted sapling. The pot is occasionally filled with water and its mouth is kept closed. In a grown up orchard, one/two/three pitchers around a tree can be buried depending on its age and spread of tree. In a 16 year old Mosambi orchard, 30 litre pitchers (2 number) per tree gave satisfactory results in terms of fruit yield and quality.

Rainwater Harvesting

In subtropical region, where moisture stress during summer appears to be the major constraints for economical yield of fruit crops under rainfed conditions. There are different types of water harvesting system used for conservation of moisture during critical stages of fruit growth and development.

In-situ Moisture Conservation

In most of the subtropical region, the rainfall is erratic and falls within a short period of time. The moisture may not be available to the crop at the critical stage of its growth. In-situ moisture conservation system can help in retaining soil moisture for longer period of time. For perennial fruit crops, Full moon, half moon, cup and saucer and trench system can be introduced for better yield and quality under rainfed conditions. This technique is most effective since the deep rooted perennial fruit trees can draw the runoff water concentrated even in the deep soil profile. However, the runoff generation depends upon several characteristics of the catchments *i.e.*, size, slope and rainfall pattern of the region.

Rainwater Harvesting Collection and Storage

Rainwater harvesting through tanks, ponds and reservoirs has been practiced traditionally in India for a long time. In this system collection of the excess runoff from the rain on the farm ponds and its utilization in later stage during critical fruit growth and development. Several indigenous methods based on local wisdom were devised to store the rainwater in dry region of Rajasthan (in the form of tanka and percolation tank), bhanndharas in Maharashtra, bandhis in MP and U.P. and ahars in Bihar.

Regulated Deficit Irrigation

It is a method that irrigates the entire root zone with less amount of water than the potential evapo-transpiration during whole or specific crop cycle periods. The principle of the Regulated Deficit Irrigation (RDI) technique is that plant sensitivity to drought is not constant during the growing season and that intermittent water deficit during specific periods of ontogenesis may increase water saving and improve fruit yield and quality. The key of the RDI strategy is the timing of the water deficit and the degree of the deficit applied to the plants. To avoid the possibility of reducing effect of RDI on fruit yield, and monitoring of soil water status is required in order to maintain a plant water regime within a certain degree of drought stress. Implementing RDI could also be difficult where there is a high water table or deep soil with a high water holding capacity. However, if RDI is managed carefully, the negative impact on crop yield could be avoided. Results of numerous field crops, tree crops and fruit crops showed beneficial effect on yield and quality under regulated deficit irrigation system.

Partial Root Drying

Partial root drying (PRD) is a method where water is manipulated over space. It is designed to maintain half of the root system in a dry or drying state, while remaining half is irrigated. The treatment is then reversed, allowing the previously well watered side of the root system to dry down while fully irrigating the previous dry side. The principal behind PRD is that irrigating part of the root system keeps the leaves hydrated and in a favourable plant water status, while drying on the other part of the root system promote synthesis and transport so called chemical signals (particularly plant hormone abscissic acid) from roots to the shoot via the xylem to induce a physiological response. The frequency is determined according to soil type, genotypes, rainfall and temperature of the region.

Effect of PRD on plant physiology are different from RDI because wet roots under PRD sustain shoot and fruit turgor that are important for plant growth. The drying roots in the PRD produce the sufficient amount of the chemical signals to maintain a physiological response to water stress. Triggering partial stomatal closure under PRD irrigation prevent excessive water loss and also the metabolic inhibition of CO_2 assimilation, that otherwise would occur in extensively development of drought stress. PRD may be applied by different techniques in the field depending on the cultivated crops or soil condition. PRD irrigation could be done by subsurface or surface drip line, furrow, micro-sprinkler or vertical soil profile methods.

Mulching

Mulching is a practice of covering orchard floor or tree basin with some materials so as to break the direct contact between orchard soil and outside atmosphere. Use of organic and inorganic mulches have greatest advantages in fruit growing on degraded lands which is not only reduce the moisture loss through evaporation, preventing weed growth and improvement of soil physical conditions but also reduce the fruit drop, improve fruit yield and quality.

Microirrigation

Water scarcity in various parts of the country has created awareness for about microirrigation system and its implementation results in significant economic and social benefits in the country. The benefit noticed in terms of improvement in crop yield, reduction in energy consumption, reduction in use of chemical fertilizers and pesticides. Microirrigation technology increases water use efficiency in fruit crops. This can be achieved by reducing conveyance losses, runoff, evaporation losses, seepage and deep percolation losses significantly. In this method application of water at low volume frequent interval under low pressure to root zone of plant. There are two type of microirrigation i) Drip irrigation ii) Sprinkler irrigation system used in subtropical horticultural crops.

Figure 5.2: Mulching in Mango.

Drip Irrigation

In drip irrigation system, the water is conveyed under pressure through pipes to the field where it drips slowly onto the soil surface through emitters located close to the plants. The system in the simplest form consist of pumping unit, water filters, PVC main and submain lines, LDPE laterals and drippers. The lateral pipes are generally 12 or 16 mm diameter and laid along the crop line in a single or double. The dripper may be on line or inline with pressure compensating to maintain uniformity. The discharge rate of drippers generally varies from 2-8 litre/hr and the operating pressure of the non-pressure compensating dripper is about 1 kg/cm^2. The frequent water application in drip irrigation maintain soil moisture in the crop root zone very near to the optimum level. The drip irrigation is a technology to grow crops under controlled conditions of soil moisture, fertilizers, salinity and pest control.

Sprinkler Irrigation

In sprinkler irrigation, water is delivered through a pressurized pipe network to sprinklers, nozzles or jet which spray the water into the air and fall to the soil surface in an artificial rain. The basic components of all sprinkler systems are a water source, a pump to pressurize the water, a plastic pipe network to distribute the water throughout the field, sprinkler system to spray the water over the surface soil layer, and valves to control the flow of water. The sprinkler system, when properly spaced, give a relative uniform application of water over the irrigation area. The system will works well at right operating pressure recommended by the manufacturer/company. The uniformity of water distribution is affected if the system is operated at higher or lower pressure and under windy conditions.

The available water must be used effectively and efficiently for the subtropical fruits so as to enhance their productivity. The relative performance of fruit crops with drip irrigation in comparison with traditional irrigation method at various locations and their performance is summarized in these fruit crops.

Mango (*Mangifera indica* L.)

Non Bearing

For establishment of new orchard, planting should be done during rainy season for better establishment. During summer or post monsoon period, young plants with shallow root system, needed water at frequent intervals (2-3 days interval). As tree grows in the age group of 2-5 years should be irrigated at 4-5 days interval. Proper moisture should be maintained in the soil for better establishment of young plants. No irrigation is required during the monsoon months unless there are long dry spells.

Figure 5.3: Drip Irrigation in Amrapali Mango under HDP.

Bearing Mango

The drip irrigation started after complete fruit set and continued till harvest. The water requirement in respective orchard was worked out on the basis of monthly mean pan evaporation and plants were irrigated daily by pressure compensated dripper (4 litre/hr) with 4 dripper connected with micro tube (4/6 mm) were placed equidistance in east, west, north, south direction at 50 per cent distance from canopy radius. The diameter of the lateral pipe is 16 mm connected with sub main pipe. The water received through natural rain is compensated in successive days.

Amount and frequency of irrigation to be given depends upon the type of soil, prevailing climatic conditions, especially rainfall and its distribution and age of the trees. No irrigation is required during the monsoon months unless there are long spells of drought. The irrigation interval could be increased to 10-15 days for 5-8 years old plants during dry season. When trees are in full bearing stage, generally 2-3 irrigations are given after the fruit set which results in increased fruit-set, decreased fruit drop, and improvement in fruit size and fruit quality. It is advisable to irrigate the mango plants in basins around them to economizing water use. The intercrops need to be irrigated independently as per their specific requirements.

Figure 5.4: Drip Irrigation and Organic Mulching in Mango.

Irrigation at the time of fruit setting (mustard size), marble stage and fruit development (before maturity) have been identified as a critical stages for optimizing fruit size and quality of mango. The research work have been conducted at ICAR–CISH, Lucknow indicated that drip irrigated orchard gave significantly higher fruit yield over basin irrigation. Drip irrigation replenishment of 60 per cent open pan evaporation gave the highest fruit yield in mango cv Dashehari and Langra. Similar work also conducted at Research and Development Farm of Jain Irrigation Systems Limited Jalgaon and found beneficial effect of drip irrigation in mango. At Central Horticultural Experiment Station-IIHR, Bhubaneswar that mango variety Arka Anmol orchard irrigated with 75 per cent Pan Evaporation Replenishment (PER) resulted better yield and at par with 100 per cent PER. Under high density planting system, water and nutrient can be applied through drip system in order to enhanced productivity and quality of mango (Kumar *et al.*, 2008). Fruit yield and quality of Alphanso mango have been improved by application of 24 litre water/day (Prakash *et al.,* 2015). Similarly organic mulching + drip irrigation @ 50 per cent PER resulted better yield and quality in Lat Sundari mango under eastern Indian conditions.

In-situ Moisture Conservation in Mango

Field experiment on *in-situ* moisture conservation techniques in mango at Central Horticultural Experiment Station (ICAR–IIHR), Bhubaneswar, India, on a new mango orchard of the variety 'Arka Neelachal Kesri' planted at 5m x 5m spacing. Results clearly indicated that rain-water harvesting enhance soil moisture availability during post monsoon periods for augmenting plant growth and fruit production.

Figure 5.5: *In-situ* Rainwater Harvesting in Almond.

Among the four in situ rain-water harvesting techniques (cup-and-plate, half-moon, full-moon, and trench) evaluated in combination with three types of mulch (no mulch, inorganic mulch, and organic mulch), the cup-and-plate system resulted in maximum annual increment in vegetative growth and fruit yield (4.67 kg/plant), while, organic mulch (paddy straw) and inorganic (black polythene, 100μ thickness) mulches improved vegetative growth, fruit yield and quality significantly over no mulch (Samant *et al.,* 2015).

Rainwater harvesting such as Cup and Plate and Full moon system in almond in Kashmir region performing well under rainfed conditions (Kumar *et al.,* 2015).

Guava (*Psidium guajava* L.)

Guava is one of the important fruit crop of tropical and subtropical region of the country. It ranks fifth most important fruit crop in production after banana, mango, citrus and papaya with a total production of 1.68 million tonnes. Irrigation is frequently needed during initial stage of plant establishment, *i.e.,* weekly or biweekly intervals. Water is essential during first few month after planting. Appropriate scheduling of irrigation increases the water use efficiency along with water saving. The work have been conducted at ICAR–CISH Lucknow indicated that drip irrigation at 80 per cent PER coupled with plastic mulching improves the fruit yield and quality of guava cultivar Allahabad Safeda. Drip irrigation and mulching provides a consistent moisture regime in the soil due to which root remains active through out the season resulting in optimum availability of nutrient and its proper absorption which favours the fruit growth and development in guava. The daily water requirement of guava varies from 15.6 to 61 litre per day per plant respectively in the month of December and April-May. Drip irrigation on alternate day or 2 days intervals is recommended during dry period of April-May month. Water and nutrient use efficiency can be improved under HDP by planting guava on raised beds at 15-20 cm height alongwith drip irrigation at 80 per cent ET + 75 per cent RDF coupled with plastic mulching improved fruit yield and quality in guava (Srivastava *et al.,* 2021).

Figure 5.6: Drip Irrigation in Guava under HDP.

Aonla (*Emblica officinalis* Gaertn)

Moisture stress affects growth and development of aonla tree during vegetative and reproductive phase and ultimately the fruit yield. In this crop, optimum availability of soil moisture can enhanced the quality of fruit. Under North Indian conditions, vegetative growth occurs during March- April and August - November, therefore, assured irrigation to aonla is very much essential. Considering the critical stage of fruit growth and development, 160-180 days of irrigation through drip is needed during this period. Maximum plant height, girth and spread were recorded with 60 per cent CPE level of irrigation at three days intervals. Mulching with various organic waste can help in conserving moisture, reducing fruit drop and increasing fruit size and quality. Drip irrigation on 20 per cent wetted area basis on alternate day has been found to improve growth and yield of aonla.

Bael (*Aegle marmelos* Correa)

Bael is one of the important hardy fruit crops of subtropical region once established can withstand even under intermittent drought conditions. Regular irrigation of plants are not required, however, young plant needs to be irrigated un till their establishment. Bearing tree needs optimum moisture regime to control the fruit drop, cracking and damaged from frost. Provision of micro-catchment around the tree trunk for runoff concentration and mulching with locally available organic waste or plastic mulch conserved more soil moisture leads better fruit yield under rainfed condition. Drip irrigation may be the ideal during critical period of growth and development of aonla.

Jamun (*Syzygium cuminii* L.)

In the early stage of orchard establishment, plants requires 8-10 irrigation in a year especially during post monsoon period. Bearing trees require 4-5 irrigation during fruit development and ripening stage. The water saving device like drip irrigation is very useful during fruit development and ripening stage. Mature plants have deep root system and draws moisture from lower strata of the soil. Once plant established, can withstand much adverse weather conditions under rainfed ecosystem. In bearing tree, irrigation is essential during April-June which help in fruit retention, development and ripening of fruits.

Litchi (*Litchi chinensis* L)

Litchi is one of the important fruit crops of tropical and subtropical region of the country. Irrigation management play an important role in quality litchi production. Flood irrigation causes significant water loss due to run off, seepage, evaporation *etc.* On the other hand, microirrigation technique provides a reduction in water consumption by 30-70 per cent with simultaneous augmentation in yield. A suitable drip irrigation level may ensure proper growth and development of trees for obtaining higher marketable yields. Application of drip irrigation at 100 per cent ER along with mulching was found to be most superior in respect to yield and quality of litchi (Tyagi 2021).

Conclusion

Among various approaches of water productivity, microirrigation has emerged as an appropriate water saving techniques for all wide spaced high value perennial fruit crops in water scare, undulating and sandy area of subtropical conditions. Since, microirrigation greatly enhances water, fertilizer and energy use efficiency and promote precision horticulture. Subtropical fruit crops have to gain much for meeting the challenges of more production with declining land and water by adoption of efficient techniques towards high water productivity for quality fruit production in the region.

REFERENCES

Descheemaeker K, Bunting SW, Bindraban P, Mathuri C, Molden D, Beveridge M, Brakel MV, Herrero M, Clement F, Boelee E and Jarvis DI, 2013. Increasing water productivity in Agriculture in book Managing Water and Agro ecosystem for Food Security (ed. E.Boelee). pp. 104-123.

Galindo A, Gonzalez JC, Grinan I, Corell M, Centeno A, Martin-Palomo MJ, Giron IF, Rodrigues P, Gruz ZN, Hemmi H, Barrachina AAC, Hernandez F, Torrecillas A and Moriana A, 2018. Deficit irrigation and emerging fruit crops as a strategy to save water in Mediterranean semi arid agroecosystems. *Agricultural Water Management*, **202**: 311-324.

Kumar D, Pandey V and Vishal Nath, 2008. Effect of organic mulching and irrigation schedule through drip on growth and yield of mango var. Lat Sundari in eastern region of India. *Indian Journal of Agricultural Sciences*,**78**(5): 385-388.

Kumar D, Pandey V and Vishal Nath, 2008. Effect of drip irrigation regimes on growth, yield and quality of mango hybrid Arka Anmol. *Indian Journal of Horticulture*, **65**(4): 409-412.

Kumar D and Ahmed N, 2015. Effect of rainwater harvesting and mulch material on soil moisture regimes, fertility status and yield of almond under rainfed conditions of north western Himalayas. *Indian Journal of Horticulture*, **72**(2): 193-200.

Malhotra SK and Das M, 2016. Augmenting water use efficiency in horticulture. Souvenir, World Aqua Congress held from 28-30 December, India Habitat Centre, New Delhi. pp. 45-60.

More TA, Sharma BD and Singh IS, 2007. Strategies for enhancing water productivity in horticultural crops. *Indian Journal of Arid Horticulture*, **2**(1): 1-8.

Prakash K, Vijaykumar RM, Balamohan TN and Singh SDS, 2005. Effect of drip irrigation and fertigation levels on yield and quality of mango cv. Alphanso under Ultra High Density Planting. ISHS Acta Horticulturae 1066. Global

Conference on augmenting production and utilizationof mango: Biotic and abiotic stresses. DOI: 10.17660/ActaHortic.2015.1066.17.

Samant D, Mandal S, Singh HS, Nath V and Kurian RM, 2015. Effect of insitu rainwater harvesting and mulching on growth, yield and fruit quality of mango var. Arka Neelachal Kesri in eastern India. *Journal of Horticultural Sciences*, **10**(1): 99-101.

Silva JRD, Rodrigues WP, Ferreira LS, Bernado WDP, Paixao JS, Patterson AE, Ruas KF, Viana LH, Sousa EFD, Bressan-Smith RE, Poni S, Griffin KL and Campostrini E, 2018. Deficit irrigation and transparent plastic covers can save water and improve grapevine cultivation in the tropics. *Agriculture Water Management*, **202**: 66-80.

Srivastava KK, Barman P, Patil P Kumar D and Sharma NK, 2021. Effect of raised bed, mulching and fertigation on productivity and quality of guava (*Psidium guajava* L.) under high density planting system. *Journal of Environmental Biology*, **42**: 1387-1394. DOI: http: //doi.org/10.22438/jeb/42/5/MRN-1642.

Tyagi M, 2021. Effect of drip irrigation and mulching on fruiting, yield and quality attributes of litchi (cv. Rose Scented) under HDP. *International Journal of Microbiology and Applied Science*, **10**(1): 1850-1858.

Chapter 6

Weed Management in Orchards

Sukhpreet Singh, Tarundeep Kaur and Gayatri Sharma

Department of Agronomy, Punjab Agricultural University, Ludhiana–141004, Punjab

The proverb "An apple a day keeps the doctor away" highlights how important fruit crops are to a person's diet. It is well known that a diet high in fruits and vegetables and low in saturated fats is beneficial and protective against cancer and cardiovascular diseases. The World Health Organization recommends a daily intake of more than 400 g of vegetables and fruits/adult (https://www.who.int, 2024). In many cultures, fruits are the primary source of vitamins, antioxidants, and dietary fibres and their health benefits are continuously highlighted by many research groups. To maintain the orchards for higher production and superior quality, proper floor management practices from the initial stage of establishment to the annual cultivation principles are very important. However, there are various biotic and abiotic threats to crop production, in addition to socio-economic and crop management-related constraints. Weeds are the most significant biotic constraint to agricultural production in both developing and developed countries. Weeds are a group of specialized plants that have evolved along temporal and spatial scales because of their large seed production, aggressive reproduction, high regenerative capacity, and large phenotypic plasticity. Weed infestation is considered as the primary obstacle in agricultural production systems specifically in orchards. Despite of providing protection against various pests and diseases, these unwanted plants compete for nutrients, moisture, space and light with the crops of interest. Infestation of weeds adversely affects the crop in terms of reducing plant growth, fruit yield, and quality along with additional costs incurred on the management practices. The yield losses caused by the weed infestation surpass the losses caused by any pest category of agricultural production system (Abouziena and Haggag, 2016). The fruit yield reduction depends upon the weed flora and its density, fruit crop species, prevailing season, *etc.* Fruit trees are poor competitors because of their low root density per unit of soil compared to weeds (Mia *et al.,* 2020). It is reported that the reduction in tree growth due to weeds ranges from 15 to 96 per cent and about 35 per cent per cent yield loss was recorded. The fruit yield reduction under weeding conditions can be up to 50 per cent in the stone fruits (Hussein *et al.,* 2016). It has been estimated that weeds result in about USD 11 billions of economic losses every year in India (Sinden *et al.,* 2004). Negi (2015) reported that weed interference affect tree growth, fruit yield, and fruit quality in nectarine by competing for light, water and nutrients and also providing a potential breeding niche for various insect-pests and diseases. Weeds are also known to have allelopathic effects on main crops by secretion of root exudates in soil. Weed management is a key factor in reducing the weed competition for water, nutrients, space, and light in avoiding a negative impact on fruit crop growth and productivity (Atucha *et al.,* 2011). In this context, this chapter highlights the types of weeds in orchards, weed impacts, various weed management practices, factors affecting the weed management decisions and future trends and technologies.

Types of Weeds in Orchards

Even though soils in this part of the country are fertile along with the availability of canal water for irrigation and sufficient rains, the orchard productivity is low due to high infestation of weeds. Table 6.1 classifies the orchard weeds according to their cotyledon character. *Chenopodium album* is an example of a cosmopolitan one-year weed, frequently occurring in the orchards of South Korea, Argentina, New Zealand, Canada, Bulgaria and Turkey (Lisek, 2014). *Setaria media, Capsella bursa-pastoris, Echinochloa crus-galli,* and *Polygonum aviculare* belong to the group of species commonly occurring in the orchards of different countries and continents. This means that the anthropogenic factor-the orchard floor management, has a significant influence on the species composition of orchard synanthropic flora, including weeds. Under Indian Punjab conditions all type of weed flora *viz.* grasses, broadleaf weeds and sedges are found. The dominant weed flora comprised of mainly grasses (*Cynodon dactylon, Sorghum halepense, Panicum maximum*), sedges (*Cyperus rotundus, Cyperus compressus*) and broadleaf weeds (*Cannabis sativa, Parthenium hysterophorus, Chenopodium album, Medicago denticulata, Rumex dentatus, Fumaria parviflora, Anagallis arvensis, Coronopus didymus* and *Malva neglecta*) (Brar *et al.,* 2020).

Table 6.1: Major Weeds of Horticultural Crops

Fruit Crop	*Monocot Weeds*	*Dicot Weeds*	*References*
Mango	*Cyperus spp., Sorghum halepense* (L.) Pers., *Digitaria ciliaris* (Retz.), *Cynodon dactlyon* L., *Cenchrus spp., Paspalum conjugatum* P.J. Bergius	*Bidens alba* L., *Ageratum houstonianum* Mill., *Phyllanthus* amarus Schumach. and Thonn., *Phyllanthus niruri* L., *Tridax procumbens* L., *Lantana camara* L., *Argemone mexicana* L., *Fumaria officinalis* L.	Kleiman *et al.* (2021) http:// ecoursesonline. iasri.res.in (2024)
Banana	*Cynodon dactylon* L., *Digitaria ciliaris* (Retz.) Koeler, *Cyperus rotundus* L., *Digitaria marginata* Link	*Mimosa pudica* L., *Mariscus alternifolius* Vahl, *Alternanthera sessilis* (L.) R.Br. *ex* DC., *Stachytarpheta cayennensis* (Rich.) Vahl, *Portulaca oleracea* L., *Asystasia gangetica* (L.) T. Anderson	Fongod *et al.* (2010) http:// ecoursesonline. iasri.res.in (2024)
Pineapple	*Cynodon dactylon* L., *Digitaria* (Retz.) Koeler, *Cyperus rotundus* L., *Megathyrsus maximus* (Jacq.) B.K. Simon and S.W.L. Jacobs, *Eleusine indica* (*L.*) *Gaertn.*	*Portulaca oleracea* L., *Mollugo pentaphylla* L., *Convolvulus arvensis* L., *Bidens pilosa L.*	Paul and Uruu (2021) http:// ecoursesonline. iasri.res.in (2024)
Grape	*Cynodon dactylon* L., *Digitaria ciliaris* (Retz.) Koeler, *Eleusine indica* (L.) Gaertn., *Setaria glauca* (L.) Beauv., *Cyperus rotundus* L.,	*Oxalis corniculata* L., *Polygonum spp.* L., *Euphorbia geniculata* L., *Portuluca oleracea* L., *Parthenium hysterophorus* L., *Commelina benghalensis* L., *Amaranthus spinosus* L.	Ramteke *et al.* (2012) http:// ecoursesonline. iasri.res.in (2024)
Papaya	*Cynodon dactylon* L., *Chloris barbata* Sw., *Digitaria ciliaris* (Retz.) Koeler, *Eleusine indica* (L.) Gaertn., *Echinochloa colona* (L.) Link, *Dactyloctenium aegyptium* (L.) Willd,	*Parthenium hysterophorus* L.,*Lagasca mollis* Cav., *EuphorbiaGeniculata* L., *Phyllanthus niruri* L., *Tridax procumbens* L., *Commelina erecta* L.	García-Cámara *et al.* (2022) http:// ecoursesonline. iasri.res.in (2024)

Fruit Crop	*Monocot Weeds*	*Dicot Weeds*	*References*
Citrus	*Cynodon dactylon* L., *Digitaria ciliaris* (Retz.) Koeler, *Eleusine indica* (L.) Gaertn., *Setaria glauca* (L.) Beauv., *Imperata cylindrica* (L.) P. Beauv., *Cyperus spp.* L., *Echinochloa crus-galli* (L.) P. Beauv., *Poa annua* L., *Paspalum distichum* L.	*Amaranthus caturus* B.Heyne ex Hook.f., *Bidens pilosa* L., *Lagasca mollis* Cav., *Euphorbia spp.*, *Amaranthus viridis* L., *Cichorium intybus* L., *Chenopodium album* L.	Nasr *et al.* (2013) http://ecoursesonline.iasri.res.in (2024)
Sapota	*Setaria glauca* (L.) Beauv., *Digitaria ciliaris* (Retz.) Koeler, *Andropogon spp.* L.	*Bidens pilosa* L., *Blumea spp.* (L.) DC., *Oxalis corniculata* L.	http://ecoursesonline.iasri.res.in (2024)
Guava	*Cynodon dactylon* L., *Digitaria ciliaris* (Retz.) Koeler, *Eleusine indica* (L.) Gaertn., *Setaria* (L.) Beauv, *Imperata cylindrica* (L.) P. Beauv., *Cyperus spp.* L., *Sorghum halepense* (L.) Pers. *Eragrostis tenella* (L.) P.Beauv. ex Roem. and Schult., *Cenchrus catharticus* Delile, *Digitaria sanguinalis* (L.) Scoop., *Commelina benghalensis* L., *Echinochloa colonum* (L.) Link, *Dactyloctenium aegyptium* (L.) Willd, *Eragrostis pilosa* (L.) P.Beauv.	*Amaranthus caturus* B.Heyne ex Hook.f., *Bidens pilosa* L., *Lagasca mollis* Cav., *Euphorbia spp.* L., *Cannabis sativa* L., *Trianthema portulacastrum* L., *Solanum nigrum* L., *Cleome viscosa* L., *Digera arvensis* Forsk.	Brar *et al.* (2017) http://ecoursesonline.iasri.res.in (2024)
Apple	*Paspalum dilatatum* Poiret, *Digitaria sanguinalis* (L.) Scop., *Echinocloa crus-galli* (L.) Beauv., *Agropyron repens* (L.) Gould, *Avena fatua* L., *Cynodon dactylon* L., *Cyperus rotundus* L., *Digitaria sanguinalis* L., *Setaria glauca* (L.) Beauv., *Sorghum halepense* (L.) Pers.	*Anagallis arvensis L.*, *Chenopodium album L.*, *Polygonum persicaria L.*, *Rumex acetosa* L., *Ageratum conyzoides* L., *Amaranthus viridis* L., *Portulaca oleracea* L., *Chenopodium album L.*, *Convolvulus arvensis* L., *Euphorbia hirta* L.	Hussain *et al.* (2020) Miñarro (2012)

Fruit Crop	*Monocot Weeds*	*Dicot Weeds*	*References*
Strawberry	*Echinochloa crus-galli* L., *Poa annua* L., *Allopecurus myosuroides* Huds., *Cynodon dactylon* L., *Cyperus rotundus* L.	*Amaranthus retroflexus* L., *Amaranthus albus* L., *Anagallis arvensis* L., *Chenopodium spp.*, *Fumaria spp.*, *Solanum nigrum* L.	Bozbuga *et al.* (2022)
Pomegranate	*Sorghum halepense* (L.) Pers., *Setaria viridis* (L.) P. Beauv.	*Amaranthus spp.*, *Portulaca oleracea* L. *Convolvulus arvensis* L., *Xanthium strumarium* L., *Chenopodium album* L., *Daucus carota* L. and *Ipomoea purpurea* L. Roth.	Muslu and Tepe (2016)

Effects of Weeds on Orchard Productivity

Competition for Nutrients, Water, Sunlight and Space

Weeds possess many growth characteristics and adaptations that enable them to successfully exploit the numerous ecological niches left unoccupied by the crop cultures. Weeds compete among themselves and with crop plants. Among the most important adaptations relevant to competitive advantage are properly synchronized germination, rapid establishment and growth of seedlings, tolerance to shading effects by the crop or by other weeds at the time of establishment, quick response to available soil moisture and nutrients, adaptation to the most severe climatic situations of the habitat, adaptations to the edaphic regime, relative immunity to post seeding soil disturbance, practices and resistance to herbicides that are used. In the initial stages of invasion by weeds into exposed orchards, only a very limited competition for resources between the crop and weed may occur, but as the establishment of the crop-weed association is completed, competition for the available resources is more obvious. Plant competition is a natural force whereby crop and weed plants tend to attain maximum combined growth and yield, with the development of each species, to some extent, at the expense of the other. It occurs when the demands of the plants for moisture, nutrients, light, and possibly carbon dioxide exceed the available supply. The weed suppresses the crop resulting in reduced yield. Weeds competing with fruit crops for light, water, and nutrients affect fruit productivity and quality. Numerous factors, including the species composition of weed communities, weed abundance, density, and ability to compete with crops, influence weed-crop competition. Total nutrient uptake by weeds depends primarily on the nutrient content of the weeds and the weed biomass, which is limited by weed control treatments.

Choi *et al.* (14) reported that South Korean persimmon orchard weeds contained 1.5-2.96 per cent N, 0.23-0.41 per cent P, and 2.85-4.46 per cent K on dry weight basis. Weed vegetation can take up 41.4 kg N, 11.9 kg P, and 89.6 kg K/ha during the year in a pear orchard in South Korea (Lim *et al.,* 2012). Weeds growing in Polish orchards contained 2.39-3.60 per cent N, 0.31-0.64 per cent P, 1.15-3.35per cent Ca and 0.12-0.23 per cent Mg on dry weight basis and significant differences among species were recorded (Lipecki, 2006). Weeds make it difficult or impossible to use a fruit harvester to harvest fruit from bushes. The presence of honey weeds blooming during the period of intensive chemical use in orchards and farm fields may poison bees and other pollinators due to the chemical residues of plant protection products on the weeds. In the case of trees which are perennial crops, the harmful effects of weeds have been observed to be shifted to the next vegetative period. Weeds might be one of the factors that cause alternate bearing in fruits (Lisek, 2010). Pome trees are particularly susceptible to weed competition during spring and the first half of summer, from April to the end of July in the northern hemisphere. Stone trees require effective weed control from April to September. Raspberry is susceptible to weed competition in spring, between May and July, and should be free from weeds in June when primo-canes are developing (Lisek, 2010). Blueberry, due to the lack of root hair, is susceptible to weed competition throughout the whole vegetative period, until October, with the greatest attention being paid to weed control from April to August (Caroll *et al.,* 2012). Weed competition is oppressive to strawberries from April to June (Pritts and Kelly, 2001). In strawberry fields planted in spring (March-April), the first weed control treatment should take place no later than between the fourth and the eighth week after the crops have been planted (Lisek, 2010). July and August are also considered to be critical periods for strawberries. Weed competition for light is normally not a problem with established trees. However, big weeds in newly planted orchards can shade and severely stunt young trees. Vining weeds like poison ivy or Virginia creeper can climb mature trees and compete for light. Conversely, trees use shading as a competitive advantage over emerging weeds. Once mature peaches leaf out, they shade a large portion of the orchard floor. Shade from trees minimizes weed seed germination and reduces the competitiveness of emerged weeds.

Table 6.2: Effects of Bermuda Grass (*Cynodon dactylon* L.) Competition on Newly Established Peach (Mitchem, 2005)

Bermuda Grass (Cynodon dactylon L.) Ground Cover (%)	*Reduction in Tree Fresh Weight (%)*	
	After 1st Year	*After 2nd Years*
0	0	0
25	19	28
50	43	44
75	64	62
100	86	87

The primary competition between weeds and peach trees is for water. North Carolina State University research showed that for every week an orchard is weed-free, yield increases by as much as 120 pounds/acre. Water is essential for maximizing fruit size. Many south-eastern peach sites are on sandy loams or coarser textured soils with limited drought tolerance. Soaking rains or irrigation offers relief for short periods of time; however, for optimum fruit size and yield, rainfall is needed every 10 days. Weeds growing in the tree rows reduce available water for trees. Insufficient soil moisture can also greatly reduce tree growth, especially in newly planted orchards, and in extreme situations can be a primary cause of tree mortality.

Allelopathic Effects

Allelopathy's possible effects on agriculture have been thoroughly examined. There has been a focus on the negative effects of plants or their waste products, which can lower crop productivity or growth. Many plants and different organs, such as leaves and fruits, contain compounds that exhibit allelopathic activity. These chemicals may have an inhibitory influence on the crops. Generally, these allelochemicals are typically referred to as secondary plant products or waste products of the primary metabolic pathways in plants. During their passage through the soil, these allelochemicals may undergo changes, metabolization, or binding to organic materials. The environmental introduction of allelopathic potential holds the capacity to greatly affect the yield and survival of crops (Figure 6.1). Specifically, autotoxicity and replanting issues, toxicity of surface residues in reduced tillage systems, issues with specific crop rotations, and significant interference from specific weed species are often brought up. Apple trees are hampered by the emission of toxins and competition from orchard sods.

Figure 6.1: Mechanism of Allelopathy to Inhibit the Growth and Development.

It has been observed that the production of toxins by potatoes interplanted in apple orchards prevented the growth and photosynthesis of apple trees. The effects of budding and grafting trees are also discussed, and it is suggested that these biochemical interactions are frequently inhibitory and may be interpreted as allelopathic reactions. The ground breaking research on ethylene derived from pome fruits shows that the same chemical may be used to promote and restrict growth. The "replant syndrome" is a significant horticultural issue that appears to include allelopathy. Replanting challenges are multifaceted and frequently stem from an accumulation of pests (such as worms or diseases). Nevertheless, soil toxins are also linked to the challenge of re-establishing numerous permanent horticultural crops, such as citrus, asparagus, apple, peach, and other ornamentals. Allelopathy is thought to have a role in the evolution of plant communities, the invasion of exotic plants, and the failure of replanting in both natural and artificial environments. Leaching or exudation through roots of plant residues leads to the distribution and subsequent accumulation of such compounds in the biotype, this activity of allelopathic substances may largely influence the dormancy of seeds and buds, the germination of seeds, resistance of plants to pathogens, vegetation pattern and plant successions.

Habitat for Pests and Diseases

Weeds integrate into field ecosystems by regulating pest populations and offering alternative hosts, with potential ramifications such as altering ecosystem processes and displacing native biodiversity. Invasive weeds may support non-native organisms, hybridize with native species, and modify gene pools (Yandoc-Ables *et al.,* 2006). Despite this, weeds serve as vital resources, providing food, shelter, and breeding sites for various organisms, including plant pathogens (Webb *et al.,* 2012), insect pests (Singh and Singh, 2016), mites (Mishra *et al.,* 2019), nematodes (Singh *et al.,* 2010), and rodents (Kumar *et al.,* 2021). The epidemiological importance of weeds as alternative hosts of phytopathogens lies in their potential to bridge infections between crop seasons and act as sources of initial inoculum for disease spread. For instance, *Heliconia* species are associated with *Ralstonia syzygii* subspp. *celebesensis*, causing banana blood disease, while *Ralstonia solanacearum*, responsible for Moko/Bugtok wilt in bananas, has a wide host range, including Solanaceous hosts (Blomme *et al.,* 2017). Thus, the removal of weeds acting as an alternative hosts is recommended to mitigate insect-pests and diseases.

Weed Management Practices

Preventive Approach

The most effective strategy involves preventing the introduction of new weeds and weed seeds into the orchard area. When purchasing nursery plants, ensure that the nursery is free from weeds, and no weed plants are present with the earth balls. It has been noted that new weeds often present in earth balls of seedlings during orchard planting. Weeds already growing in the orchard should be prevented from flowering and producing seeds. Avoid using farmyard manure that may contain weed seeds and regularly clean machinery and tractors to prevent weed seeds from entering the orchard. Keep water channels, paths, roads, and orchard boundaries free of weeds. Adhering to the principle of "one year of seeding leads to seven years of weeding," clearing weeds before they flower is necessary to prevent weed infestations in the orchard.

Weed Eradication

Complete removal of weed plants, plant parts and seeds from the orchard and its surroundings can help eradicate weeds. However, achieving total eradication of all weeds from any area is often not feasible. Efforts can still be made to achieve satisfactory results in weed control.

Weed Control

Various weed control methods have been proposed for managing orchard floors, each with its own advantages and disadvantages. This section explores sustainable weed control strategies in fruit orchards, including mechanical methods (tillage, integrated mowing, modern finger weeders), Cultural methods (living mulch, organic mulch, plastic mulch), thermal weed control (flaming, steaming), high-pressure weed control (water jet blasting), chemical weed control and precision weed control (robotic systems).

Mechanical Weed Control

Mechanical methods include various cultivation and tillage systems, integrated mowing and modern finger weeders. While traditional mechanical systems can negatively impact soil health and biodiversity, advanced integrated systems minimize these drawbacks (Granatstein *et al.,* 2010). Tillage is a primary method for weed control in perennial fruit trees, utilizing operations like hand weeding, hoeing, harrowing, and cultivators. Modern tillage equipment, like the 'Wonder Weeder' and 'Hydraweeder,' can efficiently control weeds between trees while reducing labor (Granatstein *et al.,* 2014). However, excessive tillage can harm soil quality, reducing carbon and nitrogen availability, soil organic matter, and nutrient levels (Granatstein and Sánchez, 2009). It may also inhibit tree growth, lower fruit yields, and disrupt soil microbial composition and processes. Integrated mowing, combining a brush weeder with a mower, offers a sustainable weed management option in tree rows. This system utilizes advanced equipment to effectively control weeds while minimizing soil disturbance and labor requirements. In summary, mechanical methods like tillage and integrated mowing provide sustainable weed management options for fruit orchards, though careful consideration of their effects on soil health and biodiversity is necessary for long-term sustainability. Brush weeders are equipped with a polypropylene brush powered by the tractor's power take-off. This brush bends down weeds near tree trunks without damaging them, while a mower cuts and shreds weed plants just above the soil surface, preserving the soil underneath and between trees. The shredded weeds can serve as mulching material, improving soil nutrient status and preventing erosion. Integrated weed management with brush weeders not only controls weeds but also enhances soil quality, biodiversity, and tree productivity, aligning with sustainable orchard management goals. Finger weeders are another advanced technology for weed management in fruit orchards. These tools, equipped with adjustable hydraulic widths and star-shaped discs, break soil crust and eliminate weeds between plants with minimal soil disturbance. They provide gentle weed control and are available in various versions suitable for different soil types and weed species. While finger weeders are effective in loose soil and during early weed growth stages, their efficiency decreases in compact soil and with mature weed plants, especially grasses (Pannacci *et al.,* 2017).

Cultural Weed Control

Mulching is a significant cultural method that aids in suppressing weed growth in fruit orchards and conserves soil moisture. Mulches, whether natural (living and organic mulch) or synthetic (plastic mulch), improve soil physical, chemical, and biological characteristics, enhancing orchard biodiversity (Polverigiani *et al.,* 2013). Living mulches, comprised of growing plants, offer effective weed suppression, reduce nitrate leaching, control soil erosion, and promote soil biology. However, they can compete with fruit trees for water and nutrients, potentially reducing tree growth and yield (Granatstein and Sánchez, 2009). Proper species selection and management practices are crucial for maximizing the benefits of living mulches while minimizing competition with fruit trees. Organic mulches such as wood chips, provide effective weed control but may be costly and less efficient against established weeds, particularly perennial species. They can also immobilize nutrients in the soil, affecting tree growth and yield (Granatstein *et al.,* 2014). The rapid growth and dense ground coverage provided by the crop can effectively inhibit weed growth. Introducing cover crops like clovers, oilseed radish, summer green gram, summer black gram, sunhemp, sesbania, or forages into the orchards can also help suppress weed growth. Furthermore, residues of cover crops left on the soil surface can suppress weeds by providing shade and cooling the soil. When selecting a cover crop, it's important to consider its impact on the main crop in terms of pest habitat and allelopathy. Cover crop systems are particularly effective at controlling small-seeded annual broadleaf weeds.

Thermal Weed Control

Thermal weed control methods have seen significant technological advancements for managing weeds in fruit orchards. Typically, this system involves directing heat at weed areas and the base of trees using metal flame orifices connected to propane tanks. The heat breaks down weed cuticles, leading to their desiccation. Maintaining temperatures between 60°C to 70°C is crucial, with an optimal angle of 22.5° to 45.0° C between the burner and the ground for effective weed control. Flame weeding targets broadleaf weeds in early growth stages but may be less effective against grassy weeds. It requires multiple passes and reapplication after 1 to 3 weeks, making it a temporary weed control method. Despite its potential, flaming has limitations, including fire hazards and inefficiency against certain weed types. However, integrating flaming with tillage or mulching strategies can enhance its effectiveness and sustainability, making it a promising method. Additionally, using biogas instead of fossil fuel for flaming operations can further improve its sustainability credentials. Although environmentally friendly, concerns remain regarding the social and economic sustainability of these systems due to their high initial cost.

Biological Control

In ancient times, the Chinese realized that boosting ant numbers in their citrus trees reduced the damaging populations of huge digging beetles and caterpillars. The use of a natural enemy to control a pest signalled the beginning of biological control. *Colletotrichum gloeosporioides f. sp. aeschynomene* (Collego) for management of northern jointvetch (*Aeschynomene virginica*) in rice and soybeans and *Phytophthora palmivora* (Devine) for management of strangler vine (*Morrenia odorata*) in citrus orchards, were the first commercial successes of microbial biocontrol agents (Harding and Raizada, 2015). Devine causes root rot and can remain saprophytically active for over a year, giving residual control. Table 6.3 presents some important bioherbicides thar are used in agriculture to control weeds. The bioherbicide strategy involves the enormous, typically annual, release of a biocontrol agent into selected weed-infested fields in order to infect and kill sensitive weeds. Fungi, bacteria, and even viruses can be used as bioherbicides in a variety of host-pathogen combinations. Mutagenesis, produced by irradiation or chemical therapy, can also be used to improve pathogen strains. This method appears especially promising for the selection of bioherbicide strains that are resistant to chemical pesticides.

Table 6.3: Bioherbicides and their Respective Sources, Registered Names, Target Weeds and Ecosystems

Sl.No.	*Source*	*Registered Name*	*Target Weeds*	*Ecosystem*	*References*
1	*Colletotrichum gloeosporioides*	Collego	*Aeschynomene virginica* L.	Rice, soybean	Tateno (2000);
2	*Alternaria cassiae*	CASST	*Cassia obtusifolia* L.	Soybean	Tateno (2000)
3	*Phytophthora palmivora*	Devine	*Morrenia odorata* (Hook. and Arn.) Lindl.	Citrus	Hasan *et al.* (2021)
4	*Xanthomonas campestris*	Camperico	*Poa annua* L.	Turf	Boyetchko *et al.* (2007)

Sl.No.	*Source*	*Registered Name*	*Target Weeds*	*Ecosystem*	*References*
5	*Colletotrichum gloeosporioides*	Biomal	*Malva pusila* Sm.	Horticultural crops	Hasan *et al.* (2021)
6	*Streptomyces acidiscabies*	Opportune	*Taraxacum officinale* L.	Turf	Mendes and Rezende (2014)
7	*Brassica napus*	Beloukha	*Amaranthus retroflexus* L.	Wastelands	Muñoz *et al.* (2020)
8	*Syzygium aromaticum*	Weed Slayer	*Echinochloa crus-galli* L.	Rice	Hasan *et al.* (2021)

Chemical Weed Control

Herbicides present significant potential for reducing weed control expenses across various scenarios and provide an effective alternative to cultural or mechanical weed control methods in horticultural crops. Various herbicides used in fruit crops are shown in Table 6.4. With this method, less labour is required; this allows the transferring of labour to other activities. Usage of pre-emergence herbicides assumes greater importance in view of their effectiveness from the initial stages of crop growth, which is the most critical period of weed competition (Bhullar *et al.,* 2015).

Table 4. Various herbicides used in different fruit crops.

Fruit Crop	*Herbicide*	*Dose*	*References*
Mango	Paraquat	3.0 kg *ai*/ha as post-emergence	https://www.ikisan.com/ap-mango (2024)
	Atrazine	2.0 kg *ai*/ha as pre-emergence	
	Diuron	2.0 kg *ai*/ha as pre-emergence	
Banana	Glufosinate	1-2 kg *ai*/ha as post emergence	Pattison *et al.* (2024)
	Oxyfluorfen *fb* glyphosate	0.15 kg *ai*/ha *fb* 1.64 kg *ai*/ha at 9 and 12 WAP	Shaikh and Lokhande (2005)
	Diuron *fb* glyphosate	1.0 kg *ai*/ha *fb* 1.64 kg *ai*/ha at 9 and 12 WAP	
Citrus	Glufosiante + oxyfluorfen	420 + 480 g ai/ha as (Post + Pre emergence)	Alcántara-de la Cruz *et al.* (2019)
	Glyphosate + (bromacil + diuron)	1080 + (1200 + 1200) g ai/ha as (Post + Pre emergence)	https://www.ikisan.com/ap-citrus (2024)
	Glufosinate	450 g ai/ha as post emergence	

Fruit Crop	*Herbicide*	*Dose*	*References*
	Paraquat	1.01 kg ai/ha as post-emergence	Abdulridha *et al.* (2023)
Guava	Glyphosate	4.0 L/ha as post-emergence	Brar *et al.* (2017)
	Diuron	1.6 kg/ha as pre-emergence	https://nhb.gov.in (2024)
	Oryzalin	1.67 L/ha as pre-emergence	
	Simazine	1.6 kg/ha as pre-emergence	
	Atrazine	1.6 kg/ha as pre-emergence	
Grapes	Indaziflam	-	https://www.canr.msu.edu (2024)
	Oxyfluorfen	-	
	Pendimethalin	-	
Sapota	Bromacil	2 kg/ha as pre-emergence	https://nhb.gov.in (2024)
	Diuron	2 kg/ha as pre-emergence	
Pine apple	Diuron	3 kg/ha as pre-emergence	https://nhb.gov.in (2024)
	Diuron + Bromacil	(1.5 + 2.0 kg/ha) as pre-emergence	
Strawberry	Napropamide	-	https://ag.umass.edu (2024)
	Oxyfluorfen	-	
	Sulfentrazone	-	

Weeds that emerge later also compete with the crop, diminishing its productivity and necessitating the application of post-emergence herbicides or other non-chemicals. However, relying solely on herbicides is insufficient for achieving sustained control over the diverse weed species found in fields. This underscores the importance of employing an integrated approach for long-term weed management in fruit crops.

Control of Tree Weeds in Orchards

Tree weeds create more problems often than crop weeds. It is very difficult to control tree weeds due to their extensive root system and growth. Only using herbicide spray may not be the solution. The trees like '*peepal*' and '*shahtoot*' should be repeatedly defoliated by spraying herbicides. The chemicals can be injected in tree trunks, by making deep cuts with an axe. The apical dominance can be broken by spraying herbicides or by heading back the shoot tips. Cut the trees from the soil level and treat the stumps with phenoxy-herbicides immediately after cutting down the tree. The stumps can also be given a higher dose of common salt (NaCl) or urea fertilizer to check the resprouting of the stumps.

Experimental Studies on Weed Control in Fruit Crops

A field experiment was conducted by Ustuner and Ustuner (2011). Five different weed control methods were applied in an apple orchard containing Scarlet

Spur, Granny Smith, and Red Chief apple cultivars. The treatments were, a cover of black plastic (0.15 mm thick polyethylene), a 2 cm sand layer, a 4 cm sand layer, cardboard (1.5 mm thick), wheat stem (5 cm thick) and post-emergence application of glyphosate herbicide. A total of 47 weed species from 19 families were identified in the research field. Weed densities for each method were recorded as 0.1, 19.3, 1.2, 0.2, 26.6, and 5.8 weeds per square meter, respectively. The black plastic cover method showed the highest effectiveness at 99.9 per cent, followed by the cardboard method at 99.7 per cent, the 2 cm sand layer method at 98.0 per cent per cent and the herbicide method at 90.6 per cent. Chatha and Chanana (2007) conducted an experiment to study the different pre-emergence herbicides, namely oxyfluorfen @ 0.5, 0.75 and 1.0 kg a.i./ha, oxadiazon @ 0.75, 1.0, and 1.25 kg a.i./ha, and metolachlor @ 1.0, 1.5, and 2.0 kg a.i./ha, were applied twice, in March and October. Additionally, a post-emergence herbicide glyphosate was uniformly applied @ 1.5 L/ha twice, in June and August, to all experimental plots. Manual weeding was also performed monthly. The orchard exhibited 7 monocot and 23 dicot weed species. All treatments significantly reduced weed population, weed dry weight, and nutrient depletion compared to the control (un weeded). Oxyfluorfen @ 1.0 kg/ha was the most effective treatment for controlling dicot weeds. Metolachlor @ 2.0 litres per hectare and oxadiazon @ 1.25 L/ha controlled both monocot and dicot weeds. The highest fruit weight was achieved with oxyfluorfen @ 1.0 L/ha, followed by oxadiazon @ 1.0 L/ha and oxyfluorfen @ 0.75 L/ha.

A study was conducted by Rowley *et al.* (2011) to assess the effectiveness of various mulches and organic herbicides, both individually and in combination. Mulch treatments included barley straw, pine wood chips, paper and a bare-ground control without mulch. Organic herbicide treatments consisted of lemon grass oil (14 per cent v/v), clove oil (8 per cent v/v) and two concentrations of acetic acid (10 per cent and 15 per cent v/v). Additional treatments included pelargonic acid (7 per cent v/v), glyphosate (1.06 kg ai/ha), and no-herbicide control. Each mulch significantly suppressed weeds in the first year, but they presented specific challenges over time. For instance, paper mulch became less effective over time due to the development of cracks during wet-dry cycles allowing weeds to emerge. Volunteer barley became problematic in straw mulch plots in the first year, while annual weed seed germination on top of wood chip mulch became an issue in the second year. Among the mulches, straw mulch consistently provided the highest level of weed control. Glyphosate and pelargonic acid achieved nearly 100 per cent weed control across all mulch treatments, including the non-mulched plots. Lemon oil and clove oil showed similar weed control ratings to the no-herbicide treatment when used with straw, wood chip, and paper mulches but provided 41-95 per cent weed control when applied without mulch. Shrestha *et al.* (2012) investigated a study in an almond orchard in Fresno, CA, aimed to explore sustainable weed management practices for organic orchard systems. Treatments included steam, flame and broad applications of either lemongrass oil or D-limonene, with an untreated control also included. The steam and flame treatments generally provided up to 95 per cent control at 1 weeks after treatment (WAT). However, their effects lasted only 3 to 4 weeks as new weeds emerged or the treated weeds overcame the suppressive effects of the thermal treatments. Steam and flame treatments were more effective on certain erect-growing broad-leaved weed species than on grasses.

Lemongrass oil offered minimal weed control, whereas D-limonene provided up to 95 per cent control at 1 WAT. D-limonene also resulted in lower weed biomass compared to the untreated and thermal-treated plots. Monthly applications of steam or flame, or applications of D-limonene every 5 to 6 weeks, may be necessary to adequately suppress weeds in organic almond orchards. A field study was carried out by Brar *et al.,* 2017) to check the effectiveness of various orchard floor management practices, including mulching (paddy straw, white polythene and black polythene) and different weed control methods (mechanical and chemical), for weed management in a guava orchard. The black polythene mulch effectively kept the area weed-free throughout the season, the use of paddy straw mulch showed significant results, with a reduction of 69.1 and 93.2 per cent in total weed biomass during the first and second years of application, respectively. On the other hand, the white polythene mulch performed poorly, resulting in heavy weed growth underneath and tearing of the polythene sheet. Chemical and mechanical weed management methods also reduced weed intensity, but there was a resurgence of weeds leading to significantly higher weed biomass compared to paddy straw mulch. Overall, paddy straw mulch demonstrated potential for effective weed management in guava orchards. Similarly, Brar *et al.* (2020) reported 87.1 and 91.2 per cent reduction in total weed biomass during 1st and 2nd year, respectively by use of 12.5 t/ha paddy straw mulch in ber (*Zizyphus mauritiana* Lamk.). Further, uniform spreading of paddy straw mulch at 13.75 t/ha (pear) and 11.25 t/ha (peach) (Thakur *et al.,* 2012) and plum) after fruit set and second application of second dose of fertilizers during mid-April for control of annual grasses and broadleaf weeds.

Factors Affecting Weed Management Decisions

Orchard Type

Type of orchard is a major contributing factor to weed management decisions. It includes different aspects ranging from tree species to planting density in the orchard. All these aspects affect management decisions in one or the other way. Tree species serve as the foremost basis for selecting measures for combatting weeds. Weeds growing in the vicinity of the tree could be specific to that particular species or be common for most of the species resulting in non-specific host preference. For effective management of weed species, it is imperative to know the location, distribution and abundance in the orchard. For example, in case of citrus orchards, repeated discings during the summer months can eradicate johnsongrass and bermudagrass. Trifluralin can be integrated into the soil through discing before tree planting, effectively managing germinating grasses and certain small-seeded broadleaf weeds. This practice aids in minimizing weed competition around recently planted trees (https://www.ucanr.edu, 2024). Age of trees is also a significant aspect that plays a key role in taking decisions. Mature orchards have full canopies that hinder the sunlight falling on the orchard floor. This consequently results in reduction of weed species in the area. In citrus orchards, a combination of herbicides is sometimes utilized for enhanced weed control, leveraging both their effectiveness and selective properties. Monuron and simazine, for instance, exhibit effectiveness for up to six months in newly established orchards. Simazine combined with amitrole also yields satisfactory results, whereas amitrole alone lacks efficacy. Diuron and simazine, particularly notable for their effectiveness, have been documented as highly efficient in controlling both annual and perennial weeds in citrus orchards situated on deep sandy soils (https://www.ikisan.com, 2024). Furthermore, adjusting the planting density of an orchard offers a strategy for weed management. With increased density, especially within rows, the orchard floor experiences quicker shading from the tree canopies, leading to suppression of weed growth (https://www.sciencedirect.com, 2024).

Soil Type and Fertility

The utilization of herbicides and other agrochemicals in agriculture can impact the essential functions and population dynamics of soil microorganisms and thereby soil fertility. Various soil types can significantly influence weed growth and distribution, with several key considerations. Sandy soils typically offer better drainage, inhibiting weed establishment, yet they may lack fertility. Clay soils retain moisture effectively but can be prone to compaction, impacting root growth and weed competition. Loamy soils are ideal for orchards due to their balanced qualities, including good drainage, moisture retention, and nutrient availability. Additionally, soil fertility is vital for providing essential nutrients to fruit trees. However, it also affects weed growth. For example, high nitrogen levels can encourage weed growth, necessitating careful management to maintain a balance between tree health and weed control. P and K influence crop competitiveness and excessively high levels may favor weed proliferation. Micronutrient imbalances can impact both trees and weeds, underscoring the importance of nutrient management for optimal orchard health and weed suppression (https://treefruit.wsu.edu, 2024). The efficacy of clove oil-based herbicides in weed control can vary, and their effects on soil biology and tree health may differ accordingly (Hoagland *et al.,* 2008).

Labour Availability and Cost

Historically, manual weeding has been a prevalent approach to weed control. Nonetheless, the availability of skilled labor for this method can fluctuate significantly depending on regions and seasons. Shortages in labor can result in delayed or insufficient weed removal, ultimately impacting both yield and quality. Labor expenses play a pivotal role in decisions concerning weed management. Elevated labor costs might dissuade farmers from employing labor-intensive techniques like manual weeding. Conversely, reduced labor expenses could prompt more frequent and comprehensive weed removal. Herbicides, chemical agents utilized for weed control, vary in cost depending on factors such as type, application method, and dosage.

Environmental and Regulatory Considerations

It is imperative for the agricultural community to explore methods that sustain natural resources while ensuring effective and economical weed control. Although there isn't a single solution to achieve this goal, weed scientists are investigating nonchemical weed control approaches such as harvest weed seed control, as well as leveraging technologies like machine learning, sensors and robotics. These innovations are intended to complement and address the limitations of our current chemical-dependent systems (McCauley *et al.,* 2022). Regulatory authorities establish guidelines and restrictions on herbicide usage to safeguard the well-being of humans, animals, and the environment (https://wssa.net,2024). Adherence to these regulations is crucial for sustainable weed management. It helps prevent misuse, reduces environmental impact, and increases herbicide efficacy. Through adherence to regulations and the adoption of innovative approaches, we can attain improved weed management outcomes while safeguarding our natural resources.

Future Trends and Emerging Technologies

Advances in Precision Agriculture and Remote Sensing for Weed Detection and Management in Orchards

Precision agriculture leverages automation and contemporary technological breakthroughs such as machine vision, high-resolution cameras, robotics, and nozzle technologies to boost resource utilization by means of a precise variable rate application. Machine learning and geospatial technologies are becoming increasingly sophisticated in the realm of weed control. The incorporation of robotics, artificial intelligence (AI), and diverse sensors ensures improved outcomes in executing various agricultural tasks such as weed detection and elimination. Different components covered by AI in controlling herbicide resistant weeds are presented in Figure 6.2.

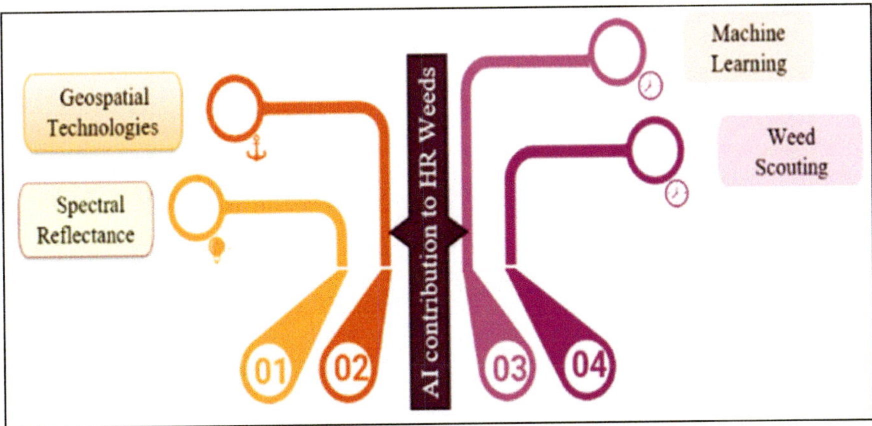

Figure 2: Role of Artificial Intelligence in Controlling HR Weeds.

Blue River Technologies, based in Sunnyvale, California (acquired by John Deere in 2017), pioneered a smart "see and spray" technology for weed identification and spraying. Employing machine learning and computer vision, they achieve precise discrimination between weeds and crops. This technology is seamlessly integrated with tractors or other agricultural machinery, enabling weeding operations at speeds of up to 19.3 km/h (https://bluerivertechnology.com,2024). Weed Detection using aerial satellite imagery and image processing software is a highly emerging and efficient technology for weed management. Based on colour and shape of plants as detected by the high-profile colour segmentation to distinguish between orchard trees and weeds can serve as a great strategical approach for site-specific management (Rasmussen *et al.,* 2019). By employing drones for image acquisition and machine learning techniques for subsequent processing, weed patches in cultivated fields can be accurately identified. Tailored algorithms can then be trained to facilitate weed removal through Autonomous Weeding Robot systems, utilizing herbicide spray or mechanical procedures (Esposito *et al.,* 2021)

After successful weed mapping, patch spraying and georeferenced application maps can be loaded into the board computer of the sprayer. In patch spraying, continuous opening and closing of boom sections and spray rate adjustment are required. With various advancements in sensor and information technologies for weed mapping and real-time weed detection, several patch spraying technologies were included in field boom sprayers (Figure 6.3 shows a drone with camera for detecting weeds). The development of precise and user-friendly systems is poised to drive broader acceptance of remote sensing data across both commercial and non-commercial Precision Agriculture (PA) prospects.

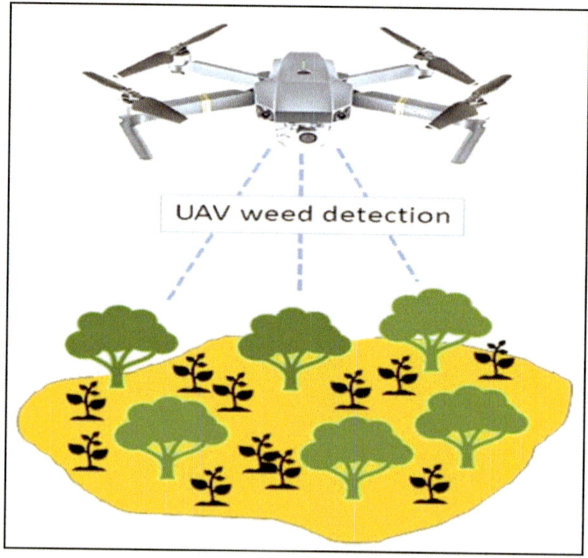

Figure 6.3: Weed Detection in Orchard Fields by Drones.

Biotechnological Approach in Weed Management

The collective endeavours in establishing genomic resources for weeds will serve as the foundation for crafting protocols in plant transformation and gene editing for the exploration of gene functions. This initiative also holds the potential to enhance scientists' comprehension of intricate traits like abiotic stresses, as well as the diverse resistance mechanisms adopted by weeds (Ravet *et al.,* 2018). Following the emergence of precise genome editing tools like CRISPR/Cas9, the effectiveness and precision of gene drive systems have been significantly enhanced, advancing their capabilities in targeting specific genes with greater efficiency. Different ways for manipulating genomics in managing weeds in orchards are:

a) Incorporating herbicide-resistant genes into trees crops (Datta and Pilli, 2019).
b) Genetic improvement of biotic agents (insects and microbes) for weed management.
c) Characterization of weeds using molecular genetics.
d) Targeting genes in crops for the production of allelochemicals to out compete weeds.

Undoubtedly, the CRISPR/Cas9 technique presents promising potential for weed control, offering two distinct approaches for utilization and advancement. One avenue involves altering the genome of crops to enhance their performance, thereby surpassing weed competition. This enhancement could occur through various mechanisms, such as heightened herbicide resistance, increased early vigour, or the implementation of allelopathic strategies to hinder weed establishment. Alternatively, gene editing technologies could be employed to manipulate weed development or enhance herbicide tolerance. Additionally, there are modified Cas9 proteins that could be utilized for these purposes (Molla *et al.,* 2021). The molecular approaches hold the potential for targeting the biosynthesis and regulatory genes responsible for producing allelochemical compounds. By employing these techniques, these can customize crop root exudate profiles, thereby enabling the exertion of negative effects on the growth of neighbouring weed species. If successfully implemented, this approach would resemble an engineered crop capable of producing its own "herbicide" to effectively control weeds. In recent years, initiatives like the International Weed Genomics Consortium have undertaken the complete sequencing of several significant weed species, including *Lolium rigidum* and *Conyza sumatrensis* (Manning, 2021). The efforts of this consortium, along with forthcoming endeavours aimed at enhancing the availability of genomic resources for various major weed species, promise numerous benefits. Not only could they facilitate the development of molecular approaches for diagnostic purposes and weed management, but they could also deepen our comprehension of weed biology, weedy traits, and the adaptive evolution of herbicide resistance (Ravet *et al.,* 2018). Conversely, enhancing the early vigour of the crop plant could result in improved growth, subsequently leading to the suppression of weed growth (De vida *et al.*, 2006). Although biotechnological approaches can create a revolutionary change in weed science, the complexity of the process need to be taken into account and all the steps involved must be done through proper studies to ensure success in this respect.

Integration of Robotic Systems for Automated Weeding

Weeding robots are transforming the way farmers manage weed control, offering a crucial mix of efficiency, precision, and environmental consciousness. Robotic weed control systems are available in diverse forms, ranging from ground-based robots to drones fitted with advanced sensors (Figure 6.4). Engineered for precision and resilience, these systems can navigate varied terrain adeptly while efficiently identifying and eradicating weeds).

Figure 6.4: AI Generated View of Robotic System for Automated and Precise Weed Control.

There has been an increased interest in moving from passive machinery-attached tractors to semi- and fully autonomous machines able to handle different tasks with minimal human supervision. One of the major distinctions among these robots is the method of weeding they employ. Few are equipped with hoes and tines (mechanical equipment's) and other are fitted with automation sensors technology to ensure precise and automated weed control. One of the first works is the one presented by Blasco *et al.* (2002). The setup consists of a cart advancing on a straight line (a crop row) and equipped with a 6 degrees of freedom (DOF) manipulator. Visual serving is facilitated by two cameras, with weed detection achieved through Bayesian discriminant analysis. Upon detection, an electrode is positioned by the robot onto the weed, capable of discharging 15kv at 30 mA for 200 ms, ensuring the removal of the weed. The AgBot II, developed by McCool *et al.* (2018) is an autonomous robot managing various weed infestations. It consists of a camera utilized for discriminating crops from infestations by leveraging different color spaces, ensuring consistent results. Additionally, the robot employs a lighting module to maintain a uniform level of illumination throughout the detection process. A robotic weed control system with multiple weeding tools, referred to as the Integrated Weed Management System, has been developed. This system aims to reduce reliance on herbicides and achieve long-term weed control by incorporating various tactics. A new robotic system developed by a team of researchers at the University of Bonn aims to assist farmers in efficiently managing weeds and monitoring crops. This robot utilizes several localization sensors based on GPS technology and odometry. It can navigate through fields to locate, classify, and count plants, while also employing various integrated tools to manage weeds directly from its body structure. The Robotic Rover for Agricultural Applications, as proposed by is a multitasking agricultural land robot designed to autonomously conduct herbicide spraying in a specialized manner within peach orchards (Veiros, 2020). The incorporation of robotics systems for automated weed control leads to substantial progression in agricultural technology, ultimately, providing precise, efficient, and sustainable solutions to the challenges of weed management. Despite hurdles in the initial adoption of this technology, the long-term advantages render these systems invaluable assets for carrying out different farming operations.

Conclusion

This chapter effectively sheds light on various aspects of weed management in orchards. It is evident that weeds pose a significant concern not only in cereal and vegetable crops but also in orchards, where they compete with and reduce yield as well as quality. A variety of host-specific and non-specific weeds hinder crop growth through mechanisms such as light, space and nutrient competition, allelopathy and acting as hosts for pests and pathogens, makes it necessary to adopt weed management practices. Weed management is influenced by factors like orchard and soil type, fertility, climate change and human and regulatory influences. Considering all these factors is crucial for making informed decisions to benefit the crops. Adopting an integrated weed management strategy is essential for a comprehensive management of weeds in a sustainable way. Cultural, mechanical, biological methods and cultural methods when combined, can provide effective weed management strategy. Leveraging advanced technologies like precision agriculture, AI, machine learning, biotechnology (genetic manipulation), and automation systems is increasingly important for precise and effective weed control. While weeds pose serious threats to orchard productivity, strategic decisions and cutting-edge technologies offer avenues to effectively manage them and ensure good crop yield and quality.

REFERENCES

Abdulridha J, Bashyal M, Ampatzidis Y and Kanissery R, 2023. Steam application with paraquat to control goat weed (*Scoparia dulcis*) in citrus orchards. *Smart Agric. Technol.* **6**: 100355.

Abouziena HF and Haggag LF, 2016. Weed control methods and effects on yield and yield attributes of common bean. *J. Plant Prod.* **7**: 1171-79.

Alcántara-de la Cruz R, Domínguez-Martínez PA, da Silveira HM, Cruz-Hipólito HE, Palma-Bautista C, Vázquez-García, JG, Domínguez-Valenzuela JA and De Prado R, 2019. Management of glyphosate-resistant weeds in Mexican citrus groves: Chemical alternatives and economic viability. *Plants* **8**: 325. doi: 10.3390/ plants8090325.

Atucha A, Merwin IA, Brown MG and Gardiazabal FE, 2011. Orchard floor management practices that maintain vegetative or biomass ground cover stimulate soil microbial activity and alter soil microbial community composition. *Plant and Soil* **348**: 379-94.

Bhullar MS, Kaur T, Kaur S and Yadav R, 2015. Weed management in vegetable and flower crop-based systems. *Indian J. Weed Sci.* 267-76.

Blasco J, Aleixos N, Roger J, Rabatel G and Moltó E, 2002. AE-Automation and emerging technologies: Robotic weed control using machine vision. *Biosyst. Eng.* **83**: 149-57.

Blomme G, Dita M, Jacobsen KS, Pérez Vicente L, Molina A, Ocimati W, Poussier S and Prior P, 2017. Bacterial diseases of bananas and enset: current state of knowledge and integrated approaches toward sustainable management. *Front. Plant Sci.* **8**: 1290.

Boyetchko S, Bailey K, Hynes R, Peng G, Vincent C, Goettel M and Lazarovits G, 2007. Development of the mycoherbicide, BioMal®. **In**: *Biological Control, a Global Perspective*; Vincent C, Goettel MS and Lazarovits G (Ed.); CAB International: Wallingford, UK, pp.274-83.

Bozbuga R, Uluisik S, Kara PA, Yuceer S, Gunacti H, Guler PG, Ince E, Yildiz HN and Tetik O, 2022. Pests, Diseases, Nematodes, and Weeds Management on Strawberries. **In**: *Recent Studies on Strawberries.* IntechOpen. http: //dx.doi. org/10.5772/intechopen.103925.

Brar JS, Gill KS, Arora NK, Gill MIS and Kaur T, 2017. Weed management in guava orchards. *Indian J. Weed Sci.* **49**: 374-77.

Brar JS, Gill KS, Kaur T and Kaur K, 2020. Appraisal of different floor management practices for weed management in ber (*Zizyphus mauritiana* Lamk.) orchards. *Indian J. Weed Sci.* **52**: 58-63.

Caroll J, Pritts MP and Heidenreisch C, 2012. Production Guide for Organic Blueberries. New York State Integrated Pest Management Publication 225, Cornell University, New York, pp.5-50.

Chatha RPS and Chanana YR, 2007. Studies on weed management in young peach orchards. *Indian J.Hort.* **64**: 300-03.

Choi ST, Kim SC, Ahn GH, Park DS and Kim ES, 2017. Nutrient release during residue decomposition of weeds mown at different times in a persimmon orchard. *J. Agric. Chem. Eviron.* **6**: 153-64.

Datta MK and Pilli K, 2019. Biotechnological approach in weed management. **In**: *Research Trends in Crop and Weed. AkiNik Publication. New Delhi, India.*

De Vida FBP, Laca EA, Mackill DJ, Fernández GM and Fischer AJ, 2006. Relating rice traits to weed competitiveness and yield: a path analysis. *Weed Sci.* **54**: 1122-31.

Esposito M, Crimaldi M, Cirillo V, Sarghini F and Maggio A, 2021. Drone and sensor technology for sustainable weed management: a review. **In**: *Chemical and Biological Technol. Agri.* **8**: 1-11.

Fongod AGN, Focho DA, Mih AM, Fonge BA and Lang PS, 2010. Weed management in banana production: the use of *Nelsonia canescens* (Lam.) Spreng as a non leguminous cover crop. *Afr. J. Environ. Sci. Technol.* **4**: 167-73.

García-Cámara I, Pérez-Brito D, Tapia-Tussell R, Magaña-Álvarez A, Cortés-Velázquez A, Martín-Mex R and Moreno-Valenzuela OA, 2022. Common weeds as alternate hosts of Mexican variant of Papaya meleira virus in papaya orchards in México. *Acta Virologica* **66**: 192-94.

Granatstein D and Sanchez E, 2009. Research knowledge and needs for orchard floor management in organic fruit system. *Intern. J. Fruit Sci.* **9**: 257-81.

Granatstein D, Andrews P and Groff A, 2014. Productivity, economics, and fruit and soil quality of weed management systems in commercial organic orchards in Washington State, USA. *Organic Agric.* **4**: 197-207.

Granatstein D, Wiman M, Kibry E and Mullinix K, 2010. Sustainability trade-offs in organic orchard floor management. *Acta Horticul.* **873**: 115-21.

Harding DP and Raizada MN, 2015. Controlling weeds with fungi, bacteria and viruses: a review. *Front. Plant Sci.* **6**: 155911.

Hasan M, Ahmad-Hamdani MS, Rosli AM and Hamdan H, 2021. Bioherbicides: An eco-friendly tool for sustainable weed management. *Plants* **10**: 1212.

Hoagland LORI, Carpenter-Boggs L, Granatstein D, Mazzola M, Smith J, Peryea F and Reganold JP, 2008. Orchard floor management effects on nitrogen fertility and soil biological activity in a newly established organic apple orchard. *Biol. Fert. Soils* **45**: 11-18.

http: //ecoursesonline.iasri.res.in/mod/page/view.php?id=12067 (Accessed on 15th May, 2024).

https: //ag.umass.edu/fruit/ne-small-fruit-management-guide/strawberries/weeds/table-24-herbicide-efficacy-against-common (Accessed on 25th May, 2024).

https: //bluerivertechnology.com/ourmethods/ (Accessed on 10th May, 2024).

https: //nhb.gov.in/report_files/guava/GUAVA.htm (Accessed on 23rd May, 2024).

https: //treefruit.wsu.edu/web-article/weed-management/ (Accessed on 11th May, 2024)

https: //ucanr.edu/blogs/blogcore/postdetail.cfm?postnum=39171 (Accessed on 10th May, 2024)

https: //wssa.net/wp-content/uploads/BCBrochure.pdf (Accessed on 12th May, 2024)

https: //www.canr.msu.edu/news/weed-management-in-vineyards-starts-with-fall-herbicide-application (Accessed on 24th May, 2024).

https: //www.ikisan.com/ap-mango-weed-management.html (Accessed on 24th May, 2024).

https: //www.ikisan.com/mh-citrus-weed-management.html (Accessed on 5th May, 2024)

https: //www.sciencedirect.com/topics/agricultural-and-biological-sciences/plant-density (Accessed on 7th May, 2024)

https: //www.who.int/news-room/fact-sheets/detail/healthy-diet (Accessed on 6th May, 2024).

Hussain S, Sharma MK, War AR and Hussain B, 2020. Weed management in apple cv. Royal Delicious by using different orchard floor management practices. *Intern. J. Fruit Sci.* **20**: 891-921.

Hussein M, Amr A and Shawky E, 2016. Effect of soil solarization, polyethylene mulch, and weed management methods on weed density, weed biomass, fruit yield, and water use efficiency of pomegranate. *Intern. J. Fruit Sci.* **16**: 404-19.

Kleiman BM, Koptur S and Jayachandran K, 2021. Weeds enhance pollinator diversity and fruit yield in mango. *Insects* **12**: 1114.

Kumar S, Bhowmick MK and Ray P, 2021. Weeds as alternate and alternative hosts of crop pests. *Indian J. Weed Sci.* **53**: 14-29.

Lim KH, Choi HS, Kim WS, Kim SG, Song JH, Cho YS, Choi JJ and Jung, SK, 2012. Nutrient contribution of green manure crops in an organic pear orchard. *Korean J. Org. Agric.* **20**: 34-48.

Lipecki, J, 2006. Weeds in orchards-Pros and contras. *J. Fuit. Ornam. Plant Res.* **14**: 13-18.

Lisek J, 2010. Wp^{3}yw terminu usuwania chwastów na wzrost i plonowanie nowo sadzonych truskawek. [Growth and yielding of newly planted strawberries after different time of weeding]. *Prog. Plant Prot/Post. Ochr. Roœlin.* **50**: 815-19.

Lisek J, 2014. Possibilities and limitations of weed management in fruit crops of the temperate climate zone. *J. Plant Prot. Res.* **54**: 318-26.

Manning A, 2021. International weed genomics consortium-collaborative effort to combat crop-threatening weeds headed by CSU Scientists. Available at: https: //agsci.source.colostate.edu/international-effort-to-combat-crop-threatening-weeds-headed-by-csu-scientists/ (Accessed on 12 May 2024).

McCauley C, Legleiter T, Herman R, Rasoulpour R, Schroeder J, Pilcher T, Meinders K and Wright T, 2022. Sustainable weed management-What is it and how are we doing? *Weed Technol.* **36**: 768-76.

McCool C, Beattie J, Firn J, Lehnert C, Kulk J, Bawden O, Russell R and Perez T, 2018. Efficacy of mechanical weeding tools: A study into alternative weed management strategies enabled by robotics. *IEEE Robot Autom. Lett.* **3**: 1184-90.

Mendes IDS and Rezende MOO, 2014. Assessment of the allelopathic effect of leaf and seed extracts of *Canavalia ensiformis* as post emergent bioherbicides: A green alternative for sustainable agriculture. *J. Environ. Sci. Health* **49**: 374-80.

Mia MJ, Massetani F, Murri G and Neri D, 2020. Sustainable alternatives to chemicals for weed control in the orchard: a Review. *Horticul. Sci.* **47**: 1-12.

Miñarro M, 2012. Weed communities in apple orchards under organic and conventional fertilization and tree-row management. *Crop Prot.* **39**: 89-96.

Mishra JS, Poonia SP, Choudhary JS, Kumar R, Monobrulla M,Verma M, Malik RK and Bhatt BP, 2019. Rice mealybug (*Brevennia rehi*): a potential threat to rice in a long–term rice–based conservation agriculture system in the middle Indo–Gangetic Plain. *Current Sci.* **117**: 566-68.

Mitchem WE, 2005. Weed management considerations for peach orchards. *Department of Horticultural Science North Carolina State University, Gastonia.*

Molla KA, Sretenovic S, Bansal KC and Qi Y, 2021. Precise plant genome editing using base editors and prime editors. *Nature Plants* **7**: 1166-87.

Muñoz M, Torres-Pagán N, Peiró R, Guijarro R, Sánchez-Moreiras AM and Verdeguer M, 2020. Phytotoxic effects of three natural compounds: Pelargonic acid, carvacrol, and cinnamic aldehyde, against problematic weeds in Mediterranean crops. *Agron.* **10**: 791.

Muslu T and Tepe I, 2016. Weeds in pomegranate orchards in Gaziantep province, Turkey. *J. Agrc. Sci.* **26**: 40-51.

Nasr N, Hajar B and Miyandeh BH, 2013. Weeds identification in west of Mazandaran Province Citrus Orchards (Iran). *Amr. J. Res. Commun.* **1**: 27-38.

Negi S, 2015. Influence of weed interference on tree growth, yield, and fruit quality of nectarine. *J. Crop Weed* **11**: 23-28.

Pannacci E, Lattanzi B and Tei F, 2017. Non-chemical weed management strategies in minor crops: a review. *Crop Prot.* **96**: 44-58.

Pattison AB, McBeath AV, Masters B and Dennis PG, 2024. Direct and indirect effects of Basta®, a glufosinate-based herbicide, on banana plantation soil microbial diversity and function. *Applied Soil Ecol.* **195**: 105225.

Paull RE and Uruu G, 2021. Major Weeds in Pineapple Fields of Hawaii. *Published by the College of Tropical Agriculture and Human Resources (CTAHR) 9.*

Peters K, Breitsameter L and Gerowitt B, 2014. Impact of climate change on weeds in agriculture: a review. *Agron. Sustain. Dev.* **34**: 707-21.

Polverigiani S, Massetani F, Tarragoni A and Neri D, 2013. Apricot root development and morphology as influenced by mulching and multispecies ground cover. *Acta Horticul.* (ISHS) **1001**: 353-59.

Pritts MP and Kelly MJ, 2001. Early season weed competition reduces yield of newly planted matted row strawberries. *HortSci.* **36**: 729-31.

Ramteke SD, Rajurkar AB, Bhange MA and Kor RJ, 2012. Chemical management of broad-leaved weeds in grapes. *Indian J. Weed Sci.* **44**: 196-99.

Rasmussen J, Nielsen J, Streibig JC, Jensen JE, Pedersen KS and Olsen SI, 2019. Pre-harvest weed mapping of *Cirsium arvense* in wheat and barley with off-the-shelf UAVs. *Precision Agric.* **20**: 983-99.

Ravet K, Patterson EL, Krähmer H, Hamouzová K, Fan L, Jasieniuk M, Lawton–Rauh A, Malone JM, McElroy JS, Merotto Jr A and Westra P, 2018. The power and potential of genomics in weed biology and management. *Pest Manage. Sci.* **74**: 2216-25.

Rowley MA, Ransom CV, Reeve JR and Black BL, 2011. Mulch and organic herbicide combinations for in-row orchard weed suppression. *Intern. J. Fruit Sci.* **11**: 316-31.

Shaikh AR and Lokhande OG, 2005. Effect of Sequential Herbicide Application on Banana and Weeds. *Indian J. Weed Sci.* **37**: 289-90.

Shrestha A, Moretti M and Mourad N, 2012. Evaluation of thermal implements and organic herbicides for weed control in a non-bearing almond (*Prunus dulcis*) orchard. *Weed Technol.* **26**: 110-16.

Sinden J, Jones R, Hester S, Odom D, Kalisch C, James R, Cacho O and Griffith G, 2004. *The Economic Impact of Weeds in Australia.* CRC for Australian Weed Management; Glen Osmond, Australia.

Singh R and Singh G, 2016. Aphids and their biocontrol. **In:** *Ecofriendly Pest Management for Food Security* (Ed.) Omkar. Academic Press. pp. 63-108.

Singh R and Singh PK, 2016. Weeds and their host insect pests in crops of Chhattisgarh plains. *Indian J. Weed Sci.* **48**: 260-64.

Singh SK, Khurma UR and Lockhart PJ, 2010. Weed hosts of root-knot nematodes and their distribution in Fiji. *Weed Technol.* **24**: 607-12.

Tateno A, 2000. Herbicidal Composition for the Control of Annual Bluegrass. U.S. Patent No. 6,162,763.

Thakur A, Singh H, Jawandha SK and Kaur T, 2012. Mulching and herbicides in peach: Weed biomass, fruit yield, size, and quality. *Biol. Agric. and Hortic.* **28**: 280-90.

Ustuner T and Ustuner M, 2011. Investigation on different mulch materials and chemical control for controlling weeds in apple orchard in Turkey. *Scient. Res. Essays* **6**: 3979-85.

Veiros AFR, 2020. Sistema Robótico Terrestre Para Apoio a Atividades de Manutenção de Solo em Pomares de Prunóideas. Master's Thesis, Interior, Covilhã, Portugal.

Webb SE, Adkins S and Reitz SR, 2012. Semi-persistent whitefly transmission of squash vein yellowing virus, causal agent of viral watermelon vine decline. *Plant Dis.* **96**: 839-44.

Yandoc-Ables CB, Rosskopf EN and Charudattan R, 2006. Plant pathogens at work: progress and possibilities for weed biocontrol. Part 1: Classical vs. Bioherbicidal Approach. doi: 10.1094/APSFeature-2006-08.

Chapter 7

Nutrient Management in Fruit Crops

O.P. Awasthi and Kripa Shankar

Division of Fruits and Horticultural Technology, ICAR–Indian Agricultural Research Institute, New Delhi–110012

India's diverse climate and topography contribute to its rich fruit diversity and have potential for feeding the growing global population. Despite the tremendous potential, challenges like the abiotic and biotic stresses due to climate change during the last two decades have resulted in productivity decline of fruit crops. Optimizing nutrient management is essential for improving both the yield and quality of fruit. This requires a balanced use of organic and inorganic fertilizers, guided by soil and leaf analysis, to ensure that nutrient levels meet the specific needs of the crops. Organic manures improve soil health, mobilize nutrient uptake and impart environmental sustainability. Green manuring and biochar also enrich soil fertility. Fruit-specific nutrient management strategies include the use of micronutrient mixtures like Banana Shakti and ATIC Mango to boost productivity. In newly established orchards, integrating *Azotobacter* and phosphorus solubilizing bacteria (PSB) with organic manures have been reported to enhance growth of different fruit trees. For proper establishment of orchards, nutrient application is adjusted based on soil nutrient content, ensuring optimal fertilization. Nutrient management research aims to optimize orchard productivity and quality through practices enhancing root nutrient acquisition and soil fertility. This comprehensive approach of nutrient management in orchards is essential for meeting the increasing demand and quality fruits amid climatic challenges.

Introduction

India's diverse fruit basket is a testament to its varied climate, ranging from tropical to temperate, hot arid to cold arid and its topographical range from plains to hills (Sankaran and Dinesh, 2020). This diversity holds substantial promise for sustaining the nation's population with top-quality fruit production. With the global population expected to approach 10 billion by 2050, India's population is expected to rise by 1.6 billion. This rise in human population with declining land resources is critical for addressing challenges such as food insecurity, abiotic stress, and pest outbreaks, exacerbated by climate change particularly in the tropical regions (Haque *et al.,* 2020). India's fruit production during the last one decade has been witnessing a a gradual decline in productivity which is thought provoking. Key fruits like apple, mango banana, citrus, papaya, and pineapple are experiencing reduced growth rates. According to the Ministry of Agriculture and Farmer's Welfare, the total fruit production increased to the tune of only 17 per cent from 2013 to 2021. A significant factor in this drop down had been the decrease in area under fruit cultivation thus witnessing a decline of 2.1 per cent from 7.12 million hectares in 2013-14 to 6.97 million hectares in 2021-22 (MoA and FW, 2023). Additionally, rising abiotic stress has led to a continuous decline in productivity of fruit crops.

For instance, in Kashmir, early flowering of fruit crops was hampered by a severe precipitation deficit 70-80 per cent below normal between March 1 and April 21, resulting in flower drop and reduced fruit production. While higher temperatures could benefit fruit growers with adequate irrigation, dry conditions and increased pest infestations are damaging crops. Effective nutrient management is critical for sustaining fruit production. The proper application of a new class of nutrients supports growth, fruit quality, and resistance to stress (Ahmad and Anjum, 2023). Sustainable productivity hinges on optimizing soil fertility and nutrient supply through the integration of different sources of nutrients, which balance soil inputs with crop requirements. Over- or underapplication of nutrients can either harm the environment or limit yields. Soil application of organic manures such as poultry manure, FYM, vermicompost and biogas slurry improves soil fertility and structure over the long term (Fageria *et al.,* 2008).

Nutrient management for perennial fruit trees focuses on optimizing orchard nutrition to enhance productivity and quality. This involves improving nutrient uptake through fertilization practices and promoting root development (Kalcsits *et al.,* 2020). The nutrient dynamics in soil and nutrient absorption through plants are the primary aspects that can be used for effective nutrient management (Day *et al.,* 2010). Biofertilizers and organic manures are environmentally safe and renewable and play a vital role in maintaining soil fertility and enhancing water retention and microbial populations. These benefits make biofertilizers crucial for improving agricultural productivity and environmental sustainability in INM systems (Malusa *et al.,* 2016). Specific nutrient solutions developed by research institutes have shown beneficial effects. For example, "Banana Shakti" from the National Research Centre for Banana in Tamil Nadu has increased fruit quality and productivity by 15-20 per cent. Similarly, the ATIC Mango Special from the Indian Institute of Horticultural Research (IIHR), Bangalore, which is enriched with micronutrients and secondary nutrients, enhances mango fruit quality and taste. The incorporation of beneficial microbes, such as nitrogen-fixing and phosphorus-solubilizing bacteria, in conjunction with mineral inputs enhances fruit quality (Bargaz *et al.,* 2018). Vermicompost, which is rich in nutrients, improves soil quality and promotes plant growth. Ultimately, sustainable and efficient nutrient management is essential for achieving long-term fruit productivity and meeting the growing demand for quality fruits. By promoting green agriculture and sustainable farming practices, the agricultural sector can support the increasing population while minimizing environmental impacts (El-Ramady, 2014).

Care Before Applying Nutrient in Orchard

Effective nutrient management in orchards depends on a thorough understanding of soil and leaf analysis, nutrient depletion from the orchard, and various plant traits, including age, variety/cultivar, performance, and scion-rootstock combinations (Rom and Carlson, 1987). Leaf concentrations of essential minerals vary throughout the growing season and are influenced by factors such as the shoot's phenological stage, fruit presence, and nutrient mobility within the plant (Yildiz *et al.,* 2022). Before flowering, leaves show high concentrations of phloem-mobile macronutrients such as N, K, Mg and P, while immobile (phloem) minerals (Ca) remain low (Smith *et al.,* 1988). From post flowering to harvest, the levels of K, P, and N in leaves decrease as these nutrients are reallocated to developing fruits and roots, with N also being diluted by leaf growth (Nachtigall and Dechen, 2006). Therefore, leaf and soil testing must be customized for each specific fruit tree species.

Soil Testing before Planting

Soil testing prior to planting is crucial for determining the lime and fertilizer needs of an orchard, particularly for nutrients such as Ca, K, O, Mg, S, Zn, B and P (Jones, 2011). This process removes uncertainty in fertilizer recommendations and enhances nutrient management efficiency. Farmers can utilize facilities provided by state departments of agriculture or universities for this purpose.

Nutrient Management in Established Orchards

For established orchards, nutrient requirements should be assessed through plant tissue and soil nutrient assessments.

Collecting Soil Samples

Accurate soil test results rely on representative soil sample collections. While soil samples can be taken at any time of the year, spring and fall are typically the most convenient (James and Wells, 1990). Soil sampling should be avoided if fields are fertilized with organic manure or fertilizers. To ensure consistency and reliable comparisons over time, soil samples should be collected at the same time each year.

Fields should be divided into uniform areas with similar cropping histories, soil textures, manure and fertilizer applications, colors and manure. For soil sampling of orchard soil, a depth of 10-12 inches is recommended for fruit crops from 15 to 20 random places in orchards (Holland *et al.,* 1967).

Leaf Sampling

Leaf sampling is an important tool for determining the nutritional status of the leaf, as the nutrient status is clearly reflected in the composition of the leaf. Leaf nutrient content changes with tree age and with varying levels of other nutrients, while leaf nutrient ratios or products remain relatively stable over time (Parent *et al.,* 2020). Therefore, sampling leaves at specific growth stages can accurately determine the nutrient status of an orchard (Sumner, 1977).

Table 7.1: Leaf Sampling Technique for different Fruit Crops

Fruit Crop	*Sampled Tissue*	*Growth Stage/Timing*
Papaya	Petiole	6th petiole from apex 6 months after planting.
Mango	Leaf + petiole	4 to 7-month-old leaves from the mid-section of the shoot (nonflowering terminal).
Sapota	Leaf	10th leaf from new growth initiated in September.
Citrus	Leaves	3 to 5-month-old leaves from new flush. 1st leaf of the shoot in June.
Grape	Petiole	Petiole from 5th leaf position (base of the shoot taken as one) 40 to 45 days after back pruning (bud differentiation stage) for yield forecast. Petiole opposite to flower at bloom time for assessing fruit quality.
Banana	Leaf lamina	Centre 20 cm^2 of leaf lamina on both sides of midrib. 3rd fully opened leaf at 16th to 20th leaves stage.
Guava	Leaves	3rd pair of recently matured leaves during bloom stage (August or December).
Pomegranate	Leaf	8th pair of leaves (counting from the growing tip) from new growth in April and August for February and June crop, respectively.

Raghupathi *et al.,* 2014.

Nutrient Management Strategies in Orchards

Effective nutrient management in orchards integrates traditional and modern agricultural practices to harness the synergistic benefits of crop-soil-environment interactions. This approach aims to optimize the use of plant nutrients to enhance soil fertility by utilizing a variety of sources, such as organic manures, crop residues, biofertilizers, nanofertilizers, specialized fruit fertilizers, crop rotations with legumes, and green manuring (Chatzistathis *et al.,* 9).

Table 7.2: Leaf Sampling for Nutrient Analysis using Index Part of different Fruits

Fruit Crop	*N (%)*	*P (%)*	*K (%)*	*Ca (%)*	*Mg (%)*
Apple	1.80–2.80	0.18–0.30	1.20–2.00	1.30–3.00	0.20–0.40
Pear	1.60–2.40	0.18–0.26	1.20–2.00	1.30–3.00	0.30–0.60
Peach	2.50–3.40	0.15–0.30	2.10–3.00	1.90–3.50	0.20–0.40
Cherry	2.30–3.30	0.23–0.38	1.00–1.90	1.00–1.90	0.49–0.65
Guava	1.63–1.96	0.18–0.24	1.31–1.71	0.67–0.83	0.52–0.65
Mango	0.95–1.45	0.03–0.12	0.40–0.77	1.74–3.45	0.22–0.75
Ber	2.77–3.32	0.18–0.32	1.69–1.96	1.07–1.36	0.39–0.50
Apple	1.50–2.80	0.10–0.40	0.80–2.90	0.70–2.60	0.21–0.64
Plum	1.74–3.19	0.22–0.40	1.80–3.60	0.56–0.80	0.76–1.33
Citrus					
Coorg mandarin	2.08–2.79	0.09–0.18	0.78–2.08	2.88–4.60	0.25–0.47
Nagpur mandarin	1.70–2.81	0.09–0.15	1.02–2.59	1.80–3.28	0.43–0.92
Malta orange	2.14–2.31	0.10–0.14	1.10–1.56	2.89–3.41	0.39–0.52
Kinnow mandarin	2.22–2.32	0.11–0.15	1.10–1.41	2.32–2.79	0.38–0.61
Sathgudi orange	2.01–2.42	0.09–0.12	1.12–1.82	1.93–2.73	0.36–0.53
Mosambi orange	1.98–2.57	0.091–0.17	1.33–1.72	1.73–2.98	0.32–0.69
Khasi mandarin	1.97–2.56	0.09–0.10	0.99–1.93	1.97–2.49	0.24–0.48
Acid lime	1.53–2.10	0.10–0.15	0.96–1.66	3.05–3.43	0.40–0.60
Grapes					
Anab-e-shahi	0.50–1.25	0.46–0.68	1.16–2.42	1.40–2.07	0.24–0.58
Thompson seedless	1.58–1.98	0.47–0.66	1.41–1.93	0.84–1.04	-
Pineapple	1.21–1.85	0.13–0.18	1.19–1.62	0.27–0.35	0.43–0.56
Sapota	1.25–2.38	0.05–0.17	0.36–0.77	1.01–2.04	0.48–0.92
Litchi	1.33–2.00	0.09–0.20	0.50–1.21	0.38–1.02	0.28–0.65
Pomegranate	0.91–1.66	0.12–0.18	0.61–1.59	0.77–2.02	0.16–0.42

Source: University of Maryland, 2006; Srivastava and Malhotra, 2017.

Organic Manures

In India, approximately 226 million tons of compost are produced annually in rural areas, with urban areas contributing approximately 6.6 million tons. Organic manures improve soil structure, decrease pH, and enhance water retention and infiltration. Promoting effective farmyard manure (FYM) and compost preparation methods is crucial. Research by Saha *et al.* (38) on the Mango cultivar Amrapali recommends applying 10 kg of FYM per plant, which significantly increases the soil organic carbon (SOC) density (193 g/m^2/year). A lower application rate of 7.5 kg/plant yielded the highest soil N (319.9 kg/ha), P (19.1 kg/ha), and K (270.8 kg/ha). Additionally, grass litter mulch quickly releases potassium within five weeks, which is essential for nutrient availability and fruit quality, particularly in preventing disorders such as bitter pit due to imbalances in K:Ca ratios (Tagliavini *et al.*, 2007).

Organic Wastages

India generates substantial crop residues, approximately 350 million tons annually. When combined with organic materials, these residues enhance soil properties, including nutrient availability, organic matter content, and microbial activity. Biochar, derived from pyrolyzed agricultural residues, is an effective soil amendment that supports biodiversity, environmental health, and carbon sequestration (Souri and Bakhtiarizade, 2019). The porous nature of biochar enhances soil physicochemical properties, making it a valuable addition to compost for improving soil health in orchards (Eyles *et al.*, 2015). Different organic fertilizers offer distinct benefits: trace elements are abundant in animal-based fertilizers but are less effective at improving soil quality, while plant-based fertilizers enhance root health and soil adsorption but are more costly (Wang *et al.*, 2024). Despite their benefits, organic fertilizers are slow to release and may not meet high nutrient demands during rapid crop growth phases.

Biofertilizers

Biofertilizers, such as *Azotobacter* and *Azospirillum,* fix nitrogen to fruit trees. Phosphate-solubilizing bacteria, such as *Bacillus* and *Aspergillus,* increase phosphorus availability from soil and rock phosphate. Other beneficial organisms, such as blue green algae and vesicular arbuscular mycorrhiza (VAM), improve water uptake and microbial activity in phosphorus-deficient soils. The use of liquid *Bacillus subtilis* microbial fertilizer (BCMF) has been found to delay the coloring of both peel and pulp and improve nutrient distribution in Tarocco blood oranges (Qiu *et al.,* 2021). In strawberries, a combination of 50 per cent FYM, 50 per cent vermicompost, *Pseudomonas* and *Azotobacter* has been reported to promote early flowering. These treatments also enhance fruit quality parameters, including vitamin C content and antioxidant capacity, and improve the soil nutrient availability of K, P and N (Negi *et al.*, 2021). The treatment combining *Azotobacter, Azospirillum,* VAM, and 2 kg of FYM resulted in the greatest increase in growth metrics in pomegranate plants, including plant height and number of fruits per tree (Dutta *et al.,* 2008). Similarly, the combination of *Azotobacter*, VAM, and 2 kg of FYM also significantly improved growth. Furthermore, this comprehensive biofertilizer treatment produced the heaviest fruits and enhanced the biochemical properties of the pomegranates. Notably, the fruits had the highest total soluble solids at 6.20 °Brix, total sugar content at 5.18 per cent and beta carotene content at 2320µg per 100 g of pulp, all of which were accompanied by the lowest acidity. Additionally, this treatment resulted in the highest bacterial population, indicating a boost in microbial activity (Dutta *et al.,* 2008).

Green Manuring

Green manuring offers a budget-friendly alternative to chemical fertilizers, preserving soil organic matter and enhancing soil fertility. *Sesbania rostrata* can fix substantial nitrogen (100-250 kg N/ha) within 45-55 days. Graminaceous species such as oat, barley, rye grass, and sorghum are suitable for preventing erosion and nutrient runoff in sloping orchards. Green manures, particularly legumes, increase soil phosphorus availability and reduce dependence on mineral fertilizers (Zhang *et al.,* 2023). In phosphorus-deficient soils, green manures significantly improve soil phosphorus pools and citrus growth (Yang *et al.,* 2024). For instance, *Desmodium ovalifolium* increases the total nitrogen content, *Grona heterocarpos* increases ammonium nitrogen, and *Stylosanthes guianensis* increases available phosphorus levels in litchi orchards (Yuan *et al.,* 2023).

Chemical Fertilizers

To achieve high yields and superior fruit quality, it is essential to apply recommended fertilizers. However, conventional chemical fertilizers typically cannot provide a sustainable, long-term nutrient supply and are often linked to nutrient loss, particularly through nitrate nitrogen (N_3O-N) leaching, ammonia (NH_3) volatilization, and N_2O emissions. These processes may pose risks of contaminating natural aquifers and contributing to greenhouse gas emissions (Wei *et al.,* 2020).

Advances in Fertilizer Technologies and Root-Stock Combinations for Enhanced Fruit Crop Production

Controlled Release and Slow Release Fertilizers (CRF_S and SRF_S)

These are engineered with nutrient elements encapsulated or coated to allow for gradual nutrient release. Research indicates that these fertilizers can significantly boost fruit crop yields and improve fruit quality while also mitigating environmental impacts (Sempeho *et al.,* 2014). Furthermore, CRFs and SRFs have been shown to enhance the composition of microbes in root zones and decrease the harmful effects of soil pathogens. Recent findings suggest that these fertilizers can also help manage citrus greening, a serious disease affecting citrus crops. Despite their higher cost, the comprehensive benefits of CRFs and SRFs make them a valuable investment in fruit crop production compared to conventional fertilizers (Wei *et al.,* 2020).

Nanofertilizers

Nanofertilizers (NFs) are emerging as a revolutionary approach for improving soil quality, plant growth, and fruit crop production. These fertilizers, including nanoscale versions of zinc, boron, and chitosan, as well as nanocomposites such as micro- and macronutrients, significantly enhanced different horticultural traits of fruit trees (Zahedi *et al.,* 2020). Nano-fertilizers are increasingly viewed as viable alternatives to synthetic fertilizers for sustainable agriculture, offering benefits such as enhanced plant development, improved stress tolerance, and better performance under adverse environmental conditions (Lopez-Vargas *et al.,* 2018). For example, a single foliar treatment involving minimal doses of nano-fertilizers (34 mg of boron and 636 mg of zinc per tree) has been shown to increase pomegranate fruit yield and improve juice pH and fruit maturity (Davarpanah *et al.,* 2016).

Root-Stock and Scion Combinations for Nutrient Absorption

The choice of rootstock is important for the vigor and productivity of fruit trees because it influences the size and structure of the root system. Vigorous rootstocks such as M.111 enhance nutrient uptake due to their extensive root networks (Ferree and Warrington, 2003). In citrus, rootstocks such as Carrizo citrange are particularly effective in alkaline soils, facilitating better absorption of essential micronutrients such as iron (Castle *et al.,* 2010).

Specific Nutrient Uptake Efficiency

Certain rootstocks are known for their ability to absorb specific nutrients, which is crucial for optimal tree growth and fruit production. For example, Carrizo citrange enhances phosphorus uptake in citrus trees, significantly impacting growth and yield (Morgan and Kadyampakeni, 2022; Shankar *et al.,* 2023). Grape rootstocks such as 1103P are recognized for their efficient nitrogen uptake, which contributes to improved vine vigor and fruit quality (Keller, 2010). In plums, *Prunus marianna* rootstocks improve the uptake of micronutrients such as iron and manganese (Layne and Bassi, 2008). Peaches grafted onto Nemaguard rootstocks benefit from resistance to root-knot nematodes, ensuring a healthy root system and better nutrient absorption (Beckman and Nyczepir, 4). Additionally, some rootstocks form beneficial associations with mycorrhizal fungi, enhancing the uptake of phosphorus and other nutrients (Smith and Read, 1988). By integrating advanced fertilizer technologies and strategically selecting rootstock-scion combinations, fruit crop production can achieve higher yields, improved quality, and greater resilience against environmental stresses and diseases.

Compatibility and Physiological Responses

The physiological compatibility between the scion and rootstock is pivotal for nutrient flow and overall tree health, significantly impacting growth and productivity (Hartmann *et al.,* 2010).

Table 7.3: Major Nutrients and their Functions, Causes and Sources

Nutrient	*Major Function and Causes*	*Source*
N	Bud fertility, delay fruit ripening, hinder fruit skin color	Mineral N
P	Flower bud differentiation	PO_4^{3-}
K	Osmotic potential, tissue dehydration. K is partitioned preferentially to the fruits.	Potassium chloride and Magnesium sulphate or sylvinite
Ca	Enzyme inactivation of polygalacturonase. Causes bitter pit and senescent breakdown (pectinase), delayed softening.	Calcium chloride
Fe	Interveinal leaf yellowing	Synthetic Fe containing (Fe-EDDHA)
B	Pollen germinability and anther fertility,	boric acid or sodium borate

Toselli *et al.,* 2020.

In apple cultivation, rootstocks such as B9 enhance water and nutrient transport, which supports better scion growth and fruit quality (Rom and Carlson, 1987). Notably, apple rootstocks such as M.26 and G.41 are efficient at absorbing nitrogen and potassium, leading to improved fruit size and quality (Robinson, 2001). In citrus crops, the Swingle citrumelo and Carrizo citrange rootstocks are known for their enhanced nutrient uptake (Bowman and Joubert, 2009). The 101-14 Mgt rootstock excels in potassium uptake and adapts well to diverse soil conditions, promoting better vine health (Keller, 2010). Rootstocks such as Lovell and Nemaguard are effective in nutrient uptake and nematode resistance, ensuring better health and productivity in peaches and plums (Layne and Bassi, 2008).

Fruit Crop Specific Nutrient Management

Integrating organic and inorganic nutrient sources based on crop- and soil-specific recommendations can optimize orchard management (Roy *et al.,* 2006). In pomegranate cultivation under semiarid conditions, a novel bioformulation based on *Penicillium pinophilum* developed by the National Research Center (NRC) for pomegranate has been shown to increase yield by up to 35.17 per cent by maintaining potassium and phosphorus levels in the soil. For banana, NRC Banana Shakti, a mixture of five micronutrients, improves fruit quality and productivity by 15-20 per cent (Uma and Kumar, 2020). The IIHR-Arka microbial consortium, which contains nitrogen-fixing microbes, phosphorus and zinc-solubilizing microbes, promotes the growth of different fruit trees. Additionally, applying seaweed extract to orchard soil under drought stress can improve the yield and quality of citrus crops. These practices enhance the rhizosphere bioflora and help fruit trees develop natural tolerance to climatic extremes (Chalise *et al.,* 2023).

Mango

Mango Orchard Fertilization and Nutrient Management

In mango orchards, it is essential to apply manures and fertilizers when soil moisture levels are adequate. These should be applied in a trench 15-20 cm deep and 1-1.5 meters from the main trunk and then covered with soil. It is crucial to avoid mixing biofertilizers with chemical fertilizers and pesticides. Zinc and iron are vital for chlorophyll production, enzyme activity, and hormone coordination in plants. Boron is important for sugar production, pollination, and preventing fruit cracking, making it necessary to address any deficiencies. The general recommendation for mango according to the age of the plant is given below.

Age of Plant	*Urea (g/plant)*	*SSP (g/plant)*	*MOP (g/plant)*	*FYM (kg/plant)*
1-3 year	100-200	250-500	175-350	5-20
4-7 year	200-400	500-750	350-700	25-50
7-10 year	400-500	750-1000	700-1000	60-90
>10 year	500	1000	1000	100

Kaur and Sharma, 2020.

Nutrient Management for Newly Planted Mango Orchards

In newly established mango orchards, the application of 50 g of *Azotobacter* and 50 g of phosphorus-solubilizing bacteria (PSB) along with 3 kg of vermicompost per plant enhances growth and development. The depths at which this combination was applied are 15-20 cm and 30 cm from the trunk at adequate soil moisture levels, after which the trees are covered with soil (Oyetunji *et al.,* 2022).

Integrated Nutrient Management for Fully Grown Mango Orchards

Fertilizer application should be based on soil tests, with doses adjusted according to soil health and nutrient content. Generally, the recommended dose of fertilizers (RDF) is applied when the soil nutrient content is intermediate. If the soil nutrient content is extremely low, 50 per cent more RDFs should be applied; if the soil nutrient content is low, 25 per cent more RDFs should be applied. Conversely, if the soil nutrient content is extremely high, 50 per cent less than the RDF should be applied; if it is high, 25 per cent less should be applied (Yadav *et al.,* 55).

Atic Mango Special (AMS)

The ATIC mango special variety, developed by the Indian Institute of Horticultural Research in Bangalore, is recommended for all mango varieties. It should be applied twice before flowering and twice after flowering at a concentration of 5 g/L, along with the addition of adjuvants and lime juice. This formulation can be mixed with any fungicide or insecticide and contains essential micronutrients. The ATIC mango specifically enhances fruit quality in terms of appearance, shelf life, and taste (ATIC Mango special, 2).

Table 7.4: Nutrition Management in Mango Orchard

Normal Condition					
	N (g/tree)	*P (g/tree)*	*K (g/tree)*	*Organic Manure (kg/tree)*	*Remarks*
	1500	500	100	40-50	
	½ dose (Source: Urea)	Full dose (Source: SSP/ DAP)	Full dose (Source: MOP)	Full dose (Source: FYM or other organic manure)	At on set of monsoon as basal dose
	½ dose	-	-	-	At flowering
Under HDP (3-year-old orchard)					
	N (g/ tree)	*P (g/ tree)*	*K (g/ tree)*	*Biofertilizer*	*Remarks*
	120	60	75	*Azotobacter* 5 ml + PSB 5 ml in 11 liter water + well decomposed FYM/ tree at onset of monsoon and at the time of flowering	At harvest
	75	60	60		At pre flowering
	60	30	75		At fruit set
	45	-	90		At fruit development
Micronutrient					
Zn (%)	*Fe (%)*	*Mn (%)*	*Cu (%)*	*B (%)*	*Remarks*
5	**2**	**1**	**0.5**	**1**	For calcareous soil

Sable *et al.,* 2023.

Banana

Banana is a heavy feeders of nutrients and require large quantities of fertilizer for their growth, development and fruit production, which is mainly exploited at very limited soil depths owing to their shallow root system. Nutrients are normally applied to the soil or by fertilization to obtain the maximum yield. However, intensive use of chemical fertilizers should be avoided because of food safety and environmental issues. Integrated nutrient management (INM), which can restore soil fertility and productivity in banana plants, is an ideal approach. In addition to INM, some newer technologies have been developed that are eco-friendly and economically viable, and the same has been discussed below in light of research carried out by different institutes/organizations.

Nutrient Pouch Technology (NPT)

The technology for banana bunch feeding was developed by the Indian Institute of Horticultural Research (IIHR), Bengaluru. This technology ensures uniform-sized bananas throughout the bunch, as typically, bananas at the upper hands are larger than those at the lower end. The method involves tying a nutrient pouch (containing 400-500 g of fresh cow dung, 13.3 g of ammonium sulphate, and 6.7 g of 0:0:50) to the lower end of the bunch after denavelling (removing the male bud after fruit set). This technique redirects nutrients from the unnecessary male bud to the banana fruits, enhancing the size and weight of the bananas. In practical applications, a slurry composed of 15 grams of nutrients combined with 500 grams of fresh dung is administered to the denaveled stalk-end of the Robusta banana bunch shortly after fruit formation. The rachis should measure between 10 and 15 cm in length to effectively secure a plastic bag containing the mixture. This method demonstrated a 67 per cent increase in bunch weight compared to that in the control group, where the male flower was retained. Field trials have shown a 22 per cent increase in bunch weight for Robusta bananas, with the technology costing around ₹2 per bunch, resulting in an additional yield of 2-3 kg for the Ney Poovan cultivar and 4-5 kg for the Robusta banana cultivar. This cost-effective technology is gaining popularity in Karnataka (Patel, 31).

Banana Shakti

It was developed in 2007 by the National Research Centre for Banana TN, is a mixture of five major micronutrients. It is widely used to improve productivity by 15-20 per cent, as well as the quality of bunches in bananas (Patel, 2020).

Grape

Nutrition has been a major focus in grapevine for increasing yield, quality and productivity. Different nutrients have been reported to affect growth, physiology and ultimately yield. Macro and micronutrients are used for obtaining higher yield and quality fruits. The application of organic manures, in addition to enhancing soil microorganisms, increases the organic carbon sink capacity and improves the soil organic matter content. Thus, the establishment of vineyards should follow a nutrient-specific program along with integrated nutrient management (INM) and research recommendations given by institutes working on grapes. The general recommendation for grape vines according to the age of the plant is given below.

Age of Plant	*Urea (g/plant)*	*SSP (g/plant)*	*MOP (g/plant)*	*FYM (kg/plant)*
1 year	400	1500	250	20
2 year	500	3500	250	35
3 Year	600	3500	500	50
4 and > 4 years	800-1000	4000-4500	650-800	65-80

Kaur and Sharma, 2020.

The fertigation schedule for NPK, which is tailored to the growth stages of Thompson Seedless vines on Dogridge rootstock, has been established as a standard by the ICAR–National Research Centre for Grapes. In mature Thompson Seedless vines on Dogridge rootstock, the currently recommended P dosage through fertigation can be reduced by 50 per cent. For Cabernet Sauvignon grapes grown on 110R rootstock, the application of farmyard manure (FYM) at 15 tons/ha (dry weight basis) combined with 100 kg K_2O/ha has proven sufficient for high-quality grape production. The application of 6-benzyladenine (6 BA) at 20 ppm and gibberellic acid (GA_3) at 20 ppm has demonstrated optimal results in terms of yield and quality production for Manjri Naveen grapes. These practices and techniques are widely adopted by grape growers to enhance the yield and quality of their crops. Additionally, the National Research Centre for Grapes has standardized the development of biocompatible nanoclay-polymer composites and nanoparticles for the retention and release of iron and zinc in grapes (*Vitis vinifera* L.). Compared with the application of conventional FeSO4 and ZnSO4 fertilizers, the application of Fe-NPs and Zn-NPs resulted in a significant increase in the iron and zinc content in both berries and leaves (Anonymous, 2022).

Citrus

The CCRI Nagpur standardizes the optimum nutrient requirements of Mosambi sweet orange. The maximum fruit yield of 25.3 kg/tree was observed with 800 g N - 200 g P_20_5-400 g K_2O/tree. The highest juice content, TSS content and TSS/acidity ratio were observed with the same treatment (Anonymous, 60). The results of studies on agrocel-based potassium schoenite (Mahalaabh) in citrus plants suggested that plants receiving 100 per cent potassium schoenite through fertigation exhibited greater vegetative growth in terms of plant height (2.78 m), N S spread, E W spread (1.58 m), canopy volume (3.54 m') and fruit yield (59.27 kg/tree or 16.53 tons/ha).

Table 7.5: Soil Application of Nutrients in different Citrus Groups

N (g/tree)	*P (g/tree)*	*K (g/tree)*	*Crop*
500	100	400	Mosambi sweet orange
125	175	100	Satsuma mandarin
2720	1810	600	Sathgudi sweet orange
160	320	480	Valencia late orange
800	170	391	Mosambi sweet orange
250	250	500	Coorg mandarins

Srivastava, 2012.

Guava

It is recommended that manures and fertilizer be applied to guava plants according to their age. However, foliar application of fertilizer is also recommended because of its lower application rates, uniform distribution and quick response. A small investment in micronutrient deficiency benefits both the yield and marketable quality of fruit. The general recommendation for a guava plant is given below.

Age of Plant	*Urea (g/plant)*	*SSP (g/plant)*	*MOP (g/plant)*	*FYM (kg/plant)*
1-3 year	150-200	500-1500	600-1000	10-20
4-7 year	300-600	1500-200	600-1000	25-40
7-10 year	750-1000	2000-2500	1100-1500	40-50
> 10 year	1000	2500	1500	50

Kaur and Sharma, 2020.

The guava orchard subjected to *Azotobacter* + 25 per cent N/tree from the FYM and 75 per cent N/tree from the inorganic fertilizer yielded the highest N, P, Ca and Mg contents in the soil. Additionally, this treatment resulted in the highest leaf N, P, Ca and Mg contents. Conversely, the maximum soil and leaf K contents were achieved through *Azotobacter* in conjunction with 50 per cent of N/tree from FYM and 50 per cent from inorganic fertilizer in guava orchards (Sharma *et al.*, 62). The Sardar variety of guava, when subjected to FYM (10 kg/tree), 250 g *Azotobacter*, N (250 g), K (250 g) and P (100 g) exhibited significant improvements in growth and yield parameters (Ram *et al.*, 2005). This treatment led to a maximum increase in plant height of 0.45 meters and an increase in plant spread, yield/tree (150.25 kg) and TSS (13.50 °Brix).

Apple

For optimal growth and fruit production in apple plants, farmyard manure (FYM) supplemented with potassium (K) and phosphorus (P) is applied from December-January. A 1/2 dose of nitrogen (N) should be applied 2-3 weeks prior to flowering and after one month, except in drought-prone areas, where a single full dose of P and K is preferable (Srivastava and Malhotra, 2017). If the full N dose is not applied initially, it may be substituted with the second dose with 1-2 foliar sprays of 0.5 per cent urea after fruit set. In heavy-bearing orchards, the spraying of 1 per cent urea postharvest and 5 per cent urea pre-fall application are recommended to enhance fruit set. For regions with boron (B) deficiency, early foliar application of blossoms aids in early bud break and fruit set, as boron uptake is limited until soils warm up. Proper calcium (Ca) nutrition, through soil or foliar application before harvesting, is essential for fruit storage and reducing metabolic breakdown. Two sprays of 0.5 per cent calcium chloride (CaCl2) 45 and 30 days before harvesting improved storability (Srivastava and Malhotra, 2017).

Table 7.6: Nutrient Management in Apple Orchard

Tree Age (years)	*FYM (kg/tree)*	*N (g/tree)*	*P (g/tree)*	*K (g/tree)*
1	10	70	35	70
2	20	140	70	140
3	30	210	105	210
4	40	280	140	280
5	50	350	175	350
6	60	420	210	420
7	70	490	245	490
8	80	560	280	560
9	90	630	315	630
10 or >10	100	700	350	700

Table 7.7: Foliar Application of Nutrients in Apple

Element	*Formulation*	*Concentration (%)*	*Time of Spray*
N	Urea	0.55.0	After petal fall Postharvest application (Oct-Nov.)
Ca	$CaCl_2$	0.5	45 and 30 DBH
Zn	$ZnSO_4$	0.5	After petal fall
Mn	$MnSO_4$	0.5	After petal fall
B	H_3BO_3	0.1	At green tip to pin bud and petal fall
Cu	$CuSO_4$	0.1	After petal fall

Conclusion

Excessive reliance on inorganic fertilizers in fruit orchards has led to significant increases in cultivation expenses and detrimental effects on soil health and surrounding ecosystems. To address this issue, it is imperative to shift toward a more efficient and sustainable utilization of biofertilizers and inorganic and organic sources. This holistic approach not only reduces production costs but also improves soil health, thereby fostering sustainable fruit production and bolstering the financial well-being of farmers. Moreover, prioritizing nutrient management offers a cost-effective and environmentally friendly strategy for sustainable fruit tree cultivation. By addressing environmental concerns and promoting balanced nutrient management practices in commercial orchards, we can build resilience and harmony within our fruit production system.

REFERENCES

Ahmad R and Anjum MA, 2023. Mineral nutrition management in fruit trees under salt stress: a review. *Erwerbs-Obstbau*. **65**: 397-405.

Anonymous, 2021. Annual report. 2021. Central citrus research institute, Nagpur. pp. 178.

Anonymous, 2022. Annual report. 2022. National research centre for grapes, Pune. pp. 200.

ATIC Mango special. https: //www.iihr.res.in/bunch-feeding-banana.

Bargaz A, Lyamlouli K, Chtouki M, Zeroual Y and Dhiba D, 2018. Soil microbial resources for improving fertilizers efficiency in an integrated plant nutrient management system. *Front. Microbil.* **9**: 1606.

Beckman TG and Nyczepir AP, 2008. Rootstock breeding and selection. Peach: botany, production and uses. CABI.

Bowman KD and Joubert J, 2009. Citrus rootstocks in Florida. University of Florida, IFAS Extension.

Castle WS, Baldwin JC, Muraro RP and Littell R, 2010. Performance of Valencia sweet orange trees on 12 rootstocks at two locations and an economic interpretation as a basis for rootstock selection. *HortSci.* **45**: 523-533.

Chalise B, Shrestha AK, Srivastava A and Tripathi KM, 2023. Effect of Plant Nutrients on Fruit Yield and Quality of Cavendish Banana (Musa sp.) cv.'Grand Naine'. *Nepalese Hort.* **17**: 86-95.

Chatzistathis T, Kavvadias V, Sotiropoulos T and Papadakis IE, 2021. Organic fertilization and tree orchards. *Agric.* **11**: 692.

Davarpanah S, Tehranifar A, Davarynejad G, Abadía J and Khorasani R, 2016. Effects of foliar applications of zinc and boron nano-fertilizers on pomegranate (Punica granatum cv. Ardestani) fruit yield and quality. *Sci. Hortic.* **210**: 57-64.

Day SD, Eric Wiseman P, Dickinson SB and Roger Harris J, 2010. Tree root ecology in the urban environment and implications for a sustainable rhizosphere. *J. Arboric.* **36**: 193.

Dutta P, Kundu S and Chatterjee S, 2008. Effect of bio-fertilizers on homestead fruit production of papaya cv. Ranchi. In II International Symposium on Papaya **851**: 385-388.

El-Ramady HR, 2014. Integrated nutrient management and postharvest of crops. *Sustain. Agric. Reviews.* **13**: 163-274.

Eyles A, Bound SA, Oliver G, Corkrey R, Hardie M, Green S and Close DC, 2015. Impact of biochar amendment on the growth, physiology and fruit of a young commercial apple orchard. *Trees.* **29**: 1817-1826.

Fageria NK, Baligar VC and Li YC, 2008. The role of nutrient efficient plants in improving crop yields in the twenty first century. *J. Plant Nutr.* **31**: 1121-1157.

Ferree D C and Warrington IJ, eds, 2003. Apples: Botany, Production and Uses. CABI, UK.

Hartmann HT, Kester DE, Davies FT and Geneve RL, 2010. Plant propagation: principles and practices. 8th ed., Pearson.

Holland DA, RC Little M Allen and Dermott Wt, 1967. Soil and leaf sampling in apple orchards. *J. Hortic. Sci.* **42**: 403-417.

James DW and Wells KL, 1990. Soil sample collection and handling: Technique based on source and degree of field variability. Soil testing and plant analysis, **3**: 25-44.

Jones Jr J B, 2011. Soil testing and plant analysis: guides to the fertilization of horticultural crops. *Hortic. Reviews.*

Kalcsits L, Lotze E, Tagliavini M, Hannam KD, Mimmo T, Neilsen D and Valverdi NA, 2020. Recent achievements and new research opportunities for optimizing macronutrient availability, acquisition, and distribution for perennial fruit crops. *Agron.* **10**: 1738.

Kaur S and Sharma Y, 2020. When and how to fertilize fruit plants. *Ind. Hortic.* **65**: 7-10.

Keller M, 2010. The science of grapevines: Anatomy and physiology. Academic Press. pp. 277.

Layne DR and Bassi D, 2008. The peach: botany, production and uses. CABI.

Lopez-Vargas ER, Ortega-Ortiz H, Cadenas-Pliego G, de Alba Romenus K, Cabrera de la Fuente M, Benavides-Mendoza A and Juarez-Maldonado A, 2018. Foliar application of copper nanoparticles increases the fruit quality and the content of bioactive compounds in tomatoes. *Appl. Sci.* **8**: 1020.

Malusa E, Pinzari F and Canfora L, 2016. Efficacy of biofertilizers: challenges to improve crop production. Microbial Inoculants in Sustainable Agricultural Productivity: **2**: 17-40.

MoA and FW, 2023. Area and Production of Horticulture Crops (2nd Advance Estimate), Ministry of Agriculture and Farmers Welfare Government of India, Directorate of Economics and Statistics.

Morgan KT and Kadyampakeni DM, 2020. Nutrition of Florida citrus trees. University of Florida Institute of Food and Agricultural Sciences: Gainesville, FL, USA, 27.

Nachtigall GR and Dechen AR, 2006. Seasonality of nutrients in leaves and fruits of apple trees. *Sci. Agric.* **63**: 493-501.

Negi YK, Sajwan P, Uniyal S and Mishra AC, 2021. Enhancement in yield and nutritive qualities of strawberry fruits by the application of organic manures and biofertilizers. *Sci. Hortic.* **283**: 110038.

Oyetunji O, Bolan N and Hancock G, 2022. A comprehensive review on enhancing nutrient use efficiency and productivity of broadacre (arable) crops with the combined utilization of compost and fertilizers. *J. Env. Man.* **317**: 115395.

Parent LE, Rozane DE, de Deus JAL and Natale W, 2020. Diagnosis of nutrient composition in fruit crops: Major developments. In Fruit Crops (pp. 145-156). Elsevier.

Patel TU, 2020. Banana mulching, sleeving and nutrient pouch feeding technology. *Ind. Hortic.* **65**: 27-30.

Qiu F, Liu W, Chen L, Wang Y, Ma Y, Lyu Q and Zheng Y, 2021. Bacillus subtilis biofertilizer application reduces chemical fertilization and improves fruit quality in fertigated Tarocco blood orange groves. *Sci. Hortic.* **281**: 110004.

Raghupathi HB, Ganeshamurthy AN and Ravishankar H, 2014. Comparison of DRIS ratio norms of selected fruit crops. *I. J. Hortic.* **71**: 168-175.

Ram RA, Bhriguvanshi SR and Pathak RK, 2005, December). Integrated plant nutrient management in guava (*Psidium guajava* L.) cv. Sardar. In I International Guava Symposium 735 (pp. 345-350).

Robinson TL, 2001. Apple rootstocks and orchard systems. Compact fruit tree, **34**: 52-55.

Rom CR and Carlson RF, 1987. Rootstocks for fruit crops, Wiley.

Roy RN, Finck A, Blair GJ and Tandon HLS, 2006. Plant nutrition for food security. Food and Agriculture organization of the United Nations, Rome.

Sable PA, Saras P, Singh P and Patel JR, 2023. Integrated nutrient management in mango. *Ind. Hortic.* **68**: 33-35.

Saha R, Nath V and Kumar D, 2010. Effects of farmyard manure on soil organic carbon stock, the pattern of fertility build-up, and plant growth in 'Mallika'mango (*Mangifera indica* L.). *J. Hortic. Sci. Biotechnol.* **85**: 539-543.

Sankaran M and Dinesh MR, 2020. Biodiversity of tropical fruits and their conservation in India. *J. Hortic. Sci.* **15:** 107-126.

Sempeho SI, Kim HT, Mubofu E and Hilonga A, 2014. Meticulous overview on the controlled release fertilizers. *Adv. Chem.* **1**: 363071.

Shankar K, Awasthi OP, Dubey AK, Singh A, Prakash J and Dolatabadian A, 2023. Rootstock mediated alteration in morphology and photosystem in sweet orange (*Citrus sinensis*) scion cv. Pusa Sharad under NaCl stress. *Ind. J. Agric. Sci.* **93**: 1103-1107.

Sharma A, Wali VK, Bakshi P and Jasrotia A, 2013. Effect of integrated nutrient management strategies on nutrient status, yield and quality of guava. *Ind. J. Hortic.* **70**: 333-339.

Smith GS, Buwalda JG and Clark CJ, 1988. Nutrient dynamics of a kiwifruit ecosystem. *Sci. Hortic.* **37**: 87-109.

Souri MK and Bakhtiarizade M, 2019. Biostimulation effects of rosemary essential oil on growth and nutrient uptake of tomato seedlings. *Sci. Hortic.* **243**: 472-476.

Srivastava AK and Malhotra SK, 2017. Nutrient use efficiency in perennial fruit crops–A review. *J. Plant Nutr.* **40**: 1928-1953.

Srivastava AK, 2012. Integrated nutrient management in citrus. Advances in citrus nutrition, 369-389.

Sumner ME, 1977. Use of the DRIS system in foliar diagnosis of crops at high yield levels. *Commun. Soil Sci. Plant Ana.* **8**: 251-268.

Tagliavini M, Tonon G, Scandellari F, Quiñones A, Palmieri S and Menarbin G, 2007. Nutrient re cycling during the decomposition of apple (Malus domestica) and mowed grasses in an orchard. *Agric. Ecosyst. Environ.* **118**: 191–200.

Toselli M, Baldi E, Cavani L and Sorrenti G, 2020. Nutrient management in fruit crops: An organic way. In Fruit Crops (pp. 379-392). Elsevier.

Uma S and Kumar PS, 2020. Banana research and development in India-challenges and opportunities. *Prog. Hortic.* **52:** 1-11.

University of Maryland, 2006. Nutrient management for tree fruits and small fruits source: University of Maryland cooperative extension.

Wang J, Lali MN, Dong Z, Xiong H, Wang Y, Zhao L and Zhang Y, 2024. Organic waste recycling for green and sustainable nitrogen management of fruit production systems in China. Agricultural Systems, **218**: 103959.

Wei X, Chen J, Gao B and Wang Z, 2020. Role of controlled and slow release fertilizers in fruit crop nutrition. In *Fruit crops* (pp. 555-566). Elsevier.

Yadav AK., Singh JK and Singh HK, 2011. Studies on integrated nutrient management in flowering, fruiting, yield and quality of mango cv. Amrapali under high density orcharding. Ind. J. Hortic. **68**: 453-460.

Yang Y, Zhang J, Chang X, Chen L, Liu Y, Xu Q and Zhang Y, 2024. Green manure incorporation enhanced soil labile phosphorus and fruit tree growth. *Front. Plant Sci.* **15**: 1356224.

Yildiz E, Yaman M, Ercisli S, Sumbul A, Sonmez O, Gunes A and Kviklys D, 2022. Effects of rhizobacteria application on leaf and fruit nutrient content of different apple scion–rootstock combinations. *Hortic.* **8**: 550.

Yuan B, Yu D, Hu A, Wang Y, Sun Y and Li C, 2023. Effects of green manure intercropping on soil nutrient content and bacterial community structure in litchi orchards in China. *Front. Environ. Sci.* **10**: 1059800.

Zahedi SM, Karimi M and Teixeira da Silva JA, 2020. The use of nanotechnology to increase quality and yield of fruit crops. *J. Sci Food Agric.* **100**: 25-31.

Zhang J, Nie J, Cao W, Gao Y, Lu Y and Liao Y, 2023. Long-term green manuring to substitute partial chemical fertilizer simultaneously improving crop productivity and soil quality in a double-rice cropping system. *Eur. J. Agron.* **142**: 126641.

Chapter 8

Balanced Fertilization in Fruit Crops and Modes of Nutrient Supply

A.K. Srivastava[1] *and D. Hota*[2]

[1]*ICAR–Indian Agricultural Research Institute, Dhemaji–787034, Assam*
[2]*Siksha-O-Anusandhan (Deemed to be University), Bhubaneswar–751003, Odisha*

Fruit crops by the virtue of extended juvenile period, long annual growth cycle and perennial woody framework are collectively known for their strong carbon sink regulating mobilization/immobilization of nutrients, thereby affecting the yield and fruit quality by ensuring the biologically active rhizosphere. Balancing nutrients supply using popular methods comprising inorganic nutrition (soil application of macronutrients and foliar fertilization of micronutrients) as per crop phenology, site specific nutrient management (tailoring nutrient supply as per nutrients removal by fruits using spatial variability in soil fertility as decision support), fertigation (supply of irrigation and nutrients through their partitioning at growth stages using either surface or subsurface drip irrigation for reduced carbon, nutrient and water footprints), organic management (considering booth soil-plant health on an on-farm module basis) and integrated nutrients supply (inorganic nutrients-organic manures-microbes synergy having multi-tier nutrients release pattern to ensure extended nutrients bioavailability) are though in practice, but witnessed periodical improvements with regard to production sustainability. All these nutrients supply modes are anchored through 4R nutrient stewardship principles, a pre-requisite for elevated nutrient-use-efficiency. Amid these scientific protocols of nutrient management, our understanding is still limited in effectively utilising the legacy nutrients stock of soil and introduction of crop nutrient holiday, if balanced fertilization is to be exercised with emphasis on production economics and nutrients fortification of fruits agronomically. Multiple objectives of balanced fertilization would only be fulfilled in fruit crops, unless a robust methods of diagnosis of nutrient constraints are developed with recurrent refinements to suit at regional agro-pedological conditions.

Background

Globally, fruit crops occupy an area of 65.29 million ha (citrus>grape>banana) with a production of 883 million tons (citrus>banana>apple) and average productivity of 13.52 tons/ha (FAO, 2020). Approximately 1.7 million (2.8%) human deaths worldwide are attributable to micronutrient deficiency induced through suboptimum consumption of fruits and vegetables, and regarded as one of the top 10 selected risks for global mortality (WHO, 2021). In the current era of climate change on continuous rampage, nutrient efficient plants will play a major role in sustaining the crop yield compared to yesteryears, amidst newer challenges like shrinking land and water resources available for fruit crop production (often comes from competitive cereal-based land uses), higher cost of inorganic fertilizer inputs, declining trends in crop yields globally, and increasing environmental concerns. Furthermore, at least 60 per cent of the world's arable lands have mineral deficiencies or elemental toxicity problems and on such soils, fertilizers and lime amendments are essential for achieving improved crop yields (Pathak and Nedwell, 2011). In the light of climate change related issuses, perennial fruit trees play an important role in carbon cycle of terrestrial ecosystems and sequestering atmospheric CO_2 (Lobell *et al.,* 2005; Guimãres *et al.,* 2014). According to Wu *et al.* (2012), net C- sink and C- storage in biomass of apple orchard ranged from 19 to 32 Tg C and from 230 to 475 Tg C in 20- years period, amounting to 4.5 per cent of total net C sink in the terrestrial ecosystems in China. In an estimate, Lakso (2010) observed that an acre of apple orchard fixed about 20 tonnes of CO_2 from the air each season, and provided over 15 tonnes of O_2, equivalent to over 5 billion BTU of cooling power. While, Bwalya (2013) showed that citrus trees carbon sequestration in biomass ranged from 23.9 tonnes CO_2/ha for young trees to 109 tonnes CO_2/ha for mature trees.

In sustaining the productivity potential of fruit crops, there will be an increasing importance of nutrient efficient cultivars (higher yields per unit of nutrients applied or absorbed than standard cultivars under similar agroecological conditions) that are higher producers under multiple nutrient stress (Srivastava, 2023a; 2023b). In the past, much research has been conducted to identify and/or breed nutrient efficient plant species or genotypes/cultivars within different fruit species, but the success in releasing nutrient efficient cultivars has been far limited (Srivastava and Mousavi, 2024). The main reasons for limited success are the genetics of plant responses to nutrients and plant interactions with environmental variables, not so well understood even today (Fageria *et al.,* 2008). Fruit crops by the virtue of their perennial nature of woody framework (nutrients locked therein), extended physiological stages of growth, differential root distribution pattern (root volume distribution), growth stages from the point of view of nutrient requirement, and preferential requirement of some nutrients by specific fruit crop, collectively make them different than the annual crops (Srivastav and Hota, 2020; Mousavi *et al.,* 2024).

Rengel *et al.* (1996) observed that the total number of bacterial colony forming units increased in the rhizosphere of Zn-efficient genotypes of wheat under Zn-deficiency and in Mn-efficient genotypes under conditions of Mn-deficiency. In contrast, a Zn-deficiency treatment acted synergistically with the number of fluorescent *Pseudomonas* in the rhizospheres. These studies provide strong cues that fruit crops exposed to varying nutrient stresses recruit plant growth promoting microorganisms, playing an important role in producing crop responses through number of mechanisms comprised of biological nitrogen fixation, growth hormone production, phosphate solubilization, siderophore production, hydrolytic enzymes production, and antagonistic activity (Bunemann *et al.,* 2006; Tailor and Joshi, 2014). Activation of each mechanism implies the production of specific compounds and metabolites, such as plant growth factors, hydrolytic enzymes, siderophores, antibiotics, and carbon and nitrogen permeases. These metabolites can be either overproduced or combined with appropriate biocontrol strains in order to obtain new formulations for their more effective applications (Tahía *et al.,* 2004). These microbial inoculants are reported to reduce the plant nutrient requirement. For example, *Azotobacter* inoculation alone could substitute up to 50 per cent nitrogen requirement of banana (Tiwari *et al.,* 1999) and 25 per cent phosphorus requirement of papaya (Padma and Kandasamy, 1990), signifying their importance in soil fertility maintenance, a pre-requisite to production sustainability (Hazarika and Ansari, 2010; Maity *et al.,* 2012). Another group of microorganisms, popularly known as arbuscular mycorrhizal fungi are equally effective in substantially improving the nutrient acquisition capacity of fruit crops, in addition to enriching the rhizosphere biologically in a much activated form (Wu *et al.,* 2013, 2017; Zhang *et al.,* 2015; Liu *et al.,* 2014; Huang *et al.,* 2014; Zou *et al.,* 2014).

Ample evidences have accrued through worldwide research that nutrient-microbe synergy plays a pivotal role in fruit crops (Hota *et al.,* 2020; 2021a; 2021b) to mobilize and accumulate the required nutrients as per the metabolic nutrient demand (Berg, 2009; Shylaja and Rao, 2012). They can facilitate mineral uptake by the plant, and can also secrete plant hormones into the rhizosphere to promote plant growth and the timing of flowering (Bottini *et al.,* 2004; Lu *et al.,* 2018). In the rhizosphere, a fierce battle is fought between plant-beneficial microbes and deleterious microbes (Berendsen *et al.,* 2018; Hota *et al.,* 2021c). Hence, the composition of the microbial community in the rhizosphere plays a crucial role in the function of plants through influences on their physiology and development (Mendes *et al.,* 2013). A still bigger question emerges, whether rhizosphere competent microbes could collectively contribute towards improved resilience of plant's rhizosphere (Wang *et al.*, 2014). And if those microbes are so successful in promoting growth response, addition of starter nutrients in such combination may further magnify the magnitude of response called nutrient-microbe synergy (Hoat *et al.*, 2023). Our earlier studies have shown that rhizosphere effective microbes have the tendency to play multiple roles (Keditsu and Srivastava, 2014; Wu *et al.*, 2017) to overcome various biotic and abiotic stresses while interacting with an environment. Mineral fertilizers on the other hand as a main stream of direct nutrient supply, their application can enhance the soil biological activity via increases in system productivity, crop residue return, and soil organic matter (Srivastava, 2012; 2013a). Therefore, a sound understanding of nutrient-microbe synergy could possibly lay a solid foundation in unlocking the productivity potential of perennial fruit crops, besides safeguarding the soil health, both physico-chemically as well as biologically (Hota *et al.*, 2022). In this background, systematic efforts were made to analyze the relationship between inorganic nutrients (macronutrients versus micronutrients) -organic manures (manures versus green manure) - microbial inoculants (single versus microbial consortium) and fruit crops as a holistic approach to address multiple nutrient stress jeopardizing the production sustainability.

Challenges of Plant Nutrition

Historically, economic development has been faster in those world regions where fertilizer use and crop yields rose in parallel. Despite bundles of scientific literature have been poured in the recent past dealing with a detailed description of one or few scientific goals like molecular genetics, plant mineral nutrients related to molecular biology, physiology processes related to plant nutrition, fertiliser application strategies, and beneficial plant–microbe interactions, yet the success of plant nutrition in ensuring the quality production of fruit crops is registered par below efficacy, and more candidly so in climate change scenario (Srivastava, 2013a; 2015). In such scenario of climate change, there are multiple expectations (from biochemical bases underlying nutritional processes described down to molecular level, involving differential selectivity of membrane transporters for nutrients or signalling functions of mineral nutrients in plant development to serving large parts of the society from farmers to policy makers, allowing further enhancements in crop yield, and quality while increasing nutrient- efficiency and reducing nutrient losses from crop production systems) from plant nutrition, thereby aimed to accomplish varied objectives (Srivastava *et al.*, 2006; Sharma *et al.*, 2021). Hence, the current understanding of plant nutrition is largely focused on monocultural situations, needs to be augmented by the interactions that occur in more complex systems. Therefore, mineral nutrients in soils and crops have important and still difficult to predict positive as well as negative interactions with global climate change (Soares *et al.*, 2019), although negative impacts of climate change appear to outweigh positive ones (St Claire and Lynch, 2010).

Plant nutrition stands on two legs, one leg anchored in plant physiology explaining the fundamental processes underlying and regulating the acquisition, allocation, and utilization of mineral elements in plants; while the other leg is deeply rooted in an applied, problem-oriented context, explaining the use and action of fertilizers and the cycling of mineral elements in agricultural plant production (Wiren, 2011). These dimensions of plant nutrition perch different challenges to different perspectives. Noordwijk and Cadisch (2002) previously pointed out the main challenge of plant nutrition as to increase the width of the domain between the access and excess frontiers, rather than to define the crop nutrient response as a single 'economic optimum' point, involving two approaches: i. the technical paradigm of precision farming and the ecological analogue approach based on filter functions and complementarity of components in mixed plant systems. Dobermann *et al.* (2022) proposed a new paradigm for responsible plant nutrition following a food systems and circular economy approach to achieve multiple socioeconomic, environmental, and health objectives involving critical actions such as: i. sustainability-driven nutrient roadmaps, ii. digital crop nutrition solutions, iii. nutritious crops, iv. nutrient recovery and recycling, v. climate-smart fertilizers, and vi. accelerated innovation. Ironically, plant responses to deficiencies of different nutrients have been studied mainly as a separate event (Kalatippi *et al.,* 2022) and only a few studies discussed the molecular basis of biological interaction amongst the nutrients, comprehensively overlooking the interaction between macro- and micronutrients at two-by-two or multi- level nutrient interactions involving transcription factor or phosphate starvation response, the master regulator of multiple nutrients homeostatis (Kumar *et al.,* 2021) using an integrative study of multiple nutrient signalling cross-talks in plants (Fan *et al.,* 2021) for ensuring better biological significance narrating higher yield- and nutrient-use-efficiency (Rietra *et al.,* 2017).

According to Brown (2022), the current definitions of essential or beneficial elements for plant growth rely on narrowly defined criteria that do not fully represent a new vision for plant nutrition and often compromise with fertilizer regulation and practice. An effective plant nutrition is supposed to ensure the level of fruit nutrition, much beyond simple minerals and vitamins, more inclined towards freshness (taste quality) as a matter of extended storability (Tatewar *et al.,* 2023). The raised CO_2 concentration in the environment reduced the overall quantity of twenty-five minerals in plants, including calcium, potassium, zinc, sulfur, copper, and iron (8 per cent on average), and also increased the ratio of carbohydrates to minerals in food plants (Loladze, 2014). A new analysis of long-term trends of the mineral content of fruits and vegetables (1940, 1991, and 2019) showed all elements except P declined in concentrations between 1940 and 2019, the greatest overall reductions during this 80-year period were Na (52%), Fe (50%), Cu (49%) and Mg (10%). Many reasons such as reductions, including changes in crop varieties and agronomic factors associated with the industrialisation of agriculture and increases in carbon dioxide could be ascribed (Mayer *et al.,* 2021).

Studies of historical fruit nutrient composition data are inherently limited, most of the other methods focus on single crops of any kind, can include any nutrient of interest and can be carefully controlled (Davis, 2009). Therefore, the proposed new definition of mineral plant nutrient is: an element which is essential or beneficial for plant growth and development or for the quality attributes of the plant or harvested product, of a given plant species, grown in its natural or cultivated environment (Brown, 2022).

While protecting plant nutrition from the effects of climate change, there is a renewed strong emphasis on enlarging the proportion of fruits, vegetables, and grains in the diet, likely to increase the intake of a wide range of minerals and phytonutrients thought to have specific nutritional benefit (Sahoo *et al.,* 2018). However, the effect of climate change on the accumulation of these phytochemicals and their biosynthetic pathways remains poorly understood. The recent developments in metabolomic analysis methods are of great help in obtaining novel insights into qualitative and quantitative changes in the composition of plant phytonutrients under different climate change scenarios (Shahzad and Rouached, 2022) in addition to linking carbon sequestration ability of crops with plant nutrition (Elbasiouny *et al.,* 2022). Amid these facts and figures, separating physiological response of nutrient deficiency versus crops responses of climatological factors becomes a more challenging exercise.

Inorganic Plant Nutrient Supply: A Complex Issue

The fertilizer requirement of fruit crops depends upon the objectivity of fertilization, whether the purpose is to grow the crop or feed the crop. Based on these objectives, two types of fertilization *viz.,* corrective and preventive are usually adopted (Srivastava and Singh, 2002). An optimum fertilizer program is one in which the cost of each unit of fertilizer applied is at least covered by an extra return of fruit yield obtained in both, the short and long term life of a citrus orchard. In a young tree care program, emphasis is placed on developing the tree canopy which will later produce large crop. Switching from a young tree program to a bearing tree, regime may result in significant reduction in applied fertilizers (Srivastava and Singh, 2008; 2009). Fertilizer programs for citrus (for example) are determined by many approaches. These are: surveys, growers' experience, following the fertilization program of high yielding orchards, replacing the amount of nutrients removed in fruits, deficiency symptoms, applying results from sand, soil culture, and field experiments, and soil/leaf analysis. Each one of these has certain advantages and limitations. Some of the difficulties in developing sound recommendation for fertilizer program may be traced in the history of developments dedicated towards determining nutrient requirements. All these studies however, aimed towards the balanced plant nutrition, a fundamental concept of any nutrition program (Srivastava *et al.,* 2021).

Optimum soil fertility induced plant nutrition has a key role to maximise yield and quality production of perennial fruit crops. The fertilizer requirement of perennial fruit crops is determined by many approaches, including surveys, growers' experience, following the fertilization program of high yielding orchards, replacing the amount of nutrients removed by fruits, deficiency symptoms, applying results from sand or soil culture and field experiments and leaf/soil analysis, with each one of these having distinct advantages and limitations (Srivastava and Malhotra, 2014).

Three approaches to fertilizer recommendations that are widely used: the deficiency correction philosophy (originates from nutrient constraints based crop response through nutrient additions to the point of maximum economic yield), maintenance concept (aims to maintain soil fertility level slightly above the point of maximum economic yield), and nutrient removal or balanced philosophy (emphasizes the return to the soil what is removed by the crop to maintain productivity, but often over recommends nutrient need, since it does not take into account for the soil's ability to supply available nutrients to the plants over time) (Srivastava 2013c; 2024).

Soil provides nearly all the nutrients essential to complete the life cycle of a plant. Different soil properties primarily determine the extent of a fertilizer response and the crop rotation on some recently published review articles changes in physico- chemical and biological properties of soil. Since the subject of fertilization in horticultural crops is so vast and complex, the readers may refer to the few review articles exclusively devoted to different issues of nutrient management in perennial fruit crops (Hazarika and Ansari, 2010; Srivastava *et al.,* 2015a; Srivastava and Ngullie, 2009; Maity *et al.,* 2012; Singh *et al.,* 1999; Srivastava *et al.,* 2022; Srivastava and Bora, 2023; Srivastava, 2024). However, the subject dealing with multipronged action to rhizocompetent microbes in combination of various nutrient sources in fruit crops is distinctly missing.

Projections of Fertiliser Requirement

Two -fold strategies is needed through expanding the acreage and improving the productivity from the present level, but both the strategies warrant for stringent vigil on balanced use of fertilisers if horticulture sector is to be laid on nutritionally secured footing where the pattern of nutrient removal will be of paramount importance. A significant amount of nutrients is removed annually by the citrus fruits.

Even within a single crop like citrus, various nutrients are reported to be removed (in kg/ha) in the proportion of : 55.5 N, 51.1 P, 53.7 K, 23.4 S, and 5.1 Mg; 89 N, 20 P, 192 K, 10 kg S, and 14 kg Mg; 270 N, 60 P, 350 K, 30 S, and 60 Mg; 100 N, 60 P, 350 K, 30 S, and 40 Mg; 68 N, 3.4 P, and 31 K in relation to an average fruit yield of 30 tonnes/ha by different citrus cultivars (Srivastava and Mousavi, 2024). The annual growth of horticultural crops needs to be accomplished at 6 per cent through 12th five year plan. On the basis of growth trend of fruit crop in last three years (2010-11, 2011-12, 2012-13), it was computed that atleast 4 per cent increase in production and 3 per cent increase in area per annum must be achieved.

The total removal of nutrients from the projected area in 2016-17 by considering the nutrient removal rate (Tiwari, 2007a), a total ($N+P_2O_5+K_2O$) will be removed by fruit crops if production and productivity have to be sustained (Table 8.2). The total nutrients (N, P and K) requirement by major fruit crops for the projected 3 per cent increase in area per year and recommended doses of nutrients given (Ghosh *et al.,* 2012a). These calculations indicated that in 2016-17, 4.89 million tonnes of fertilisers (N, P_2O_5 and K_2O) would be required, which will be 12.5 per cent higher than the fertiliser requirement for 2012-13.

Table 8.1: Projection of Nutrient (N, P_2O_5, K_2O) Requirement for Major Fruit Crops for the Year 2016-17 Based on the Recommended Doses

Crops	*Area in 2012-13 (000 ha)*	*Nutrient Require-ment (000 tonnes) in 2012-13 (N, P_2O_5, K_2O)*	*Esti-mated area in 2016-17 (000 ha)*	*Nutrient Requirement (000 tonnes) in 2016-17*			
				N	*P_2O_5*	*K_2O*	*Total (N, P_2O_5, K_2O)*
Banana	776	1240	873	440	320	636	1396
Mango	2500	358	2813	141	84	177	402
Citrus	1042	473	1172	223	122	188	533
Papaya	132	229	148	64	64	129	257
Guava	236	80	265	30	22	38	90
Apple	312	133	351	38	38	74	150
Pineapple	105	96	118	50	14	44	108
Sapota	164	43	184	18	11	19	48
Grapes	118	228	132	86	55	114	255
Litchi	83	104	93	23	37	57	117
Others	1515	1364	1705	512	341	682	1535
Total	**6983**	**4348**	**7854**	**1625**	**1108**	**2158**	**4891**

Source: Srivastava and Malhotra (2014; 2017).

Foliar Fertilization

Plants sometimes grow at rates that are faster than the ability of their roots to absorb and translocate mineral nutrients to the leaves or developing fruits. Foliar sprays are useful to maintain optimum nutrient concentration in the plant during the growing cycle by optimizing the movement of nutrients (Srivastava and Malhotra, 2014; Darshan *et al.,* 2022). Foliar fertilization means the epigean application of plant nutrient, which a plant needs for its nutrition and growth *i.e.* the non-root feeding or extra radical feeding (Table 8.4). Foliar sprays of urea (28-31 kg N ha^{-1}) in Valencia orange, multiple application at 1 per cent urea in Codoux clementine mandarin, 10 per cent KCl in Eureka lemon, 5 per cent KNO_3 with 18-20 ppm 2,4-D in Shamouti orange, and only 5 per cent KNO_3 in Valencia orange demonstrate that only two nutrients, N and K are effective through foliar application. Increase in fruit yield and size due to foliar sprays of urea further indicated an indirect effect of earliness in flowering and fruit set.

Table 8.2: Optimum Nutrient Requirement for different Fruit Crops

Fruit Crop (g/tree)	*N*	*P_2O_5*	*K_2O*	*Zn*
Apple (*Malus pumila*)	100.0	45.0	180.0	25.0
Avocado (*Persea americana*)	11.3	3.9	23.4	10.0
Banana (*Musa acuminata* L.)	250.0	60.0	1000.0	40.0
Nagpur mandarin (*Citrus reticulate* Blanco)	3.4	0.37	3.72	1.0
Khasi mandarin (*Citrus reticulata* Blanco)	1.81	0.16	2.18	1.0
Coorg mandarin (*Citrus reticulata* Blanco)	3.04	0.27	3.31	1.0
Kinnow mandarin (*C.nobilis* T. x *C. deliciosa* L.)	2.40	0.57	2.35	1.0
Acid lime (*Citrus aurantifolia* Swingle)	1.22	0.21	4.06	1.0
Durian (*Durio zibethinus* L.)	16.1	6.6	33.5	6.7
Grapes (*Vitis vinifera* L.)	170.0	60.0	220.0	20.0
Guava (*Psidium guajava* L.)	120.0	50.0	150.0	20.0
Ber (*Ziziphus mauritiana* L.)	80.0	35.0	125.0	20.0
Kiwifruit (*Actinidia deliciosa Chev.*)	165.0	50.0	273.0	40.0
Mango (*Mangifera indica* L.)	100.0	25.0	110.0	15.0
Papaya (*Caria papaya* L.)	90.0	25.0	130.0	50.0
Pecan (*Carya illinoinensis* W.)	9.7	5.3	5.4	1.2
Pistachio(*Pistacia vera* L.)	30.0	12.0	15.0	0.057
Passion fruit (*Passiflora Spp.* F.)	60.0	15.0	75.0	15.0

Note: In case of citrus, figures are expressed as g/ton fruits

Source: Tandon and Tiwari (2007), Srivastava *et al.* (2008b; 2015b), Srivastava and Pongnenger (2023), Chadha and Bhargava (1993), Ganeshamurthy *et al.* (2019)

Foliar sprays of micronutrients are more popular and, therefore, frequently used (Srivastava and Singh, 2003; Srivastava *et al.*, 2008). A large variation exists with regard to foliar recommendation of micronutrients *viz.*, Fe-EDDH (0.1%) for Valencia orange, Fe-polyflavonoid (1%) for Verna lemon, Fe (50 ppm) + $MnSO_4$ (5 ppm) + Zn (75 ppm) for Washington navel, $MnSO_4$ + $ZnSO_4$ (0.15% each) for Thompson navel, $FeSO_4$ + $CuSO_4$ (0.25% each) + $ZnSO_4$ (0.5%) for Coorg mandarin, $ZnSO_4$ (0.60 g L^{-1}) + $MnSO_4$ (1.2 g L^{-1}) for Valencia orange, Zn-EDTA (0.4%) + Cu- EDTA (0.2%) for Kinnow mandarin, borax + $MgSO_4$ (0.2%) + $ZnSO_4$ (0.1%) for Jiaogan mandarin *etc.* for different commercial citrus cultivars grown across important citrus region as an example (Srivastava *et al.*, 2015c). foliar application of 0.5 per cent borax along with GA_3 increased the vegetative, fruiting potential and qualitative aspect of guava (Darshan *et al.*, 2023a; 2023b). These recommendations are more suggestive than interpretative in nature (Srivastava and Singh, 2003). Foliar fertilization however is still not a popular choice of nutrient supply option to fruit crops due to lower nutrient-use-efficiency under the soils with acidity or alkalinity, calcareousness or contaminants/pollutants, besides crop growth stage having an impounding effect and time of application (Srivastava *et al.,* 2008).

Soil Fertilization

The main organ for absorbing water and nutrients by a plant is its roots. Climate and soil related factors such as low temperature, excessive moisture, drought *etc.* however, disturb nutrient and water uptake during plant growth, the effects of which may vary from a temporary restriction of growth to reduced yield and quality at harvest. Response of nitrogen fertilization in improving the growth, yield, and quality of different citrus cultivars is well recognized under different agroclimatic regions of the countries like Brazil, Australia, South Africa, India *etc.* Contrary to foliar fertilization, soil application of macronutrients is more efficacious (Srivastava *et al.,* 2008).

There are varied fertilization and doses schedules followed across a variety of fruit crops (Table 8.2). Some fertilization plans recommend N application since the beginning of bud break until 6 weeks after full bloom for bearing pear trees (Neto *et al.,* 2016), whereas others defend that N must be applied during the whole growth cycle considering that after harvest, trees can still improve their reserves through N uptake from soil. Some authors have studied the fertilizer N use efficiency in pears and apples (Cheng *et al.,* 2002; Chen *et al.,* 2012). Most studies were performed in pots in sand culture, comprising its applicability to field conditions. The recycling of N as a result of the decomposition of senescent leaves in soils was only addressed in one study with apple trees (Tagaliavini *et al.,* 2007; Tagliavini and Scandellari, 2013). Fertilizer N use efficiency by trees increased from the first to the third year, but was generally small (6, 14, and 33%), and estimated N losses were large (89, 46, and 53%, respectively, in the first, second, and third year). Irrigation water and soil provided more N to the trees than fertilizer N (Neto *et al.,* 2016). The fertilizer practices, therefore need to be worked out based on 4R nutrient stewardship (right source of fertilizers in right amount of fertilizers applied at right crop growth stage at right place) in fruit crops, if fertilizer-use-efficiency is to be ensured.

The optimum requirement of macronutrients for different commercial citrus cultivars suggest: 475 g N + 320 g P_2O_5 + 355 g K_2O tree^{-1} for Satsuma mandarin in Turkey, 240 g N + 40 g P_2O_5 + 100 g K_2O ha^{-1} for Dancy tangerine in Spain,1.4 kg N + 1.08 kg P + 1.1 kg K tree^{-1} for acid lime,400-1200 kg N + 200 kg P_2O_5ha^{-1} for kinnow mandarin in India, 120 kg N + 150 kg P + 75 kg S + 6 kg Cu + 0.8 kg Mo + 5.0 kg Zn ha^{-1} for Neck orange in Korea, 1.02 kg N + 0.58 kg P_2O_5 and 0.55 kg K_2O tree^{-1} for Satsuma mandarin in Georgia, 200 kg N + 140 kg P + 210 kg K ha^{-1} for Pera sweet orange in Brazil, 0.5 kg N + 0.5 kg P_2O_5 + 1.0 kg K_2Oha^{-1} for grapefruit in Greece, and 1.5 kg urea + 0.25 kg superphosphate + 1.25 kg potassium chloride + 1.1 kg magnesium sulfate + 0.10 kg zinc sulphatetree^{-1} for Jincheng orange in China (Srivastava and Singh, 2008b, Srivastava and Mousavi, 2024). Such examples testify that such information in other fruit crops if analysed provide very/little clues when spaced over different agroclimates and fitted under various land use systems.

The studies carried out world over have, therefore, shown some diversity in optimum doses of micronutrients standardized through long term field experiments. These include: Fe citrate (2.6- 6 mg kg^{-1}) + $MnSO_4$(1.3-3 mg kg^{-1}) for Satsuma mandarin, Fe + Mn + Zn-EDTA (292 g + 292 g + 315 g ha^{-1}) for Valencia orange, $MnSO_4$ (483 kg tree^{-1}) + $ZnSO_4$ (303.8 g tree^{-1}) for Valencia orange, and Zn-EDTA (30 gtree^{-1}) for Washington navel orange. The combination of two methods is also often used consisting: $ZnSO_4$+ K_2SO_4(0.5%-foliar spray) + K_2O as K_2SO_4(210 g tree^{-1}-soil application) for Kinnow mandarin and $ZnSO_4$ + $FeSO_4$ + $MnSO_4$(50 g tree^{-1} each - soil application)+ (0.50% - foliar application) for Sathgudi sweet orange. A complete fertilizers schedule using inorganic fer tilizers has been developed for three major citrus cultivars (Nagpur mandarin acid lime and sweet orange) of India (Table 8.3). Has been developed and complemented for advisory to citrus growers. Of late, there has been good success in development of customized micronutrient mixture and long term field evaluation with citrus as test crops (Srivastava and Pandey, 2021).

For many years, several authors have tested the response of different crops to application of different nutrients, especially K, with respect to yield and quality, in many crops like coffee, *Coffea arabica* L. (Mancuso *et al.*, 2014); almond, *Amygdalus communis* L. (Reídel *et al.*, 2004); pistachio, *Pistacio vera* L. (Zeng *et al.*, 2001); pecan, *Carya illinoinensis* Koch (Worley, 1994); olive, *Olea europaea* (Jasrotia *et al.*, 1994) *etc.* However, such basal fertilizer application technique is greatly conditioned by different soil properties, particularly soil moisture, which affects the mobility of the supplied nutrients. This is mainly attributed to large variation in fertilizer doses to be really effective in different crops, annual versus perennial. Such variation in optimum doses is dictated by climate, soil types, crops, and farming practices in such a way that the correct balance of nutrients necessary for one farm, may be quite different from that necessary for a farm somewhere else in the world (Sarangthem *et al.*, 2014). Therefore, determining the appropriate balance of nutrients to increase crop yield and soil fertility will require localized research.

An optimum fertilize dose once worked out through long-term fertilizer response experiment (Table 8.3) for a crop must be subjected to periodical assessment for their crop response studies, instead of assuming it as one time exercise. This has been the major drawback in our fertilizer recommendation system, thereby denying the expected magnitude of crop response.

Table 8.3: Fertilizer Schedule for different Citrus Cultivars grown in Central India (Results of long-term trails)

Fertilizer source	*Total Fertilizer (g tree^{-1} year^{-1})*	*Soil Application*					
		Ambia (g plant^{-1})			*Mrig (g plant^{-1})*		
		Apr.	*Aug.*	*Nov.*	*Sept.*	*Nov.*	*Jan.*
Nagpur mandarin *Citrus reticulata* Blanco)							
Urea	1300	433	433	434	433	433	434
Single super phosphate	1260	630	630	-	630	630	-
Muriate of potash	180	-	-	180	-	-	180
$FeSO_4$	200	100	100	-	100	100	-
$MnSO_4$	200	100	100	-	100	100	-
$ZnSO_4$	200	100	100	-	100	100	-
Acid lime (*Citrus aurantifolia* Swingle)							
Urea	1740	580	580	580	580	580	580
Single super phosphate	1260	630	630	-	630	630	-
Muriate of potash	180	-	-	180	-	-	180
$FeSO_4$	200	100	100	-	100	100	-
$MnSO_4$	200	100	100	-	100	100	-
$ZnSO_4$	200	100	100	-	100	100	-
'Mosambi' sweet orange (Citrus sinensis Osbeck)							
Urea	1740	580	580	580	580	580	580
Single super phosphate	1260	630	630		630	630	-
Muriate of potash	180	-	-	180	-	-	180
$FeSO_4$	300	150	150	-	150	150	-
$MnSO_4$	300	150	150	-	150	150	-
$ZnSO_4$	300	150	150	-	150	150	-
Fertilizer Source	*Total Fertilizer (g tree^{-1} year^{-1})*	*Foliar Application*					
		Ambia (g plant^{-1})			*Mrig (g plant^{-1})*		
		Apr.	*Aug.*	*Nov.*	*Sept.*	*Nov.*	*Jan.*
Nagpur mandarin (*Citrus reticulata* Blanco)							
Urea	1300	-	-	-	-	-	-
Single super phosphate	1260	-	-	-	-	-	-
Muriate of potash	180	-	-	-	-	-	-
$FeSO_4$	200	0.5 per cent	-	0.5 per cent	0.5 per cent	-	0.5 per cent

Fertilizer Source	*Total Fertilizer (g tree^{-1} year^{-1})*	*Foliar Application*					
		Ambia (g plant^{-1})			*Mrig (g plant^{-1})*		
		Apr.	*Aug.*	*Nov.*	*Sept.*	*Nov.*	*Jan.*
$MnSO_4$	200	0.5 per cent	-	0.5 per cent	0.5 per cent	-	0.5 per cent
$ZnSO_4$	200	0.5 per cent	-	0.5 per cent	0.5 per cent	-	0.5 per cent
Acid lime (*Citrus aurantifolia* Swingle)							
Urea	1740	-	-	-	-		-
Single super phosphate	1260	-	-	-	-	-	-
Muriate of potash	180	-	-	-	-	-	-
$FeSO_4$	200	0.5 per cent	-	0.5 per cent	0.5 per cent	-	0.5 per cent
$MnSO_4$	200	0.5 per cent	-	0.5 per cent	0.5 per cent	-	0.5 per cent
$ZnSO_4$	200	0.5 per cent	-	0.5 per cent	0.5 per cent	-	0.5 per cent
'Mosambi' Sweet orange (*Citrus sinensis* Osbeck)							
Urea	1740	-	-	-	-	-	-
Single super phosphate	1260	-	-	-	-	-	-
Muriate of potash	180	-	-	-	-	-	-
$FeSO_4$	300	0.5 per cent	-	0.5 per cent	0.5 per cent	-	0.5 per cent
$MnSO_4$	300	0.5 per cent	-	0.5 per cent	0.5 per cent	-	0.5 per cent
$ZnSO_4$	300	0.5 per cent	-	0.5 per cent	0.5 per cent	-	0.5 per cent

In order to prepare 0.5 per cent of $FeSO_4$ solution, dissolve 500 g $FeSO_4$ in 100 liters of water and spray on the plant till drench, likewise prepare the solutions of other micronutrients

Caution: The solution of $FeSO_4$ and $MnSO_4$ can be mixed together, but avoid applying $ZnSO_4$ along with $FeSO_4$ and $MnSO_4$

Source : Srivastava *et al.* (2017; 2021).

Fertigation

Fertigation (application of nutrients through the irrigation using principles of 4R for fertilizers and 4R for irrigation water to be combined together) has still produced better results in improving the tree growth, fruit yield, quality, the reserve pool of soil nutrients, and consequently, the plant nutritional status. Besides the mobility of nutrients, fertigation has several advantages over broadcast application of granular fertilizers with respect to effective placement of nutrients (Shirgure *et al.,* 2021) and flexibility in application frequency, development of uniform root distribution, an important pre-requisite for better fertilizer-use-efficiency, and improvement in fruit quality (Srivastava *et al.,* 2003; Srivastava and Singh, 2016; Jeyabaskaran *et al.,*2021). The importance of ground coverage of orchard floor by fertigation reported that the treatment having 37 per cent coverage of ground and 82 per cent of canopy area produced fruit yield higher than the broadcast fertilizer treatment covering 100 per cent of soil surface and 53 per cent canopy area (Shirgure *et al.,* 2003a; 2003b). Other studies in central India showed far superior results with fertilizers applied through drip irrigation (fertigation) over basal fertilizer application using basin/flood irrigation. Irrigation at 20 per cent depletion of available water content combined with fertilizer treatment of 500 g N + 140 g P + 70 g K tree^{-1}year^{-1} produced a significantly higher magnitude of fruit yield m^{-3} of canopy in addition to higher nutrient status and fruit quality parameters in 14-year-old Nagpur mandarin (*Citrus reticulata* Blanco) on an alkaline calcareous Lithic Ustochrept soil type (Shirgure *et al.,* 2001a; 2001b; 2001c).

Large number of studies have, however, demonstrated much better fertilizer-use-efficiency with fertigation in crops like guava (Ramniwas *et al.,* 2012), banana (Reddy *et al.,* 2017), apple (Banyal and Sharma, 2011), kiwifruit (Chauhan and Chandel, 2008), sweet cherry (Ahmed *et al.,* 2010), litchi (Dey *et al.,* 2010), sapota (Khot *et al.,* 2012), mango (Singh *et al.,* 2009), banana (Pawar *et al.,* 2014), citrus (Srivastava *et al.,* 2014, Shirgure *et al.,* 2001a; 2001b; 2001c), pomegranate (Haneef *et al.,* 2014) *etc.* using various bases of drip irrigation scheduling and NPK-based fertilizers, but without much success with micronutrient fertigation. These studies have provided a wealth of information with unanimous result that fertigation reduced both irrigation and nutrient requirement by 30-50 per cent compared to conventional split application within plant basin. Open hydroponics (Shirgure *et al.,* 2000; Rubio-Asensio *et al.,* 2020; Velazquez-Gonzalez *et al.,* 2022) and variable rate fertilizer linked site specific nutrient management (Johnston *et al.,* 2009; Srivastava *et al.,* 2014) in fruit crops like citrus, olive, avocado, coconut *etc.* have also started showing their utility in improving fertilizers use efficiency to various dimensions, using a variety of microbial inoculation, collectively called bio-fertigation.

Site Specific Nutrient Management

It is a dynamic concept based on plant nutrient uptake, however finds application using nutrient removal pattern by fruits as an alternative to conventionally used nutrient uptake in field crops. The concept advocates that every time, a crop is grown, all the nutrients should be applied in a particular proportion. Rather fertilizer application should be tailored according to the crops' need keeping in view the capacity of these soils to fulfill various demands. To achieve this, it is necessary to keep an overall nutrient balance in relation to total crop load (Srivastava and Singh, 2016). This may indicate the need for the application of different nutrients at specific times, in a particular order to derive the maximum benefit from the application of a given quantity of nutrients.

However, findings from the long term fertilizer trials revealed that: i. intensive cropping with only N input is a short-lived phenomenon, ii. omission of limiting macro- or micro-nutrient leads to its progressive deficiency due to heavy removals, iii.sites initially well supplied with P, K or S become deficient when continuously cropped using N alone, and iv. fertilizer rates considered optimum still resulted in nutrient depletion at high productivity levels, if continued, become sub-optimum rates (Srivastava *et al.,* 2014).

Tailoring of fertilizer requirement*:* Exploiting soil spatial variability is a pre-requisite to tailor fertilizer requirement within an orchard using variable rate application technique. In this regard, attempts were made to tailor fertilizer requirement of citrus without increasing the total dose of fertilizers.In a long term experiment on evaluation of differential fertilizer treatments on two contrasting soil types in an orchard showed that with the same level of NPK, application of micronutrients (Zn + Fe + Mn) produced comparatively higher increase in canopy volume and fruit yield on shallow soil (Typic Ustorthent) than deep soil (Typic Haplustert). However, the magnitude of response of micronutrients on fruit yield was almost similar at 600 g N + 400 g P + 600 g K tree^{-1} versus 1200 g N + 600 g P + 600 g K tree^{-1}. These two soil types showed a differential behaviour with respect to fruit quality as well. Higher application of K at the rate of 900 g tree^{-1} alongwith 600 g N+ 500 g P tree^{-1} produced much higher acidity on Typic Haplustert which developed more green color of fruits, thereby, took comparatively longer time for the color break and attain harvest maturity (Srivastava *et al.,* 2006). The effect of K was more pronounced in combination with 600 g N + 400 g P tree^{-1} than with 1200 g N+ 600 g P tree^{-1} irrespective of soil type. The fruits on Typic Ustorthent attained an early maturity compared to those on Typic Haplustert. Increasing the application of K from 0 to 900 g tree^{-1} with 600 g N + 400 g P tree^{-1} or K from 600 to 1500 g tree^{-1} with 1200 g N + 600 g P tree^{-1} produced concurrently a higher Zn concentration in leaf and greater available Zn in both the soils. Such a synergism between Zn and K can be befittingly exploited with the objectivity of FUE vis-à-vis quality citrus production (Srivastava and Singh, 2013b). These results suggested that the same orchard could produce a differential response of fertilization, and fertilizer application needs to be tailored even within an orchard instead of practising umbrella kind of recommendation.

Organic Nutrient Supply

Organic nutrient supply compared instant mode of inorganic nutrient supply is considered relatively slow nutrient responsive strategy. For example, growing citrus organically is considered as the modern day necessity. We have developed an effective protocol of organic citrus. The application of vermicompost loaded with microbial consortium (100 per cent N-equivalent basis) + IPM_2 (foliar application of Horticulture Mineral oil (2%) followed by *Beauveria bassiana* @ 5g l^{-1} and Azadirachtin (1%) @ 4 ml l^{-1}) + IDM_1 – Bordeaux paste ($CuSO_4$: Lime : Water = 1:1:10) as pre monsoon/post monsoon trunk application along with *Trichoderma harzianum*native antagonistic strain, NRCfBA44 (100 g $plant^{-1}$) with carrier material of FYM (1kg) as soil application at root zone) recorded maximum soil and plant nutrients, plant height, canopy volume, fruit yield, number of fruits, fruit weight and lower incidence of insect pests and disease as compared to other organic treatments (Table 8.4). Organic nutrient supply through liquid formulation has also proved to be effective in sustaining the plant health through combating biotic and abiotic stresses and maintain the microbial niche of rhizosphere (Solanki *et al.,* 2015). The organic culture of fruit crops has consistently not been able to provide that expected magnitude of success in fruit crops (tropical fruits or temperate fruits, despite far better suitability of such fruit crops under organic farming), unless developed on on-farm principles. The half-life period of carbon (called carbon recalcitrance) added through different organic manures/composts is also far less (2-3 years only an dcarbon deposited in acyive pool of soil organic carbon stock) compared to more prudent sources like mycorrhizas (glycoproteins added through mycorrhizal hyphal networking, 10-15 years) or biochars (75-100 years, adding carbon into passive pool of soil organic carbon). The studies on these lines would surely perch organic culture of fruits on a more sound footing. Until then, sustainability of organic culture of fruits would continue to be under big question mark.

Table 8.4: Scheduling of different Components Involved in Organic Management of Nagpur Mandarin

Components	*Dose*	*Time of Application*
Nutrient Management		
Vermicompost	15 kg $tree^{-1}$	Three splits with each split containing 5 kg $tree^{-1}$ vermicompost to be applied in the month of April, August and November
Microbial consortium	150 ml $tree^{-1}$	Three splits with each split having 50 ml microbial consortium $tree^{-1}$ in the months of April, August and November
Insect Pest Management		
Horticultural mineral oil (HMO)	2 per cent (20 ml L^{-1})	Application at the time of new flush emergence
Beauveria bassiana (BB)	5 g L^{-1}	Application after 15 days of new flush emergence

Components	Dose	Time of Application
Azadirachtin	1 per cent (4 ml L^{-1})	Application at new flush emergence and may be repeated at 10-15 days interval
Disease Management		
Bordeaux paste	$CuSO_4$: Lime: water in 1:1:10	Pre-monsoon (middle of June) and post-monsoon (first week of October) application
Trichoderma (NRCFBA44) inoculation	100 g $plant^{-1}$	100g culture mixed with 1kg vermicompost FYM and to be applied in July and second application in the month of September

Source: Srivastava *et al.* (2002; 2008).

Specialty Fertilizer, a Tool for Site Specific Nutrient Management (SSNM)

Conventionally designed long term fertilizer trials revealed that: i. omission of limiting macro- or micronutrients led to their progressive deficiencies due to heavy removals; ii. Sites initially well supplied with P, K or S become deficient when continuously cropped using N alone and iii. Fertilizer rates considered optimum still resulted in nutrient depletion at higher productivity levels, if continued, become sub-optimum rates (Srivastava and Pandey, 2021; Dong *et al.,* 2023). There is a strong necessity to keep overall nutrient balance in relation to total crop load (Tiwari 2007b). Most of the existing fertilization guidelines have been derived from experimental data on yield responses to individual nutrients after controlling for other soil conditions by way of randomization. But, soil conditions in farm fields, former being more heterogeneous in soil fertility, is distinctly very different from those in the experimental fields, and soil data collected from a farm fields cannot be analyzed using the statistical method developed for experimental data (Srivastava *et al.,* 2017).

Knowing the nutrients required for all stages of growth, and understanding the soil's ability to supply those needed nutrients, are critical to profitable crop production. The recommendations on fertilizer application may not, however, produce the same magnitude of results when practiced on an orchard of large area, because of its inability to accommodate large spatial variation in soil fertility (Srivastav *et al.,* 2006). Application of uniform single rate of nutrients, hence, most often result in over-application of nutrients at some sites and under-application at other sites, cumulatively leading to reduced fertilizer use efficiency (Srivastava *et al.,* 2009; 2014). Under such circumstances, SSNM as a dynamic concept of nutrient management exploits indigenously and spatially available nutrients within bigger orchards requiring differential fertilizer treatments in patches so as to improve the orchard efficiency (average yield of specified trees in relation to average orchard yield) in ultimate terms. The success of SSNM during the last 15- years has been prominently realized on a number of crops *viz.*, cereals, black gram, (Biradar *et al.*, 2006), black gram (Gupta *et al.*, 2007) coconut intercropped with fodder (Lakshmi *et al.*, 2007), avocado (Salaza-Garcia and Lazcano-Ferrat 2003), citrus (Srivastava *et al.*, 2006, 2009) *etc.* to cite few success stories.

Färe *et al.* (2012) introduced a model to answer three questions pertaining to SSNM in production of perennial crops: which input factors of crop production are limiting yield; what action should be taken to remove the limiting factors; and what is the potential gain in revenue from taking the action. The suggested model captured the essence of the law of the minimum in yield and revenue increase only if the limiting nutrients are appropriately adjusted. In the past, various aspects of SSNM have been highlighted in reviews (Nielsen *et al.*, 1973, Wibawa *et al.*, 1993, Pinter *et al.*, 2003, Corwin and Lesch 2005, Elprince 2009, Johnston *et al.*, 2009, Zhang *et al.*, 2010, Chan 2012, Srivastava 2013c), a number of reviews, but there is hardly any comprehensive coverage addressing the perennial crops. In this background, the present review attempts to highlight certain pertinent issues on the necessity of developing SSNM concept as rationale of fertilizer use to harness better fertilizer use efficiency with emphasis on perennial crops.

SSNM is usually carried out with the objectives: i. identification and quantification of the variability of soil physical and chemical properties; ii. understanding the impact of soil variability on crop growth, yield profitability and iii. management of soil variability to improve production, profits, and reduce environmental impact (Srivastava *et al.*, 2010). The SSNM aims to apply nutrients at optimal rates and times to achieve high yield and yield efficiency of nutrient use leading to high economic return per unit of fertilizer invested (Dobermann *et al.*, 2003a, 2003b). The SSNM also adjusts the fertilizer use to optimally bridge up the deficit between the nutrient needs of a crop and the nutrient supply from naturally occurring indigenous sources such as soil, organic amendment, crop residues, manures and irrigation water (De La Cruz 2008). There are five steps through which the SSNM is accomplished. These are: i. establishment of yield target; ii. estimation of actual yield responses to fertilizer N, P, and K; iii. selection of fertilizer N, P, and K rates based on expected yield responses to fertilizer application considering agronomic efficiencies and nutrient balances; iv. application of fertilizer to meet the crop demand for nutrients at critical growth stages and v. optimization nutrient use (Hach and Tan, 2007) efficiencies. The following steps are involved in arriving at fertilizer requirement based on the principles of SSNM (Srivastava 2013c): step i.: estimating target yield; step ii. estimating nutrient requirement to get target yield, step iii. estimating indigenous nutrient supply from soil (based on our earlier studies dealing with progressive nutrient response experiment), step iv : calculating fertilizer requirement. Working out the nutrient requirement : defined as the amount of nutrients added for producing the target yield minus the amount of indigenous nutrient (soil and other sources)and fertilizer recovery (%) defined as the percentage of nutrients absorbed by a crop out of the total amount of fertilizers applied

Future gains in productivity and input- use- efficiency will require soil and crop management technologies that are knowledge-intensive, and are tailored to specific characteristics of individual farms or fields to manage the variability that exists between and within them (Tiwari, 2007a). The SSNM approach is one such option that has been tried successfully in India using different approaches. A plant-based SSNM approach developed in the early 1990s by the International Rice Research Institute in collaboration with partners has been adopted by IPNI, India. The scientific principles of SSNM were compiled into a practical into a practical guidebook to nutrient management for rice (Dobermann *et al.,* 2003a, 2003b). Large variation in tree canopy and subsequently, the tree-to-tree yield difference are common in many of the large sized citrus orchards. Knowing the required nutrients for all stages of growth, and understanding the soil's ability to supply those needed nutrients are critical to profitable crop production. The recommendations on fertilizer application may not, however, produce the same magnitude of yield response when practised in an orchard of large area, because of its inability to accommodate variation in soil fertility status.

In subsequent sections, we discuss the principles and strategies that could be effectively used developing customized fertilizers use strategy, leading to development of specialty fertilizers. We developed a specialty micronutrient mixtures for Nagpur mandarin grown in central India and evaluated during 2012-20 in Vertic Ustochrept Soil facing multiple nutrient deficiencies in form of N, P,Fe, MN, Zn and B. The treatments involving a total of four treatments *viz.*,$T_{1:}$ Micronutrient mixture(Citrus special, IIHR Bangalore, T_2 : Macronutrients (Soil application) + Micronutrients (Foliar application); T_3: Micronutrient mixture-1(Soil application) + Macronutrients (Soil application) and T_4 : Micronutrient mixture-2 (Soil application) + Macronutrients (Soil application) were tested in field with 6 replications. The responses were evaluated in terms of vegetative growth, changes in soil fertility status and nutrient composition of index leaves to identify the best treatment during the fifth year of evaluation. The results obtained are as follows:

Response on Vegetative Growth and Quality

Vegetative growth response using different growth contributing parameters were evaluated and expressed as canopy volume, responded significantly to the different treatments, as in preceding years (Figure 8.1). Among all the treatments, the most striking increase in canopy volume over 2018-19 was observed with the treatment T_4 (0.96 m^3) followed by the treatment T_3 (0.92 m^3), T_2 (0.85 m^3) and T_1 (0.46m^3). These responses in canopy volume suggested the superiority of both the micronutrient mixtures (T_3 and T_4) over conventionally used fertilizer (T_1 and T_2) application. Similarly, the yield response with treatments T_4 (58.2 kg tree or 16.2 tons/ha) and T_3 (46.3 kg/tree or 12.9 tons/ha) were of much higher magnitude over T_2 (42.4 kg/tree or 11.8 tons/ha) and T_1 (40.4 kg/tree or 11.1 tons/ha), suggesting that our prepared mixtures have out-yielded the conventionally used fertilizer doses.

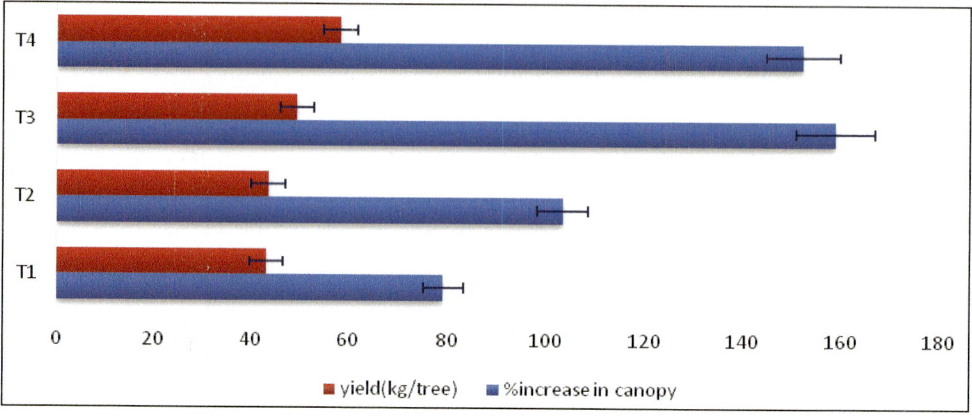

Figure 8.1: Response of different Micronutrient Mixtures on Changes in Canopy Volume and Fruit Yield of Nagpur Mandarin (2007-2019)

T_1 – Micronutrient mixture (Citrus special, IIHR Bangalore); T_2 -Macronutrients (Soil application) + Micronutrients (Foliar application; T_3- Micronutrient mixture-1(Soil application) + Macronutrients (Soil application); T4- Micronutrient mixture-2 (Soil application) + Macronutrients (Soil application).

Changes in Soil Fertility and Plant Nutrition

The soil fertility and plant nutrition was significantly available nutrients, with an exception of DTPA-Cu a affected by different treatments (Figure 8.5). Amongst plant available nutrients, with the exception of DTPA-Cu, all the nutrients *viz.*, $KMnO_4$-N, Olsen-P, NH_4OAc-K, DTPA-Fe, DTPA-Mn, and DTPA-Zn showed a signification response of different treatments. The treatments T_4 observed significantly higher plant available macronutrient than treatment

T_3 compared to treatments $T_{2.}$ These observations suggested the far superior response of micronutrient mixtures (Treatment T_3 and T_4) than treatments involving T_1 and $T_{2.}$ The treatment T_4 and T_3 registered a significantly higher plant available micronutrient than T_2 and T_1. Similar responses were observed in leaf nutrient status, where T_4 treatment showing the best response over T_3, T_2 or T_1. Our observations strongly lend support shortly support towards better efficacy of micronutrient mixtures compared to conventionally used fertilizers (Fig 2) Hence our long term studies showed that micronutrient mixture developed by simply mixing straight micronutrient fertilizers such as ferrous sulphate, manganese sulphate, zinc sulphate and borax in ratio of 1:1:0.50:0.20 that stoichiometrically coincide with the similar concentration ratio of micronutrients present in fruits. Interesting, our developed micronutrient mixture at the rate of 100g/plant overtaken the response of these micronutrients when applied at the rate of 100g ferrous sulphate, 100g manganese sulphate, 100g zinc sulphate and 50g borax/ plant/year (total micronutrients at the rate of 350g/plant proved far inferior over 100g customized micronutrients mixture).

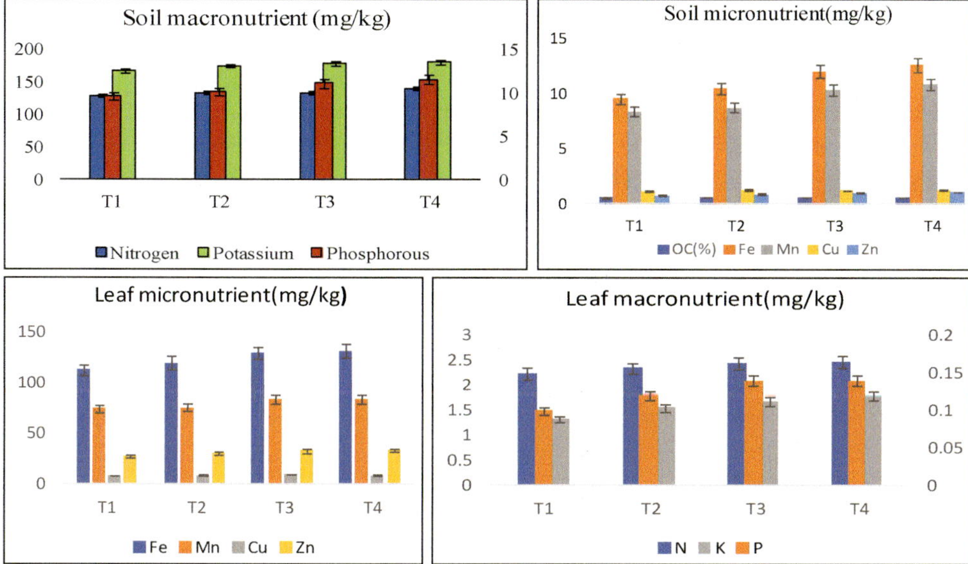

Figure 8.2: Response of different Micronutrient Mixtures on Changes in Plant Available Nutrients and Leaf Nutrient Composition of Nagpur Mandarin (Pooled data : 2007-2019)

T_1 - Micronutrient mixture (Citrus special, IIHR Bangalore); T_2 -Macronutrients (Soil application) + Micronutrients (Foliar application; T_3- Micronutrient mixture-1(Soil application) + Macronutrients (Soil application); T4- Micronutrient mixture-2 (Soil application) + Macronutrients (Soil application).

The fall out of a generalized fertilizer s recommendation over large areas of small-scale farming very frequently leads to the possibility of over or under-application of nutrients, by and large, coupled with its economic and environmental consequences (Srivastava *et al.,* 2009). The more apparent consequences of falling productivity and nutrient efficiency, multi-nutrient deficiencies, increasing pace of nutrient mining, and falling farm income are highlighted by earlier researchers (Srivastava *et al.,* 2006). The environmental impacts are not very apparent yet, probably because of the generally low nutrient application rates, except few crops of very high commercial importance. The SSNM using crop-customized fertilizers on the other hand, is an approach for feeding crops with nutrients, as and when they are needed, considering inherent spatial variability associated with fields under crop production. The most important step towards the calibration of site specific fertilizer requirement is the estimation of the indigenous nutrient supplies, which we define as the cumulative amount of nutrient originating from all indigenous sources (Srivastava and Das, 2017). The major challenges for SSNM research and extension in future will be two-folds: i. to retain the demonstrated potential of the approach and ii. to build upon what has already been achieved while reducing the complexity of the technologies as it is disseminated to farmers (Johnston *et al.,* 2009). However, Zhang *et al.* (2010) suggested that the regionalized maps are practical alternative to site-specific soil nutrient management approaches in areas where it is not practical to implement SSNM due to small field size or other constraints to use intensive soil sampling and chemical analyses. These strategies would go a long way in offering some fruitful solution in mitigating multiple nutrient deficiencies using specialty fertilizers, with an eye on the soil agroecology playing a pivotal role in ensuring a sustained productivity in the backdrop of climate change as well (Srivastava *et al.,* 2010).

Optimising Integrated Nutrient Supply

The long-term experiments conducted around the world indicated that chemical fertilizer alone is not enough to improve or maintain soil fertility at high levels and the soil acidification problem caused by over-application of synthetic N fertilizers can be reduced if more fertilizer N is applied as NO3 relative to ammonium- or urea-based N fertilizers (Pathak and Nedwell, 2011). Organic fertilizers can improve soil fertility and quality, but long-term application with high rates can also lead to more nitrate leaching, and accumulation of P, if not managed well. Well- managed combination of chemical and organic fertilizers can overcome the disadvantages of applying single source of fertilizers and substantially achieve higher crop yields, improve soil fertility, alleviate soil acidification problems and increase nutrient-use efficiency compared with only using chemical fertilizers (Miao *et al.,* 2011). Organic amendments comprising manures have helped in increasing the fruit yield coupled with quality, effectively replacing mineral fertilizers in the nutrient management of commercial fruit tree orchards through associated changes in soil C pool, microbial pool and available nutrient pool of soil (Baldi *et al.,* 2010; Montanaro *et al.,* 2010; 2016). Nutrient (nutrient as well as inorganic source)-microbe as tripartite association on the other hand is better known in fruit crops (Singh and Banik, 2011; Srivastava, 2020; 2023a;2023b; Srivastava and Ngullie, 2009; Khehra and Bal, 145) with respect to both agronomic response as well as soil health. An array of fruit crops have been reported to respond to the synergies originated through combination of organic nutrient- microbe-inorganics (Table 8.5). And such associations have invariably witnessed substantially higher productivity than any single component alone. However, there is a greater need to expand such plant response advantages using more rhizocompetent microbes preferably in consortium mode, plant response as well as soil health response both have to be sustained on a long term basis (Srivastava and Kohli, 1999; Srivastava, 2020).

Our long term study entitled "Integrated nutrient management in relation of sustained quality production of Nagpur mandarin (*Citrus reticulata* Blanco) was evaluated under field condition during 2007-16) initiated in 2007 with a total of five treatments *viz.,* Module I (T_1)- 100 per cent RDF (600 g N – 200 g P – 300 g K - 200 g $ZnSO_4$ – 200 g $FeSO_4$ – 200 g $MnSO_4$/tree/year), Module II(T_2) – 75 per cent RDF + 25 per cent Vermicompost, Module III (T_3)-75 per cent RDF + 25 per cent Vermicompost + Microbial Consortium, Module IV(T_4)-50 per cent RDF +

50 per cent Vermicompost, and Module V (T_5) – 50 per cent RDF + 50 per cent Vermicompost + Microbial Consortium. These modules were evaluated in terms of growth parameters, fruit yield, fruit quality (including micronutrient composition of fruit juice), plant available pool of nutrients in soil, soil carbon stock, soil microbial load, CO_2 emission rate and spectral properties of soil on a black clay soil (Typic Ustocherpt). The pooled results obtained during 2007-16 (10 years) are summarized (Table 8.6).

Table 8.5: Different Components of Integrated Nutrient Management Recommended for different Fruit Crops

Crop	*Nutrient-Microbe Combination*	*References*
Acid lime (*Citrus aurantifolia Swingle*)	50 per cent RDF + 50 per cent through goat manure	Lal and Dayal (2014)
Aonla (*Emblica officinalis* Gaertn.)	50 per cent NPKS (105 kg N – 7.20 kg P_2O_5 – 125.25 kg K_2O/ha) – Biofertilizers (*Azotobacter sp* – *Azospirillum sp* – Phosphate solubilising bacteria) – FYM (2 tons/ha)	Yadav *et al.* (2007)
Aonla (*Emblica officinalis* Gaertn.)	100 g N – 25 g P_2O_5 – 150 g K_2O/plant – FYM 10 kg/plant – Phosphate solubilising bacteria 50 g/plant	Mandal *et al.* (2013)
Aonla (*Emblica officinalis* Gaertn.)	Azotobacter + 50 per cent N through Urea + 50 per cent through FYM	Kumar *et al.* (2017)
Apple (*Malus domestica*)	70 kg/tree FYM + 19 kg/tree vermicompost + 165g/tree N + 178 g/tree P_2O_5 + 548 g/tree K_2O	Das *et al.* (2016)
Apricot (*Prunus armeniaca* L.)	50 per cent N (CN) + 50 per cent N (Urea) + Azotobacter + PSB + Vermicompost and 0 per cent N (CN) + 50 per cent N (Urea) + Azotobacter + PSB + Vermicompost	Kumar *et al.* (2010)
Apricot (*Prunus armeniaca* (Linn.)	75 per cent RDF – 25 per cent FYM	Shah *et al.* (2014)
Banana (*Musa acuminata* L.)	FYM 12 kg/plant – *Azospirillum sp* 50 g/plant - Phosphate Solubilising Bacteria 50 g/plant *T.harzianum* 50 g/plant	Hazarika and Ansari (2010)
Banana (*Musa acuminata* L.)	50 per cent RDF- FYM 20 kg/plant – *Azotobacter sp* 50 g/plant – Phosphate solubilising bacteria 50 g/plant –VAM 250 g/plant	Patil and Shinde (2013)
Banana (*Musa acuminata* L.)	100 per cent RDF – 40 per cent Wellgrow organic manure	Kuttimani *et al.* (2013)
Banana (*Musa acuminata* L.)	100 per cent RDF – FYM 10 kg/plant – *Azospirillum sp* 25 g/pant Phosphate solubilising bacteria 250 g/plant	Bhalerao *et al.* (2009)
Guava (*Psidium guajava* L.)	FYM 50 kg/plant – *Azotobacter sp* 50 g/plant - *Azospirillum sp* 50 g/plant – *Sesbania sp* as green manure	Ram and Rajput (2000)

Crop	*Nutrient-Microbe Combination*	*References*
Guava (*Psidium guajava* L.)	488 g N – 244 g P_2O_5 - 281 g K_2O/plant – FYM 50 kg/plant – *Azotobacter* 250 g/plant – phosphate solubilising bacteria 25 g/plant	Barne *et al.* (2011)
Guava (*Psidium guajava* L.)	50 per cent RDF (250 g N – 100 g P_2O_5 - 250 K_2O g/plant) - FYM 25 kg/plant – vermicompost 5 kg/plant	Dwivedi (2013)
Guava (*Psidium guajava* L.)	236 g N – 66 g P_2O_5 – *Azospirillum sp* 30 g/plant – VAM 30 g/plant	Dutta *et al.* (2009)
Guava (*Psidium guajava* L.)	250 g N – 100 g P – 250 g K_2O/plant – *Azotobacter sp* 250 g/plant	Shukla *et al.* (2009)
Guava (*Psidium guajava* L.)	50 per cent RDF (225 g N – 195 g P_2O_5 – 150 g K_2O/plant)- FYM 50 kg/plant – *Azospirillum* 250 g/plant	Goswami *et al.* (2012)
Guava (*Psidium guajava* L.)	2/3 of T_1+25 kg FYM+250g Azospirillum+250g Azotobacter/Plant	Meena *et al.* (2017)
Lemon (*Citrus limon* L.)Burm.f.)	N 525 g/plant – FYM 150 kg/plant – *Azotobacter sp* 18 g/plant	Khehra and Bal (2014)
Litchi (*Litchi chinensis* Sonn.)	500 g N – 250 g P_2O_5 – 500 g K_2O/plant – FYM 50 kg/plant – *Azotobacter sp* 150 g/plant – VAM 100 g/plant	Dutta *et al.* (2010)
Mango (*Mangifera indica* L.)	500 g N - 250 g P_2O_5 – 250 K_2O g/plant –50 kg FYM/plant – *Azospirillum sp* 250 g/plant	Singh and Banik (2011)
Mango (*Mangifera indica* L.)	250 N – 425 P_2O_5 – 1000 K_2O – *Azospirillum sp* 250 g/plant – PSB – 250 g/plant – $ZnSO_4$ 100 g/ plant – Borax 100 g/plant	Hasan *et al.* (2013)
Mango (*Mangifera indica* L.)	NPK_{100}+VAM+ Azotobacter	Kumari *et al.* (2019)
Mosambi (*Citrus sinensis* Osbeck)	300 g N – 250 g P_2O_5 – 300 g K_2O – AMF 10 g/ plant - *Azospirillum sp* 25 g/plant	Patel *et al.* (2009)
Papaya (*Caria papaya* L.)	50 per cent RDF (100 N – 100 P_2O_5 – 125 K_2O g/ plant)- *Azotobacter sp* 50 g/plant – Phosphate solubilising bacteria 2.5 g/m^2	Singh and Varu (2013)
Papaya (*Carica papaya L.*) *Cv.* Taiwan Red Lady	25 per cent RDN through bio compost + 25 per cent RDN through castor cake + 50 per cent RDN through inorganic fertilize	Tandel *et al.* (2014)
Papaya (*Caria papaya* L.)	Vermicompost 20 kg/plant – rhizosphere culture 50 g/plant- 150 N – 200 P_2O_5 – 200 K_2O g/plant (75 per cent RDF)	Kirad *et al.* (2010)
Peach (*Prunus persica* (L.)Stokes)	75 per cent RDF - 25 per cent N equivalent FYM	Shah *et al.* (2014)
Peach (*Prunus persica* Batsch)	75 per cent RDF +VM 15kg	Solanki *et al.* (2020)

Crop	*Nutrient-Microbe Combination*	*References*
Plum (*cv. Kala Amritsari*)	75 per cent of N + 12.5 per cent N through vermicompost + 12.5 per cent N through FYM+ biofertilizers.	Solanki *et al.* (2020)
Pomegranate (*Punica granatum* L.)	400 g N- 100 g P_2O_5 – 300 g K_2O/plant – FYM 20 kg/plant,	Ghosh *et al.* (2012b)
Pomegranate (*Punica granatum* L.)	300 g N/plant – neem cake 1 kg/plant	Ray *et al.* (2014)
Pomegranate (*Punica granatum* L.)	10 kg vermicompost + 5 kg neem cake + 25 per cent recommended dose of NPK + 20 g PSB	Meena *et al.* (2019)
Sapota (*Achras zapota* L. Syn)	2/3rd part of RDF (RDF: 1000g: 500g: 500g :: N:P:K) + 50 kg FYM + 250 g Azospirillum + 250 g Azotobacter plant–	Baviskar *et al.* (2011)
Sapota (*Manilkara zapota* (L.) P. Royen)	Vermicompost 12.5 kg tree-1 + RDF (1000 g N + 1000 g P + 1500 g K tree-1) + EM (1:250 dilution)	Sheik *et al.* (2019)
Sapota (*Achras zapota* L.)	75 per cent RDF + 25 per cent RDF equivalent vermicompost	Hebbarai *et al.* (2006)
Strawberry (*Fragaria ananassa* Duches)	75 per cent N as RDF – 25 per cent N as FYM – *Azotobacter sp*	Umer *et al.* (2009)
Sweet Orange (*Citrus sinensis* (L.) Osbeck)	60 per cent nitrogen of recommended dose of fertilizer + 40 per cent organic manure (FYM)	Kumar *et al.* (2021)

RDF and PSB stand for recommended doses of fertilizers and phosphate solubilising bacteria predominantly (*Pseudomonas fluorescens*), respectively

FYM and AMF stand for farmyard manure, and arbuscular mycorrhizal fungi, respectively

Plant Growth, Fruit Yield and Quality Response

All the ISFM-based treatments displayed significant response on changes in canopy volume (Table 8.7). Maximum cumulative increase in canopy volume was observed with module-III (18.67 m^3) followed by module-IV(17.16), Module-V(16.93 m^3), on part with module-II (16.70 m^3) and canopy volume with either module-III or module-V which were better than either module-II or module-IV compared to module-I which comprised of only inorganic fertilizers. Module I- T_1 (11.50 m^3). Incorporation of microbial consortium either with module-III or with Module –Vinvariably induced higher canopy volume suggesting better response on Fruit yield was observed to be significantly affected by different ISFM-based treatments (Table 8).The maximum fruit yield of 88.8 kg/tree was observed with treatment module-V which was better than 83.2 kg/tree with T_4 or 80.5 kg/tree with module-III than 71.0 kg/tree with treatment Module-II However, all these Modules treatments were far superior in magnitude of response when compared with 100 per cent RDF as module-I (67.7 kg/tree). Hence, different ISFM-based modules (71.0-88.8 kg/tree) were much better than exclusive inorganic fertilizer treatment like module (67.7 kg/tree).

Table 8.6: Growth Attributing Parameters in Response to different Vermicompost-based Modules of ISFM (Pooled data :2007-16)

ISFM Modules	*Plant Height (m)*	*Tree spread (m)*		*Canopy Volume (m³)*	*Cumulative Increase in Canopy Volume Over 2007-08 (m³)*
		E-W	*N-S*		
Module I	3.82 (2.02)	2.52 (1.27)	2.50 (1.20)	13.04 (1.54)	11.50
Module II	5.09 (2.10)	2.98 (1.39)	2.89 (1.25)	18.51 (1.81)	16.70
Module III	3.87 (2.13)	3.00 (1.27)	3.22 (1.23)	20.28 (1.61)	18.67
Module IV	3.76 (2.10)	3.73 (1.30)	2.93 (1.30)	19.78 (1.62)	17.16
Module V	3.86 (1.87)	2.97 (1.18)	2.96 (1.22)	18.33 (1.40)	16.93
CD (*P=0.05*)	0.49	0.38	0.51	1.18	1.32

RDF stands for recommended doses of fertilizer (600 g N – 200 g P – 300 g K - 200 g $ZnSO_4$ – 200 g $FeSO_4$ – 200 g $MnSO_4$ tree/year).

Vm stands for vermicompost (Nutrient composition : 2.38 per cent N, 0.09 per cent P, 1.42 per cent K, 1072 ppm Fe, 116 ppm Mn, 39 ppm Cu, and 46 ppm Zn).

MC stands for microbial consortium developed by isolating the native microbes as a (mixture of *Bacillus pseudomycoides* (MF113272), *Acinetobacter radioresistens* (MF113273), *Micrococcus yunnanensis* (MF113274), *Paenibacillus alvei* (MF113275) and *Aspergillus flavus* (MF113270).

Figures in parenthesis indicates the value obtained in 2007-2008.

Source: Srivastava *et al.* (2015b; 2019).

Different fruit quality parameters, except peel thickness (Table 8.7) displayed significant response in relation to different treatments. There was much high fruit weight with different ISFM-based treatments modules I-V 101.4-114.6 g compared to inorganic RDF module-I (101.4), showing that incorporation of both organic manure as well as microbial cultures improved the efficiency of organic fertilizers. Similar observations were obtained with respect to other three fruit quality related parameters such as juice content, acidity and TSS. While on the other hand, acidity observed a significant reduction with those superior treatment, *e.g.* module-III (0.86%) and module-V (0.80%) compared to module-II (0.93%) and module-IV(0.91%) highlighting the favorable changes in different fruit quality changes in response to different ISFM-based treatments.

ISFM: Response on Soil Health

Rational soil use practices must allow economically and environmentally sustainable yields, which will only be reached with the maintenance or recovery of the soil health. Thus, a healthy soil has been defined as "The continued capacity of soil to function as a vital living system, within ecosystem and land-use boundaries, to sustain biological productivity, promote the quality of air and water environments, and maintain plant, animal and human health" (Doran and Safley, 2000). Soil health refers to the ecological equilibrium and the functionality of a soil and its capacity to maintain a well-balanced ecosystem with high biodiversity above and below surface, and productivity (Ngullie *et al.,* 2015).

Table 8.7: Fruit Yield and Quality Parameter in Response to different Vermicompost based Modules of ISFM-Treatments (Pooled data: 2007-16)

ISFM Modules	*Yield (kg/ tree)*	*Fruit Weight (g/fruit)*	*Peel Thick-ness (mm)*	*Fruit Quality Parameters*			
				Juice Content (%)	*TSS (°Brix)*	*Acidity (%)*	*TSS/Acid Ratio*
Module I	67.7	101.4	4.1	40.2	8.4	0.96	8.92
Module II	71.0	104.1	3.0	39.6	8.6	0.93	9.23
Module III	80.5	106.2	3.4	41.5	9.5	0.86	10.05
Module IV	83.2	109.1	3.4	42.7	8.4	0.91	9.56
Module V	88.8	114.6	3.0	44.2	9.3	0.80	11.62
CD(*P=0.05*)	2.1	1.4	0.50	2.1	0.30	0.10	1.10

RDF stands for recommended doses of fertilizer (600 g N – 200 g P – 300 g K - 200 g $ZnSO_4$ – 200 g $FeSO_4$ – 200 g $MnSO_4$ tree/year).

Vm stands for vermicompost(Nutrient composition : 2.38 per cent N, 0.09 per cent P, 1.42 per cent K, 1072 ppm Fe, 116 ppm Mn, 39 ppm Cu, and 46 ppm Zn).

MC stands for microbial consortium developed by isolating the native microbes (mixture of *Bacillus pseudomycoides* (MF113272), *Acinetobacter radioresistens* (MF113273), *Micrococcus yunnanensis* (MF113274), *Paenibacillus alvei* (MF113275) and *Aspergillus flavus* (MF113270).

Source: Srivastava and Mousavi (2024); Srivastava *et al.*(2015a;2015b; 2015c).

To understand and use soil health as a tool for sustainability, physical, chemical, and biological properties must be employed to verify which respond to the soil use and management within a desired timescale. Attributes with a rapid response to natural or anthropogenic actions are considered good indicators of soil health. However, most of soil health-related parameters generally have a slow response, when compared to the biological ones, such as microbial biomass C and N, biodiversity, soil enzymes, soil respiration, *etc.*, in addition to macro- and meso-fauna. Thus, a systemic approach based on different kinds of indicators (physical, chemical and biological) in assessing soil health would be safer than using only one kind of attribute (Cardoso *et al.,* 2013). The soil microbial load in terms total bacterial count and fungal count in response to different ISMF-based treatments showed variable higher soil microbial load with Module III and Module V compared to either Module II or Module IV including I. In our study changes in soil fertility indices with regard to available macro- as well as micronutrients were observed highly significant, but of variable nature in response to different treatments (Table 8.8).

Table 8.8: Changes in Soil Fertility Status in Response to different Modules of ISFM-based Treatments (Pooled data: 2007-16)

ISFM Modules	*Available Nutrients (mg/kg)*						
	Macronutrients			*DTPA-Micronutrients*			
	$KMnO_4$-N	*Olsen-P*	*NH_4O Ac-K*	*Fe*	*Mn*	*Cu*	*Zn*
Module I	140.0	9.32	185.2	10.35	10.54	2.2	0.98
Module II	1435	9.27	196.6	12.57	11.54	2.3	1.00
Module III	155.7	9.15	203.6	15.18	12.45	2.7	1.10
Module IV	161.0	9.57	209.2	16.56	11.67	2.8	1.18
Module V	178.5	9.55	212.6	19.50	12.93	2.5	1.32
CD(*P=0.05*)	6.2	NS	1.3	1.32	1.01	NS	0.46

RDF stands for recommended doses of fertilizer (600 g N – 200 g P – 300 g K - 200 g $ZnSO_4$ – 200 g $FeSO_4$ – 200 g $MnSO_4$/tree/year).

Vm stands for vermicompost (Nutrient composition: 2.38 per cent N, 0.09 per cent P, 1.42 per cent K, 1072 ppm Fe, 116 ppm Mn, 39 ppm Cu, and 46 ppm Zn).

MC stands for microbial consortium developed by isolating the native microbes (mixture of *Bacillus seudomycoides* (MF113272), *Acinetobacter radioresistens* (MF113273), *Micrococcus yunnanensis* (MF113274), *Paenibacillus alvei* (MF113275) and *Aspergillus flavus* (MF113270).

Source: Srivastava *et al.* (2015a; 2015b).

Amongst macronutrients, $KMnO_4$-N and Olsen-P showed significant responses, without any significant response on NH_4OAc-K. The treatments such as module-III (155.7 and 203.6 mg kg^{-1} $KMnO_4$-N and Olsen-P, respectively) and module-V(178.5 and 212.6 mg/kg $KMnO_4$-N and Olsen-P, respectively) were comparatively higher than module-II (193.5 and 196.6 mg kg^{-1} $KMnO_4$-N and Olsen-P, respectively) and module-IV(161.0 and 209.2 mg/kg $KMnO_4$-N and NH_4OAc-K, respectively). These observations suggested that superiority of INM-based treatments than those treatments without microbial consortium. Module-I involving exclusive inorganic fertilizers registered lowest test values as against treatment like module-V registering maximum values, validating the supremacy of those treatments which carry all the three components of ISFM. The soil properties such as soil pH and soil EC were not affected by any of the ISFM-based treatments (Table 8.9). While, amongst different fractions of soil carbon *viz.*, organic-C (SOC), inorganic-C (SiC) and total-C (TC) only SOC and TC were significantly affected. These observations showed that changes in soil carbon stock are more governed by organic fraction than inorganic fraction. Maximum SOC and TC of 7.43 g/kg and 9.14 g/kg were observed with module-V. Likewise, module-III much SOC and TOC than module-II, displaying the significant role of microbial consortium in improving the carbon sink capacity of soil. Different ISFM-based modules were also observed to aid in improving the total soil N stock, being maximum with treatment like module-III-V(0.741-0.748%) compared to rest of the treatments like module-I, module-II or module-IV (0.721-0.738%). However, soil C:N ratio in the range of 12.00-12.32, without displaying significant changes in response to different treatments.

Table 8.9: Soil Carbon Fractions and C:N Ratio in Response to different Modules of ISFM (Pooled data: 2007-16)

ISFM Modules	*pH*	*EC (dS/m)*	*Soil carbon (g/kg)*			*Soil Total N (g/kg)*	*Soil C:N Ratio*	*Soil microbial load (x 10^3 cfu/g)*	
			SOC	*SiC*	*TC*			*Bacterial count*	*Fungal count*
Module I	7.7	0.168	6.61	1.71	8.32	0.721	11.54	26	12
Module II	7.8	0.174	6.70	1.71	8.41	0.732	11.48	36	18
Module III	7.7	0.163	7.02	1.74	8.76	0.741	11.82	54	22
Module IV	7.6	0.173	7.11	1.75	8.86	0.738	12.00	44	26
Module V	7.8	0.174	7.43	1.81	9.24	0.748	12.32	86	42
CD (*P=0.05*)	NS	NS	0.08	NS	0.09	0.006	NS	8.2	4.8

RDF stands for recommended doses of fertilizer (600 g N – 200 g P – 300 g K - 200 g $ZnSO_4$ – 200 g $FeSO_4$ – 200 g $MnSO_4$/tree/year).

Vm stands for vermicompost (Nutrient composition: 2.38 per cent N, 0.09 per cent P, 1.42 per cent K, 1072 ppm Fe, 116 ppm Mn, 39 ppm Cu and 46 ppm Zn).

MC stands for microbial consortium developed by isolating the native microbes (mixture of *Bacillus seudomycoides* (MF113272), *Acinetobacter radioresistens* (MF113273), *Micrococcus yunnanensis* (MF113274), *Paenibacillus alvei* (MF113275) and *Aspergillus flavus* (MF113270).

SOC, SiC, and TC stand for soil organic carbon, soil inorganic carbon and total carbon, respectively.

Source: Srivastava *et al.* (2015a; 2015 c; 2019).

In endo-sphere, both the forms of phosphatase enzyme as acid phosphatase (EC : 3.1.3.2) and alkaline phosphatase (EC : 3.1.3.1) expressed significant changes in response to different ISFM-based treatments (Figure 8.3). Acid phosphatase (EC : 3.1.3.2) enzyme activity was statistically on par with each other when compared with inorganic fertilizers (Module-I) versus (Module-II) and Module-III. But, with Modules-II or Modules-V vermicompost, acid phosphatase (EC : 3.1.3.2) enzyme displayed significant changes. While alkaline phosphatase (EC : 3.1.3.1) activity was lowest in inorganic fertilizers treatment as module-I (1888.88 units/g) and highest with Module-III (6861.11 units/g). While, peroxidase (EC : 1.11.1.7) activity was minimum with module-I (2.76 units/g) and it increased further with Module-II(4.97 units/g) and Module-III(9.99 units/g). While, replacing further 50 per cent RDF with vermicompost either without microbial consortium (module-IV) or with microbial consortium induced no further significant changes. The succinate dehydrogenase (EC : 1.3.99.1) enzyme activity was observed minimum with Module-I (91.66 μ mol TTZ reduced/g) and maximum with Module-V (799.99 μ mol TTZ reduced/g). It was interestingly observed a concurrent increase in activity of all the three enzymes with addition of vermicompost and microbial consortium under different ISFM-based modules.

ISFM: Changes in Rhizosphere CO_2 Emission

All the ISFM-based modules influenced the CO_2 emission to varying proportions across various seasons in a year (Figure 8.3). The appraisal on CO_2 emission rate in response to differential treatments showed comparatively higher CO_2 emission in the morning was observed over evening time, irrespective of the season. While, much higher CO_2 emission was observed during summer season (3127.2-4530.6 mg C^2/hr) compared to either rainy season (1858.9-2371.4 mg C/m^2/hr) or winter season (1212.1-2052.9 mg C/m^2/hr) with Module-I involving exclusive use of inorganic fertilizers. With replacement of 25 per cent of RDF with vermicompost (Module-II) CO_2 emission rate was slowed down to 1297.0 – 1959.7 mg C/m/hr 1,381.7-2075.8 mg C/m/hr and 746.3-1189.8 mg C/m/hr during summer, rainy and winter season, respectively, irrespective of time of sampling. Incorporation of microbial brought down the CO_2 emission rate to 980.5-1030.6 mg C/m/hr, 408.6-528.8 mg C/m^2/hr and 988.3-1061.6 mg/C/m^2/hr, respectively, during summer, rainy and winter season, respectively, irrespective of time of sampling. Combined use of 50 per cent RDF + 50 per cent vermicompost + Microbial consortium (Module-V) further brought down the CO_2 emission rate to 724.6-938.2 mg C/m^2/hr, 685.5-864.9 mg C/m/hr and 697.2-1007.1 mg C/m^2/hr, respectively during summer, rainy and winter season, respectively. On the other hand, Module-IV showed comparatively hither CO_2 emission, 2205.0-2635.0 mg C/m^2/hr, 1108.4-1670.0 mg C/m/hr, and 831.3-1402.7 mg C/m^{-2} hr^{-1} during summer, rainy, and winter season, respectively. These observations showed better carbon accredition with those ISFM modules having all the three components

Futuristic Viewpoints

Despite many cutting edge technologies, addressing a variety of core issues on role of soil management-based nutrient -use- efficiency amidst multiple nutrient deficiencies in raising the productivity of perennial fruits is the major research and developmental issue. Microbes are the most diverse soil organisms, yet very little is known about them. Until recently, research has focused on those organisms that are culturable. However, a wealth of information is now being collected from both culturable and as yet, unculturable organisms.

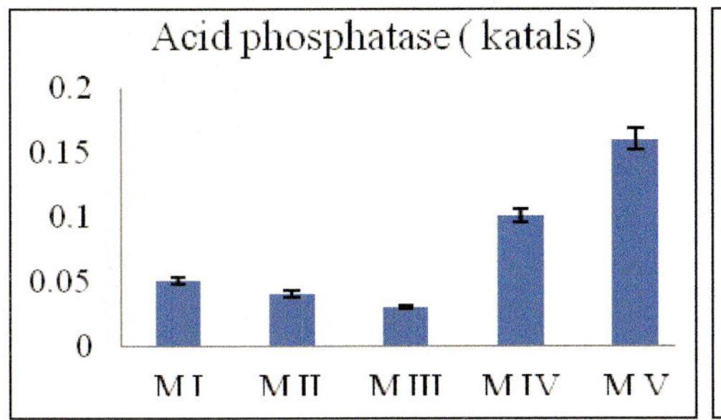

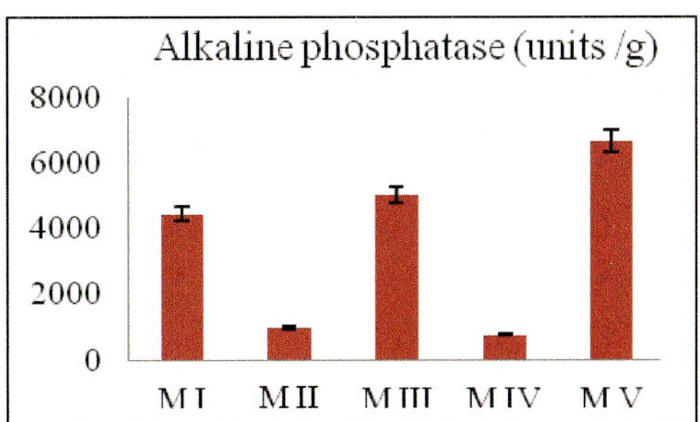

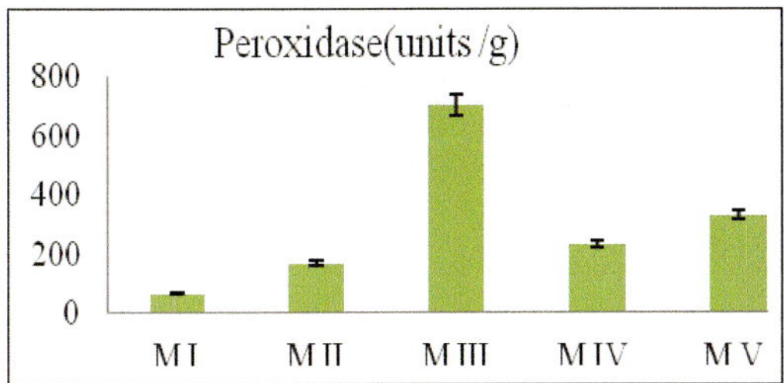

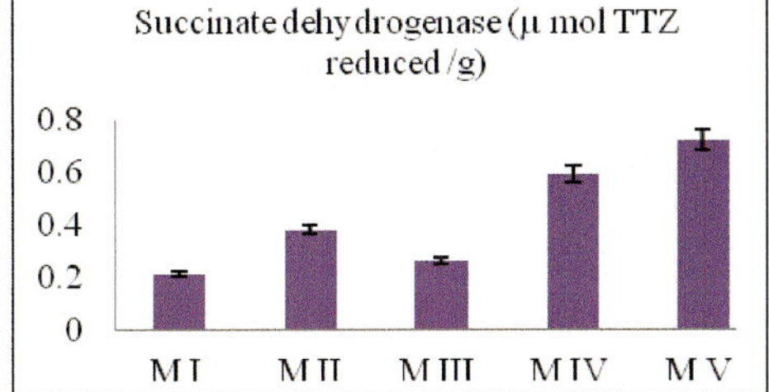

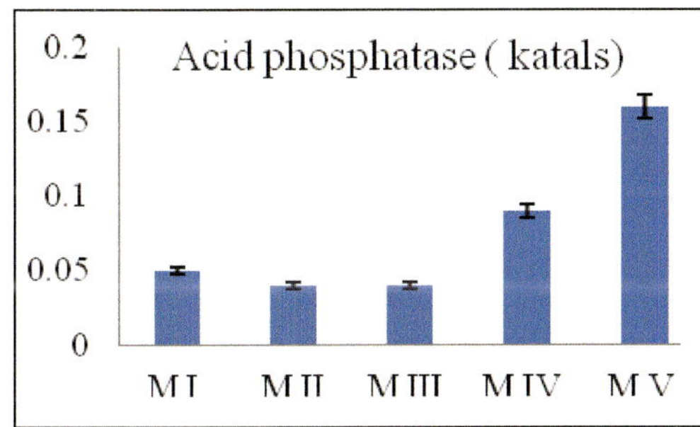

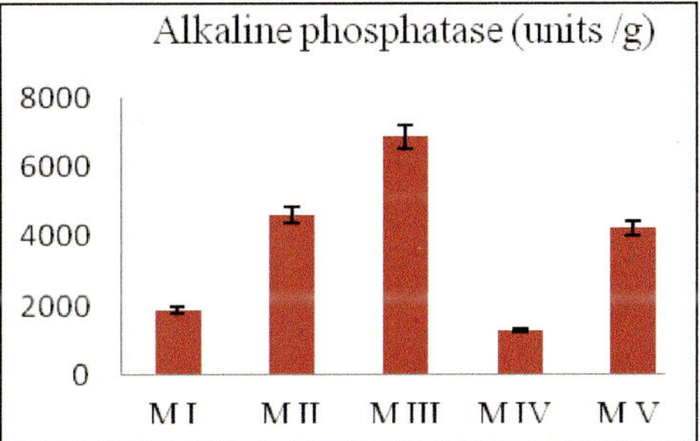

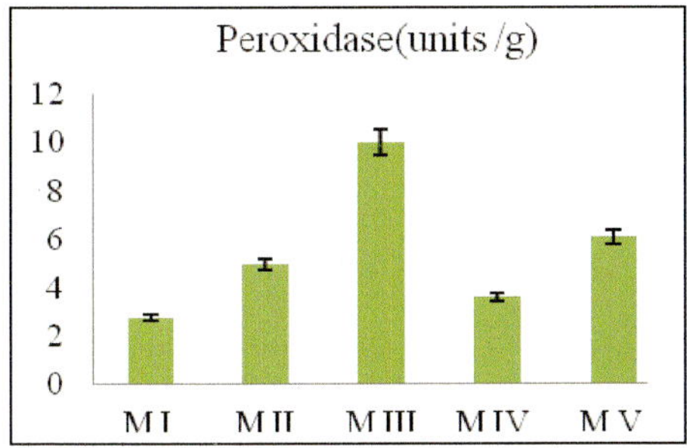

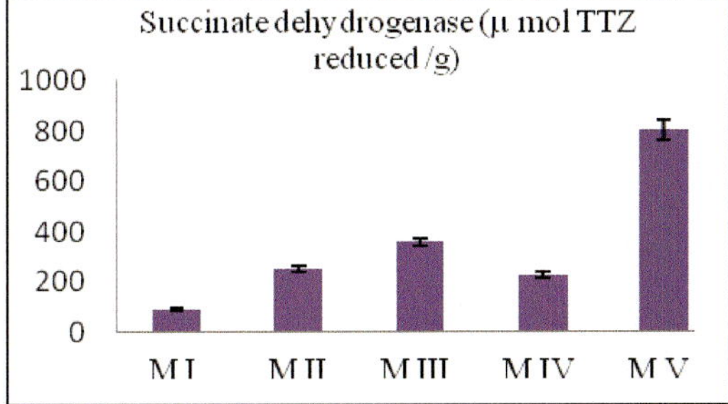

Figure 8.3: Profile of Enzymatic Activities in Rhizosphere in Response to different Modules of ISFM (upper four figures represent rhizosphere profile and lower four figures represent endosphere profile) (Source: Srivastava *et al.*, 308).

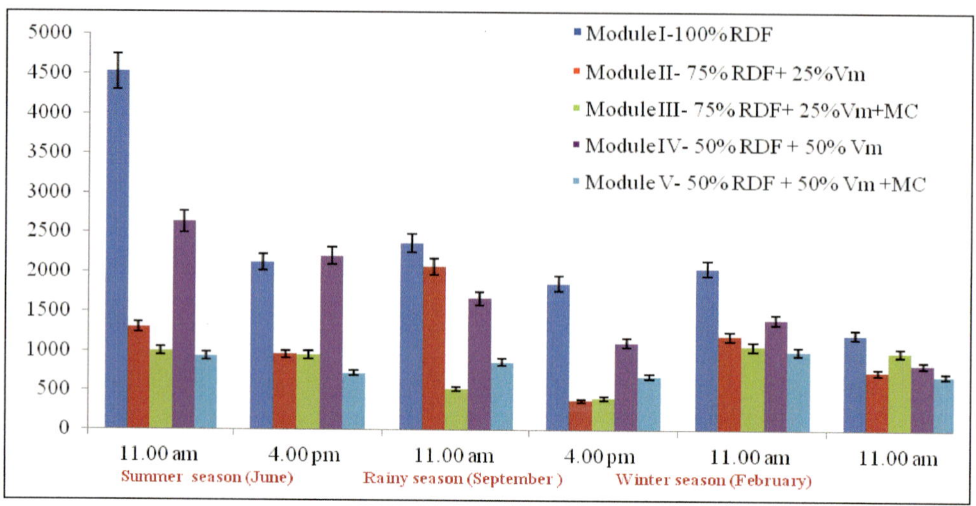

Module I (M I) - 100 per cent RDF; Module II (M II) - 75 per cent RDF + 25 per cent Vm; Module III (M III) - 75 per cent RDF + 25 per cent Vm + MC; Module IV (M IV) - 50 per cent RDF + 50 per cent Vm; Module V (M V) - 50 per cent RDF + 50 per cent Vm+ MC

Figure 8.4: ISFM: Reduction in CO_2 Emission Rate (mg C/m^2/hr) in Response to different Modules of IFSM (*Source*: Srivastava *et al.,* 2015c).

The microbial biomass is one of the biological properties of soil that undergoes immediate change in response to fertilizers like input. Studies, therefore, need to be undertaken with a view to explore the possibility of which soil microbial property could be used as a potential tool for finding out soil health -related constraint, instead of concentration of available nutrients in soil using some indicator fruit crop(s). While the genetic, functional and metabolic diversity of soil microorganisms within the rhizosphere of wide range of fruit crops is important, the capacity of soil microbial communities to maintain functional diversity of those critical soil processes could ultimately be more important to ecosystem productivity and stability than mere taxonomic diversity. In this context, it remains to be assessed how nutrient-microbe synergism is associated with productivity of perennial fruits. New research methods involving molecular techniques will extend our understanding of taxonomic and functional diversity microbes in fruit -based agro-ecosystems.

With the availability of more technical know-how on combined use of organic manures, prolonged shelf life of microbial bio-fertilizers, and more balanced use of inorganic chemical fertilizers, an understanding on nutrient acquisition and regulating the water relations would help switch orchards to CO_2 sink (expanding carbon capturing capacity of rhizosphere), so that a more sustainable fruit-based integrated crop production system catering biotic as well as abiotic stress could be evolved. Role of mycorrhizas in providing an additional resilience to rhizosphere's ability of carbon accredition within rhizosphere and associated development of plant's antioxidant profile as a defense mechanism should divert the research studying strong mycorrhizal dependency of fruit crops. Rhizosphere specific AMF-based microbial consortium (mass multiplication technology, however still a lot desired to be addressed) would add a new dimension in providing newer options for raising the productivity potential of fruit crops. The framework on soil biodiversity effects from field to fork comprises: i. recognizing both direct and indirect mechanisms of soil biodiversity effects on crops properties, ii. identifying postharvest processes that affect biodiversity legacy effects on crop properties; and iii. pinpointing biodiversity-related crop properties that influence the efficacy and success of operations occurring in the agri-food chain.

Nutrient dynamics is another neglected and under-utilized area where limited attempts have been made using fruit crops. Amongst different nutrients, Zn has attracted worldwide investigation from various angles. The changes in rhizosphere bring different simultaneous changes in microbial diversity vis-a-vis C_{mic}, N_{mic}, P_{mic} and nutrient regime, especially for diffusion- limited nutrients like P, Zn, Fe, Mn, *etc.* has to find serious considerations in any nutrient management program that involves INM-based corrective treatments. Additionally, the conditions under which fruit trees are most likely to respond to corrective Zn-treatments are still not fully understood. The role of Zn in flowering, fruit set, fruit quality (external and internal) and shelf life of fruits; models defining the critical periods of Zn-supply to assure sustained response and its uptake for helping the management decision under different fruit-based cropping systems; and devising means for improved Zn-uptake efficiency need to be attempted to unravel many of the complexities involved with Zn-nutrition under INM-based production management.

An eye should be kept on long term changes in total carbon pool of soil to arrive at the logistic conclusion that sequestration of carbon through improved production level could rejuvenate the lost productivity potential of nutritionally depleted soil. However, it remains to be further established that any change in microbial diversity within the rhizosphere is brought about with different sources of substrate, and if there is any, how the nutrient dynamics is associated with orchard productivity.

Fertilizer applications are currently managed to protect environmentally sensitive areas by using controlled release fertilizers (use of organic manures, a befitting option), frequent low concentration fertigation, multiple applications, and variable rate application technology in order to improve fertilizer- use-efficiency. However, using newly emerging techniques of nutrient management and site specific nutrient management on the principles of INM could be worked out accommodating soil's nature and properties. Simultaneously, concerted efforts would be required to develop INM-based yield monitors and soil quality indicators in order to develop a comprehensive system, whereby the concept of soil security could effectively be brought into a reality with an emphasis on development of minimum data set to define fruit-based Soil Health Card, still a long pending issue. Amidst these important issues, the molecular approach to breeding of mineral deficiency resistance and mineral efficiency would facilitate to produce nutritionally efficient biotypes in order to maximize the quality production of fruit crops on sustained basis. The work related to microbial inoculants for mass production, formulation coupled with innovative marketing, interaction and signaling with the plant and soil environment need further redressal to reorient nutrient-microbe synergy in fruit science.

REFERENCES

Ahmad M, Feza Samanta A and Jabdeen A, 2010. Response of sweet cherry (*Prunus avium*) to fertigation of nitrogen, phosphorous and potassium under Karwa land of Kashmir valley. *Indian J Agri Sci.* **80**: 212-216.

Baldi E, Toselli M, Marcolini G, Quarieri M, Cirillo E, Innocenti A and Marangoni B, 2010. Compost can successfully replace mineral fertilizers in the nutrient management of commercial peach orchard. *Soil Use Manag* **26**: 346-353.

Barne Varsha G, Bharad SG, Dod VN and Baviskar MN, 2011. Effect of integrated nutrient management on yield and quality of guava. *Asian J. Hort.* **6**: 546–548.

Banyal SK and Sharma SK, 2011. Effect of fertigation and rootsocs on yield and quality of apple under high density plantation. *Indian J Hort.* **68** : 419-424.

Baviskar MN, Bharad SG, Dod vn and Barne V, 2011. Effect of integrated nutrient management on yield and quality of sapota. *Pl Archives.* **11**(2): 661-663.

Berendsen RL, Visman G, Yu K, Song Y, Jongede R, Burgman M, Burmollie M, Herschend J, Bokrer PAHM and Pieterse CMJ, 2018. Disease induced assemblage of a plant beneficial bacterial consortium. *Microb. Ecol.* **12**: 1406-1409.

Bhalerao NM, Patil NM, Badgujar CO and Patil DR, 2009. Studies on integrated nutrient management for tissue cultured Grand Naine banana. *Indian J. Agri. Res.* **43**: 107–12.

Bhargava BS and Chadha KL, 1993. Leaf nutrient guide for fruit crops. **In** Advances in Horticultural Fruit Crops,K.L. Chadha and O.P. Pareek (ed.), pp. 973-1030. Malhotra Publishing House, New Delhi

Biradar DP, Aladakatti YR, Rao TN and Tiwari KN, 2006. Site specific nutrient management for maximization of crop yields in northern Karanataka. *Better Crops.* **90**(3): 33–5.

Bottini R, Cassan F and Piccoli P, 2004. Gibberellin production by bacteria and its involvement in plant growth promotion and yield increase. *App. Microbiol. Biochem.* **65**: 497-503.

Brown HB, Zhao FJ and Dobermann A, 2022. What is plant nutrient ?. Changing definitions to advanced science and innovation in plant nutrition. *Pl Soil.* **476**: 11-23.

Buneman E, Schwenka GD and Van Zwiten L, 2006. Impact of agricultural inputs on soil organisms: a review. *Aust J Soil Res.* **44**: 379-408.

Bwalya JM, 2013. Estimation of net carbon sequestration potential of citrus under different management systems using life cycle approach. *Res Thesis* (submitted on Feb. 2, 2013), University of Zambia, Lusaka, Zambia.

Chan CW, 2012. Variable rate application of fertilizer in citrus. (In) Advances in Citrus Nutrition, pp 257–69. Srivastava A K (Ed). Springer-Verlag, The Netherlands.

Cheng L, S Dong S Guak and Fuchigami LH, 2001. Effects of nitrogen fertigation on reserve nitrogen and carbohydrate status and regrowth performance of pear nursery plants. *Acta Hortic.* **564:** 51–56.

Cardoso EJBN, Vasconcellos RLF, Bini D, Miyauchi MYH, Alcantara C, Alves PRL, Paula A, Nakatani AS, Pereira J de M and Nogueira M, 2013. Soil health; looking for sustainable indicators. What should be considered to assess the effects of use and management on soil health. *Scien. Agricola.* **70**(4): 219-303.

Chauhan N and Chandel JS, 2008. Effect of fertigation on growth,yield fruit quality and fertilizer use efficiency of kiwifruit (*Actinidia deliciosa*). *Indian J Agri Sci.* **78**: 389-393.

Corwin DL and Lesch SM, 2005. Characterizing soil spatial variability with apparent soil electrical conductivity I. Survey protocols. *Comput. Electron. Agric.* **46**: 103-33.

Darshan D, Hota D, Yadav S and Kumar V, 2023a. Foliar application of growth regulators and nutrients for better quality aspects of guava cv Lalit. *Environ Ecol.* **41**(3): 1383-1387.

Darshan D, Shukla AK, Hota D, Singh V and Kumar V, 2023b. Foliar application of growth regulators and nutrients for enhanced yield attributes of guava cv Lalit. *Emergent Life Sci Res.* **9**(1): 72-76 DOI: 10.31783/elsr.2023.917276

Darshan D, Hota D, Devi R and Shukla JK, 2022. Micronutrients and plant growth regulators affecting the yield and quality of fruit crops: A review. *Emergent Life Sci Ress.* **8**(2): 92-103.

Das B, Harekrishna Ranjan JK, Pragya Agmed N and Attri JK, 2016. Integrated nutrient management and mulching for higher productivity of spur type apple (*Malus domestica*) cultivar. *Indian J Agri Sci.***86**: 1016-1023.

Davis DR, 2009. Declining fruits and vegetables nutrient composition: what is the evidence. *Hort Sci.* **44**(1): 15-19.

De La Cruz RT, 2008. A site specific and integrated nutrient approach to growing corn in the Philippines. *Bur Agri Res Chronicle.* **9**: 1-3.

Dey P, Rai M, Gangopadhyay KK, Das B, Nath V and Reddy NN, 2010. Effct of phosphorous on growth, yield and nutrient use efficiency of litchi grown on Alfisols. *Indian J Hort.* **67**: 394-395.

Dobermann A, Bruulsema T, Cakmak J, Gerard B, Majumdar K, McLaughlin L, Reidsma P, Vanlauwe B, Wollenberg L, Zhang F and Zhang X, 2022. Responsible plant nutrition to support food system transformation. *Global Food Sec.* **33**: 100636 doi.org/10.1016/j.gfs-2022.10636.

Dobermann A, Witt C, Abdulrachman S, Gines HC, Nagarajan R, Son TT, Tan PS, Wang GH, Chien NV, Thoa VTK, Phung CV, Stalin P, Muthukrishnan P, Ravi V, Babu M, Simbahan GC and Adviento MAA, 2003b. Estimating indigenous nutrient supplies for site-specific nutrient management in irrigated rice. *J. Agron.* **95**: 924-35.

Dobermann A, Witt C, Abdulrachman S, Gines HC, Nagarajan R, Son TT, Tan PS, Wang GH, Chien NV, Thoa VTK, Phung CV, Stalin P, Muthukrishnan P, Ravi V, Babu M, Simbahan GC and Adviento MAA, 2003a. Soil fertility and indigenous nutrient supply in irrigated rice domains of Asia. *J. Agron.* **95**: 913-23.

Dong ZH, Srivastava AK, Tan QL, Raiz M, Lv XL, Liu XD, Gao Y and Chen M, 2023. Effect of different summer and spring fertilization proportions on fruit yield, quality and nutrient-use-efficiency of Nehwall orange (*Citrus sinensis* Osbeck) orchard. *J. Plant Nutr.* **46**(9): 2144-2154.

Dorans SF and Zeiss MR, 2000. Soil health and sustainability: managing the biotic components of soil quality. *Appl. Soil Ecol.* **15**(1): 3-11.

Dutta P, Kundu S and Biswas S, 2010. Integrated nutrient management in litchi cv Bombai in new alluvial zone of West Bengal. *Indian J. Hort.* **67:** 181–4.

Dutta P, Maji SB and Das BC, 2009. Studies on the response of bio-fertilizer on growth and productivity of guava. *Indian J. Hort.* **66:** 39–42.

Dwivedi V, 2013. Effect of integrated nutrient management on yield, quality and economics of Guava. *Ann. Plant Soil Res.* **15**: 149–51.

Elbasiouny H, El-Ramady H, Elbehiry F, Rajput V.D, Minkina T and Mandzhieva S, 2022. Plant nutrition under climate change and soil carbon sequestration. *Sustainability* **14**: 914. https: //doi.org/10.3390/su14020914.

Fageria NK, Baligar VC and Li YC, 2008. The role of nutrient efficient plants in improving crop yields in the twenty first century. *J. Plant Nutr.* **31**: 121-157.

Fan X, Zhou X, Chen H, Tang M and Xie X, 2021. Cross -talk between macro- and microutrients uptake and signaling in plants. *Front Pl Sci.* **12**: doi.org/16.3389/ fpls.2021.663477.

FAO, 2020. Statistical Year Book of Food and Agricultural Organization, FAOState div. Metalink: P3.Reu.FAO, ESS.FRD.AHSci.p.108-113.www.fao.org.docrep/017/ i3138e/i3138e05.pdf.

Färe R, Wang C and Seavert C 2012. A model of site-specific nutrient management. *App. Econ.* **44**: 4369-380.

Ghosh SN, Bora B, Roy S, Kundu A, 2012b. Integrated nutrient management in pomegranate grown in laterite soil. *Indian J. Hort.* **109**: 1-7.

Goswami Kumar A, Shant L and Misra KK, 2012. Integrated nutrient management improves growth and leaf nutrient status of guava cv Pant Prabhat. *Indian J. Hort. Sci.* **69**: 168–72.

Guimaraes VD, Gonzaga MIS and Neto JOM, 2014. Manangemnt of soil organic matter and carbon storage in tropical fruit crops. *Rev. Bras. Eng. Agric. Ambient.* **18**: 83-93.

Gupta BR, Tiwari R, Tiwari TP and Tiwari KN, 2007. Maximizing yield, nutrient use efficiency, and profit in summer black gram. *Better Crops.* **91**(3): 22–3.

Hach CV and Tan PS, 2007. Study on site-specific nutrient management (SSNM) for high-yielding rice in the Mekong Delta. *Omonrice.* **15**: 144-52.

Haneef Mohd, Kaushik RA, Sarolia DK, Mordia A and Dhakar M, 2014. Irrigation scheduling and fertigation in pomegranate cv Bhagwa under high denisty planting system. *Indian J Hort.* **71**: 45-48.

Hasan MA, Manna M, Dutta P, Bhattacharyya K, Mandal S, Banerjee H, Ray SK and Jha S, 2013.Foliar nutrient content in mango as influencd by organic an dinorganic nutrients and their correlative relatinship with yield and quality. *Acta Hortic.* **992**: 201-206.

Hazarika BN and Ansari S, 2010. Effect of integrated nutrient management on growth and yield of banana cv. Jahaji. *Indian J. Hort.* **67**: 270–3.

Hebbarai M, Ganiger VM, Masthana RBG and Joshi VR, 2006. Integrated nutrient management in sapota (*Manilkara zapota*) using vermicompost to increase yield and quality. *Indian J. Agric. Sci.* **76**: 587–90.

Hota D, Kumar V and Singh IP, 2021a. Agronomic performance of acid lime in response to microbial fortification of rhizosphere. *Indian J Agri Sci.* **91**(1): 146-149.

Hota D, Kumar V and Singh IP, 2021b. Pre-evaluation response of rhizosphere hybridization in acid lime. *Indian J Agri Sci.* **91**(1): 159-163.

Hota D, Srivastava AK, Dahat S and Dubey S, 2020. Rhizosphere engineering through microbes enhances agronomic performance of acid lime. *Prog Hort.* **52**(2): 120-123.

Hota D, Sharma LD, Sahu R and Dahat S, 2021c. Rhizosphere hybridization: A better alternative for exploiting citrus microbiome. *Int J Plant Biotechnol.* **7**(2): 10-17.

Hota D, Kalatippi AS, Dahat S, Rathore JP and Jatav R, 2023. Role of rhizosphere hybridization-mediated biometric response in citrus. **In**: Singh, S.B., Haldhar, S.M., Singh, R.K.D., Thaochan, N. and Murkute, A.A. Citrus Crop: Production and Management in NEH Region. Scientific Publishers, New Delhi, India. Pp: 187-202.

Hota D, Srivastava AK, Kumar V and Kalatippi AS, 2022. Rhizosphere hybridization: sustaining citrus production through microbiome changes. *Indian Horticulture.* **67**(5): 37-39.

Huang YM, Srivastava AK, Zou YN, Ni QD, Han Y and Wu QS, 2014. Mycorrhizal -induced calmodulin mediated changes in antioxidant enzymes and growth response of drought -stressed trifoliate orange. *Front Microbiol.* **5**: 682-628.

Jasrotia A, Singh RP, Singh JM and Bhutani VP, 1999. Response of olive trees to varying levels of N and K fertilizers. *Acta Hortic.* **474**: 337-339.

Jeyabaskaran KJ, Shirgure PS, Srivastava AK, Pandey V and Uma S, 2021. Fertigation in horticulture: a guarantee to economized quality production. *Indian J Fertilizers.* **17**: 364-383.

Johnston AM, Khurana HS, Majumdar K and Satyanarayana T, 2009. Site specific nutrient management – concept, current research and future challenges in Indian agriculture. *J. Indian Soc. Soil Sci.* **57**: 1-10.

Kalatippi AS, Pandey SK, Huchche AD, Hota D, Kuldeep DK and Bhooriya MS, 2022. Effect of foliar applied nitrogen with growth regulators (2,4 D, GA_3, Amrut-pani) and soil applied potassium on the management of the rough-skin disorder of Nagpur mandarin (*Citrus reticulata* Blanco). *Indian J Agric Res.* **A**-6003.

Keditsu R and Srivastava AK, 2014. Substrate dynamics: Developments and issues. *Ann. Plant Soil Res.* **16**(1): 1–8.

Khehra S and Bal JS, 2014. Influence of organic and inorganic nutrient sources on growth of lemon (*Citrus limn* (L.) Burm.) cv Baramasi. *J Exp Biol Agric Sci.* **2**: 126–9.

Khot AB, Ashoka P, Neelkanth JK, Rajkumar S and Gundlur SS, 2012. Performance on growth period of sapota (*Achras sapota* L.) to drip method of irrigation and fertigation on Vertisol. *Environ Ecol.* **30**: 1513-1515.

Kirad KS, Barche S and Singh DB, 2010. Integrated nutrient management in papaya cv. Surya. *Acta Hortic.* **1**: 377–80.

Kumar P, Murugan P, Murkute A and Singh SB, 2010. A carbon sequestration strategy involving temperate fruit crops in the trans-Himalayan region. *J Hortic Sci Biotechnol.* **85**(5): 405–409.

Kumar S, Kumar S and Mohapatra T, 2021. Interaction between macro- and micronutrients in plants. *Front Pl Sci.* **12**: doi.org/10.3389/fpls.20211665583.

Kumar A, Sharma N, Sharma CL and Singh G, 2017. Studie son nutrient management in apple cv Oregon Spur II under cold desert region of Himachal Pradesh of India. *Int J Agri Res.* **51**: 161-166.

Kumari R, Kundu M, Das A, Rakshit R, Sahay Sengupta S and Ahmad Md Feza, 2019. Long term integrated nutrient management improves carbon stock and fruit yield in a tropical mango (*Magnifera indica* L.) orchard. *J Soil Sci Pl Nutri.* **20**: 725-731.

Kuttimani R, Velayudham Somasundram E and Muthukrishnan P, 2013. Effect of integrated nutrient management on yield and economics of banana. *Global J Biol Agri Health Sci.* **2**: 191–5.

Lakso AL, 2010. Estimating the environmental footprint of New York apple orchards. *New York Fruit Quarterly*. **18**: 26-28.

Lakshmi S, George A, Pillai GR and Rao TN, 2007. Nutrient needs of coconut-based fodder production systems in homesteads of Kerala. *Better Crops*. **91**(4): 28–9.

Liu CY, Srivastava AK and Wu QS, 2014. Effect of auxin inhibitor and AMF inoculation on growth and rot morphology of trifoliate orange (*Poncirus trifoliata*) seedlings. *Indian J Agric Sci.* **84**(11): 342-346.

Lobell DB, Field CB, Chahill KN and Bonfils C, 2005. Impacts of future climatic changes on California crop yields: Model projections with climate and crop uncertainties. *Agric. For. Meteorol.* **141**: 208-218.

Lal G and Dayal H, 2014. Fffect of integrated nutrient management on yield and quality of acid lime (*Citrus aurantifolia* Swingle). *Afr J Agri Res.* **9**(4) : 2985-2991.

Loladze I, 2014. Hidden shift of the ionome of plants exposed to elevated CO_2 depletes minerals at the base of human nutrition. *eLife* **3**: e02245.

Lu T, Ke M, Lavoie M, Jin Y, Fan X, Zhang Z, Fu Z, Sun L, Gillings M, Penuels J, Qian H and Zhu YZ, 2018.Rhizosphere microorganisms can influence the timing of plants flowering. *Microbiome*. **6**: 231-238.

Maity A, Jadhav VT and Ram Chandra, 2012. Exploration of microbial wealth for sustainable horticultural production. *Int. J. Stress Manag*. **3**: 489-500.

Mancuso MAC, Soratto RP, Crusciol CAS and Casstro GSA, 2014. Effect of potassium sources and rates on Arabica coffee yield, nutrition and nutrient export. *R Bras a Solo*. **38**: 14448-1456.

Mandal KK, Rajak A, Debnath S and Hasan MA, 2013. Integrated nutrient management in aonla cv A-7 in the red lateritic region of West Bengal. *J. Crop Weed*. **9**: 121–123.

Mayer AMB, Trenchard L and Rayns F, 2021. Historical changes in the mineral content of fruits and vegetables in the UK from 1940 to 2019 : a concern for human nutrition and agriculture. *Int J Food Sci Nutri* **73**(2): doi.org/10.1080 109637486.2021.1981831.

Meena KR, Maji SMS and Meena SC, 2017. Use of shoot pruning for crop regulation and quality production of guava (*Psidium guajava* L.). *Int J Agri Res.* **13**(2): 184-191.

Meena HR, Somasundaram J, Kaushik RA, Sarolia DK, Singh RK, Kala S and Meena GL, 2019. Integrated nutrient management affects fruit yield of sapota (*Achras zapota* L.) and nutrient availability in a Vertisol, *Commun Soil Sci Pl Anal.* **50**: 2848-2863

Mendes R, Garbeva P and Raajmakers JM, 2013.The rhizosphere microbiome: significance of plant beneficial, plant pathogenic and human pathogenic microorganisms. *FEMS Microbiol. Rev.* **37**(7): 634-663.

Miao Y, Bobby AS and Zhang F, 2011. Long-term experiments for sustainable nutrient management in China. *Agron. Sustain. Dev.* **31**: 397-414.

Montanaro G, Celano G, Dicho B and Xiloyannis C, 2010. Effects of soil protecting agricultural practices on soil organic carbon and productivity of fruit tree orchards. *Land Degrad Develop* **161**: 46-54.

Montanaro G, Tuzio AC, Xylogiannis E, Kolimenakis A and Dichio B, 2016. Carbon budget in a mediterranean peach orchard under different management practices. *Agri Ecosys Environ* **238**: doi.org10.1016/j.agee.2016.05031.

Mousavi SM, Srivastava AK and Raiesi T, 2024. Citrus nutrition in Iran: learning lessons from calcareous soils. *J. Plant Nutr.* **47**(2): 1-26.

Neto Mário de O, Rebouças Azeved de, Thales V de, Viana A, Mesquita José BR de, Carvalho Marco A R de and Carvalho Lilian CC de, 2016. Potassium fertilization via fertigation and conventional application on quality of tomato fruits. *Revista Brasileira de Engenharia Agrícola e Ambiental.* **20**: 913-917.

Ngullie E, Singh AK, Sema A and Srivastava AK, 2015. Citrus growth and rhizosphere properties. *Commun Soil Sci Pl Anal.* **46**(12): 1540-1550.

Nielsen DR, Biggar JW and Erh KT, 1973. Spatial variability of field-measured soil-water properties. *Hilgardia* **42**(7): 215–59.

Noordwijk MV and Cadisch G, 2020. Access and excess problems in plant nutrition. *Pl Soil* **247**(1): 25-39.

Padma TMR and Kandasamy D, 1990. Effect of interaction between VA-mycorrhizae and graded levels of phosphorous on the growth of papaya (*Carica papaya*). *Proceedings of the National Conference on Trends in Mycorrhizal Research,* February 14-16, Haryana Agricultural University, Hisar, Haryana India, pp 133–134.

Panwar DD, Kale KD and Dingre SK, 2014. Nutrient use an ddynamics under different fertigation scheduling for banana(*Musa paradisiaca*) cultivation in Vertisol. *Indian J Agri Sci.* **84**: 1159-1165.

Patel VB, Singh SK, Asrey R, Nain Lata, Singh AK and Singh L, 2009. Microbial and inorganic fertilizers application influenced vegetative growth, yield, leaf nutrient status and soil microbial biomass in sweet orange cv Mosambi. *Indian J Hort.* **66**: 163–8.

Pathak H and Nedwell DB, 2011. Nitrous oxide emission from soil with different fertilizer, water levels, and nitrification inhibitors. *Water Air Soil Pollution* **129**: 217-228.

Patil VK and Shinde BN, 2013. Studies on integrated nutrient management on growth and yield of banana cv. Ardhapuri (Musa AAA). *J. Hortic. For.* **5**: 130–8.

Pinter JPJ, Hatfiel JL, Schepers JS, Barner EM, Moram HS, Daughtry CST and Upchurch DR, 2003. Remote sensing for crop management. *Photogram Eng Remote Sens.* **69**(6): 647–94.

Ram RA and Rajput MS, 2000. Role of biofertilizers and manures on production of guava (*Psidium guajava* L.) cv. Allahabad Safeda. *Haryana J Hort Sci.* **29**: 193–4.

Ramniwas RA, Kaushik DK, Sarolia DK, Pareek S and Singh V, 2012. Effect of irrigation and fertigation scheduling on growth and yield of guava (*Psidium guajava* L.) under meadow orcharding. *Afr J Agri Res.* **7**: 6350-6356.

Ray Dutta SK, Takawale PV, Chatterjee R and Hanmte V, 2014. Yield and quality of pomegranate as influenced by organic and inorganic nutrients. *Bioscan* **9**: 317–620.

Reddy AGK, Kumar JS, Maruthi V, Venkatasubbaiah K and Rao CS, 2017. Fruit production under climate changing scenario in India: a review. *Environ Ecol* **35**(2B): 1010-1017.

Reídel EJ, Brown PH, Duncu RA, Heerema RJ and Weinbaum SA, 2004. Sensitivity of yield determinants to potassium deficiency in 'Nonpareil' almond. *J Hort Sci Biotechnol.* **79**: 906-910.

Rengel Z, Gutteridge R, Hirsch P and Hornby D, 1996. Plant genotype, micronutrient fertilization, and take-all infection influence bacterial populations in the rhizosphere of wheat. *Pl Soil.* **183**: 269-277.

Rietra RPJJ, Heinen M, Dimkpa CO and Bindraban PS, 2017. Effects of nutrient antagonism and synergism on yield and fertilizer use efficiency. *Commun Soil Sci Pl Anal* **48**(6): 1895-1920.

Rubio-Asensio JS, Parra M, Intrigliolo D, 2020.open filed hydroponicsin fruit crops : developements and challenges. In : A.K.Srivastava and Chenxiao Hu, Fruit Crops : Diagnosis and Management of Nutrient Constraints, Elseviers, pp. 419-439.

Sahoo T, Kaluram Chander S, Hota D and Adhikari B, 2018. Response of perennial fruit crops to climate change. In: Naresh, R.K. *Research trends in agriculture science*. Akinik Publication, Delhi, India. Volume 4 pp53-58.

Sarangthem I, Sharma LD and Srivastava AK, 2014. Development of nutrient diagnostic techniques for khasi mandarin (*Citrus reticulata* Blanco) grown in Manipur. *J Indian Soc Soil Sci.* **62**: 118-125.

Shah AS, Mohammad W, Shah MS, Ali ER and Haroon AAB, 2014. Integrated effect of organic and inorganic nitrogen on peach fruit yield and orchard fertility. *Res. J. Agric. Sci.* **4**: 78–82.

Shahzad Z and Rouached H, 2022. Protecting plant nutrition from the effects of climate change. *Curr Biol* **32**(13): R725-R727.

Sharma LD, Sadhukhan R and Hota D, 2021. Neutralising climate change through fruit crops. *J Crop Sci Technol.* **10**(3): 28-36.

Sheik RT, Kamalakannan S, Rajeswari R and Sudhagar R, 2019. Effect of integrated nutrient management on fruit quality in sapota (Manilkara zapota(L) P. Royen. *Anal Pl Soil Res.* 21(1): 67-70.

Shirgure PS, Srivastava AK and Singh S, 2000. Water management in citrus: a review. *Agri Rev.* **21**(4): 223-230.

Shirgure PS, Srivastava AK and Singh S, 2001a. Growth, yield and quality of Nagpur mandarin (*Citrus reticulata* Blanco) in relation to irrigation and fertigation. *Indian J Agric Sci.* **71**(8): 547-550.

Shirgure PS, Srivastava AK and Singh S, 2001b. Effect of nitrogen fertigation and band placement fertilizer application in soil-leaf nutrient build-up and incremental growth of acid lime. *J. Soil Water Conserv.* **45**(3- 4): 176-181.

Shirgure PS, Srivastava AK and Singh S, 2001c. Fertigation and drip irrigation in Nagpur mandarin (*Citrus reticulata* Blanco). *South Indian Hort.* **49/50**: 95-97.

Shirgure PS, Srivastava AK and Singh S, 2003a. Irrigation scheduling and fertigation in acid lime (*Citrus aurantifolia* Swingle) *Indian J Agric Sci.* **73**(7): 363-367.

Shirgure PS, Srivastava AK and Singh S, 2003b. Evaluating microirrigation systems in Nagpur mandarin (*Citrus reticulata* Blanco) under sub-humid tropical climate. *Trop. Agri.* (Trinidad) **80**(2): 91-96.

Shirgure PS, Srivastava AK and Hota D, 2021. Fertigation in citrus, a toolkit for sustainable quality production. *Int. J. Innov. Hort.* **10**(1): 35-46.

Shukla AK, Sarolia DK, Kumari B, Kaushik RA, Mahawer LN and Bairwa HL, 2009. Evaluation of substrate dynamics for integrated nutrient management under high density planting of guava cv. Sardar. *Indian J. Hort.* **66**: 461–4.

Shylaja M and Rao MS, 2012. In vitro compatibility studies of *Trichoderma harzianum* with inorganic fertilizers. *Nematologia Mediterranea.* **40**: 51-54.

Singh SK, Singh CP and Dhingre SK, 2014. Response of fertigation and plastic mulch on characteristics of young Dashehari mango. *Indian J Hort.* **66**: 390-392.

Singh JK and Varu DK, 2013. Effect of integrated nutrient management in papaya (*Carica papaya* L.) cv Madhubindu. *Asian J. Hort.* **8**: 667–70.

Singh S, Srivastava AK and Shivankar VJ, 1999. Citrus-based cropping system: a review. *Agri Rev* **20**(2): 123-128.

Singh SR and Banik BC, 2011. Response of integrated nutrient management on flowering. *Fruits* **42**: 151–154.

Salaza-Garcia S and Lazcano-Ferrat I, 2003.Site-specific fertilization increased yield and fruit size in 'Hass' Avocado. *Better Crops Int.* 17(1): 12–6.

Soares JC, Santos CS, Carvalho SMP, Pintado MM and Vasconcelos MW, 2019. Preserving the nutritional quality of crop plants under a changing climate: importance and strategies. *Pl Soil* **443**: 1-26.

Solanki SPS, Sharma NC, Chandel JS and Hota D, 2020. Effect of integrated nutrient management on fruit yield and quality of peach (*Prunus persica* Batsch) cv July Elberta. *Int. Res. J. Pure App. Chem.* **21**(10): 152-160. DOI: 10.9734/irjpac/2020/v21i1030217

Solanki SPS, Telkar SG, Hota D, Kant K and Dey JK, 2015. Folk liquid manures for sustainable horticulture. *Int J Econ Plant.* **2**(4): 175-177.

Srivastava AK, 2012. Integrated nutrient management in citrus. **In**: Srivastava A K. (Ed). *Advances in Citrus Nutrition,* pp 369-89. Springer-Verlag, The Netherlands.

Srivastava AK 2013a. Nutrient deficiency symptomology in citrus: An effective diagnostic tool or just an aid for post–mortem analysis. *Agricultural Advances.* **2**: 177-94.

Srivastava AK, 2013b. Site specific nutrient management in citrus. *Scientific Journal of Agricultural* (*Agricultural Advances*). **2**: 1-15.

Srivastava AK, 2013c. Nutrient diagnostics in citrus: Are they applicable to current season crop. *Agrotechnology.* **2**, 104-05. Doi.10.4172/2168P-9881.1000e104

Srivastava AK, 2015. Soil fertility dynamics vis-à-vis climate change in citrus. **In**: Choudhary, M.L., Patel, V.B., Siddiqui, M.W. and Verma, R.B. (eds) *Climate dynamic in horticultural science.* Apple Academic Press, USA, pp 1-34.

Srivastava AK, 2020. Climate-smart integrated soil fertility management in fruit crops : an overview. In: Srivastava, A.K. and Hu, C. (ed) *Diagnosis and management of nutrient constraints in fruit crops.* Elseviers USA pp 521-540.

Srivastava AK, 2023a. Multiple nutrient stresses accentuating citrus decline in northeast India. *Indian Horticulture.* **68**(6): 40-44.

Srivastava AK, 2023b. Integrating natural farming with agroecology for soil health care under fruit production system. *Ann. Plant Soil Res.* **25**(4): 524-533.

Srivastava AK, 2024. Challenges of plant nutrition and climate change: Focus on fruit crops. **In**: Prado, R.D., Etesami, H. and Srivastava A.K. (Ed.). *Silicon Advances for Sustainable Agriculture and Human Health,* Springer Switzerland, pp. 1-40.

Srivastava AK and Bora P, 2023. Multiple dimensions of agroecology in sustaining agriculture. *Indian Farming.* 73(6): 35-37.

Srivastava AK and Das SN, 2017. Nutrient optima-based productivity zonality and rationale of fertilizer use in citrus. **In**: *Sustainable Management of Land Resources.* Apple Academic Press, USA. pp. 717-738.

Srivastava AK and Hota D, 2020. Fruit crop under nutrient capped scenario: A timeless journey. *Curr Hort* **8**(2): 14-17.

Srivastava AK and Kohli RR, 1999. Agroecological approaches for land use planning of citrus. *Agri Rev* **20**(1): 41-47.

Srivastava AK and Malhotra SK, 2014. Nutrient management in fruit crops -review. *Indian J Fertilizers* **10**(12): 72-88.

Srivastava AK and Malhotra SK, 2017. Nutrient use efficiency in perennial fruit crops -review. *J. Plant Nutr.* **40**(13): 1928-1953.

Srivastava AK and Mousavi SM, 2024. Sustaining fruit production under multiple nutrient stresses linked to water deficit stress. **In**: Etesami, H. and Chen, Y. (Ed). *Sustainable Agriculture under Drought Stress*, Academic Press, Elseviers, USA. pp.141-169.

Srivastava AK and Ngullie E, 2009. Integrated nutrient management: Theory and practice. *Dynamic Soil, Dynamic Plant.* **3**: 1-30.

Srivastava AK and Pandey V, 2021. Speciality fertilizer centric site specific nutrient management in fruit crops: principles and achievements. *Indian J Fertilizers.* **17**(5): 458-474.

Srivastava AK and Pongnenger A, 2023. Asia citrus-a meta-analysis on soil fertility and plant nutrition constraints. *Ann. Plant Soil Res.* **25**(30): 345-357.

Srivastava AK and Singh S, 2002. *World citrus: climate and soil analysis.* International Book Distribution Company, Lucknow, Uttar Pradesh, India. pp. 1- 458.

Srivastava AK and Singh S, 2008. Analysis of citrus orchard efficiency in relation to soil properties. *J Plant Nutr.* **30**: 2077-090.

Srivastava AK and Singh S, 2009. Citrus decline: soil fertility and plant nutrition. *J Plant Nutr.* **32**(2): 197-245.

Srivastava AK and Singh S, 2016. Site–Specific nutrient management in Nagpur mandarin (*Citrus reticulate* Blanco) raised on contrasting soil types. *Commun Soil Sci Pl Anal.* **47**(4): 447-45.

Srivastava AK, Das AK, Jagannadham PTK, Bora P, Ansari FA and Bhate R, 2022. Bioprospecting soil and plant health management amidst huanglongbing threat in citrus: A review. *Front Pl Sci* **13**: 858842.

Srivastava AK, Das SN, Malhotra SK and Majumdar K, 2014. SSNM-based rationale of fertilizer use in perennial crops: A review. *Indian J Agri Sci.* **84**: 3-17.

Srivastava AK, Huchche AD, Ram L and Singh S, 2015a. Performance of irrigate mono-cropped versus intercropped Nagpur mandarin orchard on basalt derived tropical soils. *Trop Agri.* **82**(3): 159-161.

Srivastava AK, Malhotra SK and Krishna Kumar NK, 2015b. Exploiting nutrient-microbe synergy in unlocking productivity potential of perennial fruits: A review. *Indian J Agric Sci* **85**(4): 459-81.

Srivastava AK, Marathe RA and Singh S, 2002. Organic citrus; soil fertility and plant nutrition. *J. Sustainable Agriculture.* **19**(3): 5-29.

Srivastava AK, Paithankar DH, Ventataramana KT, Hazarika B and Patil P, 2019. INM in fruit crops: sustaining quality production and soi health. *Indian J Agric Sci* **89**(3): 379-395.

Srivastava AK, Shirgure PS and Singh S, 2003. Differential fertigation response of Nagpur mandarin (*Citrus reticulata* Blanco) on an alkaline Inceptisol under sub-humid tropical climate. *Trop Agri* (Trinidad) **80**(2): 97-04.

Srivastava AK, Shirgure PS, Deshmukh S and Bhoyar P, 2017. Soil fertility and soil health care in citrus: A review. *Anal Pl Soil Res* **19**(2): 127-136.

Srivastava AK, Singh S and Albrigo LG, 2008. Diagnosis and remediation of nutrient constraints in citrus. *Hort Rev.* **31**: 277-363.

Srivastava AK, Singh S and Huchche AD, 2015c. Evaluation of INM in citrus on Vertic Ustochrept: Biometric response and soil health. *J Plant Nutr* **38**(5): 1-15.

Srivastava AK, Singh S, Das SN and Singh H, 2010. Delineation of productivity zones in mandarin orchards using DRIS and GIS. *Better Crops – South Asia.* **4**: 13-15.

Srivastava AK, Singh S, Diware VS and Singh H, 2009. Site-specific nutrient management in 'Mosambi' sweet orange. *Better Crops – India* **3**: 10-11.

Srivastava AK, Singh S, Tiwari KN, 2006. Site specific nutrient management for Nagpur mandarin (*Citrus reticulata* Blanco). *Better Crops.* **88**: 22-25.

Srivastava AK, Wu QS, Mousavi SM and Hota D, 2021. Integrated soil fertility management in fruit crops: An overview. *Int J Fruit Sci* **21**(1): 413-439.

St Claire SB and Lynch JP, 2010. The opening of Pandora box: climate change impacts on soil fertility and crop nutrition in developing countries. *Pl Soil* **335**: 101-115.

Tagliavini M and Scandellari F, 2013. Methodologies and concepts in the study of nutrient uptake requirements and partitioning in fruit crops. *Acta Hortic.* **984**: 47-56.

Tagliavini M, Tanon G, Scandellari F, Quinones A, Palmieri G, Giaacchini P and Masia A, 2007. Nutrient recycling during the decomposition of apple leaves (*Malus domestica*) and mowed grasses in an orchard. *Agri Ecosys Environ.* **118**: 191-200.

Tahi B, Ana MR, Limon MC and Antonio CD, 2004. Biocontrol mechanisms of Trichoderma strains. *Int. Microbiol. J.* **37**: 249-260.

Tailor AJ and Joshi BH, 2014. Harnessing plant growth promoting rhizobacteria beyond nature: a review. *J Pl Nutri.* **37**: 1534-1571.

Tandel BM, Patil BN and Patel BB, 2014. Effect of integrated nutrient management on growth and physiological parameters on papaya cv. Taiwan Red Lady. *Trends in Biosci.***7**(16): 2175-2178.

Tandon HLS and Tiwari KN, 2007. Fertilizer use in Indian agriculture: An eventful half century. *Better Crops India.* **1**(1): 15-19.

Tatewar AV, Hota D, Mohanty S, Nayak SS, Karna AK and Mishra S, 2023. Nutrient dynamics in sweet orange as affected by plant density and nutrient source. *Int J Innov Hort* **12**(2): 179-183.

Tiwari DK, Hasan MA and Chattopadhyay PK, 1999. Leaf nutrient and chlorophyll content in banana (Musa AAA) under influence of *Azotobacter* and *Azospirillum* inoculation. *Environ Ecol.* **17**: 46–50.

Tiwari KN, 2007a. Breaking yield barriers and stagnation through site-specific nutrient management. *J Indian Soc Soil Sci.* **55**: 444-454.

Tiwari KN, 2007b. Reassessing the role of fertilizers in maintaining food, nutrition and environmental security. *Indian J Fertilizers.* 3(1): 33-48.

Umer I, Wali VW, Kher R and Jamwal M, 2009. Effect of FYM, urea and *Azotobacter* on growth, yield and quality of strawberry cv. Chandler. *Not Bot Horti Agrobo.* **37**: 139–43.

Velazquez-Gonzalez RS, Garcia-Garcia AL, Venture-Zapata E, Barceinas-Snchez JDO and Sosa-Savedra JC, 20222. A review on hydroponics and the technologies associated for medium and small scale farmers. *Agriculturae.* **12** (5): 646.

Wang S, Srivastava AK, Wu QS and Fokom R, 2014.The effect of mycorrhizal inoculation on the rhizosphere properties of trifoliate orange (*Poncirus trifoliata* L. Raf.). *Sci. Hortic.* **170**: 137-142.

Wibawa WD, Dludlu DL, Swenson LJ, Hopkins DG and Dahnke WC, 1993. Variable fertilizer application based on yield goal, soil fertility, and soil map unit. *J Prod Agri.* **6**: 255–62

Wiren NV, 2021. Grand challenges in plant nutrition. *Front Pl Sci* **2**: 4 doi.org 10.3389/fpls.2011.00004.

WHO, 2021. World Health Organisation, United Nations Organisations.Rome, Italy.

Worley RE, 1994. Long term performance of pecan trees when potassium application is based on prescribed threshold concentration in leaf tissue. *J Am Soc Hort Sci.* **119**: 434-434.

Wu QS, Srivastava AK and Zhou YN, 2013. AMF-induced tolerance to drought stress in citrus: A review. *Sci. Hortic.* **164**: 77-87.

Wu QS, Sun P and Srivastava, AK, 2017. AMF divsersity and in citrus rhizoshere. *Indian J Agric Sci.* **87**(5): 653-656.

Wu T, Wang Y, Yu C, Chiarawipa R, Zhang X, Han Z and Wu L, 2012. Carbon sequestration by fruit trees-chinese apple orchards as an example. *PlosOne* **7**(6): e38883. Doi.org/10.1371/journal.pone0038883.

Yadav R, Singh BH, Singh HK and Yadav AL, 2007. Effect of integrated nutrient management on productivity and quality of aonla (*Emblica officinalis* Gaertn.) fruits. *Plant Archives* (**1 and 2**): 881–3.

Zeng Q, Brown PH and Holtz BA, 2001. Potassium fertilization affects soil K, leaf concentration, and nut yield and quality of mature pistachio trees. *HortScience*. **36**: 85-89.

Zhang ZZ, Srivastava AK, Wu Q.and Li GH, 2015. Growth performance and rhizosphere traits of peach (*Prunus persica*) in response to mycorrhization on replant versus non-replant soil. *Indian J Agric Sci* **85**(10): 125-130.

Zou YN, Srivastava AK, Wu QS and Huang YM, 2014. Increased tolerance of trifoliate orange (*Poncirus trifoliata*) seedlings to waterlogging after inoculation with arbuscular mycorrhizal fungi. *J Animal Pl Sci.* **24**(5): 415-420.

Chapter 9

Insect Pest Management in Fruit Nurseries

H.S. Singh and S. Routray

Division of Crop Protection, ICAR–Central Institute of Subtropical Horticulture, Rehmankhera, PO Kakori, Lucknow– 226001, Uttar Pradesh

To feed the growing population, more food will need to be produced. To ensure that this goal can be accomplished, it might be necessary to adopt new technologies and modify the systems and technologies that are currently in use. India is renowned for its diverse range of fruits, from Kerala's succulent bananas to Ratnagiri's sweet mangoes. Despite being a leading producer of fruits like papayas, pomegranates, bananas, and mangoes, India struggles to meet the demand for its fruit exports and pests are a major bottleneck in attaining this goal. This challenge is largely due to issues with the health and quality of its horticultural products. Factors such as diseases, pests, and health issues may diminish the appeal of Indian fruits in international markets. Additionally, stringent quality standards set by importing countries have further complicated India's export efforts.

There have been many encouraging developments that give hope for the future. To boost export and fruit production potential, India has launched the *Atamanirbhar* Clean Plant Programme. This initiative aims to establish ten Clean Plant Centers across the country, focusing on increasing the availability of high-quality, disease-free planting materials for crops such as avocados, blueberries, apples, almonds, and grapes by 2030. The main objective of this programme is to enhance the quality and health of horticultural plants, thereby overcoming the barriers to fruit exports.

Healthy planting material is essential to the horticultural sector and linking research with practical farming for supplying both commercial growers and home gardeners with healthy, diverse fruit plants will offers significant economic and environmental benefits.

Nurseries often face challenges with insect pests, especially since cultivated plants are more susceptible than their wild counterparts. Large numbers of the same species planted close together can facilitate rapid pest spread. Commercial fruit nurseries may create an environment conducive to pest proliferation. To manage this, an effective cleanliness system is crucial. Identifying potential pest issues and understanding their life cycles are essential for maintaining hygiene. Understanding how environmental factors impact crops and pests in nurseries is vital. Addressing pest problems on-site is the most cost-effective approach, as it prevents pests from spreading to larger production areas. The primary goal is to minimize the number of pests in the nursery and prevent their spread to healthy regions.

Weeds, arachnids, nematodes, animals, and birds attack plants and inflict damage in a variety of ways. There are various groups into which pests are divided. Among them are pests that feed on roots, defoliators, gall-forming, sap-sucking, shoot-boring, stem-boring, and cutting pests. These pests mostly harm plantations, but they can also be found in nurseries to some degree and can cause large losses if nursery owners ignore them. This chapter addresses various insect pests found in fruit crop nurseries and their management strategies.

Introduction

India's fruit production is estimated to exceed 102 million metric tons, demonstrating consistent growth over the years. The total value of fruit production is projected to hit $69.75 billion in 2024, with an annual growth rate of 2.34 per cent (CAGR from 2024 to 2029), potentially reaching $78.31 billion by 2029. The export value of fruit is expected to be $0.9 billion in 2024, growing at an annual rate of 5.62 per cent (CAGR from 2024 to 2029). In 2023, fruit production was approximately 107 million metric tons, including bananas, mangoes, papayas, and apples. As a major producer of affordable fruits and vegetables, India has a substantial export market, with key exports including mango pulp, fresh mangoes, dried walnuts, and fresh grapes.

Fruit nurseries play a crucial role in maintaining agricultural health by providing high-quality, pest-free planting materials essential for productive orchards. They employ rigorous pest management techniques, such as using resistant rootstocks, certified virus-free materials, and integrated pest management (IPM) strategies, to minimize pest susceptibility in young plants.

Nurseries also focus on preserving genetic diversity and producing high-yielding varieties of true to type, significantly contribute to plant health. By adhering to strict quality control and certification standards, nurseries ensure that only healthy plants are supplied to growers, reducing disease spread risk. Additionally, nurseries function as educational centers, sharing best practices in plant health management that support sustainable agriculture. By promoting the use of healthy plants that require fewer chemical inputs, nurseries aid in environmental sustainability, benefiting both the ecosystem and the agricultural community.

Common Insect Pests of Fruit Nursery

Common insect pests in fruit crop nurseries include aphids, leaf miners, mealy bugs, scales, weevils, thrips, mites, whiteflies, termites, and other non-insects like nematodes *etc.* (Table 9.1). These pests are more prone to be dispersed via planting materials.

Table 9.1: Insect and Nematode Pests likely to Spread via Nursery

Sl.No.	*Insect Pests*	*Scientific Name*	*Host Crops*
1	Scale insects	*Aonidiella aurantii* (Maskell) *Aulacaspis tubercularis* (Newstead) *Aonidiella orientalis* (Newstead)	Mango, Guava
2	Nematodes	*Meloidogyne enterolobii* Yang and Eisenback, 1983	Guava
3	Leaf miner	*Phyllocnistis citrella* Stainton	Citrus
4	Citrus psylla	*Diaphorina citri* Kuwayama	Citrus
5	Mealy bugs	*Pseudococcus maritimus* (Ehrhorn) *Ferrisia virgata* (Cockerell, 1893) *Maconellicoccus hirsutus* (Green)	Mango, guava, citrus
6	Thrips	Various species	Mango, guava, citrus
7	Aphids	*Aphis gossypii* Glover *Aphis pomi* De Geer*Anuraphis* sp	Guava, Apple and other temperate fruits
8	San Jose scale	*Quadraspidiotus pernicious* Comstock	Apple
9	Mites	*Tetranychus* sp. *Panonychus ulmi* Koch. *Tetranychus urticae* Koch.	Guava, Citrus, Apple
10	Black fly	*Aleurocanthus woglumi* Ashby, 1915	Citrus
11	Aonla aphid	*Schoutedonia emblica* Patel and Kulkarni	Aonla

Insect Pests of Temperate Fruit Nurseries

The temperate fruit-growing regions in India, such as Himachal Pradesh, Jammu and Kashmir, Uttarakhand, and other eastern hill states, are known for producing fruits like apples, pears, peaches, plums, apricots, almonds, and

Figure 9.1: Infestation of White Mango Scale, *Aulacaspis tubercularis*.

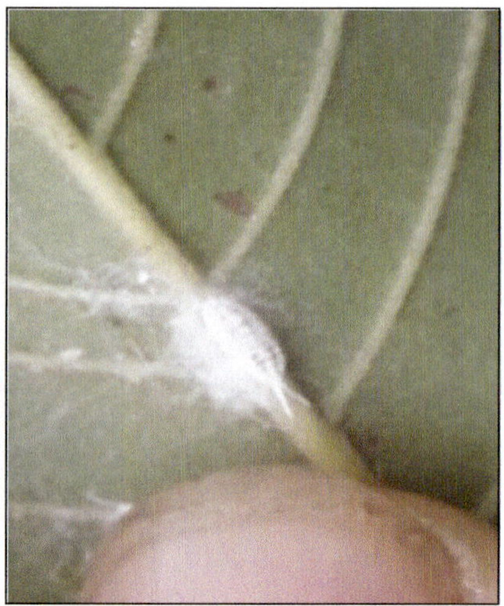

Figure 9.2: Infestation of Two Tailed Mealy Bug on Guava.

Figure 9.3: Colonies of Guava Aphids, *Aphis gossypii* on Terminal Shoots.

Figure 9.4: Infestation of Mango Mealy Bugs.

walnuts. The fruit tree ecosystem has been severely impacted by the expansion of high-yielding cultivars, monoculture practices, dense planting, intensive cropping systems, high fertilizer applications, frequent irrigation, broad-spectrum pesticide use, and the increased biotic potential of pest species. Temperate fruits are highly export-oriented and hold significant potential for generating foreign exchange. However, temperate fruit nurseries have been found to be infested with various insects. Altaf *et al.,* 2019 reported fourteen insect-pests of four different orders in apple fruit nursery; among these 4 insects belonged to Hemiptera (*Quadraspidiots pernicious, Aphis pomi, Anuraphis spp, Halyomorpha hayla*), 7 to Coleoptera (*Altica spp, Chaetocnema pulicaria, Melolontha furcicauda, Brahmina* spp *Holographic longipennis, Oryctes* spp, *Protaettia speciosa*) and 2 each to Lepidoptera

(*Choristoneura roseceana, Rhopobota navena*) and Homoptera (Typhlocyba pomaria), respectively. Apart from these, 14 insect species 2 non-insect pests (mite species) belonging to order 2 Acrina (*Panonychus ulmi, Tetranychus urticae*) were recorded from the site. All the parts of the plant like leaves flowers twigs and roots were damaged by foliage feeders while sucking insect pests sucked the plant sap. Among these San Jose scale and leaf rollers caused serious damage to the nurseries.

Table 9.2: Temperate Fruit Crop Nursery Insect Pests

Sl.No.	*Common Name*	*Scientific Name*	*Order: Family*
1	San Jose scale	*Quadraspidiotus perniciosus* Comstock	Hemiptera: Diaspididae
2	Codling moth	*Cydia pomonella* (Linnaeus)	Lepidoptera: Tortricidae
3	Archips moth	*Archips pomivora* Meyrick	Lepidoptera: Tortricidae
4	Black headed fire worm	*Rhopobotana evana* (Hubner)	Lepidoptera: Tortricidae
5	Wooly apple aphids	*Eriosoma lanigerum* (Hausmann, 1802)	Hemiptera: Aphididae
6	European Red Mite	*Panonychus ulmi* Koch.	Trombidiformes: Tetranychidae
7	Two spotted spider mite	*Tetranychus urticae* Koch.	Trombidiformes:Tetranychidae
8	Aphids	*Aphis pomi* De Geer	Hemiptera: Aphididae
9	Aphids	*Anuraphis* sp	Hemiptera: Aphididae
10	Brown marmo-rated stink bug	*Halyo morphahalys* Stal	Hemiptera: Pentatomidae
11	Flea beetles	*Altica* sp.,	Coleoptera: Chrysomelidae
12	White grub	Various spp	Coleoptera: Scarabaeidae
13	Leaf roller	*Choristoneura roseceana* (Harris, 1841)	Lepidoptera: Tortricidae
14	Pear psylla	*Cacopsylla pyricola* (Foerster)	Hemiptera: Psyllidae
15	White apple leaf hopper	*Typhlocyb apomeria* McAtee	Hemiptera: Cicadellidae

San Jose Scale (*Quadraspidiotus perniciosus*)

Several insect pests have been identified as threats to apple nurseries (Shahida *et al.*, 2019), with the San Jose scale being particularly problematic. This sucking insect poses a serious risk to apples, pears, peaches, and plums. It can infest trees from the flowering stage through to harvest, causing damage to fruits, twigs, and limbs. Without proper management, this pest can be fatal, especially to young trees. The San Jose scale is widespread in nearly all apple-growing regions and remains active from April to December in temperate areas. The toxins it injects result in localized discoloration of the plant, while the nymphs and adults continuously feed on cell sap, weakening and potentially killing young plants. Affected leaves, twigs, and fruits, and sometimes even the entire bark, may become covered with ashy-grey scales, which can be scraped off to reveal the orange pests underneath. Fruits often develop pink areas around the scales, reducing their market value (Sherwani *et al.,* 2016).

To manage this pest, avoid using infested nursery plants, buds, and graft materials. Pruning and burning heavily infested branches during the dormant period can help control pest buildup. The introduction of the parasitoid *Encarsia perniciosi* has proven effective in managing the overwintering population of this pest (Sherwani *et al.,* 2016). Green lacewing larvae and other insects are aggressive predators of the San Jose scale. Applying a dormant spray during the late dormant stage (from green tip to tight cluster stage) is crucial for controlling the scale (Paul, 2008; Sofi, 2017). Augmentative and inoculative releases of *Encarsia perniciosi* (Tower) and *Aphytis proclia* (Walker) or *Aphytis diaspidis* (Howard), at a rate of 2,000 per infested tree, have shown significant results in scale management (Rao *et al.,* 1971). In Kashmir, releasing *E. perniciosi* and *A. proclia* has increased parasitism from 8.9 to 64.3 per cent.

Ahmed Sofi and Khan (2021) observed 82.69 per cent mortality in San Jose scale with the application of 2.5 per cent Arbofine Extra (Horticultural mineral oil) and about 80.71 per cent mortality with 2.0 per ent Code-204. The activities of the parasitoids *Encarsia perniciosi* and *Aphytis proclia*, along with the predator *Chilocorus infernalis*, have been observed on infested twigs. In some cases, horticultural mineral oil at 4 liters per 200 liters of water has been applied at the half-inch green tip stage against San Jose scale. Severe infestations may require additional summer sprays with Oxy demeton methyl at 200 ml per 200 liters of water. Muneer (2023) reported 71.98 per cent mortality of San Jose scale with a 0.035 per cent concentration of Chlorpyrifos. The combination of horticultural oil and Lorsban (organophosphates) has been effective in managing scales, while Pyriproxyfen (Esteem) and Spirotetramat (Ultor) are also used for effective scale control.

Leaf Rollers

Various leaf rollers from the Tortricidae family are known to infest plants from April to October, with the highest infestation occurring during the first two weeks of July. Rolled leaves at the tips of plants indicate feeding within the terminal portions. In the temperate region of Jammu and Kashmir, 64 species of tortricid moths have been recorded, with only two causing damage to apple crops: the codling moth (*Cydia pomonella*), which is a significant issue in Ladakh, and the Archips moth (*Archips pomivora*), which occasionally appears in the Kashmir Valley (Bhagat *et al.,* 2014). Recently, the black-headed fireworm (*Rhopobota naevana*) has emerged as a serious problem, particularly affecting young apple trees in orchards.

Leaf rollers on fruit trees are typically managed effectively with dormant oil sprays. Under an alfalfa cover crop, several parasitoids have been found effective in controlling leaf rollers in apples, including the eulophid *Colpoclypeus florus*, two species of *Apantales*, two species of *Bracon*, and the tachinid parasitoid *Nemorilla pyste* (Mullinix *et al.*, 2011).

Woolly Apple Aphid (*Eriosoma lanigerum*)

Woolly apple aphids are a major pest affecting both apple nurseries and mature trees. These globally significant pests primarily feed on Malus species and have spread from eastern North America to nearly all apple-growing regions worldwide. Infestations can occur on various parts of apple trees, including roots, trunks, limbs, shoots, and occasionally fruit. Colonies of woolly apple aphids form on the above-ground parts of the tree and overwinter on the roots. In summer and fall, nymphs migrate up and down the trunks of infested trees, feeding on cell sap from tender plant parts and roots. This feeding causes swellings or knots on the roots, disrupting normal plant functions. The loss of cell sap leads to twig shriveling, and young nursery plants, which are particularly vulnerable, may die rapidly (Atwal and Dhaliwal, 1997). Severe infestations also result in undersized, poor-quality fruits.

Regular monitoring of this pest in nurseries and orchards is crucial for effective management. Pruning and applying dormant sprays can help reduce pest incidence. The exotic parasitoid *Aphelinus mali* has been found to be highly effective against woolly apple aphids (Paul, 2008). In cases where woolly apple aphids are a significant problem, using rootstocks like M111 or M106 is recommended. Additionally, applying a 0.04 per cent Chlorpyrifos solution or drenching the basin area with a 0.08 per cent Chlorpyrifos solution after harvest can be effective in managing woolly apple aphids.

Phytophagous Mites (European Red Mite, *Panonychus ulmi* and Two Spotted Spider Mite, *Tetranychus urticae*)

Young mite nymphs primarily feed on the undersides of leaves, particularly around the veins and midrib, while adults feed on both the upper and lower leaf surfaces. Mild infestations result in speckling on leaves, whereas severe infestations cause bronzing. Two-spotted mites are known for creating delicate webs on their host plants. Heavy infestations can impede fruit growth and cause premature fruit drop in certain varieties.

To manage these pests, horticultural mineral oils should be applied when buds begin to show green tissue but before they reach the pink bud stage. This timing is critical for effectively removing overwintered eggs. Encouraging the presence of various mite predators is also beneficial. The use of acaricides such as Spiromesifen, Hexythiazox, Fenpyroximate, Etoxazole, or Propargite helps keep mite populations in check. Additionally, *Stethorus punctum* and *Phytoseiulus persimilis* are effective predators of plant-feeding mites.

Pear Psylla (*Cacopsylla pyricola*)

Pear psylla is a major pest affecting pear trees (*Pyrus* species). This small insect feeds on the sap of pear leaves and shoots, with both nymphs and adults causing damage by extracting sap. This feeding weakens the tree, diminishes fruit quality and yield, and can lead to distorted growth, premature leaf drop, and reduced photosynthesis. Additionally, pear psylla can transmit pathogens that cause pear decline and other diseases. It is a significant pest in all regions of India and the Pacific Northwest pear production areas (Mahal *et al.,* 2011). Managing pear psylla is economically crucial for growers (Jenser *et al.,* 1999).

Effective management strategies for pear psylla generally include cultural practices such as sanitation, which involves removing and destroying infested plant parts. Predatory plant bugs, including *Deraeocoris, Anthocoris,* and *Campylomma,* play a key role in controlling pear psylla populations. Other natural predators include lady beetles in both their adult and larval stages, lacewings, and earwigs. Parasitic wasps such as *Trechnites* species also help suppress psylla populations. Maintaining these natural enemies within orchards is essential for successful biological control of pear psylla (NASS, 2013).

Pear psylla often quickly develops resistance to chemical control methods. Therefore, it is important for growers and advisors to alternate between different chemical formulations and modes of action to manage this pest effectively. Insecticides like Esfenvalerate (Asana XL) and Fenpropathrin (Danitol) are commonly used, but horticultural oils are recommended to delay egg laying. Additionally, Kaolin (Surround) has also been found effective.

Insect Pests of Tropical and Sub-Tropical Fruit Nurseries

In Indian context, subtropical and subtropical fruits include avocado, citrus, litchi, pomegranate, banana, carambola, guava, jackfruit, sapota, mango, papaya and pineapple. Many of these fruit trees are propagated through nursery and infested by insect pests. Mango being the major crop in subtropical has been given more attention for their pest occurrence. Patel *et al.* (1997) studied the insect pests infesting mango root stocks in nursery and their relative occurrence in south Gujarat and reported its nursery pests as thrips (*Pantachaetothrips* sp., *Selenothrips rubrocinctus* Giard, *Caliothrips impurus* Priesner, *Scirtothrips dorsalis* Hood), shoot borer (*Chlumetia transversa* Walker), mango hopper (*Amritodus atkinsoni* Lethierry), gall midge (*Procontarinia matteiana* Kieffer and Cecconi), leaf eating caterpillars (*Chelasia spathota* Meyrick, *Penicillaria jocosatrix* Guenee), leaf eating weevils *Deporaus marginellus* Faust and *Crinorrhinus crassirostris* Faust, leaf miner (*Acrocercops syngramma* Meyrick), scales (*Eucalymnatus tessellatus* Signoret and *Ceroplastes actiniformis* Green), mites (*Tetranychus cinnabarinus* Boisduval and *Oligonychus mangiferus* Rahman and Sapra), leaf mining weevil (*Rhynchaenus mangiferae* Marshall) and mealy bug (*Drosicha mangiferae* Green). The list of some of nursery insect pests is given in Table 9.3.

Table 9.3: Tropical and Sub-Tropical Fruit Crop Nursery Insect Pests

Sl.No.	*Common Name*	*Scientific Name*	*Order: Family*
1	Lemon butterfly	*Papilio demoleus* Linnaeus	Lepidoptera: Papillionidae
2	Citrus leaf miner	*Phyllocnistis citrella* Stainton	Lepidoptera: Gelechiidae
3	Citrus psylla	*Diaphorina citri* Kuwayama	Hemiptera: Psyllidae
4	Mealy bugs	*Pseudococcus maritimus* (Ehrhorn), *Ferrisia virgata* (Cockerell, 1893), *Maconellicoccus hirsutus* (Green)	Hemiptera: Psyllidae
5	Flea beetle	*Scelodonta strigicollis* (Olivier)	Chrysomelidae: Coleoptera
6	Leaf eating cater-pillar	*Spodoptera* sp	Noctuidae: Lepidoptera
7	Red mites	*Tetranychus* sp	Acari: Tetranychidae
8	Thrips	Various species	Thysanoptera: Thripidae
10	Scale insects	Various species	Coccidae: Hemiptera
11	Mango shoot borer	*Chlumetia transversa* Walker	Lepidoptera: Euteliidae
12	Guava shoot borer	*Microcolona technographa* Meyrick	Lepidoptera: Agonoxidae
13	Aphids	Various sp	Hemiptera: Aphididae
14	Guava nematode	*Meloidogyne enterolobii* Yang and Eisenback, 1983	Tylenchida: Heteroderidae
15	Litchi Erinose mite	*Aceria litchi* (Keifer)	Acari: Eriophyiidae
16	Weevil	Various sp	Coleoptera
17	Whiteflies	Various sp	Hemiptera: Aleyrodidae
18	Black fly	*Aleurocanthus woglumi* Ashby, 1915	Hemiptera: Aleyrodidae

Figure 9.5: Aggregation of Citrus Black Fly Colonies.

Figure 9.6: Infestation of Lemon Butterfly, *Papilio demoleus* on Citrus Foliage.

Lemon Butterfly (*Papilio demoleus*)

The lemon butterfly, also known as the citrus swallowtail butterfly (*Papilio demoleus*), is a major pest in citrus cultivation. Its caterpillars are particularly damaging, as they feed on young foliage in nurseries and on new growth in mature trees. These larvae consume leaf tissue aggressively, often leaving only the midrib. Severe infestations can lead to complete defoliation of the trees. Initial control measures should involve the mechanical collection and destruction of caterpillars. Alternate host plants that harbor the pest should be removed. Plants that attract predators include members of the Apiaceae (Umbelliferae), Asteraceae (Compositae), as well as buckwheat, alfalfa, corn, and shrubs. It is also beneficial to plant resistant cultivars at a distance from repellent plants and include nectar-rich plants with small flowers like caraway, parsley, anise, sunflower, mustard, cowpea, and buckwheat to attract braconid wasps. Natural enemies such as the yellow wasp (*Polistes hebreus F.*) and praying mantis (*Creobroter gemmatus*) prey on young caterpillars.

Maintaining parasitoids such as *Trichogramma evanescens*, egg parasitoids like *Telenomus* species, larval parasitoids including *Brachymeria* species, *Cotesia*, and *Distatrix papilionis* (which can achieve up to 30 per cent parasitism in India), and pupal parasitoids like *Pteromalus* species is crucial. Installing bird perches near nurseries, at a rate of 10 per acre, can help attract predatory birds. Releasing egg parasitoids like *Trichogramma chilonis* and *Telenomus* species can lead to more than 75 per cent parasitism. Promoting the activity of larval parasitoids such as *Pristomerus testaceus*, *Cremastus flavoorbitalis*, *Apanteles lunatus*, and *Pteromalus puparum* is effective against both lemon butterfly larvae and pupae.

In cases of severe infestation, applying Neem Seed Kernel Extract (NSKE) at a 0.5 per cent concentration twice, at an 8-day interval, is recommended due to its strong antifeedant and repellent properties against citrus butterflies. Botanical treatments such as Azadirachtin at 0.5 per cent concentration and Nimbecidine at 0.03 per cent (4.5 ml per liter of water) are also effective against citrus butterfly caterpillars. If pest levels exceed the Economic Threshold Level (ETL), insecticides such as Neem oil at 1500 ppm (1-1.5 ml per liter) or fungal agents like *Beauveria bassiana* and *Metarhizium anisopliae* at 5-10 ml per liter should be applied starting one month after planting, at 15-day intervals (Kumar *et al.*, 2022).

Citrus Leaf Miner (*Phyllocnistis citrella*)

The citrus leaf miner (CLM) is a persistent pest affecting nurseries, young citrus plantations, and tender shoots in groves. Its impact is particularly severe in nurseries, where it causes significant economic losses and disruption if not controlled. CLM larvae infest young leaves, creating serpentine mines within the leaf epidermis, which trap air and give the leaves a silvery appearance. Infected leaves turn pale yellow, become distorted, crumple, and eventually dry up and die. Additionally, the presence of CLM can exacerbate the development of citrus canker disease.

In citrus nurseries, generalist predators play a crucial role in reducing pest populations (Wise, 1993; Romeldiet *et al.,* 2012). Various predators of *Phyllocnistis citrella* have been identified, with *Chrysoperla* species being the primary contributors to its control. For instance, a single larva of *Chrysopa boninensis* has been reported to potentially consume up to 49 *P. citrella* larvae throughout its lifetime (Chen *et al.,* 1989). Spiders also contribute significantly to pest control (Amalin *et al.,* 2002). Parasitoids such as *Citrostichus phyllocnistoides* and *Ageniaspis citricola* have been effective in managing *P. citrella* in specific environments (Argov and Rossler, 1996). *Citrostichus phyllocnistoides* (Narayanan) is a solitary ectoparasitic species that targets 2nd and 3rd instar larvae and pupates within the leaf miner mines (Ding *et al.,* 1989; Pomerinke, 1999).

Studies show that combining *Chrysoperla* with Diver (a mineral oil) at concentrations of 15 + 0.15 per cent and 10 + 0.1 per cent resulted in 87 per cent and 67 per cent control of *P. citrella*, respectively. Additionally, combinations of *Chrysoperla* with emamectin at 15 + 0.15 per cent and 10 + 0.1 per cent achieved 50 per cent control. These findings suggest that effective control of *P. citrella* in nurseries can be achieved without relying on more toxic pesticides, thereby preserving natural enemies, benefiting the environment, and reducing control costs (Ahmed *et al.,* 2013).

Alongside natural enemies, mineral oils are highly recommended for managing *P. citrella* infestations (Khalid *et al.,* 2012). Mineral oils are commonly used in nurseries, home gardens, and orchards and are as effective as organophosphate and carbamate insecticides (Beattie and Smith, 1993). They function by reducing female oviposition when sprayed on leaf surfaces (Beattie *et al.,* 1995; Liu *et al.,* 2001). Mineral oils are safer for beneficial fauna compared to conventional pesticides, making them compatible with biological control agents (Cranshaw and Baxendale, 2005).

In orchard trials comparing various insecticides, *Thiamethoxam* 25 WG (0.06%) proved most effective, followed by *Spinosad* 45 SC (0.03%), *Imidacloprid* 17.8 SL (0.06%), *Diflubenzuron* 25 WP (0.08%), *Acetamiprid* 20 SP (0.04%), and *Diafenthiuron* 50 EC (0.05%) (Shinde *et al.,* 2017). These insecticides significantly reduced leaf miner infestations compared to untreated controls when applied as foliar sprays. In studies on lemon seedlings, *Thiamethoxam* 25 WG, *Abamectin* 1.9 EC, and *Spinosad* 48 SC showed the most notable results over a two-week period (Sharma and Bhatti, 2012). Another field study found *Imidacloprid* 0.005 per cent to be more effective than other insecticides (Patil, 2013; Perovic and Malidzan, 2001). *Acetamiprid* also demonstrated significant efficacy against *P. citrella* infestations (Zi *et al.,* 2007; Singh and Saravanan, 2008; Lad, 2006; Shinde *et al.,* 2015).

Citrus Psylla (*Diaphorina citri*)

The citrus psylla (*Diaphorina citri*) poses a significant threat to citrus nurseries by feeding on the sap of leaves, buds, and branches while injecting a toxic substance.

This feeding behavior results in raised swellings on the upper surfaces of leaves and pits or hollows on the undersides where the nymphs have fed. In severe cases, the leaves may curl, distort, and eventually fall off, leading to complete defoliation. Both the adults and nymphs of this pest feed on tender plant parts, such as buds, leaves, and branches, causing similar damage. Furthermore, the citrus psylla is a known vector for spreading 'greening' disease.

Ladybird beetles and other coccinellids are natural predators of citrus psylla nymphs. Effective management of this pest can be achieved by applying Parathion at a concentration of 0.025 per cent. Additionally, insecticides such as Imidacloprid and Thiamethoxam have proven effective against citrus psylla (Qasim and Hussain, 2015).

Mango Thrips

In India, several thrips species have been reported to infest mango trees, including *Aeolothrips collalis* (Priesner), *Anaphothrips sudanensis* (Trybom), *Caliothrips indicus* (Bagnall), *Rhipiphorothrips scruentatus* (Hood), *Neoheegeria mangiferae* (Priesner), *Haplothripsganglbaueri* (Schmutz), *Selenothrips rubrocinctus* (Giard), *Ramaswamia hiella subnudula* (Karny), *Frankliniella schultzai* (Karny), and *Scirtothrips dorsalis* (Hood). Of these, the first four species primarily feed on leaves, while the remaining five target the inflorescences.

Figure 9.7: Thrips Infested Tender Mango Twigs.

Monitoring infestation levels with sticky traps placed at regular intervals is recommended. Munj *et al.* (2020) suggest applying three weekly sprays of *M. anisopliae* (IIHR oil formulation at 0.5 ml/l) starting from the panicle initiation stage, with additional treatments during the pea and lemon-sized stages. Visalakshy and Krishnamoorthy (2016) recommend using *M. anisopliae* at a concentration of 1×10^9 spores/ml, combined with adjuvants like sunflower oil and Triton-X. For prevention, neem soap at 10 g/l or neem-based pesticides can help deter thrips establishment. Conservation of natural predators such as predatory thrips, predatory mites, anthocorid bugs (minute pirate bugs), ground beetles, lacewings, hoverflies, and spiders is also advised. Spraying Neem Seed Kernel Extract (5%) or Neem oil (2%) has proven effective in controlling early stages of thrips. In cases of severe infestation, insecticides like Tolfenpyrad 15 per cent EC, Thiamethoxam 25 per cent WG (0.3 g/l), Imidacloprid 17.8 per cent SL (0.3 ml/l), or Spinosad 45 per cent SC (0.4 g/l) can be used.

Scale Insects

Scale insects are highly damaging pests affecting both mango nurseries and mature trees. In nurseries, these pests feed on the cell sap from leaves and twigs, significantly impacting the quality of young plants. Their sedentary nature, along with the resulting chlorotic spots on leaves, twigs, and subtle blemishes on fruit skins, can often be overlooked by farmers, particularly in regions like Kenya (Noriega-Cantú *et al.,* 2016; Arias de López *et al.,* 2001). Young mango trees, especially during hot and dry conditions, frequently experience leaf loss and twig mortality due to scale infestations (El-Metwally *et al.,* 2011). Severe infestations can cause the death of small mango plants in nurseries (Abo-Shanab, 2012), while even mild infestations can delay growth in hot and dry seasons (Daneel *et al.,* 2009). The feeding of scale insects impairs photosynthesis, leading to leaves changing color from green to pale yellow (Miller *et al.,* 2005).

Figure 9.8: Infestation of Scale Insects on Mango Foliage.

Natural enemies such as Aphelinidae and Encyrtidae wasps have been observed preying on White Mango Scale (WMS) (Sayed, 2012). However, the effectiveness and distribution of these parasitoids vary by region. In South Africa, Egypt, and

Mexico, *Encarsia citrina* Crawford (Hymenoptera: Aphelinidae) has been observed parasitizing WMS, though control levels have often been below Economic Threshold Levels (ETL) (Hamdy, 2016). At times, the parasitism rate by *E. citrina* can reach up to 80 per cent (Daneel, 2009), and it is also documented in WMS populations in Andalusia, Spain (LaSalle, 1991). Recognized for its efficacy, *E. citrina* has been introduced in various countries for WMS management (Myartseva, 2007). Additionally, the ectoparasitoid *Aphytis sp.* (Hymenoptera: Aphelinidae) has been observed feeding on WMS in Australia and Egypt (Hamdy, 2020).

Pteroptrix koebelei is a native biocontrol agent known for its effectiveness against WMS, capable of reducing populations by up to 90 per cent in India (Mani, 1997). Various arthropod predators, including those from orders Neuroptera (Chrysopidae), Coleoptera (Coccinellidae and Cybocephalidae), Thysanoptera (Phlaeothripidae), and Diptera (Cecidomyiidae), as well as mites from families Cheyletidae and Stigmaeidae, have been reported preying on WMS. Notably, Cybocephalid beetles have been observed feeding on both Diaspidid scales and WMS in mango orchards (Smith and Cave, 2006).

Among the predators, *Cybocephalus rufifrons flaviceps* Reitter was identified as a major predator, accounting for 35 per cent of predation, followed by *Chrysoperla carnea* (Stephens) (Neuroptera: Chrysopidae) at 25.5 per cent, and *Chilocorus bipustulatus* (L.) (Coleoptera: Coccinellidae) at 21.4 per cent. Mite predation was less common, representing only 11.8 per cent (Sayed, 2012). Additionally, various lady beetle species from the genus *Chilocorus* (Coleoptera: Coccinellidae) have been observed preying on *A. tubercularis* on mango trees in South Africa (Labuschagne and De Beer, 1995), Australia (Klein *et al.,* 2012), India (Mani *et al.,* 1997), and other countries.

To manage scales populations, conventional broad-spectrum insecticides, including organophosphates, pyrethroids, and neonicotinoids, are commonly used worldwide. Soil drenching is also recognized as an effective control method (Miranda *et al.,* 2016). For example, Fenvalerate has been shown to control 85.3 per cent of WMS populations (Handa and Dahiya, 1999). Early stages of scales can be effectively managed with sprays of dimethoate, monocrotophos, neem oil, and quinolphos. In Kenya, the pyrethroid insecticide deltamethrin has been utilized, while in South Africa, the application of Thiamethoxam via drip irrigation has proven to be highly effective (Le Lagadec, 2004). Application rates of Thiamethoxam 25 per cent WG at 6 g/tree and 12 g/tree efficiently suppressed scale insect populations, though results showed no significant differences (Qureshi *et al.,* 2011).

Mango Shoot Borer

Several borer species affect mango shoots, including *Chlumetia transversa* Walker, *C. alternans* Moore, *Gatesclarkeanaerotias* Meyrick, *Anarsia melanoplecta* Meyrick, *A. lineatell* most common and inflicts severe damage on young plants in nurseries. In addition to mango, this borer also targets litchi leaves. It is widely distributed throughout India and has also been reported in Bangladesh, Sri Lanka, China, Java, and the Philippines (Butani 1979; Srivastava 1997).

To manage infestations, remove and destroy affected shoots. Apply quinalphos at 2 ml/l or profenophos at 1.5 ml/l, along with neem oil at 5 ml/l, twice at fortnightly intervals, and alternate the chemicals as new flushes emerge.

Figure 9.9: Damage Symptoms of Mango Shoot Borer.

Guava Shoot Borer (*Microcolona technographa*)

Guava shoot borer infestations have been noted in Punjab, with slightly higher levels observed in nurseries compared to the main fields. In West Bengal, the infestation of the shoot borer *M. technographa* typically starts from late January to mid-February, coinciding with rising temperatures and new flushes. The pest reaches its peak activity from mid-July to the second week of August (Devi and Jha, 2017). The apical buds of mature trees are more susceptible than axillary buds, leaf petioles, or leaf midribs of nursery plants. In Punjab, the pest is active from mid-April to November, with peak infestation occurring between mid-July and mid-August. Abiotic factors such as mean temperature and relative humidity significantly affect the population growth of the shoot borer (Sharma *et al.,* 2004).

To manage the pest, it is recoded to prune infested shoots. Varieties such as Red Flesh, Sardar Guava, Dharwar, and Annu Ishkwala show moderate resistance to this pest. In cases of severe infestation, spraying with Chlorpyriphos at 0.05 per cent or Quinalphos at 0.05 per cent has proven effective.

Guava Aphid (*Aphis gossypii*)

This pest mainly targets new flushes, damaging the leaves and growing shoots by sucking sap. Common signs of infestation include wilting and drooping of leaves, and ants may also be seen accompanying the pest. The affected areas often develop a black coloration due to the growth of sooty mold on honeydew secreted by the insect. The pest population is naturally controlled by its predators, particularly syrphid and coccinellid beetles. To manage the pest, clip and dispose of infested leaves and shoots at the early stages of infestation. In cases of severe infestation, it may be necessary to spray Dimethoate at a concentration of 0.06 per cent.

White Flies (*Bemesi tabaci, Aleurodicus disperses, Aleurodicus rugioperculatus*)

Whiteflies cause damage by piercing the leaves and extracting sap, which can lead to early plant death when their populations are high. Additionally, sooty mold, which grows on the honeydew produced by whiteflies and their nymphs, can also cause damage. This mold weakens the leaves by blocking sunlight, impairing their normal growth.

Spiraling whiteflies (*Aleurodicus dispersus*) are particularly active in warm, dry conditions and reproduce rapidly, making them challenging to manage. They lay their eggs on the undersides of leaves, where both nymphs and adults feed on sap, leading to the development of sooty mold. This results in yellowing and dropping of affected leaves.

Rugose spiraling whitefly (*Aleurodicus rugioperculatus*), believed to have originated from Central America, has been recorded in various Indian states, including Tamil Nadu, Karnataka, Kerala, Andhra Pradesh, Goa, and Assam. It affects a wide range of plants, such as coconut, banana, mango, sapota, guava, cashew, maize, and ornamental plants. The nymphs and adults feed on the undersides of leaves, causing significant honeydew excretion that leads to sooty mold growth on the upper leaf surfaces and understory crops. The waxy, spiraling patterns produced by the adults are also noticeable on plant parts like leaf petioles and tender nuts.

Figure 9.10: Infestation of Rugose Spiraling Whitefly (*Aleurodicus rugioperculatus*).

Whitefly populations decline during cool, rainy periods but surge when conditions are warm and dry. They can become especially problematic during droughts when natural predators may be less active. Whiteflies spread through the movement of crawler nymphs and winged adults, and their high reproductive and dispersal rates pose significant threats to banana, tropical fruit trees, vegetables, and ornamental plants. There are no specific measurements of yield losses due to whiteflies.

The wasp parasite *Encarsia haitiensis* lays eggs in immature whiteflies, killing them. Lacewings and ladybird beetles also target all stages of whiteflies. In the absence of *Encarsia*, spiraling whiteflies can reproduce rapidly, but the presence of this parasitoid offers excellent control. Horticultural oils (petroleum-based), white oils (vegetable oil-based), or soap solutions may not eliminate all whiteflies but can suppress their populations enough to allow predators and parasites to establish control. Frequent applications (every 3-5 days) with thorough coverage are necessary, particularly on the undersides of leaves and terminal buds, where whiteflies congregate. Spraying is best done between 4 and 6 pm to reduce the risk of leaf sunburn. Commercial horticultural oil can also be effective.

For control, field sanitation, removal of host plants, and installation of yellow sticky traps are recommended. During heavy infestations, applying imidacloprid 200SL at 0.01 per cent, triazophos 40EC at 0.06 per cent, neem oil 3 per cent, or NSKE 5 per cent can be effective. Additionally, releasing predators such as *Coccinellid* beetles and *Cryptolaemus montrouzieri*, as well as parasitoids *Encarsia haitiensis* and *E. guadeloupae*, has been recommended.

Guava Thrips (*Selenothrips rubrocinctus* and *Rhipiphorothrips cruentatus*)

Thrips are a major pest of guava, causing significant damage. *Selenothrips rubrocinctus* is commonly found in large numbers on the leaves of various plants, including cashew, guava, mango, and pear. This pest poses a serious threat, particularly in nurseries, and can also damage mature trees. The nymphs feed by extracting sap from the undersides of leaves, appearing yellow with a broad band around their abdomen. Adults are slender, fragile insects, dark brown in color, approximately 1-1.5 mm long, and have dark-grey heavily fringed wings. The life cycle of this thrip ranges from 23 to 35 days. They attack fruits during winter, from fruit setting until harvest in April, with populations typically low in January but peaking in March and April. Severe infestations can reduce fruit quality and yield. The sap extraction by *S. rubrocinctus* leads to leaf withering and shedding, while heavy infestations of *Rhipiphorothrips cruentatus* can cause fruit scarring and premature dropping (Harmit Singh Dahiya, 2001).

Effective control of thrips involves practicing clean cultivation and promptly removing infested plant parts at the initial signs of infestation. Conserving natural enemies like *Ciranisus maculatus* and *Tetrastichus rhipiphorothripsidis* by limiting the use of harmful chemicals or choosing safer insecticides is also beneficial. Additionally, using yellow sticky traps or yellow pan water traps can help monitor and manage these pests effectively.

Guava Nematode (Several species in various crops)

Obtaining robust seedlings or grafts is essential for achieving optimal plant density, ensuring healthy crop development, and maximizing yields. Weak seedlings or grafts, often the result of significant nematode and pathogen infestations, exhibit poor root development. Stunted roots hinder proper establishment after transplanting, leaving the plants vulnerable to secondary infections by soil-borne bacteria and fungi. Nematode damage also reduces the plant's resistance to pathogenic fungi. Furthermore, infected seedlings contribute to the spread of nematodes across larger areas, exacerbating the problem. Thus, producing seedlings or grafts free from nematode and pathogen infestations is crucial.

Nematodes cause severe root necrosis, leading to several economic issues: inefficient use of fertilizers, extended planting and harvesting periods, reduced fruit or bunch weights, compromised fruit quality, and a need for replanting every two to three years due to declining plant numbers. Nematode-damaged roots also facilitate fungal invasion, accelerating root deterioration. *Meloidogyne enterolobii* feeds on woody plants, including guava, and causes symptoms such as stunted growth, leaf yellowing, leaf shredding, branch dieback, reduced vigor, lower productivity, and plant wilting. This nematode worsens overall disease susceptibility, making plants more prone to secondary diseases like Fusarium wilt. Major nematodes affecting fruit crops include the citrus nematode (*Tylenchulus semipenetrans*), burrowing nematode (*Radopholus similis*), and root-knot nematode (*Meloidogyne incognita*), which can cause yield losses of 30 to 40 per cent.

Nematodes also contribute to an increased incidence of fungal infections, sometimes overwhelming the plant's resistance to pathogenic fungi. Infested planting materials are the primary vector for nematode spread. For instance, in bananas, the common practice of removing or clipping suckers before planting often fails to eliminate deeply penetrating viruses. Additionally, nematodes can spread through recycled irrigation water from infected areas.

Soil associated with tools, vehicle tires, and workers' footwear can transfer nematodes to new locations. The highest concentrations of nematodes are typically found at depths of 20 to 40 cm and distances of 25 to 50 cm from the plant base. Grafts and seedlings of papaya, guava, acid lime, and pomegranates are usually grown in substrate mixtures within polythene bags. These mixtures often contain nematodes and other harmful fungi and bacteria, as nurserymen rarely treat the soil.

Consequently, nematode infections can lead to the introduction of dangerous fungi and bacteria into seedlings or grafts, which subsequently spread to the farmer's crops, rendering the soil less productive and eventually unfit for cultivation. Nematodes such as *Meloidogyne incognita*, *M. javanica* (root-knot nematodes), *Rotylenchulus reniformis* (reniform nematode), and *Heterodera* sp. (cyst nematodes) infect the roots of banana seedlings during the hardening process due to poor substrate quality, which also hosts harmful bacteria and fungi.

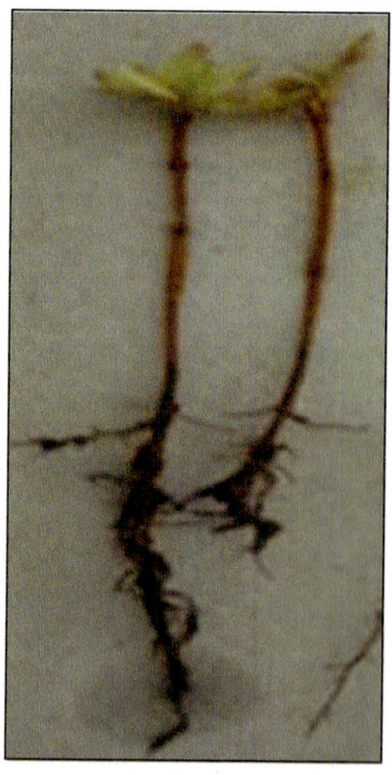

Figure 9.11: Guava Saplings Damaged by Nematodes *Meloidogyne enterolobii.*

To address nematode issues in banana seedlings, the IIHR, Bengaluru, has developed a bio-pesticide formulation containing *Trichoderma harzianum* and *Pseudomonas fluorescens*. This organic mixture includes nematicidal agents such as *Paecilomyces lilacinus* 1 per cent W.P., *Pochonia chlamydosporia* 1 per cent W.P., *Trichoderma harzianum* 1 per cent W.P., and *Trichoderma viride* 1.5 per cent W.P. The IIHR recommends this formulation for producing healthy banana seedlings and rootstocks, as well as grafts of various fruit crops.

Litchi Erinose Mite (*Aceria litchi*)

Litchi mite infestations are first detected in March and can persist until June or July. In Bihar, severe infestations occur from March to September, with a decrease in mite populations from November to February. However, mites remain active throughout the year in various stages. From January to October, they are visible on litchi trees, while from November to December, they hibernate as adults within the hairy, velvety growth known as erineum, found on the underside of leaves. Adult mites begin to proliferate by the end of March, reaching peak activity in July. Female mites lay their eggs singly on the lower leaf surfaces, usually at the base of leaf hairs. These eggs hatch within two to three days, with the emerging nymphs feeding on tender leaves.

Both nymphs and adults damage leaves, flowers, and immature fruits by piercing and lacerating leaf tissues with their robust rostrum and feeding on cell sap. Affected leaves exhibit abnormal growth of epidermal cells on the underside, forming velvety, chocolate-brown hair-like structures. Occasionally, mites induce galls, depressions, or swellings on the upper leaf surface, resembling warts. Infested fruits may develop green patches on their outer skin. Young plants and seedlings in nurseries are especially susceptible, and severe leaf drop can lead to plant death. Erineum mites typically spread between orchards through old infested litchi leaves or by wind dispersal.

Preventive measures are essential for managing litchi mites, as complete eradication once established is nearly impossible. Depending on the severity of the infestation, the following steps are recommended:

- Use only healthy, non-infested plants for layering. Treat saplings with a 0.05 per cent dimethoate solution before planting, repeating this treatment twice at 10-14 day intervals. Regularly inspect leaves for symptoms during summer and autumn. As not all trees will be infested simultaneously, cut and burn any infested branches.
- Remove infested branches after the June harvest. In September-October, treat trees just before vegetative flushing with a 0.05 per cent dimethoate solution, alone or combined with 0.12 per cent dicofol. Repeat spraying two weeks later and then monthly until new growth is free of infestation. Collect and either burn or deeply bury infested leaves.
- In December-January, remove affected shoots and spray with a 0.15 per cent kelethane solution just before the appearance of flush or flower buds. In February, perform two sprays: one before flowering and one after, which has proven effective.

Leaf Cutting Weevil (*Myllocerussp, Apoderus blandus*)

These pests primarily target leaves, shoots, and flowers. Adult weevils tend to gather on tender leaves, causing irregular holes or sometimes consuming the entire leaf, leaving only the midrib. This damage is most severe during the emergence of new shoots, as they favor fresh leaves, making newly established orchards and nurseries particularly susceptible to leaf damage

To control adult weevils, shake a branch vigorously over an open, inverted umbrella to collect the weevils, then transfer them into a container filled with soapy water. Additionally, two applications of Chlorfenapyr 10 EC at 3 ml/l, Buprofezin 25 SC at 1.5 ml/l, or Imidacloprid 17.8 SL at 0.5 ml/l have been shown to be effective against weevils.

Figure 9.12: Infestation of Leaf Cutting Weevil on Mango.

Insect Pests of Arid Fruit Nurseries

India's diverse climate and geography allow for the cultivation of various fruit crops, even in arid and semi-arid regions. Among the horticultural crops, ber (*Ziziphus mauritiana*), date palm (*Phoenix dactylifera*), lasoda (*Cordia myxa*), aonla (*Emblica officinalis*), bael (*Aegle marmelos*), jamun (*Syzygium cumini*), pomegranate (*Punica granatum*) and, khejri (*Prosopis cineraria*) are the main arid and semi-arid fruit crops. The major pests of these crops are ber fruit fly, *Carpomyia vesuviana*; ber stone weevil, *Aubeus himalayanus*; fruit borers; aphids; thrips; lemon butter fly; date palm scale, *Parlatoria blanchardi*; fruity fly, *Bactrocera cucurbitae*; leaf miner, *Liriomyza trifolii*; white flies, *Bemisia tabaci*; thrips, *Frankliniella occidentalis*; aphids, *Aphis gossypii* and ash weevil, *Myllocerus subfasciatus* are major pests which cause the considerable economic loss and increasing the cost of arid and semi-arid horticultural crops (Haldhar and Maheshwari, 2018). Various insect pests attacking in nursery stage in arid crops are enlisted in Table 9.4.

Lime Swallowtail (*Papilio demoleus*)

The larvae of *P. demoleus* present a significant threat to bael nursery stock, particularly affecting trees 1-2 feet tall and other young citrus trees. They have the potential to defoliate entire nursery groves and may also feed on young foliage of more mature trees. Infestations of *P. demoleus* in bael trees are typically observed from August to February, with peak egg and larval populations recorded as 12.75 eggs and 6.15 larvae per 10 branches, respectively, and a plant infestation rate of 65.0 per cent (Haldar and Maheswari, 2018).

Table 9.4: Arid Fruit Crop Nursery Insect Pests

Sl.No.	*Common Name*	*Scientific Name*	*Order: Family*
1	Lime Swallowtail	*Papilio demoleus* Linnaeus	Lepidoptera: Papillionidae
2	Aonlashoot gall maker	*Betousastylophora* (Swinhoe)	Lepidoptera: Thyrididae
3	Aonla mealy bug	*Nipaecoccus vastatar* Newstead	Hemiptera: Pseudococcidae
4	Aonla aphids	*Schoutedonia emblica* (Patel and Kulkarny)	Hemiptera: Aphididae
5	Termite	*Odontotermes obesus* (Rambur)	Isoptera: Termitidae
6	Pomegranate leaf eating caterpillar	*Achea Janata* (Linn.)	Lepidoptera: Noctuidae
7	Pomegranate aphid	*Aphis punicae* Passerini	Hemiptera: Aphididae
8	Anar thrips	*Scirtothrips dorsalis* (Hood) *Rhipiphorothrips cruentatus* (Hood)	Thysanoptera: Thripidae
9	Root-Knot Nematode	*Meloidogyne incognita* Chitwood 1949	Tylenchida: Heteroderidae
10	Date palm white scale	*Parlatoria blanchardii* Targioni-Tozzetti	Hemiptera: Coccidae

Manual removal of larvae from nursery stock is recommended. Natural predators of these caterpillars include the yellow wasp (*Polistes hebreus*), praying mantis (*Creobrator gemmatus*), and various spiders. Additionally, the pupal parasitoid *Pteromalus* sp. has been observed parasitizing them. *Trichogramma chilonis* and *Telenomus* sp. act as egg parasitoids for lemon butterflies, with egg parasitism rates of up to 75.9 per cent reported for both *P. demoleus* and *P. polytes* species.

Plants that effectively reduced the number of *P. demoleus* larvae by 52.8 per cent include Pipal (*Ficus religiosa*), Beshram (*Manchoria hastifolia*), *Parthenium hysterophorus*, Neem (*Azadirachta indica*), and Datura (*Datura stramonium*). Moreover, plant products such as Allitin, Replin, Margosol, and Neem Guard at a 1 perr cent aqueous extract concentration achieved over 75 per cent pest control in citrus. Allium sativum clove extracts, along with leaf extracts from *Eucalyptus globulus* and *Ageratum conyzoides*, cause morphological defects in fifth-instar larvae, making them valuable for pest management.

For chemical control, Bacillus thuringiensis applied at a concentration of 1 gram/liter has been used to manage citrus butterflies. Novaluron demonstrated a mortality rate of 94.8 per cent, followed by neem (59.4%), chlorpyriphos (59.4%), and emamectin benzoate (84.5%) (Haldar and Maheswari, 2018). Murthy *et al.* (2009) found that emamectin benzoate, chlorpyriphos, and neem at 1500 ppm were effective against *P. demoleus*, with mortality rates of 75.0 per cent and 65.0

per cent, respectively. Quinalphos 25 EC and Chlorpyriphos 20 EC also showed significant reductions in caterpillar populations (Pawandeep *et al.*, 2020).

Aonla or Indian Gooseberry Shoot Gall Maker (*Betousa stylophora*)

Haldar *et al.* (2019) report that *B. stylophora* caterpillars infest both nursery and mature Aonla plants, leading to the formation of galls on newly developed shoots. This pest is notably impactful because extensive infestations can slow tree growth, potentially affecting their fruit and flower production. The pest is active from June to December, with a preference for the rainy season, particularly when new growth begins during the monsoon (Bharpoda *et al.*, 2009). The larvae expel reddish frass through a hole at one end of their body. Once fully developed, the caterpillar enlarges this hole to exit the gall and pupates on the apical region of a branchlet, which is webbed with three to five leaves.

The gall appears as an uneven, spindle-shaped swelling or localized tumescence on the tender shoot, resembling a snake charmer's flute. *B. stylophora* is a pest exclusive to Aonla and has not been observed on other hosts. While it is generally a minor, intermittent pest, Aonla gall disease is notably prevalent in India.

The relationship between climatic factors and pest activity remains unclear. Malvia (2006) found that weather conditions did not significantly affect the occurrence of the shoot gall maker in Aonla. Conversely, Mahor (2007) identified a significant correlation between pest occurrence and average temperature, but found no relationship with relative humidity.

To manage this pest, avoid overcrowding Aonla branches during the rainy season. Remove and destroy galled shoots along with the pests after harvest. Early treatments with quinalphos (0.05%) or chlorpyriphos 20 EC (0.05%) are recommended. Research on Aonla genotypes showed that varieties such as NA-7, Kanchan, Chakaiya, Krishna, LS-2, LS-3, and LS-4 exhibited resistance to *B. stylophora* and had reduced pest incidence (Malshe *et al.*, 2016). Thanlass *et al.* (2018) and Halder *et al.* (2019) reported that Chakaiya among the tested Aonla varieties showed a 34 per cent damage rate.

For recurring infestations, initiate control measures with dimethoate (30 EC) or chlorpyriphos (20 EC) at the start of the monsoon season, repeating treatments every two weeks if necessary. Muthiah and Indiragandhi (2021) noted the effectiveness of dimethoate at 2 ml/l and chlorpyriphos at 2 ml/l for controlling shoot gall makers. Additionally, azadirachtin at 1000 ppm (1 ml/l), neem oil, or NSKE 5 per cent may be used as alternative treatments for managing Aonla shoot gall makers.

Aonla Mealy Bug (*Nipaecoccus vastatar*)

Mealybugs, once regarded as minor pests, have now become major threats due to their wide host range, high reproductive capacity, and short life cycle. Their impact has been exacerbated by prolonged drought conditions and their rapid spread through wind, seeds, and planting materials. High populations of mealybug nymphs can lead to leaf drying and dropping, which adversely affects sapling growth.

To manage mealybug infestations, it is recommended to practice clean cultivation, maintain tree health and vigor, and promptly prune and destroy affected parts at the early stages of infestation. The entomopathogenic fungus *Beauveria bassiana* has proven effective in controlling mealybug nymphs. Various predatory spiders, such as *Neoscona sp.*, *Peucetia sp.*, *Argiope sp.*, and *Oxyopes sp.*, also help reduce mealybug populations.

Frequent applications of neem oil or insecticidal soap (every couple of weeks) can effectively manage mealybug numbers. For severe infestations, the following treatments are recommended: chlorpyriphos 20 EC at 2 ml/l, imidacloprid 70WS at 0.4 ml/l, profenophos 50 EC at 2 ml/l, or Spinosad 45 SC at 0.25 ml/l.

Aonla Aphids (*Schoutedonia emblica*)

The occurrence of this pest is primarily observed from July to October, with peak infestations in September. The pest targets new shoots at their growing points, where both nymphs and adult females feed on the sap. Severe infestations can impede the growth and vitality of saplings, causing leaves to turn yellow and dry out, and shoots to become bent and twisted. The presence of ants often indicates aphid infestations, which typically occur on the new flushes at the growing points of aonla plants.

To manage the pest, it is recommended to clip and remove affected leaves and shoots. Two species of coccinellid beetles, *Menochilus sexmaculatus* and *Coccinella septempunctata*, have been observed preying on aphids. Effective treatments include spraying with neem oil at a concentration of 0.5 per cent (50 ml neem oil + 10 gm detergent in 10 liters of water), or using Dimethoate 30 EC at 0.06 per cent, Spinosad 45 SC at 0.25 ml/l, or Imidacloprid 17.8 SL at 0.40 ml/l when aphids first appear.

Termite (*Odontotermes obesus*)

Termites are a significant pest in Ber nurseries, affecting a wide range of host plants, including pomegranates, date palms, ber, aonla, lasora, and phalsa, due to their polyphagous nature. They thrive in environments with poor drainage and aeration. In Ber nurseries, termite damage can range from 40 to 50 per cent, causing plants to lose their internal parts, which become contaminated with soil and eroded at the stem height of one to two feet. Termites also affect orchards planted on virgin soil. Damage is particularly severe in nurseries and newly planted orchards, where entire populations of seedlings or saplings may dry up and die.

Heliphobic termites can build mud galleries on tree trunks or remain underground, feeding on roots before moving upwards to hollow out the trunk. The formation of clay galleries on tree trunks decreases in March and increases in

September as temperatures rise. Termites exhibit distinct castes, including workers, soldiers, reproductive, and royals.

To manage termites, maintain nursery areas clean and free of residual plant materials. Regularly remove dead and decaying wood, as well as previously infested plant sections. Eliminate mud galleries from tree trunks and treat them with either 2 per cent methyl parathion dust or a spray of 1.5 ml/l of malathion. Irrigation and finely crushed mahua cake can help repel termites. For effective control, rake and dust the soil around affected plants with 2 per cent methyl parathion at a rate of 22 to 27 kg/ha. Additionally, applying 1 ml/l of chlorpyriphos to the soil at the base of the trees is recommended.

Pomegranate Leaf Eating Caterpillar (*Achea janata*)

Numerous studies have shown that this insect is polyphagous, affecting a wide range of crops such as citrus, cocoa, coconut, pomegranate, grapevine, guava, rose, and ber. In Rajasthan, India, it has been noted for causing significant damage to ber crops by feeding on the leaves and young plants in grafted ber and pomegranate nurseries.

A female typically lays around 400 eggs on tender leaves, with an incubation period of three to five days. The larval stage can last between nine to twenty-three days, depending on environmental conditions. Pupation occurs either in leaf folds, soil, or fallen leaves, with the pupal stage lasting from seven to twenty-six days. After pupation, the insect takes six to twenty-one days to begin oviposition. The entire life cycle spans 48 to 50 days, with an average of five to six overlapping generations per year. The caterpillars feed on young fruits, delicate petioles, and leaves. Heavy infestations can lead to complete loss of foliage, leaving only the leaf veins.

In small-scale farming, caterpillars are often collected and destroyed manually. For pest management, it is recommended to spray Chlorpyriphos 20 EC at 2.0 ml/L as a preventative measure or in response to infestations. Additionally, spraying Indoxacarb 14.5 per cent SC at 1.0 mL, Spinosad 45 per cent SC at 0.5 mL, or Cypermethrin 25 per cent EC at 1.0 mL per liter of water may also effectively control the pest.

Pomegranate Aphid (*Aphis punicae*)

Aphis punicae is a major pest of pomegranate orchards. These small yellowish-green aphids, both winged and wingless, primarily infest the upper surfaces of pomegranate leaves. Both adults and nymphs form colonies and feed on plant sap by piercing leaves, stems, and even reproductive parts. This feeding causes discoloration and deformation of the affected plant parts. Additionally, the aphids excrete large amounts of honeydew, which promotes the growth of sooty mould and impairs the plant's photosynthesis. Severe infestations can lead to extensive sooty mould development due to sap extraction.

To manage this pest, it is recommended to collect and destroy damaged plant parts and use yellow sticky traps. Botanicals such as neem oil, azadirachtin, neem alkaloid, and neem seed kernel extract have proven effective for quick knock-down control. For more effective management, spraying with Imidacloprid 17.5 SL at 0.4-0.6 ml/l or Dimethoate 30 EC at 2 ml/l is advised.

Thrips (*Scirtothrips dorsalis* and *Rhipiphorothrips cruentatus*)

Thrips, including *Rhipiphorothrips cruentatus* and *Scirtothrips dorsalis*, are widely distributed polyphagous pests. While they are generally considered minor pests, they can occasionally become severe. Both nymphs and adults feed on the underside of leaves by rasping the surface and sucking the oozing cell sap, leading to browning, curling, drying, and shedding of leaf tips.

To manage these pests, it is recommended to remove and destroy affected plant parts and use blue sticky traps at a rate of 1 trap per 10 plants. Spraying with Acetamiprid 20 SP at concentrations of 0.005 to 0.01 per cent (25 to 50 g/100 L), Spinosad 45 SC at 0.25 ml/L (25 ml/100 L), or using NSKE 5 per cent or *Verticillium lecani* (2×10^8 cfu/g) at 200 g/100 L, starting before flowering and repeating at 10-day intervals, has been found effective.

Mite (*Tenuipalpus punicae*)

The pomegranate mite, *Tenuipalpus punicae*, is recognized as a pest in Punjab, Haryana, and Rajasthan. This mite affects all parts of the pomegranate plant, including leaves, stems, fruits, and flowers. It is primarily found on the underside of pomegranate leaves, particularly around the mid-rib and petiole. Persistent feeding by these mites leads to defoliation of the plant. The mite is present on pomegranate stems year-round, with peak populations occurring between May and August.

To manage mite infestations, regular monitoring is essential. Frequent irrigation during the summer can help reduce mite populations. Recommended treatments include spraying with Dicofol 18 EC at 2 ml/liter, Ethion 50 EC at 2 ml/liter, or using Wettable Sulphur 80 WP at 2 g/liter applied with a hand sprayer. Alternatively, Spiromesifen 240 SC at 0.4-0.7 ml per liter of water can be used.

Root-Knot Nematode (*Meloidogyne incognita*, Race-II)

The root-knot nematode, *Meloidogyne incognita*, is a significant pest causing substantial yield losses in pomegranate cultivation. Recent reports indicate issues with leaf yellowing, stunting, and reduced tree productivity, often due to severe infestations of root-knot nematodes as confirmed by soil and root sample analyses. In addition to direct damage, these nematodes are known for complicating disease interactions with fungi, bacteria, viruses, mycoplasmas, insects, and other nematodes. Due to their polyphagous nature and endoparasitic feeding habits, controlling them, especially in established crops, is challenging.

To manage root-knot nematodes, avoid introducing infected soil or plant materials. Ensure proper light and water drainage, use resistant rootstocks, and apply farmyard manure at 20 tonnes/ha and neem cake at 250 g/tree. Following this, the bioagent *Paecilomyces lilacinus* at 25 g/plant is recommended. Several nematicides, including Nemagon, Mocap, Dasanit, Nemacur, and Vydate, have proven effective against root-knot nematodes in pomegranates.

Date Palm White Scale (*Parlatoria blanchardii*)

The white scale insect, *Parlatoria blanchardii*, primarily infests the leaves of date palms, sucking sap and covering the leaves with white scales, which eventually cause them to dry out. Infestation begins in December and peaks in October.

Variety Tamoda has shown tolerance to this pest, while Brakawi and Jaw exhibit moderate resistance, and Mishrig and Gondaila are highly sensitive. Effective management includes implementing cultural practices that provide undergrowth to protect the natural predators of this pest. Removing infested lower leaves of the date palm is an effective method to reduce pest numbers. The predatory mite *Hemisarcoptes coccophagus* has been identified as a natural enemy of white scale. To support biological control, maintaining a less toxic environment in date palm nurseries enhances the activity and survival of these natural predators. Additionally, oil-based chemicals, such as petroleum oil and Dimethoate (30 EC), applied topically, have proven effective in controlling pests in their early stages.

Conclusion

Fruit nurseries play a crucial role in maintaining plant health and ensuring the success of fruit orchards. They provide pest-free planting materials, implement effective pest management strategies, preserve genetic diversity, and adhere to high-quality control standards. However, understanding pest dynamics within nurseries remains a challenge, and heavy reliance on synthetic insecticides can drive up production costs. Adopting biorational pest management techniques is essential for advancing sustainable agriculture. These methods offer environmentally friendly alternatives to conventional pesticides, enhancing ecosystem health and human well-being. A comprehensive approach to pest management in nurseries not only produces healthy saplings and boosts economic returns for growers but also supports the development of resilient and thriving orchards.

REFERENCES

Altaf Shahida Raina, Ishtiyaq Maqbool and Showkat, 2019. Insect pest complex of apple nurseries in North Kashmir. 697-700.

Abo-Shanab A, 2012. Suppression white mango scale, *Aulacaspis tubercularis* Newstead (Hemiptera: Diaspididae) on mango trees in El-Beheira Governorate, Egypt. *Acad. J. Biol. Sci.* **5**: 43-50.

Ahmad Sofi M and Hussain Khan Z, 2021. Management of san jose scale, *Quadraspidiotus perniciosus* Comstock in apple orchards of Kashmir through horticultural mineral oils. *Jour. of Eco-Friend. Agri.* **16**: 32–35.

Ahmed S, Shakir MM and Younis A, 2013. Integrated Management of *Phyllocnistis citrella* Stainton (Lepidoptera: Gracillariidae) through Natural Enemies,

Mineral Oil and Insecticide in a Citrus Nursery of Faisalabad, Punjab, Pakistan. *Thai J of Agri. Sci.***46**: 135-140.

Amalin DM, Peña JE, Duncan RE, Browning HW and Mcsorley R, 2002. Natural mortality factors acting on citrus leafminer, *Phyllocnistis citrella*, in lime orchards in South Florida. *Biocontrol.* **4**: 327-347.

Argov Y and Rossler Y, 1996. Introduction, release and recovery of several exotic natural enemies for biological control of the citrus leafminer, *Phyllocnistis citrella*, in Israel. *Phytoparasitica.* **24**: 33-38.

Arias de López M, Jines Carrasco A, Carrera C, Bustos Nivela P, Plúas M and Gutiérrez K, 2001. *Bioecología, DinámicaPoblacional, Muestreo, Nivel de Daño y Alternativas Para elManejoSostenible de Aulacaspis tubercularis en Mango de Exportación*; INIAP, Estación Experimental Boliche: Guayaquil, Ecuador.

Atwal AS and Dhaliwal GS, 1997. Pests of temperate fruits. In: Agricultural pests of South Asia and their management. Kalyani publishers, Ludhiana. Pp. 274-286.

Beattie GAC and Smith D, 1993. Citrus Leafminer. Agfact H2.AE.4, 2nd ed. NSW Agric. Sydney.

Beattie GAC, Liu ZM, Watson DM, Clift AD and Jiang L, 1995. Evaluation of petroleum spray oils and polysaccharides for control of *Phyllocnistis citrella* Stainton (Lepidoptera: Gracillariidae). *J. Aus. Entom.* **34**: 349-353.

Bhagat D, Bhardwaj S, Sharma AK and Chopra V, 2014. Warning system for European red mite using Beta Model. *Int. J. of Sci. and Res. Pub.* **4**: 1-5.

Bharpoda TM, Koshiya DJ and Korat DM, 2009. Seasonal occurrence of insect-pests on aonla *Emblica officinalis* (Geartn) and their natural enemies. *Karn. J. of agri. Sci.* **22**: 314-318.

Butani DK, 1979. Insects and fruits. Periodical Expert Book Agency, Delhi

Chen RT, Chen YH and Huang MD, 1989. Biology of green lacewing, *Chrysopa boninensis* and its predation efficiency to citrus leafminer, *Phyllocnistis citrella*, pp. 96-105. In Studies on the Integrated Management of Citrus Insect Pests, Beijing, China Academic Book and Periodical Press.

Cranshaw W and Baxendale B, 2005. Insect Control: Horticultural Oils. Colorado State University Extension Service. Publication No. 5.569.

Daneel M and Joubert P, 2009. Biological control of the mango scale *Aulacaspis tubercularis* Newstead (Coccidae: Diaspididae) by a parasitoid *Aphytis chionaspis* Ren (Hymenoptera: Aphelinidae). *Acta Hortic.* **820**: 567-574.

Devi AR and Jha S, 2017. Incidence pattern of guava shoot borer, *Microcolona technographa* Meyrick (Lepidoptera: Agonoxidae) in relation to weather conditions in West Bengal, India. *Pest Manag. Hort. Eco.* **23**: 1-6.

Ding Y, Li M and Huang MD, 1989. Studies on the Integrated Management of Citrus Insect Pests, Beijing, China Academic Book and Periodical Press.

El-Metwally MM, Moussa SF and Ghanim NM, 2011. Studies on the population fluctuations and distribution of the white mango scale insect, *Aulacaspis tubercularis* Newstead within the canopy of the mango trees in eastern of Delta region at the north of Egypt. *Egypt. Acad. J. Biol. Sci. A Entomol.* **4**: 123-130.

Gomes VM, Souza RM, Mussi–Dias V, Silveira SF and Dolinski C, 2011. Guava decline: A complex disease involving *Meloidogyne mayaguensis* and *Fusarium solani. Jour of Phyto.* Jan; **159**(1): 45-50

Gomes VM, Souza RM, Silva MM and Dolinski C, 2008. Caracterização do estado nutricional de goiabeiras em declínio parasitadas por *Meloidogyne mayaguensis. Nematologia Brasileira.* Jun; **32**(2): 154-60.

Haldhar S, Ramesh K and Saroj PL, 2019. Integrated pest management in aonla crop. In book: Integrated Pests Management of Arid Fruit and Vegetable crops Publisher: ICAR–CIAH, Bikaner

Haldhar SM and Maheshwari SK, 2018. Insect-pests management in arid and semi-arid horticultural crops. Technical Bulletin No: 64, ICAR–CIAH, Bikaner

Haldhar SM, Chet R and Singh D, 2019. Biotic stress (insect) of aonla (*Emblica officinalis*) in arid region of India: A review. *J. Agri. and Eco.* **7**: 16-26.

Hamdy NM, 2016. Some ecological aspects on mango white scale, *Aulacaspis tubercularis* and associated natural enemies infesting mango trees in *Qalyubiya govrnorate* (Hemiptera: Sternorrhyncha: Diaspididae). *J. Plant Prot. Pathol.* **7**: 377-383.

Hamdy NM, 2020. Life Table and Morphometric Studies of *Aulacaspis tubercularis* Infesting Mango Trees in Egypt (Diaspididae: Hemiptera). *J. Plant Prot. Pathol.* **11**: 613-620.

Handa S and Dahiya KK, 1999. Chemical management of scales. *Agric. Sci. Dig. Karnal.* **19**: 112-114.

Harmit Singh Dahiya KK, 2001. Studies on morpho-chemical attributes for resistance against thrips, *Rhipiphorothrips cruentatus* Hood in guava, *Psidium guajava. Har. J. of Hort. Sci.* **30**: 62-65.

Jenser G, Balázs K, Erdélyi C, Haltrich A, Kádár F, Kozár F, Markó V, Rácz V and Samu F, 1999. Changes in arthropod population composition in IPM apple orchards under continental climatic conditions in Hungary, *Agr. Ecosyst. Environ.* **73**: 141-154.

Khalid M, Malik A, Saleem B, Khan A and Javed N, 2012. Horticultural mineral oil application and tree canopy management improve cosmetic fruit quality of Kinnow mandarin. *Afric. J. Agric. Res.* **7**: 3464-3472.

Klein AM, Vaissiere B, Cane JH, Steffan-Dewenter I, Cunningham SA, Kremen C and Tscharntke T, 2007. Importance of pollinators in changing landscapes for world crops. *Proc. R. Soc. B Biol. Sci.* **274**: 303-313.

Kumar A, Kumar P, Mishra PK, Aman AS and Tripathi P, 2022. Integrated Approach for Management of Citrus Butterfly (*Papilio demoleus* Linnaeus). **4**: 32-35.

Labuschagne T and De Beer M, 1995. First successful import of parasitoids of the mango scale (*Aulacaspis tubercularis* Newstead) to South Africa. *S. Afr. Mango Grow. Assoc. Yearb.* (*S. Afr.*). **15**, 106-108.

Lad DV, 2006. Seasonal Incidence and Management of *Phyllocnistis citrella* Stainton. M.Sc. Agri. Thesis (unpub).

LaSalle J, 1991. Armoured scale insects: Their biology, natural enemies and control. *Bull. Entomol. Res.* **81**: 497.

Le Lagadec M, 2004. The control of mango scale in commercial orchards through the use of the predatory beetle, *Cybocephalus binotatus. Acta Hortic.* **645**: 509–515.

Liu Z, Hodgkinson M, Rose H and Jiang L, 2001. Influence of petroleum-derived spray oil aromaticity, equivalent n-paraffin carbon number and emulsifier concentration on oviposition by citrus leafminer, *Phyllocnistis citrella* Stainton (Lepidoptera: Gracillariidae). *Aust. J. Entomol.* **40**: 193-197.

Mahal MS, Singh B, Sarao PS and Singh S, 2011. Role of insect pest forecasting in integrated pest management In: Arora R, Singh B and Dhawan A K(eds), Theory and Practice of Integrate Pest Management, Scientific Publishers, Jodhpur, India. 248-261.

Mahor B, 2007. Studies on non-hemipteron insect pests of aonla, *Phyllanthus emblica* Linneus., M.Sc. (Ag.) Thesis (Unpublished), JNKVV, College of Agriculture, Gwalior (M.P.).

Malshe K, Salvi B and Gawankar M, 2016. Evaluation of different varieties of Aonla (*Emblica officinalis* Gaertn) under hard lateritic rocky conditions of South konkan coastal zone of Maharashtra. *New Agricul.* **27**: 135-138.

Malvia TN, 2006. Studies on insect pest complex of aonla. M.Sc. (Ag.) Thesis (Unpublished), College of Agriculture, JNKVV, Jabalpur (M.P.).

Mani M, Krishnamoorthy A and Verghese A, 1997. Suppression of the soft green scale *Coccus viridis* (Green) (Coccidae: Homoptera) on acid lime in India. In Proceedings of the First National Symposium on Pest Management in Horticultural Crops: Environmental Implications and Thrusts, Bangalore, India, 15–17 October 1997; pp. 210–212.

Miller DR and Davidson JA, 2005. *Armored Scale Insect Pests of Trees and Shrubs* (Hemiptera: Diaspididae); Cornell University Press: Ithaca, NY, USA, 2005.

Miranda MP, Yamamoto PT, Garcia RB, Lopes JP and Lopes JR, 2016. Thiamethoxam and imidacloprid drench applications on sweet orange nursery trees disrupt the feeding and settling behaviour of *Diaphorina citri* (Hemiptera: Liviidae). *Pest Manag. Sci.* **72**: 1785-1793.

Mullinix K, Brunner JF and Isman MB, 2011. Apple Leafroller (Lepidoptera: Tortricidae) Populations and Parasitism in an Orchard Managed With Either a Grass or Alfalfa Cover and Without Insecticides Over Four Growing Seasons. *Int. J. of Fruit Sci.* **11**: 99-110. https: //doi.org/10.1080/15538362.2011.578509.

Muneer A, 2023. Management of San Jose Scale (*Quadraspidiotus perniciosus*) by HMOs and Insecticides in Apple Orchards of Kashmir, India. *Int. J. of Plant Soil Sc.***35**: 1621-1627. 10.9734/ijpss/2023/v35i183434.

Munj AY, Reddy PV and Irulandi S, 2020. Efficacy of entomopathogen formulations against thrips, *Scirtothrips dorsalis* hood on mango: A multilocation study. *Pest Manag.Hortic. Ecosyst.* **26**: 179-183.

Murthy KSRK, Reddy AR and Yogi K, 2009. Efficacy of Certain Eco-friendly Pesticides against citrus butterfly, *Papilio demoleus. Ind. J. Plant Prot.* **37**: 46-49.

Muthiah C and Indiragandhi P, 2021. Efficacy of bio rational insecticides against major pests of aonla under arid zone farming ecosystem. *J. Ento. and Zool. Stud.* **9**: 228-231.

Myartseva S and González-Hernández A, 2007. *Encarsia citrina* (Craw) (Hymenoptera: Aphelinidae), un parasitoide de las escamas armadas (Homoptera: Diaspididae) en México. *Folia Entomol. Mex.* **46**: 101-106.

NASS, National Agricultural Statistics Service, U.S. Department of Agriculture, 2013.

Noriega-Cantú D, Urias-Lopez M, Gonzalez-Carrillo J and López-Guillén G, 2016. Seasonal abundance of white mango scale, *Aulacaspis tubercularis* Newstead, in Guerrero, Mexico. *Southwest. Entomol.* **41**, 845–853.

Patel JR, Patel MB, Radadia GC, Shah AA and Pandya HV, 1997. Insect pest management in mango nursery. *Journal of Applied Horticulture Navsari,* **3**(1/2): 125-128

Patil SK, 2013. Evaluation of insecticides against citrus leaf miner, *Phyllocnistis citrella* Stainton in acid lime. *Pest manag. Horti. Eco.* **19**: 237-239.

Paul A, 2008. Insect pests and their management. In: Phylogeny and classification of Rosaceae. Potter D, Eriksson T, Evans RC, *et al., Plant Syst. and Evo.* **266**: 5-43.

Paul A, 2008. Insect pests and their management. In: Phylogeny and classification of Rosaceae. Potter D, Eriksson T, Evans RC, *et al.,* 2007. Plant Systematics and Evolution, **266**: 5-43.

Pawandeep Singh K and Singh A, 2020. Management of *Papilio demoleus* L. using Insecticides on Kinnow nursery plants at Talwandi Sabo, Punjab. *J. Pharmac. Phytochem.* **9**: 3368-3369.

Perovic TS and Malidzan, 2001. Chemical control of citrus leaf miner *Phyllocnistis citrella* Stainton (Lepidoptera: Gracillariidae, Phyllocnistinae) [Serbian]. *Jugoslovensko Vocarstvo*. **35**: 175-182.

Pomerinke M, 1999. Biological control of citrus leafminer, *Phyllocnistis citrella* (Lepidoptera: Gracillariidae) in Southwest Florida. PhD, University of Florida, Gainesville, Florida, USA.

Qasim M and Hussian D, 2015. Efficacy of insecticides against citrus psylla (*Diaphorinacitri*Kuwayama) in field and laboratory conditions. *Cercetări Agronomiceîn Moldova*. **48**. 10.1515/cerce-2015-0033.

Qureshi MS, Thistleton B, Syeda SS, Hearnden M and Qureshi M, 2011. Managing mango leafhoppers and other associated species affected through systemic insecticides in mango orchards at Darwin. Australia. *Pak. J. Entom.* **26**: 81-87.

Rao VP, Ghani MA, Sankaran T and Mathur KC, 1971. A Review of the biological control of insects and other pests in South-East Asia and the Pacific region. *Commonw. Inst. Biol. Control Tech. Commun.,* **6**: 1-149.

Romeldi F, Schneider MI and Ronco AE, 2012. Short and Long-term effects of endosulfan, cypermethrin, spinosad, and methoxyfenozide on adults of *Chrysoperla externa* (Neuroptera: Chrysopidae). *J. Econ. Entom.* **105**: 1982-1987.

Sayed AMM, 2012. Influence of certain bio-agents and climatic changes on the population density of the white mango scale insect, *Aulacaspistuber cularis* Newstead. *Egypt. J. Agric. Res.* **90**: 607-624.

Shahida A, Ishtiyaq R and Showkat M, 2019. Insect pest complex of apple nurseries in North Kashmir. 697-700.

Sharma DR, Bhatti IS, 2012. Bio-Efficacy of Neonicotinoid Insecticides against *Phyllocnistiscitrella* Stainton in Drip-Irrigated Kinnow Plants. *Ann. of Plant Prot. Sci.* **20**: 334-337.

Sherwani A, Mukhtar M and Wani AA, 2016. Insect pests of apple and their management. Insect pest management of fruit crops. New Delhi: Biotech Books, pp. 295-306.

Shinde SR, Tambe VJ, Kumar G, Sarvan D and Rama A, 2015. Evaluation of Newer Molecule of Insecticides Against Citrus leaf miner (*Phyllocnistis citrella*) Sanitation in nursery. *J. Tre. in Bio-sci.* **7**: 2552-2555.

Shinde SS, Neharkar PS, Dhurve NG, Sawai HR, Lavhe NV and Masolkar DS, 2017. Evaluation of different insecticides against citrus leaf miner on Nagpur mandarin. *J. of Ento. and Zool. Stud.* **5**: 1889-1892.

Singh H and Saravanan L, 2008. Seasonal incidence and management of pea leaf miner *Phytomyza horticola* Goureau infesting pea. *Int. J. of Plant Prot.* **1**: 33-37.

Smith TR and Cave RD, 2006. Pesticide susceptibility of *Cybocephalus nipponicus Rhyzobius lophanthae* (Coleoptera: Cybocephalidae: Coccinellidae). *Fla. Entomol.* **89**: 502-507.

Sofi MA, 2017. Major insect pests of apple grown under temperate conditions of Kashmir and their integrated pest management. https: //www. researchgate. net.

Sood AK, 2010. Integrated Pest Management under Protected Environment: Principles and Practices. *Agropedia*. **29**: 13-21.

Srivastava RP, 1997. Mango insect pest management (No. Ed. 1). International Book Distributing Co, Lucknow

Thanlass N, Hafeez A, Uma S and Mahesh K, 2018. Population dynamics of fruit borer, *Deudorix Isocrates* of aonlafabr. (Lycaenidae: Lepidoptera) in relation to abiotic factors. *J. Ento. and Zool. Stud.* **6**: 1790-1793.

Visalakshy GPN and Krishnamoorthy A, 2012. Comparative field efficacy of various entomopathogenic fungi against *Thrips tabaci*: prospects for organic production of onion in India, *Acta Hort.* **933**: 433-437.

Wise DH, 1993. Spiders in Ecological Webs. Cambridge University Press, Cambridge; New York.

Zi C, Hu J, Wei H and Xia, 2007. Evaluation on the controlling effects of eight insecticides on the natural population of citrus leaf-miner. *Chi. J. of Eco Agri.* **15**: 109-112.

Chapter 10

Management of Nursery Diseases of Fruit Crops

M.L. Supriya, G.Y. Rakhonde and S. Sriram

Division of Crop Protection; ICAR–Indian Institute of Horticultural Research, Bengaluru–560089, Karnataka

Nursery diseases in fruit crops pose significant challenges to productivity and economic viability, impacting plant health, yield potential, and market value. The management of nursery diseases has become an increasingly critical aspect of horticultural practices due to the growing recognition of its impact on crop yield, quality, and economic viability. Nursery diseases, caused by a diverse range of pathogens including fungi, bacteria, viruses, and nematodes, leads to significant losses during early stages of plant development. The emerging importance of nursery disease management, emphasizing necessity for integrated approaches that combine cultural practices, chemical treatments, and biological controls was highlighted. Advances in diagnostic techniques, including molecular tools, have improved early detection and precise identification of pathogens, enabling more effective disease management strategies. Additionally, role of environmental factors and climate change in influencing prevalence and severity of nursery diseases is also important. The integration of sustainable practices and development of Integrated disease management (IDM) is essential for long-term disease management. This chapter underscores the urgency of addressing nursery diseases comprehensively to ensure the health and productivity of crops, ultimately contributing to food security and agricultural sustainability.

Introduction

Nursery plays a crucial role in propagating plants for fruit, vegetable, forestry, and ornamental species, forming the backbone of many horticultural production systems. The foundation of a successful fruit production system begins in the nursery, where young plants are propagated and nurtured. Ensuring the health and vigor of these plants is crucial stage for the entire lifecycle of fruit crops, and therefore hold immense economic importance. However, nurseries are highly susceptible to plant diseases, causes significant economic repercussions within and beyond nursery production system, especially if diseased stock is not detected before distribution. The introduction of infected plant material, including live plant stocks, into nurseries is believed to account for approximately 69 per cent of pathogen outbreaks globally (Liebhold *et al.,* 2012). Several reports document spread of diseases from greenhouses to agricultural, horticultural, and forestry lands (Liebhold *et al.,* 2012; Parke *et al.,* 2014). The resulting losses can be substantial, as seen in the US state of Georgia, where plant disease-related losses in 2015 were approximately $800 million, up from $647.2 million in 2006 (Martinez-Espinoza, 2013). Pathogen outbreaks are often linked to the movement of nursery plants. Unlike agricultural monocultures, nurseries typically host a diverse range of plants within a confined area, increasing variety of potential pathogens and complicating disease detection and management. Pathogen infections in plants may manifest as visible symptoms or latent, making it challenging to identify infected plants without formal diagnostic testing.

Traditional methods of managing nursery diseases have primarily relied on chemical treatments. While effective in the short term, these methods pretense environmental risks and lead to the development of fungicide resistance, making disease control more challenging task. Consequently, there is an increasing need for innovative approaches that integrate advancements in technology, biology, and sustainable agricultural practices. Biological control methods offer sustainable alternatives to chemical treatments. The use of beneficial microorganisms and biocontrol agents can suppress pathogens while promoting overall plant health. Practices such as soil solarization and strict sanitation measures are crucial in preventing the spread of pathogens in nurseries. IDM is a holistic approach that combines cultural, biological, and chemical methods to manage diseases sustainably.

Understanding nursery diseases and their management is crucial for enduring research and development efforts. By studying these diseases, researchers can develop new resistant cultivars, improve diagnostic techniques, and create more effective management strategies. This supports continuous improvement of nursery practices and overall resilience of fruit production systems. Nursery diseases of fruit crops are of vital significance due to their far-reaching impacts on economic viability, plant health, and environmental sustainability. Effective management of diseases at nursery stage is indispensable for ensuring long-term productivity and quality of fruit crops. By integrating cultural, biological, chemical, and biotechnological approaches, growers can protect their investments, meet regulatory requirements, and contribute to sustainable agricultural practices.

Furthermore, educational and extension services play a vital role in promoting adoption of novel approach. By current research and practical insights, these services aim to guide nursery managers, researchers, and policymakers in adopting sustainable and effective management practices. Ensuring production of healthy and robust fruit crops not only secures economic benefits but also contributes to overall sustainability of agricultural systems. By disseminating knowledge and best practices, these services ensure that nursery managers and growers are equipped to implement effective disease management strategies.

Importance of Nursery Diseases of Fruit Crops

Nursery diseases lead to significant economic losses for growers. Infected plants often suffer from stunted growth, poor vigor, and reduced survival rates, leading to increased costs for replanting and additional treatments. The sale of diseased plants results in customer dissatisfaction and loss of market reputation, impacting future business prospects. Nurseries are starting point for lifecycle of fruit crops, and health of nursery plants directly influences performance in field. Diseases affecting young plants can compromise development, reducing their ability to endure environmental stresses and affecting their overall yield and quality once transplanted. Diseases originating in nurseries can spread to orchards and plantations, becoming more complicated to manage. Pathogens can be transported *via* infected plant material, soil, and tools, leading to widespread outbreaks affecting large areas and multiple growers. The quality of planting material is crucial for successful fruit production. Disease-free nursery plants ensure that growers start with a strong foundation, leading to healthier crops and higher yields. Conversely, infected nursery plants can introduce pathogens into new environments, undermining growers' efforts to maintain healthy orchards. Effective management of nursery diseases is an essential component of IDM strategies. By controlling diseases at nursery stage, growers can reduce need for chemical treatments later in crop's lifecycle, promoting more sustainable and environmentally friendly practices. Early intervention and prevention of nursery diseases are key elements of successful IDM programs (Valatin *et al.,* 2022).

Many countries have stringent regulations and certification programs to ensure health and quality of nursery plants. Compliance with these standards is necessary for sale and distribution of plants, both domestically and internationally. Managing nursery diseases efficiently helps growers meet these regulatory requirements and access broader markets. The use of chemical treatments to control diseases can negatively impact environment, including soil health, water quality, and non-target organisms. By managing nursery diseases through integrated approaches that emphasize cultural and biological controls, growers can reduce their reliance on chemicals, leading to more sustainable farming practices and reduced environmental impact.

Major Nursery Diseases of Fruit Crops

Nurseries of fruit crops are vulnerable to a range of diseases that can significantly impact plant health and productivity. Fire blight, caused by the bacterium *Erwinia amylovora*, affects apples, pears, and other members of the *Rosaceae* family, leading to blossom blight and shoot dieback. Powdery mildew, caused by fungi such as *Podosphaera* spp. and *Erysiphe* spp., affects apples, grapes, strawberries, and stone fruits, manifesting as white, powdery fungal growth on leaves and stems. Downy mildew, caused by oomycetes like *Plasmopara* spp. and *Peronospora* spp., affects grapes and cucurbits, resulting in yellowing leaves and mold on the undersides of leaves. Root rot, caused by pathogens such as *Phytophthora* spp. and *Fusarium* spp., can devastate citrus, avocados, and strawberries by causing root decay and stunted growth. Leaf spot diseases, caused by fungi like *Cercospora* spp. and *Alternaria* spp., lead to necrotic spots on leaves, which can result in defoliation in crops like strawberries and bananas. Rust diseases, caused by fungi such as *Puccinia* spp. and *Gymnosporangium* spp., affect apples, pears, and stone fruits, presenting as orange to brown pustules on leaves and stems. Verticillium wilt, caused by the fungus *Verticillium* spp., affects strawberries and various fruit trees, leading to yellowing and wilting of leaves. Bacterial blight, caused by *Xanthomonas* spp., impacts stone fruits like peaches and cherries, causing water-soaked lesions and cankers. Virus diseases, such as Cucumber mosaic virus (CMV), affect melons, leading to mottling. Lastly, anthracnose, caused by *Colletotrichum* spp., affects mangoes, avocados, and bananas, causing dark, sunken lesions on various plant parts. Effective management of these diseases involves regular monitoring, adopting cultural practices to reduce disease pressure, using resistant varieties, and applying appropriate control measures (Parke and Grünwald, 2012; Agrios, 2005; Chen *et al.*, 2021).

Phytophthora root rot (PRR) has been a significant problem in nursery industry for nearly a century. *Phytophthora* infestations are common in nurseries, leading to average losses of 10 to 20 per cent, with some production blocks experiencing total losses (Weiland *et al.*, 2021). A large European survey found that 91.5 per cent of nurseries across 23 countries were infested with *Phytophthora* species (Jung *et al.*, 2016), and similar infestation rates have been reported in the USA and Australia (Rooney-Latham *et al.*, 2019). Various *Phytophthora* species, *P. cactorum, P. cambivora, P. cinnamomi, P. citrophthora, P. drechsleri, P. lateralis, P. megasperma, P. nicotianae, P. niederhauserii, P. occultans, P. pini, P. plurivora, P. pseudocryptogea,* and *P. tentaculata* responsible for PRR in nurseries. Most of these species reside in soil, infecting and killing roots, leads to above-ground symptoms like chlorosis, anthocyanosis, stunting, wilting, and plant death. Often, multiple *Phytophthora* species are involved in PRR on a single host, and many of these species have wide host ranges. This is particularly concerning because nurseries typically grow multiple crop species, with many nurseries cultivating dozens or even hundreds of plant genera. This diversity increases likelihood that any *Phytophthora* species will find a suitable host within a nursery. Additionally, national and international trade of nursery stock plays a crucial role in spreading *Phytophthora* globally, posing a threat not only to other nurseries but also to commercial tree plantations, agricultural fields, and natural areas, where they can cause significant economic and ecological damage (Weiland, 2021).

Phytophthora species affecting citrus, leads to substantial production losses, particularly in high rainfall subtropical regions. This includes major citrus production areas in São Paulo, Brazil, and Florida, USA. In seedbeds, *Phytophthora* causes damping-off; in nurseries, leads to foot rot and root rot; and in groves, it results in foot rot, fibrous root rot, and brown rot of fruit. The pathogen can spread to adjacent fruit in packing boxes, exacerbating losses. The most serious disease caused by *Phytophthora* is foot rot, also known as gummosis. In drier climates, water-soluble gum produced by infection is not washed away by rainfall. The infection typically occurs near ground level, creating lesions that extend down to bud union on resistant rootstocks or up trunk into the major limbs of the tree. This damages the cambium and inner bark, and as the lesions spread around the trunk's circumference, they girdle the cambium, ultimately killing the tree. Nursery trees and young grove trees with small trunk circumferences can be rapidly girdled and killed by the infection. *Phytophthora* also infects emerging seedlings, causing damping-off, and can decay the root cortex, leading to the decay of fibrous roots. Root rot is particularly severe on susceptible rootstocks in infested nursery soil or on young nursery trees planted in infested soil. In small trees, significant root loss can result in tree death. In fruit-bearing groves, root rot occurs on susceptible rootstocks, typically not killing the tree but causing a decline in vigor and fruit production. Infection of citrus fruit by *Phytophthora* results in brown rot, characterized by a light brown and leathery rind on the affected fruit (Feichtenberger, 2001: Graham and Feichtenberger, 2015).

Grapevine fungal trunk diseases (GTDs) are a key issue for grapevines globally. The primary ways GTD pathogens spread is through nursery vines infected during propagation. In Uruguay, nursery grapevine plants have a high incidence of GTDs, regardless of scion/rootstock combination. Throughout all propagation stages, typical brown to black streaks and sectorial wood necrosis were observed, with the incidence markedly increasing as the nursery process progressed, reaching nearly 100 per cent in grafted rooted vines ready for planting. The main GTDs identified were *Botryosphaeria* dieback, Petri disease, and black-foot disease. *Botryosphaeria* dieback and Petri disease pathogens infect materials from early stages, with a significant increase towards end of plant production process. In contrast, black-foot disease pathogens were found exclusively in vines ready to plant. Diaporthe dieback pathogens were present at all stages but in low proportions. Notably, *Phaeoacremonium oleae* and *Diaporthe terebinthifolii* were recorded as new pathogens on grapevines worldwide. The locally produced vines are a significant means of GTD pathogen dissemination, highlighting the need for a nursery sanitation program to reduce the incidence of these diseases (Carbone *et al.,* 2022). Various fungi, bacteria, and viruses affecting nursery fruit crops are detailed in Table 10.1.

Table 10.1: Fungi, Bacteria and Viruses Affecting the Nursery Fruit Crops

Sl.No.	Disease	Causal Organism	Host	References
1	Root and collar rot	*P. aphanidermatum*	Papaya	Koffi *et al.* (2010).
2	Rotting of mango stones and wilting of root stocks	*Macrophomina phaseolina, R. solani*	Mango	ICAR Complex Goa, (2009)
3	Apple replant disease (ARD)	*Pythium rregulare, Pythium ultimum, Pythium sylvaticum Phytopythium littorale*	Apple	Moein *et al.* (2019).
4	Crown and root diseases	*Phytophthora cactorum and Verticillium spp.*	Strawberry	Villarino *et al.* (2021)
5	Angular leaf spot (ALS)	*Xanthomonas fragariae*	Strawberry	Turechek *et al.* (2021)
6	Anthracnose	*C. gloeosporioides and C. acutatum*	Strawberry	Wu *et al.* (2022)
7	Black root rot and crown rot	*Pestalotiopsis spp., Rhizoctonia fragariae and Cylindrocarpon destructans.*	Strawberry	Lieten *et al.* (2021)
8	Citrus variegated chlorosis	*Xylella fastidiosa*	Citrus	Coletta-Filho *et al.* (2020).
9	Banana Bunchy Top Disease (BBTD)	*Banana bunchy top virus*	Banana	Simbare *et al.* (2020).
10	Mango Malformation	*Fusarium moniliforme* var. *subglutinans*	Mango	Kumar *et al.* (2021).
11	Almond band canker and prune Cytospora canker	*Cytospora spp. B. dothidea, Diplodia spp., Lasiodiplodia spp., Neofusicoccum spp.,* and *Phomopsis spp.*	Almond	Luo *et al.* (2024)
12	Black foot disease (BFD)	*Campylocarpon fasciculare* and *Campylocarpon pseudofasciculare; Dactylonectria alcacerensis, Dactylonectria macrodidyma, Dactylonectria novozelandica, Dactylonectria pauciseptata* and *Dactylonectria torresensis* as well as *Ilyonectria liriodendra*	Grapes	van Jaarsveld *et al.* (2021)
13	Grapevine trunk diseases	*F. annulatum*	Grapes	Bustamante *et al.* (2024)
14	Root rot	*Globisporangium ultimum, Phytopythium mercuriale,* and *Phytopythium helicoides*	Apple and peach	Mannai *et al.* (2024)

Emerging High Throughput Diagnostic Techniques

Early detection of pathogens in nurseries is crucial in preventing spread of disease beyond nursery environment. This necessitates use of rapid, specific, sensitive, and cost-effective diagnostic techniques that can routinely screen for notorious pathogens before plants are sold or transported for large-scale production. Traditional methods for detecting nursery diseases involve assessing symptoms, host range, and identifying isolated organisms through microscopy or biochemical tests. These methods are complex, require trained professionals, exhibit low sensitivity, are time-consuming, and are ineffective for unculturable organisms (Keremane *et al.*, 2015). Serological methods, such as ELISA, are widely used in plant nursery diagnostics due to their ability to detect protein molecules using polyclonal or monoclonal antibodies. ELISA-based methods include direct tissue blot immunoassay (DTBIA), double antibody sandwich (DAS) ELISA, and tissue-print (TP) ELISA. The dipstick ELISA test is particularly popular in commercial nurseries due to its ease of use at point-of-care (POC) (Cambra *et al.*, 2000). Although suitable for on-site, specific detection of many plant pathogens, these methods may be less sensitive and have potential for cross-reaction with non-target pathogens, especially in multiplexed reactions (Djelouah *et al.*, 2014). Several molecular approaches based on PCR have been developed to improve sensitivity and specificity over ELISA methods. These PCR variants include reverse transcription PCR (RT-PCR), real-time PCR (qPCR), immunocapture PCR (IC-PCR), and nested PCR. These methods have shown enhanced sensitivity and specificity in detecting bacteria, viruses, and fungi but require sophisticated laboratory equipment and skilled operators. Additionally, they can have issues with specificity and are prone to false-positive results due to detection of even minimal contamination in reaction mixture (Yamane *et al.*, 2008). New high-throughput diagnostic techniques are available and being applied in other fields but have yet to be adopted for nursery disease diagnostics (Hariharan and Prasannath, 2021; Ahale *et al.*, 2024).

Nursery Management Practices

Nursery diseases in fruit crops encompass a diverse range of fungal, bacterial, viral, and nematode pathogens that affect seedlings, grafts, and young plants during their early growth stages. These diseases can lead to stunted growth, reduced vigor, and even plant death, impacting subsequent orchard productivity. Effective disease management in nurseries is critical to ensure the establishment of healthy plants and to minimize need for costly interventions in mature orchards.

Preventive Measures

Sanitation

Sanitation plays a crucial role in effectively managing diseases in fruit crop nurseries, especially where environmental conditions can accelerate disease spread. Maintaining strict sanitation practices is essential. Regular cleaning and disinfection of tools, containers, and nursery facilities are key measures to prevent introduction and spread of pathogens. Additionally, removing and properly disposing of plant debris, fallen leaves, and weeds helps eliminate potential reservoirs for disease-causing organisms. This proactive approach not only reduces availability of disease inoculum but also disrupts lifecycle of pathogens, thereby significantly reducing disease pressure. Sanitation practices complement other disease management strategies such as cultural controls and chemical treatments, enhancing their overall effectiveness. In nurseries focused on sustainable production and healthy crops, sanitation serves as a foundational practice to safeguard plant health and optimize yields (Galanti and Lutgen, 2021).

Pathogen-Free Planting Material

In fruit crop nursery disease management, ensuring pathogen-free planting material is critical to mitigate disease outbreaks and maintain plant health. Various plant parts, such as buds and branches, are used for vegetative propagation, sourced from designated "mother plants." These mother plants must be meticulously maintained to produce healthy, true-to-type planting material. The mother plant block houses progeny trees that are true-to-type, healthy, free from diseases and insect pests, and exhibit high yield potential. In addition to progeny block, an adjacent area maintains disease-free rootstocks suitable for propagation. Protective fungicides play a vital role in preventing the germination, growth, and penetration of pathogens. To effectively use these fungicides, it is crucial to select the appropriate type and apply it in proper dosage, at right times, and using proper application techniques. Ensuring accurate fungicide application–avoiding both insufficient amounts that fail to control disease and excessive amounts that may harm plants is paramount. Nursery managers and applicators must strictly follow usage instructions to achieve optimal results. By adhering to these practices, nurseries can enhance their ability to produce healthy planting material and sustain crop health effectively.

Quarantine and Inspection

Quarantine and inspection are essential strategies in fruit crop nursery management to prevent introduction and spread of plant diseases, particularly in regions with diverse agricultural activities and international trade that heighten the risk of pathogen transmission. Quarantine regulations establish stringent protocols for monitoring and controlling movement of plant materials, including seeds, seedlings, and grafts, into nurseries. Incoming plant stock is isolated in designated quarantine areas to detect and eliminate diseased or pest-infested plants before they can disseminate pathogens to healthy crops. Comprehensive inspection procedures involve meticulous visual assessments and, at times, laboratory testing to verify the health status of plant material. By promptly identifying and removing potential sources of infection, quarantine and inspection protocols effectively mitigate risk of disease outbreaks. This proactive approach reduces necessity for costly disease management interventions later in growing season. Overall, these measures are crucial for safeguarding plant health and promoting sustainable agricultural practices. By ensuring use of healthy planting material and minimizing reliance on chemical treatments, quarantine and inspection contribute to long-term crop productivity and economic viability in fruit crop nurseries.

Cultural Practices

Cultural practices exert significant influence on plant disease development by modifying surrounding environment to create conditions favorable for crop plants while discouraging pathogens. These practices target key stages of a pathogen's life cycle: survival between crops, production of inoculum for primary infection, and actual inoculation. To manage soil-borne organisms, crop rotations involving non-host species effectively reduce pathogen populations. Catch crops are also used to control nematodes and other soil-borne pathogens. Biological methods, such as promoting antagonistic organisms through practices like green manuring and appropriate soil additives, contribute to pathogen management. For soil-borne parasites, cultural practices such as deep ploughing, flooding, frequent cultivation, and fallowing can disrupt their overwintering stages and reduce pathogen populations. Additionally, therapeutic control methods are employed for diseases caused by organisms that survive as parasites within perennial hosts or seeds of annual plants.

Proper Water Management

Effective water management is critical in regions prone to heavy rainfall and high humidity that can exacerbate fungal diseases. Proper drainage and irrigation practices are essential to prevent water logging, creating ideal conditions for root rot and other fungal infections. In nurseries, ensuring irrigation systems deliver water efficiently and uniformly are crucial for maintaining optimal soil moisture levels without inducing waterlogged conditions. Well-drained soils are essential to reduce risk of fungal pathogens that thrive in moist environments. Monitoring soil moisture levels and adjusting irrigation schedules based on weather conditions and plant requirements further helps mitigate disease risks. Furthermore, utilizing weather prediction models enhances effectiveness of management strategies by providing advance information (Rakhonde *et al.,* 2024). By implementing these effective water management strategies, nursery operators can minimize root rot diseases and fungal infections, promoting healthier plant growth and reducing reliance on chemical interventions (Redekar *et al.,* 2020; Purushotham *et al.,* 2024).

Optimal Plant Spacing

Optimal plant spacing is crucial as it influences air circulation, light penetration, and overall plant health. In regions prone to high humidity and where dense planting conditions can heighten disease susceptibility, proper spacing is essential to mitigate these risks effectively. Adequate spacing between plants enhances air circulation around foliage, reducing humidity levels. This adjustment creates less favorable conditions for fungal pathogens such as powdery mildew and leaf spot diseases, which thrive in moist environments. Optimal plant spacing also ensures sufficient sunlight exposure, minimizing duration of leaf wetness that promotes fungal spore germination and infection. Furthermore, proper spacing facilitates easier access for monitoring and management practices, including pruning, pest and disease scouting, and timely treatments. This proactive approach allows growers to detect disease symptoms early and take prompt actions to prevent disease spread effectively.

Soil Management

Soil management is essential pillars of effective disease management strategies. Different crop species exhibit varying susceptibility to diseases and pests, making crop rotation an effective strategy to break disease cycles and minimize reliance on chemical treatments in seed beds used in nursery. By rotating fruit crops with non-host or less susceptible crops, growers can reduce pathogen survival and spread, thereby promoting healthier soils and crops on nursery beds. Soil management practices, such as incorporating organic matter, employing appropriate tillage methods, and adjusting pH levels, also play crucial roles in disease dynamics. Organic amendments improve soil health and stimulate microbial activity, suppress pathogens through competitive exclusion or biological antagonism. Effective tillage techniques help bury crop residues and disrupt pest habitats, further lowering disease risks. Soil management practices contribute to sustainable disease management in fruit crop nurseries by fostering soil health, reducing pathogen pressures, and enhancing crop resilience. Integrating these practices into nursery management plans allows growers to effectively mitigate disease risks and sustain high-quality fruit production over time (Parajuli *et al.,* 2022).

Chemical Control

Chemicals used to control plant diseases are applied in various ways, depending on specific pathogen and conducive conditions. Systemic fungicidal chemicals may also be used therapeutically. Oxathiin derivatives are employed therapeutically to combat fungi that infect plant embryos. Benomyl, known for its systemic action against powdery mildews and other leaf-infecting fungi. Additionally, volatile fungicides are often employed as soil-fumigating chemicals with eradication capabilities.

Seed Treatments

Seed treatments can effectively control plant pathogens that affect seeds, both inside and outside the seed coat and the propagating material. Therapeutic seed treatments target bacteria or fungi that infect embryos, cotyledons, or endosperms within the seed and propagating material. Eradicative treatments focus on killing fungal spores that contaminate surfaces. Protective seed treatments aim to prevent soil-borne fungi from penetrating seedling stems. Seed treatments are classified into two types: physical and chemical. Physical treatments include methods like hot-water treatment and solar-heat treatment. Chemical treatments involve use of fungicides and bactericides. Carbendazim fungicide is applied to mango seeds using various methods. Seeds are treated in small batches using simple seed-treaters. Another method involves preparing a fungicide suspension in water and dipping the seeds into it for a specified period. Overall, seed treatments play a crucial role in preventing the spread of plant diseases by targeting pathogens at different stages of seed development and germination (Villarino *et al.,* 2021; Parajuli *et al.,*2022).).

Soil Treatments

Chemical treatments of soil eradicate plant pathogens and facilitate rapid reclamation of infested soils, ensuring uninterrupted use. Pre-planting chemical treatments of nursery soils are commonly employed to control nematode-induced diseases and fumigate seedbeds and greenhouse soils with methyl bromide to eradicate weeds, insects, and plant pathogens. Soil-treatment chemicals like formaldehyde or captan are applied against sclerotia-producing fungi responsible for seedling blights, stem rots, and root rots in nursery seedlings. Other soil-treatment fungicides such as vapam and "Vorlex" are also utilized. Field applications of soil-treatment chemicals for fungal control are typically limited to furrow treatments. Soil treatments applied at planting time are most effective against early-season parasitic attacks (Villarino *et al.,* 2021; Parajuli *et al.,*2022).

Integrated Nursery Disease Management

Soil or any planting medium used in the nursery may be contaminated by various pests. The presence of pests in the medium causes huge losses to the crop in the nursery, and the infection caused by the pests may be carried to the field through seedlings or adhering medium on the roots. It is, therefore, advocated that the medium used for the nursery must be free from infections. The different methods adopted for soil treatment are as follows.

Soil Solarization

It is an environment-friendly method to control soil-borne plant pathogens, including bacteria, fungi, nematodes, insect-pests and weeds. Solar energy increases the temperature of the soil, which helps control various soil-borne pathogens. The most appropriate time for soil solarization is May–June when the temperature reaches 47°C or above. This treatment causes physical, chemical and biological changes in the soil.

Soil Treatment by Fungicide

- Fungicides like captan or thiram @ 5 g/m^2 are used to control soil-borne pathogens.
- These fungicides can also be used as soil drench by preparing a solution of 2.5–3 per cent and drenching @ 4–5 litre/m^2.

Use of Bio-agents

- Certain biological agents like trichoderma are used to control soil-borne pathogens.

Table 10.2: Practices in Nursery Disease Management in Fruit Crops

Sl.No.	*Disease*	*Pathogen*		*Symptoms*	*Management*
1	Bacterial blight/ canker	*Pseudomonas syringae* pv. *syringae*	Apple, lilacs, flowering cherries and Oriental pears	Blackening (or browning) of new shoots and tip dieback Blossom blast on pear and leaf spots and vein blackening	✰ Protect frost-sensitive plants from cold temperatures and ice formation. ✰ Control weeds around susceptible nursery stock. ✰ Prune trees in January or February, or in mid-summer when weather is dry and minimize pruning wounds. ✰ Provide optimum nutrition, good drainage and growing conditions ✰ *Pseudomonas* cankers with a hand-held propane burner have been shown to limit spread of cankers and prevent girdling of cankered branches. ✰ Rootstocks and varieties of some flowering cherry and oriental pear species have shown resistance to *Pseudomonas syringae*.
2	Crown Gall	*Agrobacterim tumefaciens*	Blueberry, Raspberry, *Pyrus*, Raspberry, *Vitis*	Galls on the crown and roots, or in some cases on shoots and branches. Galls usually soft, spongy and white at first, but later turn hard and brown. Infected plants often show symptoms of nutrient deficiency, such as yellowing of leaves, followed by a general decline and stunting.	✰ Purchase diseases free new stock only from a reputable supplier. ✰ Inspect new stock for pathogen before planting and avoid plant any with gall symptoms. ✰ Avoid injury to roots and bark in crown area when planting or cultivating. ✰ If only a few plants are affected, remove and destroy these to prevent spread of infection. Remove all roots and soil from around infected plants and take to a landfill. Do not compost infected plant debris or soil. ✰ Sterilize pruners in 10 per cent bleach (corrosive to metals), Lysol, 70 per cent alcohol (2 minute exposure) or other disinfectant between cuts.

Sl.No.	Disease	Pathogen		Symptoms	Management
					☆ Do not take cuttings or propagate from diseased plants. ☆ Do not replant susceptible species into previously infected soil for at least 2 years.
3	Fire Blight	*Erwinia amylovora*	*Malus, Prunus, Pyrus*	Foliar and blossom blight, twig dieback and branch cankers. The succulent tips of blighted shoots often droop, forming a "shepherd's crook", and turn brown to black. Leaves remain attached to dead twigs. A cream or tan coloured ooze may be present at the edge of cankers and on young infected shoots. Entire trees can sometimes be killed and rootstocks can also be infected.	☆ Cut out and burn or bury infected foliage and branches. ☆ Eliminate old infected trees from vicinity of new plantings and nursery stock. ☆ Space nursery stock to provide good air circulation. Provide good drainage. ☆ Avoid overhead irrigation as much as possible or water early in morning so foliage can dry off quickly. ☆ Use moderate amounts of nitrogen and avoid fertilizer after mid-summer to prevent excessive succulent shoot growth. ☆ Control aphid, leafhopper and other insect infestations if this can be done without harming bees. ☆ Use of resistant varieties and cultivars
4	Damping-Off	*Cylindrocarpon, Fusarium, Rhizoctonia, Pythium, Phytophthora* and other fungi.		Pre-emergence form of this disease causes rotting of ungerminated seeds or germinants before they emerge Post-emergence from causes rotting of seedling	☆ Sow stratified seed. ☆ Sow at optimum temperature and apply a sterile sand cover. ☆ Irrigate in the morning. ☆ In severe cases, a captan treatment may be necessary.

Sl.No.	*Disease*	*Pathogen*		*Symptoms*	*Management*
5	Powdery Mildew	*Podosphaera leucotricha*	Apple, Blueberry	The presence of white powdery (ash like) coating in patches on both sides of the leaves, and young shoots. The affected leaves turn pale and curl up. Affected shoots remain weak and immature. In case of severe infection, leaf fall and pre-mature fruit drop may occur	☆ Use of disease-resistant hybrids or cultivars. ☆ Rake up and burn or bury fallen leaves. In most cases, composting will destroy powdery mildew spores. ☆ Reduce humidity by spacing plants for good air circulation and by adjusting irrigation practices. ☆ Avoid planting host plants in shaded areas. ☆ In nurseries, spraying young seedlings with Bayleton (500 ppm) at an interval of 7 days to control disease. ☆ Spraying 0.2 per cent Dithane Z-78 or Bordeaux mixture 0.1 per cent.
6	Root Rot	*Armillaria* spp.	Apple, Rubus (blackberry, raspberry), Avocado	Dark brown to black, shoestring-like strands of fungal mycelium called "rhizomorphs" may be visible in the mat or scattered around the base of the plant. Leaf yellowing, wilting, plant decline and dieback	☆ Water deeply when needed. Avoid surface watering around the crown and main trunk. ☆ Remove infected trees, including roots, and soil in the root ☆ When new sites are being cleared, avoid *Armillaria* root rot by: Girdling large trees before removal so roots will decay. ☆ Leaving the prepared site fallow for at least one year
7	Root, Stem and Crown Rot	*Fusarium* and *Rhizoctonia*	Apple, *Populus*, *Prunus*	Fusarium causes Pre- and post-emergence damping-off. *Rhizoctonia* cause damping-off, foliar lesions, and stem cankers.	☆ Imbibe seed in running water, sanitize all seeding equipment and growing containers, reduce early heat and water stress, and encourage rapid germination.

Sl.No.	Disease	Pathogen		Symptoms	Management
8	Root Rots Foliar Blight	*Phytophthora* *Pythium* *Phytophthora*	Papaya, Strawberry	Water soaked spot in stem at ground level which enlarge and griddle stem. The diseased area turns brown or black and rot. Yellowing, leaf drop, wilting and general decline	☆ Soil solarisation ☆ Disease-free stock plants should be used for propagation at all times. ☆ Use strict hygiene in propagation, media preparation and potting. ☆ If plants become infected, do not move plants that are susceptible to *Phytophthora* into that bed until the drainage has been improved and fresh ground cloth has been put down. ☆ Do not plant susceptible cultivar in soil contaminated with *Phytophthora.* ☆ Thermotheraphy : nursery transplants exposed to aerated steam in a closed chamber at 37°C (98.6°F) for 1 hour followed by 44°C (111.2°F) for 4 hours were much less likely to die from *Phytophthora* crown rot ☆ Seed treatment with Thiram or Captan 4 g/kg or Chlorothalonil. ☆ 9. Drenching with Copper oxychloride 0.25 per cent or Bordeaux mixture 1 per cent or Metalaxyl 0.1 per cent.
9	Verticillium Wilt	*V. dahliae* and *V. albo-atrum*	Raspberries and Strawberries	Yellowing, wilting and dieback of young twigs and branches	☆ Remove infested roots and soil and replant with a tolerant or resistant species. ☆ Avoid drought stress or flooding on mature landscape trees, as stress will accelerate *Verticillium* damage.

Sl.No.	Disease	Pathogen		Symptoms	Management
					☆ Mature trees where dieback is just starting can be pruned to remove dead wood and may show only intermittent symptoms for many years. Pruners should be disinfected with 10 per cent bleach before being used on other trees ☆ Purchase propagating stock only from reliable sources.
10	Malfor-mation	*Fusarium moniliforme* var. *subglutinans*	Mango	In bunchy top phase in nursery at 4-5 months, bunching of thickened small shoots, bearing small rudimentally leaves. Shoots remain short and stunted giving a bunchy top appearance. Vegetative malformation induces excessive vegetative branches of limited growth in seedlings. They are swollen with short internodes forming bunches of various size and top of seedlings shows bunchy top appearance	☆ Plant disease-free seedlings, but if symptoms appear, remove and destroy the affected plants. ☆ Select budwood and scions only from healthy trees. ☆ Use of disease free planting material ☆ Incidence reduced by spraying 100-200ppm NAA during October. ☆ Pruning of diseased parts along the basal 15-20 cm apparently healthy portions followed by spraying of Carbendazim (0.1%) or Captafol (0.2%).

Sl.No.	Disease	Pathogen		Symptoms	Management
11	Mango sudden death	*Cerastocystis fimbriata*	Mango	Symptoms not distinct in nursery but evident in field-planted trees. Leaves become dull and wilting spreads throughout tree. Gummosis can occur through bark on trunk and branches and there may be vascular discoloration when the bark is removed.	☆ Make sure nursery is located remotely from existing orchards. ☆ Take budwood only from healthy trees. ☆ Avoid wounding trees and roots during field planting. ☆ Sterilise pruning equipment. ☆ Use effective methods to control insects, especially bark beetles.
12	Wilt	*Macrophomina phaseolina* and *Rhizoctonia solani*	Mango	Sudden wilting of plants Shredding of stem portion near ground level Discolouration and death of secondary roots	☆ Soil solarisation ☆ Treat soil with talc based formulation of *Trichoderma* spp. @of 50g/50kg of soil before placing nuts. ☆ Apply talc based formulation of *Pseudomonas fluorescence* or *Bacillus* spp @ 10g/graft during planting.
13	Bacterial black/leaf spot	*Xanthomonas campestris* pv. *Mangiferae indicae*; *Xanthomonas fragariae*	Mango, Strawberry,	On leaves, slightly raised, angular, black lesions appear On stems the lesions appear as elongated, black lesions with greasy margins	☆ Bacterial black spot infections can be avoided by not raising rootstocks or grafted trees under infected mango trees, keeping leaves dry by using drip irrigation and by selecting a nursery site that is protected from the wind. ☆ In wet weather, fortnightly applications of a protectant copper fungicide may be needed.

Sl.No.	Disease	Pathogen		Symptoms	Management
					☆ Destroy infection sources by pruning infected stems and leaves. ☆ Use certified plants, use drip irrigation instead of overhead irrigation
14	Huang-longbing	*Candidatus Liberibacter* sp.	Citrus	Asymmetric blotchy mottle on leaves, pattern of yellow and green is different on either side of midrib, yellowing across leaf with no green, green islands or chlorotic patterns resembling nutrient deficiency (*e.g.* zinc). Vein corking and veins thicken and turn brown.	☆ Use healthy HLB-free propagation material to produce nursery trees. ☆ Trained staff should detect infection or vector populations early. ☆ Remove and burn diseased trees; use chemical or biological control methods to manage the insect host.
12	Canker	*Xanthomonas citri* subsp. *citri*	Citrus	Canker lesions form on leaves, stems and fruit. The young lesions start as tan-coloured pinpoint spots and become raised as they develop. Older lesions are brown or grey, corky and crater-like with a raised margin and sunken centre.	☆ Reduce spread by planting windbreaks; controlling leaf miner populations and removing heavily infected trees. ☆ Copper sprays may be used to protect young leaves against infection but this protection is lost as the leaf grows and new tissue is exposed.

Sl.No.	Disease	Pathogen		Symptoms	Management
13	Scab	*Elsinoe* sp.	Citrus	Lesions are raised, irregular and wart-like on leaves, twigs and fruit. Lesions are grey to pink and darken over time.	✰ Remove infected trees and apply fungicide. Use different fungicides on separate spraying occasions so that single chemicals are not overused in the nursery. This gives the best chance for controlling the fungus.
14	Viruses and viroids	Citrus psorosis virus, Citrus tristeza virus (CTV), Citrus leaf blotch virus, Citrus exocortis viroid (CEV) and Citrus cachexia viroid	Citrus	Graft-transmissible diseases can cause severe stunting in trees, reduced yield, tree decline, and in some cases, death. Other specific symptoms include stem-pitting, gum pockets and bark-scaling	✰ There is no cure for graft-transmissible diseases but they can be managed by preventing them from being introduced to plants. ✰ Use disease-free budwood and rootstock seed; disinfect all cutting tools in bleach (1.25% sodium hypochlorite) and remove infected trees from the nursery.
15	Nematodes	*Meloidogyne* sp.	Guava	Aboveground symptoms include chlorosis, and reduced yield, growth and leaf size. Below ground, these pathogens reduce fineroot densities and severely distort roots	✰ Exclusion at the nursery stage by using non-infested planting materials and disinfested soil (Lim and Khoo, 1990). ✰ Fernandez *et al.* (1987) reported that 50 g of 98 per cent methyl bromide m−2 and 100 g of dazomet m−2 controlled *Meloidogyne* spp. for 5 months and resulted in 100 per cent healthy seedlings.

- Bio-agents @ 10–25 g/m^2 are mixed in the soil, and after 2–3 days, the seeds are sown.

General IDM Practices to be followed

- Selection of apparently healthy seeds/propagules for seedling production
- Seed dressing with 0.2 per cent Carbendazim/Methyl thiophanate/ Benomyl/Thiram
- Sowing in sterilized/fumigated, clean beds and adequate watering
- Using sterilized budding knife, secateurs, and scissors during budding and grafting
- Transplanting seedling after root dip for 3-5 min in 0.02 per cent Carbendazim solution
- Healthy planting material maintenance by keeping them under proper sunlight, watering and clean environment
- Frequent examination of seedling health and removal of diseased stocks
- Foliar spray of 0.2 per cent Carbendazim/Dithane M-45 at regular interval

Conclusion

Nurseries are essential for plant propagation in horticultural systems but face significant challenges from plant diseases having severe economic impacts if not managed well. Infected plant introductions are a major cause of global pathogen outbreaks, underscoring need for rigorous disease detection and management. While traditional chemical treatments are effective in the short term, posing environmental risks and contributes to fungicide resistance, prompting search for sustainable alternatives like biological controls. IDM, combining cultural, biological, and chemical methods, offers a holistic approach to sustainable nursery disease control. Research into nursery diseases is vital for developing resistant plants, improving diagnostics, and refining management strategies. Education and extension services play crucial roles in disseminating knowledge and enhancing nursery resilience, thereby safeguarding plant health and promoting sustainable agricultural practices.

REFERENCES

Agrios GN, 2005. *Plant pathology*. Elsevier.

Ahale S, Rakhonde G, Bhateja S, Kuppuraj J and Mishra S, 2024. Disease and Pest Control Through Advance Technology. In *Artificial Intelligence and Smart Agriculture: Technology and Applications* (pp. 437-459). Singapore: Springer Nature Singapore.

Bagshaw J, 2008. In: Rogers, M. (Ed.), BioSecure HACCP : Guidelines for Managing Biosecurity in Nursery Production/[developed by John Bagshaw, Matt Rogers and John McDonald. Nursery and Garden Industry Australia, Sydney.

Bustamante MI, Todd C, Elfar K, Hamid MI Garcia JF, Cantu D and Eskalen A, 2024. Identification and Pathogenicity of Fusarium Species Associated with Young Vine Decline in California. *Plant Disease*, **108**(4): 1053-1061.

Cambra M *et al.,* 2000. Routine detection of citrus tristeza virus by direct immunoprinting- ELISA method using specific monoclonal and recombinant antibodies. In: Proceedings of the 14th Conference of the International Organization of Citrus Virologists, Campinas, São Paulo State, Brazil, 13-18 September 1998. International Organization of Citrus Virologists

Carbone MJ, Gelabert M, Moreira V, Mondino P and Alaniz S, 2022. Grapevine nursery propagation material as source of fungal trunk disease pathogens in Uruguay. *Frontiers in Fungal Biology*, **3**: 958466.

Chen L, Zhu H, Horst L, Wallhead M, Reding M and Fulcher A, 2021. Management of pest insects and plant diseases in fruit and nursery production with laser-guided variable-rate sprayers. *HortScience*, **56**(1): 94-100.

Coletta-Filho HD, Castillo AI, Laranjeira FF, de Andrade EC, Silva NT, de Souza AA and Lopes, JR, 2020. Citrus variegated chlorosis: an overview of 30 years of research and disease management. *Tropical Plant Pathology*, **45**: 175-191.

Djelouah K, Frasheri D, Valentini F, D'ONGHIA AM and Digiaro M, 2014. Direct tissue blot immunoassay for detection of *Xylella fastidiosa* in olive trees. *Phytopathologia Mediterranea*, 559-564.

Feichtenberger E, 2001. Doenças incitadas por Phytophthora em citros. In: Luz EDMN, Matsuoka K, Santos AF, Bezerra JL, editors. Doenças causadas por Phytophthora no Brasil. Campinas. Livraria Rural. p. 283-342.

Galanti R and Lutgen H, 2021. Greenhouse and nursery sanitation. Tools, equipment, workers, and visitors. *Coll. Trop. Agric. Hum. Resour. Univ. Hawaii Manoa. Extension Publication OF-54*, 1-7.

Graham J and Feichtenberger E, 2015. Citrus Phytophthora diseases: management challenges and successes. *Journal of Citrus Pathology*, **2**(1).

Hariharan G and Prasannath K, 2021. Recent advances in molecular diagnostics of fungal plant pathogens: a mini review. *Frontiers in Cellular and Infection Microbiology*, **10**: 600234.

ICAR Complex Goa, 2009. Management of seedling mortality in mango nursery. Extension folder no. 35.

Jung T, Orlikowski L, Henricot B, Abad–Campos P, Aday AG, Aguín Casal O and Peréz–Sierra A, 2016. Widespread Phytophthora infestations in European nurseries put forest, semi–natural and horticultural ecosystems at high risk of Phytophthora diseases. *Forest Pathology*, **46**(2): 134-163.

Keremane ML *et al.,* 2015. A rapid field detection system for citrus huanglongbing associated *'Candidatus Liberibacter asiaticus'* from the psyllid vector,

Diaphorina citri Kuwayama and its implications in disease management. *Crop Protect.* **68**: 41–48.

Koffi CN, Diallo HA, Kouadio JY, Kelly P, Buddie AG and Tymo LM, 2010. Occurrence of *Pythium aphanidermatum* root and collar rot of papaya (*Carica papaya* L.) in Côte d'Ivoire. Fruit, Vegetable and Cereal Science and Biotechnology, **4**(Special Issue 1): 62-67.

Kumar A, Bhuj BD, Ram S and Singh CP, 2021. Mango malformation: Etiology and preventive measures. *Annals of the Romanian Society for Cell Biology,* **25**(6): 8025-8058.

Liebhold AM, Brockerhoff EG, Garrett L J, Parke JL and Britton KO, 2012. Live plant imports: the major pathway for forest insect and pathogen invasions of the US. *Frontiers in Ecology and the Environment,* **10**(3): 135-143.

Lieten P, Van Kerckhove S and Van Hemelrijck W, 2021. *In vitro* evaluation of fungicides against *Cylindrocarpon destructans, Rhizoctonia fragariae* and *Pestalotiopsis* spp. in strawberries. In IX International Strawberry Symposium **1309**: 795-800.

Luo Y, Niederholzer F, Camiletti BX and Michailides TJ, 2024. Survey on Latent Infection of Canker-Causing Pathogens in Budwood and Young Trees from Almond and Prune Nurseries in California. *Plant Disease,* **108**(3): 550-557.

Mannai S, Benfradj N and Boughalleb-M'Hamdi N, 2024. Occurrence of *Globisporangium* and *Phytopythium* species associated with apple and peach seedlings decline in Tunisian nurseries. *Journal of Plant Pathology,* 1-13.

Martinez-Espinoza A, 2013. Georgia Plant Disease Loss Estimation. University of Georgia Extension, pp. 102–106.

Moein S, Mazzola M, Ntushelo NS and McLeod A, 2019. Apple nursery trees and irrigation water as potential external inoculum sources of apple replant disease in South Africa. *European Journal of Plant Pathology,* **153**: 1131-1147.

Nerida Donovan, Ian Bally and Tony Cooke Susan House, 2016. Nursery Manual for Citrus and Mango, Australian Centre for International Agricultural Research (ACIAR)

Parajuli M, Panth M, Gonzalez A, Addesso KM, Witcher A, Simmons T and Baysal-Gurel F, 2022. Cover crop usage for the sustainable management of soilborne diseases in woody ornamental nursery production system. *Canadian Journal of Plant Pathology,* **44**(3): 432-452.

Parke JL and Grünwald NJ, 2012. A systems approach for management of pests and pathogens of nursery crops. *Plant Disease,* **96**(9): 1236-1244.

Parke JL, Knaus BJ, Fieland VJ, Lewis C and Grünwald NJ, 2014. Phytophthora community structure analyses in Oregon nurseries inform systems approaches to disease management. *Phytopathology,* **104**(10): 1052-1062.

Purushotham P, Sahana G, Rakhonde G, Karunakar Reddy N, Nayak AM and Pravalika KM, 2024. Pathways for Intensified Agriculture Through Sustainable Water Management: Need of an Hour. In *Artificial Intelligence and Smart Agriculture: Technology and Applications* (pp. 79-101). Singapore: Springer Nature Singapore.

Rakhonde GY, Ahale S, Reddy NK, Purushotham P and Deshkar A, 2024. Big Data Analytics for Improved Weather Forecasting and Disaster Management. In *Artificial Intelligence and Smart Agriculture: Technology and Applications* (pp. 175-192). Singapore: Springer Nature Singapore.

Redekar NR, Bourret TB, Eberhart JL, Johnson GE, Pitton BJ, Haver DL and Parke JL, 2020. The population of oomycetes in a recycled irrigation water system at a horticultural nursery in southern California. *Water Research*, **183**: 116050.

Rooney-Latham S, Blomquist CL, Kosta KL, Gou YY and Woods PW, 2019. Phytophthora species are common on nursery stock grown for restoration and revegetation purposes in California. *Plant Disease*, **103**(3): 448-455.

Simbare A, Sane CAB, Nduwimana I, Niyongere C and Omondi BA, 2020. Diminishing farm diversity of East African highland bananas in banana bunchy top disease outbreak areas of Burundi–the effect of both disease and control approaches. *Sustainability*, **12**(18): 7467.

Turechek WW, Myhrene O, Slovin J and Peres NA, 2021. The use of aerated steam as a heat treatment for managing angular leaf spot in strawberry nursery production and its effect on plant yield. *Phyto. Frontiers™*, **1**(2): 104-119.

Valatin G, Price C and Green S, 2022. Reducing disease risks to British forests: An exploration of costs and benefits of nursery best practices. *Forestry: An International Journal of Forest Research*, ***95***(4): 477-491.

van Jaarsveld WJ, Halleen F, Bester MC, Pierron RJ, Stempien E and Mostert L, 2021. Investigation of *Trichoderma* species colonization of nursery grapevines for improved management of black foot disease. *Pest Management Science*, **77**(1): 397-405.

Villarino M, Larena I, Melgarejo P and De Cal A, 2021. Effect of chemical alternatives to methyl bromide on soil–borne disease incidence and fungal populations in Spanish strawberry nurseries: A long–term study. *Pest Management Science*, **77**(2): 766-774.

Weiland JE, 2021. The challenges of managing Phytophthora root rot in the nursery industry. *Plant Health Progress*, **22**(3): 332-341.

Weiland JE, Scagel CF, Grünwald NJ, Davis EA, Beck BR, Foster ZS and Fieland VJ, 2020. Soilborne Phytophthora and Pythium diversity from rhododendron in propagation, container, and field production systems of the Pacific Northwest. *Plant Disease*, **104**(6): 1841-1850.

Wu J, Hu S, Ye B, Hu X, Xiao W, Yu H and Zhang C, 2022. Diversity and Resistance to Thiophanate-Methyl of *Colletotrichum* spp. in Strawberry Nursery and the Development of Rapid Detection Using LAMP Method. *Agronomy*, **12**(11): 2815.

Yamane K, Oyama K, Iuchi E, Ogawa H, Suzuki T and Natsuaki T, 2008. RT–PCR Detection of Odontoglossum ringspot virus, Cymbidium mosaic virus and Tospoviruses and Association of Infections with Leaf–Yellowing Symptoms in Phalaenopsis. *Journal of Phytopathology*, **156**(5): 268-273.

Chapter 11

Nematode Management in Fruit Nurseries

M. Devindrappa and R. Umamaheswari

Division of Crop Protection, ICAR–Indian Institute of Horticultural Crops, Bengaluru–560089, Karnataka

Agriculture and Horticulture are vital sciences as they meet the fundamental need for food for human beings. Qualitative and quantitative food can be essentially produced from healthy plants, which are in turn cultivated from vigorous and healthy seedlings/saplings. Consequently, nurseries are the basic requirement of horticulture. India has diverse climatic and soil conditions conducive to fruit production. Fruits are a staple in the regular Indian diet. Fruit cultivation has become a lucrative choice among small and marginal farmers in India. Fruit crops are estimated to cover around 7.46 million hectares, producing 11.26 million metric tons in India (2023-24, Department of Agriculture and Farmer Welfare, Government of India). The major fruit-producing states of India are Maharashtra, Karnataka, Andhra Pradesh, Tamil Nadu, Uttar Pradesh, Gujarat, and Bihar. Plant parasitic nematodes continue to threaten horticultural crops throughout the world, particularly in tropical and sub-tropical regions. Mainly nematodes as pathogens by themselves, and secondarily as predisposing microbes for easy entry of soil borne fungi and bacteria into the host plants, nematodes cause huge economic losses quantitatively and qualitatively. Overall average annual yield loss of the world's major horticultural crops due to damage by plant parasitic nematodes is estimated at 13.54 per cent (Reddy, 2012). All India Coordinated Research Project on Nematodes assessed the monetary losses in 19 horticultural crops at ₹50,224.98 million (Kumar *et al.*, 2020). Among the various nematodes identified in the fruit crops, the most devastating genera affecting fruits are *Meloidogyne* spp., *Radopholus similis, Pratylenchus* spp., and *Tylenchulus semipentrans* (Table 11.1). *Meloidogyne enterolobii,* the highly pathogenic guava root knot nematode was first reported from Tamil Nadu in 2016 and subsequently in 11 other states in India, mainly spread through nematode infested planting material (Poornima and Walia, 2021).

Table 11.1: Important Nematodes Pest of Fruit Crops and Yield Loss (Kumar *et al.*, 2020)

Fruit Crops	*Nematode Pest*	*Yield Loss (%)*	*Monetary Loss (₹ in millions)*
Banana	*Radopholus similis* + *Meloidogyne incognita*	15 per cent	9710.46
Citrus	*Tylenchulus semipenetrans*	27 per cent	9828.22
Grapes	*Meloidogyne incognita*	30 per cent	3940.44
Papaya	*Meloidogyne incognita* + *Rotylenchulus reniformis*	30 per cent	2516.64
Guava	*Meloidogyne* spp.,	28 per cent	2350.88
Pomegranate	*Meloidogyne* spp.,	23 per cent	3023.44

There is an urgent need for nematode free seedlings for sustainable crop production to meet the demands of both domestic and international fruit markets. Hence, there is a discernible necessity for implementation of integrated nematode management packages by judicial blend of several environment-friendly tactics like use of resistant varieties, biopesticides, botanicals, grafting with resistant root stocks, physical measures like soil solarisation, hot water treatment *etc.* for suppressing the nematode population. ICAR– Indian Institute of Horticultural Research, Bengaluru, is the pioneering Institute in India for its frontier research and commercialization of biopesticides all over India. It has registered five biopesticides products, *viz., Purpureocillium lilacinum* (=*Paecilomyces lilacinus*) 1 per cent W.P., *Pochonia chlamydosporia* (=*Verticillium chlamydosporium*) 1 per cent W.P., *Trichoderma viride* 1.5 per cent, *Trichoderma harzianum* 1 per cent W.P., and *Pseudomonas fluorescens* 1 per cent W.P. with the Central Insecticides Bureau and Registration Committee, Faridabad under section 9(3b) of the Insecticides Act, 1968 for managing nematode problems and associated disease complexes in horticultural crops in India (Rao *et al.*, 2016).

Nematode Spread

Nematodes primarily spread from one area to another through infested planting materials. Additionally, soil adhering to implements, tyres of machineries, and shoes of workers may also contribute to the spread of nematodes between areas. Furthermore, nematodes can be disseminated by irrigation water especially furrow irrigation systems.

Typical Symptoms of Nematode Infested Fruit Crops

Root-Knot Nematode, *Meloidogyne* spp.

Generally, above ground symptoms of root knot nematode infected plants are revealed through stunted growth and chlorotic yellow leaves. Below ground, characteristic galls are prominent on the roots and the size of gall varies from crop to crop (Figure 11.1).

Figure 11.1: Root Knot Nematode (*Meloidogyne* spp.) Damage in Guava Nurseries and Orchards.

Life Cycle of the Root-Knot Nematode, *Meloidogyne* spp.

Once the nematode penetrates into the root and finds a suitable site, usually behind the root cap, the juveniles start feeding on epidermal cells. They become sedentary during feeding and enlarge in cross-section. The juveniles continue to feed except during the 3rd and 4th stages. The third and fourth molts occur in quick succession, leading to the development of pear-shaped, white adult females. The males are elongated and vermiform. Reproduction is typically by parthenogenesis, though sometimes by amphimixis. Each adult female lays about 400-500 eggs in a gelatinous matrix (Figure 11.2). Under optimal conditions (temperature 27-30°C), the total time for one life cycle is 3-4 weeks, depending on the host and other weather conditions, nematode allowing for several generations per year (Rao *et al.,* 2015a).

Symptoms Due to Citrus Nematode, *Tylenchulus semipenetrans*

T. semipenetrans is an important pest responsible for slow decline disease of citrus. Infestation level depends on the age of trees and time of infection, susceptible rootstock and the population density of nematode. Affected trees exhibit reduced vigour, chlorosis and falling of leaves, twig dieback and consequently, reduced fruit production. Decline of the tree is gradual and persists until the crop is so small that tree maintenance may become uneconomical. Heavily infested roots

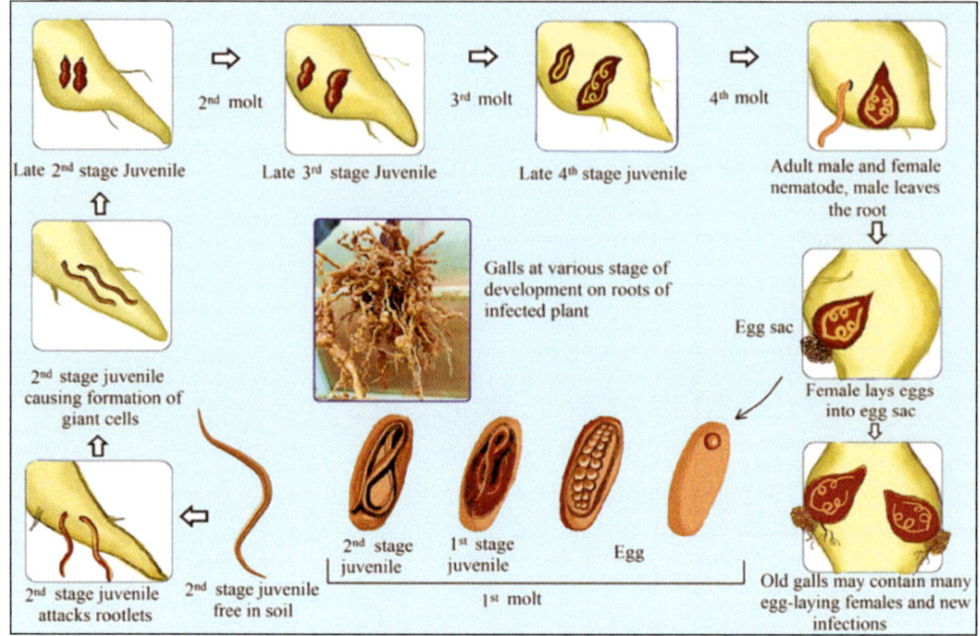

Figure 11.2: Life Cycle of Root-Knot Nematodes of the Genus *Meloidogyne* spp. (Pranaya *et al.*, 2023).

are darker in colour with rootlets shortened, swollen and irregular in appearance than in normal ones.

Life Cycle of the Citrus Nematode

The citrus nematode, *Tylenchulus semipenetrans*, exhibits sexual dimorphism and reproduces sexually, occasionally using facultative parthenogenesis without the need for males. From the egg, which contains the first-stage juvenile (J_1), the second-stage juvenile (J_2) emerges and searches for host roots. The motile and vermiform J_2 female molts into the vermiform J_3 and J_4 stages, and finally into the sedentary adult. The immature female penetrates deeply into the cortical layers of the root, becomes sedentary, and establishes a permanent feeding site consisting of specialized cells called 'nurse cells,' which are the main source of nutrients. Upon maturation, the posterior portion of the female's body swells and protrudes from the root surface while its elongate neck and head remain embedded in the cortex. Mature females produce eggs embedded in a gelatinous matrix. The length of the female life cycle from egg to egg ranges from four to eight weeks. The development of the J_2 male into an adult is completed in seven days and does not require feeding. Citrus nematode-infected roots become thicker, darker, decayed, and show a dirty appearance due to the large number of females sticking to the infected roots (Figures 11.3 and 11.4). The infected root systems lose the ability to absorb enough water and nutrients for normal growth due to the nematode damage.

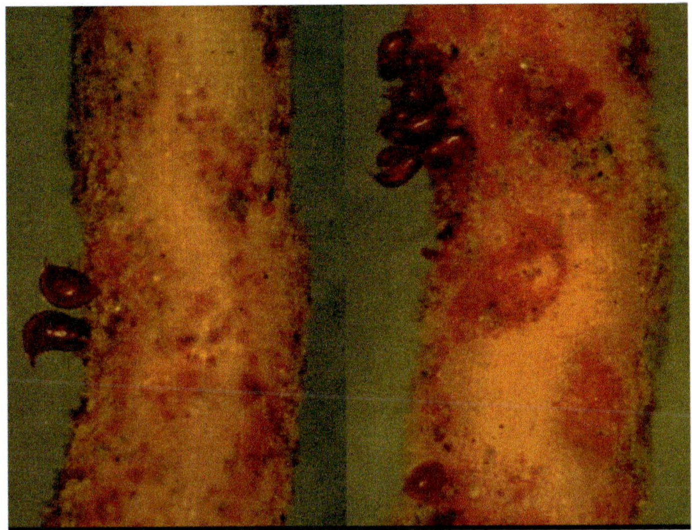

Figure 3: Citrus Nematodes, *Tylenchulus semipenetrans* in Citrus.

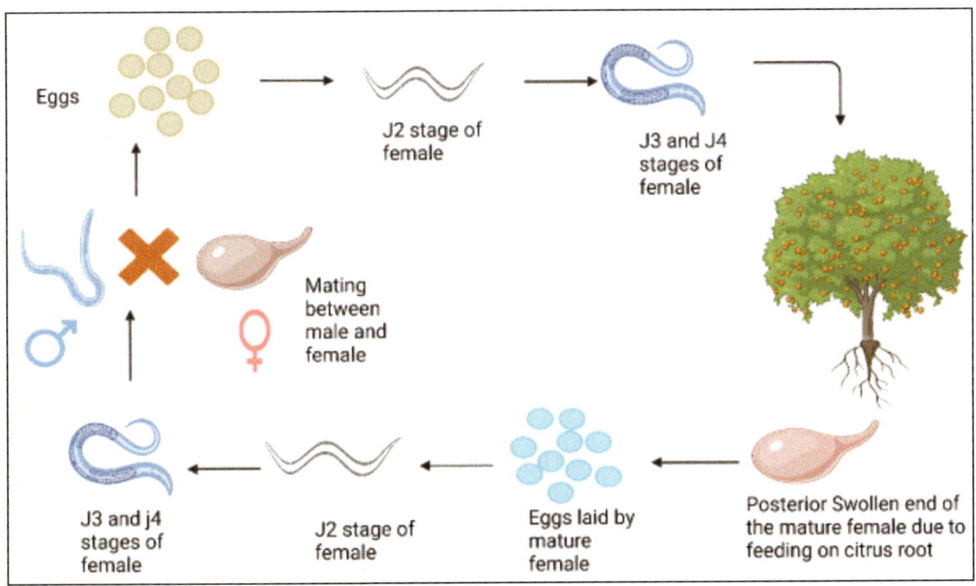

Figure 11.4: Life Cycle of *Tylenchulus semipenetrans* (Ankit Choudhary *et al.*, 2023).

Symptoms Due to Burrowing Nematode, *Radopholus similis*

When the burrowing nematodes wound the roots, they usually induce reddish-brown cortical lesions, which are a diagnostic symptom of the disease in bananas. Root lesions are clearly visible when an affected root is split longitudinally and examined immediately. Root and rhizome necrosis, stunted growth, leaf yellowing, and falling of mature plants are observed in nematode infected plants (Figure 11.5). The lesions on the primary roots, along with the girdling and death of these anchor roots, make the plant prone to 'tip over or topple' due to wind action.

Figure 11.5: Symptoms Due to Burrowing Nematode, *Radopholus similis* in Banana.

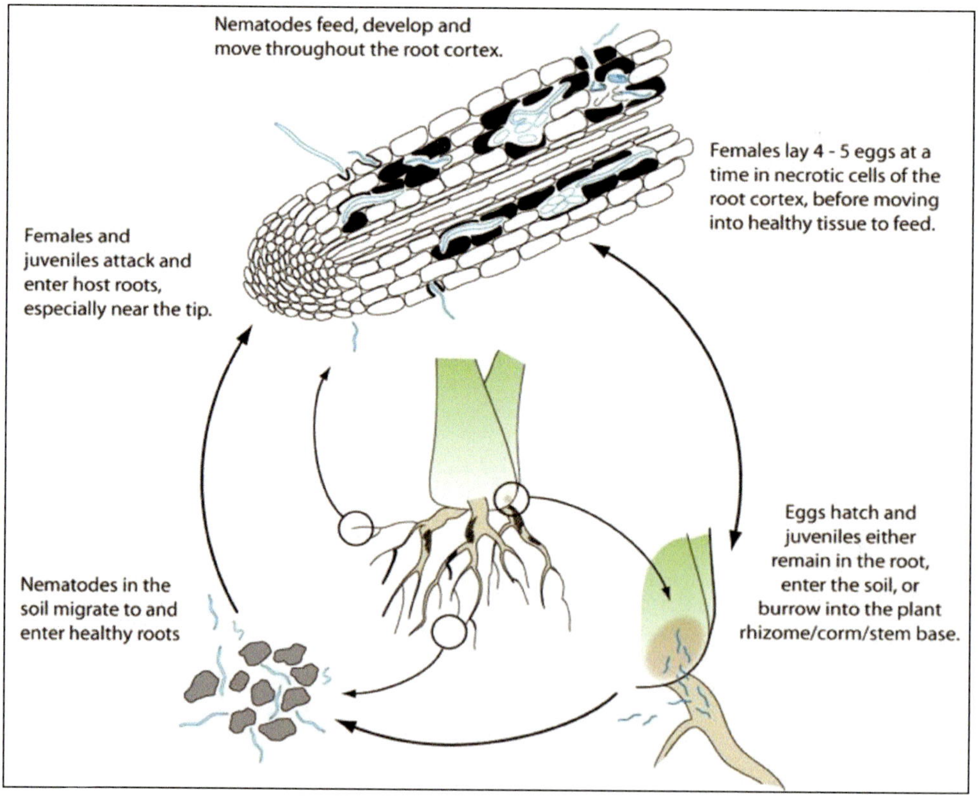

Nematodes feed, develop and move throughout the root cortex.

Females lay 4 - 5 eggs at a time in necrotic cells of the root cortex, before moving into healthy tissue to feed.

Females and juveniles attack and enter host roots, especially near the tip.

Eggs hatch and juveniles either remain in the root, enter the soil, or burrow into the plant rhizome/corm/stem base.

Nematodes in the soil migrate to and enter healthy roots

Figure 11.6: Life Cycle of Burrowing Nematode, *Radopholus similis* (www.apsnet.org/edcenter/intropp/lessons/Nematodes/Pages/Burrowingnematode.aspx)

Life Cycle of the Burrowing Nematode

Burrowing nematode, *Radopholus similis* exhibits a migratory endoparasitic habit. Despite all stages remaining vermiform, sexual dimorphism is evident, with adult males being somewhat degenerate and probably non-parasitic. Eggs are typically laid in infested tissue over 7-8 days at a rate of about four eggs per day. The life cycle from egg to egg takes 20 to 25 days, with eggs taking 8-10 days to hatch and larvae taking 10-13 days to mature (Figure 6).

Symptoms Due to Reniform Nematode, *Rotylenchulus reniformis*

Shedding of leaves, formation of malformed fruit and seeds, stunted plant growth are generally observed. Roots system becomes impaired with necrotic discoloration and decay.

Integrated Nematode Management Practices in Fruit Nurseries

Seedlings of papaya, lime, guava and pomegranate grafts are produced generally in substrate mixture in polythene bags. Many a times substrate mixture (sand + soil + FYM or any organic manure) harbour the above mentioned nematodes and other pathogenic fungi and bacteria. Hence, production of nematode-free healthy planting materials has to be guaranteed by commercial nurseries which can be accomplished through the following measures.

Sanitation

- ☆ Use of nematode free saplings for planting material.
- ☆ Removal and destruction of nematode infected saplings or trees.
- ☆ Maintaining the orchard free from weeds and alternate hosts.

Adoption of Suitable Propagation Techniques

- ☆ Tissue culture propagation techniques are promising for production of nematode and disease free planting materials. However, these techniques are available in selected fruit crops such as banana, citrus and pomegranate and not available in guava.
- ☆ Use of air layering techniques especially in guava propagation. Soil-less rooting medium can be used in layers and rooted cuttings can be transferred to sterilized medium for further raising of plants.

Use of Soil-less Medium

Use of soil-less medium (sterilized coir pith, cocopeat or vermiculite) instead of native soil which usually harbours nematodes, for raising or hardening of rooted planting materials.

Use of Nematode Free Soil

- ☆ Soil solarization by using linear low density transparent polythene sheets for about 4-6 weeks during peak summer months is effective and may be used as a routine practice for suppressing disease pathogens. Double solarization can also be done by covering with polythene sheets after filling soil in polybags

- Use of virgin soil

Use of Biopesticides

1. **Treatment of soil mixture used for rootstocks:** A ton of soil mixture has to be mixed with 50 kg of neem cake or pongamia cake enriched with the bio-pesticides such as *Paecilomyces lilacinus, Pseudomonas fluorescens* and *Trichoderma harzianum*. Enrichment of neem cake/pongamia with bio-pesticides can be done by mixing 2-3 kg each of *Paecilomyces lilacinus/ Pochonia chlamydosporia, Pseudomonas fluorescens* and *Trichoderma harzianum/T. viride* in one ton of neem cake/pongamia cake and leaving it under shade for 20 days at 25 – 30 per cent moisture for multiplication of beneficial microbes.
2. **Soil application in main field:** Apply 5 kg of FYM or 2.5 kg of vermicompost enriched with *Pseudomonas fluorescens* + *Trichoderma harzianum* + *Purpureocillium lilacinum* per pit before planting the fruit crop seedlings. For application in standing crop, mix 20 kg of bio-pesticide enriched neem cake/pongamia cake in 200 lit water, leave it for two days. This can be used for drenching at 2–3 lit/plant or filter it thoroughly and use it for sending along the drip, once in an interval of 30 days. (Rao *et al.,* 2015b).

Use of Resistant Varieties/Root Stocks

- Dogridge, Black Champa, 1613, Salt Creek, Cardinal, Banquabad are grape root stocks resistant to root knot nematodes, *Meloidogyne* sp.
- Trifoliate orange (*Poncirus trifoliata*) and its hybrids (Citrumello), rootstocks by crossing Rangpur lime (*Citrus limonia*) with *P. trifoliata viz.,* CRH-3, CRH-5 and CRH-41 are highly resistant to the citrus nematodes (Reddy, 2012).
- *Psidium cattleianum* and *P. friedrichstalianum* are the wild guava germplasm identified highly resistant to root knot nematodes, *M. enterolobii* and *M. incognita*. Grafting on these resistant germplasm as rootstocks needs to be explored.

Hot Water Treatment

- Dipping the bare roots of citrus seedling in hot water at 45°C for 25 min or 46.7°C for 10 min controls citrus nematodes.
- Dipping of grape rooted cuttings for 10min. at 50°C or 30min. at 47° reduces nematode damage due to root knot nematodes.

Psidium cattleianum

Psidium friedrichstalianum

Figure 11.7: Nematode Resistant Wild Germplasm of Guava.

Chemical Nematicides

- ✰ Soil application of Furadan (Carbofuran 3G) @ 30 to 40 g per plant in established orchards; 3 g/plant in nursery rootstocks

Conclusion and Future Thrusts

Production of clean planting material is of utmost importance in the current Indian scenario as horticultural nurseries are playing a major role in dissemination of plant parasitic nematodes especially in fruit crops. As most of the nematodes feed on plant roots and underground plant parts, farmers, nurserymen and orchardists remain ignorant about their infestations and disseminations through planting material. Unawareness about nematode damage potential and its diagnosis by the field functionaries is one of the main reasons behind non adoption of the technologies available to manage nematodes. As an outcome of research trials through All India Coordinated Research Project on Nematodes, Poornima and Walia (2021) recommended the following measures to avoid nematode infestation and spread through horticultural nurseries.

- ✰ Training should be mandatory for State Horticultural Officials and nurserymen in nematode disease diagnosis and remedial measures in horticultural nurseries.
- ✰ Mandatory use of sterilised substrate media (planting media) for propagation of horticultural planting materials should be practiced by the nursery growers. Soil-less medium or air layering can be used for production of planting materials and clones.

- ☆ Strict compliance is necessary for regular monitoring of nursery materials by specially trained staff of state governments and destruction of all nematode infected planting materials/soil.
- ☆ Nursery laws/regulations should be suitably amended to make provisions for issuing mandatory Health certificate by the concerned District Horticultural Officer to the registered nursery owner/carrier at the time of transporting planting materials from these nurseries.
- ☆ Researchers can focus their future line of work into *in-vitro* production of resistant plants through somatic embryogenesis and tissue culture techniques that would help in avoiding nematode infestation.

REFERENCES

Ankit Choudhary, Sunil Kumar, Anirudh Jaswal, Jitender Singh Sankhla, Ravi Kant, 2023. Infestation of *Tylenchulus semipenetrans* on citrus plantations and their management. *International Journal of Multidisciplinary Research and Development,* **10**(6): 67-73.

Kumar V, Khan MR and Walia RK, 2020. Crop Loss Estimations due to Plant-Parasitic Nematodes in Major Crops in India. *Natl. Acad. Sci. Lett.,* **43:** 409–412. https: //doi.org/10.1007/s40009-020-00895-2

Poornima K, Walia RK, 2021. Horticultural Nurseries and Nematode dissemination. *Current Science,* **120**: 2, 25

Pranaya Pradhan, Ponnam Naresh, Satyaprakash Barik, Gobinda Chandra Acharya, Ramakrushna Bastia, Anil Kumar Adamala, Mahendra Prasad Das, 2023. Breeding for root knot nematode resistance in fruiting Solanaceous vegetable crops: a review. *Euphytica,* **219**: 71. https: //doi.org/10.1007/s10681-023-03204-2.

Rao MS, Umamaheswari R, Priti K, Rajinikanth R, Grace GN, Kamalnath M, Prabu P, Kumar RM, Chaya MK, 2016. Role of Biopesticides in the Management of Nematodes and Associated Diseases in Horticultural Crops. In: Plant, Soil and Microbes. Springer International Publishing, pp 117-148.

Rao MS, Umamaheshwari R, Chaya MK, Manojkumar R, Priti K, Grace GN, Vidyashree, Narayanaswamy B, 2015a. Management of Nematodes in Fruit crops. IIHR technical bulletin No.45, ICAR– IIHR, Bengaluru.

Rao MS, Umamaheshwari R, Priti K, Rajinikanth R, Vidyashree, Prabu P, Kamalnath M and Narayanaswamy B, 2015b. Nematode Management In Vegetable Crops. IIHR technical bulletin No. 47. ICAR– IIHR, Bengaluru.

Reddy PP, 2012. Bio-intensive nematode management in horticultural crops. Scientific Publishers, Jodhpur, India, 300p.

Chapter 12

Drip Irrigation in Orchard Management

Rakesh Sharda and Nav Prem Singh*

Project Coordinator (AICRP on PEASAM), CIPHET, Ludhiana–141004, Punjab

**Department of Fruit Science, Punjab Agricultural University, Ludhiana–141004, Punjab*

In India, fruit cultivation spans 7.04 million hectares. The majority of exported raw or processed fruits, including grapes, pomegranates, mangoes, bananas, and oranges, are sent to various countries such as the USA, Bangladesh, China, UAE, Netherlands, Nepal, Malaysia, Iraq, UK, Sri Lanka, Iran, Oman, and Saudi Arabia. India's fruit production reached 112.62 m MT in 2023-24.

Irrigation is considered one of the major inputs in increasing the agricultural yield in arid and semi-arid areas. Water is available to us through various sources. However, proper water management is needed. Lack of proper water management sometimes results in over- and under-irrigation. An optimum solution to this is drip irrigation, which is one way to shift towards a disciplined water management strategy.

Efficient water usage can be achieved through drip irrigation. A properly designed drip irrigation system eliminates water loss due to runoff, evaporation, or deep percolation. By minimizing water contact with the crop's leaves, stems, and fruit, drip irrigation may create less favourable conditions for disease proliferation. Furthermore, precise irrigation scheduling can be implemented to meet crop requirements, potentially leading to improved yield and quality. It is an irrigation system with high-frequency application of the desired quantity of water in and around the root zone of the plant system. It consists of a network of pipes and a suitable emitting device called an emitter. Drip irrigation adoption in fruits increases water use efficiency (upto 90%), saves water (40-60%), reduces fertilization requirement (20-30%) through fertigation, produces better quality crops and increases yield (25-40%) as compared with conventional irrigation. A drip irrigation system uses every drop which is given as irrigation. The total amount of water which the plant needs is exactly made by the drip irrigation system. A drip irrigation system irrigates the area where water is needed, whereas conventional methods irrigate all cropped areas.

Advantages of Drip Irrigation

a. Water saving.
b. Enhanced plant growth and yield.
c. Uniform and better quality of produce.
d. Efficient and economical use of fertilizers.
e. Less weed growth.
f. Also suitable to wastelands.
g. Possibility of using saline water.
h. No soil erosion.
i. Flexibility in operation.
j. Easy installation.
k. Labor saving.
l. Suitable for all types of land terrain.
m. Saves land as no bunds, *etc.*, are required.
n. Minimum diseases and pest infestation.
o. Saves electricity

Layout of Drip Irrigation Systems

The different planting densities are adopted in the orchards–the conventional planting densities of 12 mx12m, 8mx8m, 6mx 6m, *etc.* High-density orchards are also practiced with spacing such as 6mx3m, 3mx2m, *etc.* In such a system, the primary question is what type of lateral layout has to be selected. If the plant-to-plant distance is less than 2m, go for inline laterals placed at least 0.5 m away from the trunk on both sides. The dripper spacing will depend upon the type of soil. The farther the soil is, the farther the spacing; if the soil is coarser, it will be vice versa. In the case of plant-to-plant spacing of more than 2m, go for a ring system of lateral with the idea that the ring has to be small during the initial stages and increased as the tree matures.

Planning and Design of Drip Irrigation System

The practical implementation of a drip irrigation system necessitates meticulous planning and design to ensure precise water delivery. This process begins with gathering essential information about the installation site, including farm maps, soil characteristics, water quality, crop rotation patterns, and topographical features. Subsequently, the peak water demand for the summer crop within the rotation must be calculated. Various methods exist for determining water requirements, such as utilizing soil and plant indicators, employing soil water budgeting techniques, or analyzing weather data. However, the most straightforward and commonly used approach involves pan evaporimeter data. After calculating the peak water requirement, the next phase involves designing and selecting the various components of the drip irrigation system. To achieve uniform water distribution across all plants in the field, it is crucial to design the system to maintain consistent hydraulic pressure throughout the distribution network. This design process aids in selecting appropriate emitters, laterals, manifolds, sub-mains, main pipelines, pumping units, and filtration systems.

General Information

General information on water source, crop, topographic condition, type and texture of the soil and climatic data are obtained and is essential for the design of the drip irrigation system.

Crop Water Requirement

The daily crop water requirement is estimated by using the equation

Irrigation water requirement = $E_p \times K_p \times K_c \times A \times W_p$ (1)

Net daily irrigation water required (2)

Daily operating hours of the system (T) = $\frac{V_n}{N_e \times N_p \times q}$ (3)

where,

V = Volume of water required in liter (L day^{-1})

E_p = Mean pan evaporation for the month in mm day^{-1}

K_c = Crop coefficient

K_p = Pan coefficient (FAO irrigation and drainage paper No. 24)

A = Plant to plant x Row to row spacing (m^2)

R_e = Effective rainfall (mm)

W_p = Wetting fraction (varies from 0.3 to 1.0)

N_e = Number of emitters per plant

N_p = Number of plants

q = Emitter discharge (Lh^{-1})

The crop coefficient (K_c) varies with crop growth stage and season. The crop coefficient (K_c) is considered for the development of a crop for the design of a drip irrigation system.

Hydraulic Design of System

The ideal microirrigation system is one in which all emitters (orifices) deliver the same volume of water in a given time. From the practical point of view, this ideal performance is almost impossible to achieve. However, the hydraulic design can control the flow variation of water pressure.

Flow carried by each lateral line

Q_l = Discharge of one dripper × No. of drippers per lateral (4)

Flow carried by each sub-main line

$Q_s = Q_l \times$ No. of lateral lines per sub-main (5)

Flow carried by mainline (Q) = $Q_s \times$ No. of sub-main line (6)

The size of the mains, submains, laterals, and pump is decided based on the desired flow rate and pressure head in the system. The pressure drop due to friction can generally be evaluated using the William Hazen or Dacy-Weishbach equation. This section gives the procedure for computing head loss in lateral, sub-main, and main pipes and the horsepower requirement of the pump.

Head Loss in Laterals

The pipes used in microirrigation systems are made of plastic and are considered smooth pipes. The pressure drop due to friction can be evaluated using the Hazen -William empirical equation, as given below.

$$H_f(100) = K\left(\frac{Q}{C}\right)^{1.852} \times D^{-4.871} \times F \qquad (7)$$

As the length of the pipe increases, the discharge in the pipe decreases due to emission outlets and hence the total energy drop is less than as given by the above equation. For this reason, a reduction factor F is introduced.

where,

$H_f(100)$ = head loss due to friction per 100 meter of pipe length, m/100m

Q = flow of water in pipe, lps

D = Internal diameter of pipe, cm

L = length of the pipe, m

C = Hazen - William constant (140 for PVC pipe)

F = Reduction Factor due to multiple openings in pipe, which can be computed by equation.

$$F = \frac{1}{m+1} + \frac{1}{2N} + \frac{\sqrt{m-1}}{6N^2}$$

m = 1.852

N = number of outlets on the lateral

The design criteria for lateral pipe are to keep pressure and discharge variations within the prescribed limit. For lateral design, the discharge and operating pressure at the emitter are required to be known. Based on this, the allowable head loss can be calculated using the above formula. The diameter of the lateral pipe is usually selected such that the difference in discharge between emitters operating simultaneously will not exceed 10 per cent. Pressure head difference should not exceed 10 to 15 per cent of the operating pressure. For the discharge variation of 10 per cent, the emission uniformity has to be more than 90 per cent.

Head Loss in Sub Mains

The hydraulic properties of sub-main lines bear a resemblance to those of lateral lines. One can calculate the sub-main hydraulic characteristics by treating the laterals as equivalent to emitters on lateral lines. Generally, sub-main and mainline pipes are considered hydraulically smooth, as they are typically made of PVC or HDPE materials. The Hazen Williams roughness coefficient is commonly assigned a value between 140 and 150. To determine the energy loss in the sub-main, one can employ techniques akin to those utilised for lateral calculations.

Head Loss in Main Line

Typically, pressure regulators or adjustments are located at the entrance of the sub-main. As a result, energy losses in the main pipeline should not impact system uniformity. The main line lacks outlets, hence the reduction factor is not applied. The calculation of frictional head loss in the main pipeline employs the same Darcy-Weisbach or Hazen-William's equations.

Horsepower Requirement of Pump

The horsepower requirement of pump is computed by using flowing equation.

$$\text{Horsepower required (hp)} = \frac{H \times Q_m}{75 x \eta_p \ x \eta_m} \tag{8}$$

where,

H = Total pumping head ($H_f + H_e + H_S$), m

H_f = Total head loss due to friction (Friction head loss in mains + Friction head loss in sub mains + Friction loss in laterals + Head loss in accessories), m

H_e = Operating pressure head required at the dripper, m

H_S = Total static head, m

Q_m = Discharge of main, lps

η_p = Efficiency of pump, fraction

η_m = Efficiency of motor, fraction

Selection and Size of the Filtration Unit

The filtration unit is selected based on the type of water source. If the water source is a water storage tank, then an assembly of sand filters and screen filters is used. If the water source is a tube well and there is a chance of sand coming out along with pumped water, select a hydro-cyclone filter in conjunction with the disc filter. If there is no sand in the tube well water, then only a disc filter can be used. The size of the filtration assembly is selected based on the quantity of water used per section.

Irrigation Schedule

The conventional method of irrigation for fruit crops is flood irrigation. This method is highly inefficient, with around 40 per cent efficiency. Drip irrigation schedules have been recommended and are available in the packages of practices of the various institutes. The following drip irrigation schedules were developed for Kinnow and guava based on the age and month of the crop.

Table 12.1: Drip Irrigation Water (litre/day/plant) Schedule for Kinnow (Monthly basis) (6m × 6 m)

Month	*Age of Plants (Years)*				
	0-2	*3-5*	*5-6*	*7-8*	*9 and Above*
January	3	6	9	12	15
February	6	12	18	24	30
March	9	18	27	36	45
April	13	25	39	52	65
May	16	32	48	64	80
June	17	34	51	68	85
July	13	26	39	52	65
August	12	24	36	48	60
September	11	22	33	44	55
October	8	16	24	36	48
November	5	10	15	20	25
December	3	6	9	12	15

Table 12.2: Drip Irrigation Water (litre/day/plant) Schedule for Kinnow (Monthly basis) (6m × 3 m)

Month	*Age of Tree (Years)*				
	1	*2*	*3*	*4*	*5*
January	-	-	-	-	-
February	0.8	1.5	3.8	5.9	13.6
March	1.7	2.2	6.3	10.2	22.5

Month	*Age of Tree (Years)*				
	1	*2*	*3*	*4*	*5*
April	3.3	5.4	13.6	33.1	40.9
May	4.1	9.6	16.3	39.2	53.3
June	4.4	8.4	15.8	38.1	54.3
July	3.2	6.5	10.1	27.8	35.8
August	2.6	4.4	9.8	21.4	29.3
September	1.9	4.9	7.8	19.6	26.1
October	1.7	3.5	6.4	15.2	18.8
November	1.1	1.9	4.0	5.7	12.5
December	0.6	1.4	2.5	7.0	7.5

Table 12.3: Drip Irrigation Water (litre/day/plant) Schedule for Guava (Monthly basis) (6m × 5 m)

Month	*Age of Trees*								
	1	*2*	*3*	*4*	*5*	*6*	*7*	*8*	*9*
January	0.5	0.8	2.3	2.8	5.3	5.7	7.0	8.0	8.4
February	0.6	2.3	4.5	5.7	9.6	10.2	11.4	12.5	13.2
March	1.2	3.4	7.5	9.9	15.9	17.0	18.4	19.2	19.8
April	2.4	8.5	16.0	17.9	28.8	30.5	32.5	33.6	34.5
May	3.0	14.9	19.2	23.0	37.5	39.5	41.5	43.4	44.0
June	3.3	13.1	18.6	27.0	47.8	49.2	51.4	52.6	53.0
July	2.4	10.0	11.9	19.7	28.1	30.5	32.4	33.5	33.8
August	1.9	6.9	11.6	14.9	20.6	22.5	24.3	24.8	25.0
September	1.4	7.6	9.7	14.0	18.4	20.2	21.5	22.7	23.0
October	1.3	5.4	8.5	10.8	13.2	13.8	15.0	16.2	16.8
November	0.8	3.0	5.4	4.0	8.8	9.2	10.4	11.1	12.1
December	0.4	2.1	3.3	5.0	5.2	6.0	6.8	7.2	7.8

Fertigation

Fertigation is the method of application of soluble fertilizer with irrigation water. Fertigation is a prerequisite for drip irrigation. Since the wetted soil volume is limited, the root system is confined and concentrated. The nutrients from the root zone are depleted quickly and a continuous application of nutrients along with the irrigation water is necessary for adequate plant growth. Fertigation offers precise control on fertilizer application and can be adjusted to the rate of plant nutrient uptake.

Advantages of Fertigation

Several distinct advantages of fertigation in comparison with conventional application methods are as follows:

1. Nutrient management can be more precisely controlled and observed.
2. The distribution of nutrients can be more uniformly spread across the soil profile and root area.
3. Throughout the growing season, nutrients can be supplied in increments to match the crop's nutritional needs.
4. When field conditions prevent the use of conventional machinery, nutrients can still be applied to the soil.
5. The absence of heavy machinery in the field eliminates soil compaction issues.
6. Unlike traditional chemical application methods, this approach avoids damaging crops through root pruning, leaf breakage, or plant bending.
7. The chemical application process typically demands less energy and labour for supervision.

The equipment used for fertigation can be either fertilizer tank or ventury.

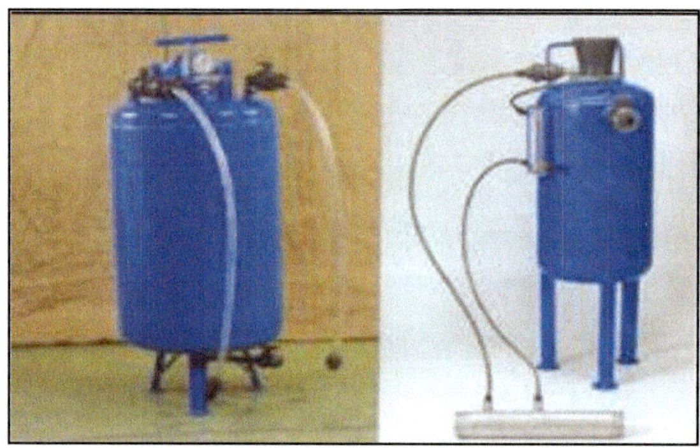

Figure 12.1: Fertilizer Tank

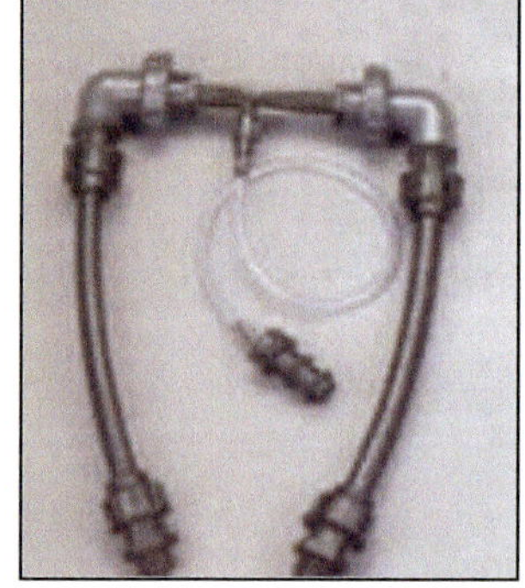

Figure 12.2: Venturi.

Urea is the nitrogen source, urea phosphate is the source of phosphorous, and muriate of potash (white) is the source of potash.

Table 12.4: Fertigation Schedule for Kinnow

Age of Trees (Year)	*February-April (15 splits) (grams)*			*July-Mid*	*September (12 splits) (grams)*	
	Urea	*UP*	*MOP*	*Urea*	*UP*	*MOP*
1	115	-	-	77	-	-
2	230	-	-	154	-	-

Age of Trees (Year)	*February-April (15 splits) (grams)*			*July-Mid*	*September (12 splits) (grams)*	
	Urea	*UP*	*MOP*	*Urea*	*UP*	*MOP*
3	346	-	-	230	-	-
4	370	240	80	247	160	120
5	466	300	80	310	200	120
6	558	360	80	373	240	120
7	647	420	80	431	280	120
8	742	480	240	494	320	160

Table 12.5: Fertigation Schedule for Guava

Age of Trees (Year)	*March-May (12 splits) (grams)*				*July-September (12 splits) (grams)*	
	Urea	*UP*	*MOP*	*Urea*	*UP*	*MOP*
1	33	72	40	33	72	40
2	16	145	100	16	145	100
3	-	218	160	-	218	160
4	40	218	240	40	218	240
5	88	256	320	88	256	320
6	132	292	400	132	292	400
7	192	292	440	192	292	440
8	226	328	520	226	328	520
9	248	346	560	248	346	560

Solar Photo Voltaic Pumping System for Drip Irrigation

Solar pumping systems are expensive to install but have almost nil operative costs. The cost of a solar pumping system becomes at par or even less than providing a new electricity connection, especially in remote areas far away from the existing electrical transmission network. The cost of installing a new electricity line is quite high.

Solar pumping systems are very effective with drip irrigation systems. Popularizing and promoting solar pump systems, especially micro irrigation systems, can lead to water savings and substantial savings in electricity (fuel). The 2-3.0 hp pump is sufficient for a 5 ha area under drip irrigation.

Maintenance of Drip Irrigation System

Current or Daily Maintenance

1. Check emitter functioning, wetting pattern, leakage of pipes, valve fittings, *etc.*
2. Inspect the dripping and ensure that water is reaching all the corners of the plot/field. If water is not dripping correctly at some portion, find the cause and correct it.
3. Always visit the field to check twisting, folding, cutting, punching, *etc.* in the lateral pipe, which may cause faulty system operation. Correct it immediately.
4. Flush all the laterals by opening end plugs 1 to 5 in a series; then close them 1-5 in the same sequence, allowing flushing for 3 minutes until clean water starts flowing.
5. Flush each sub-main at the end of every section (shift) till dirt-free clear water starts flowing.
6. Check inlet and outlet filter pressures. Remove slurry from the hydrocyclone, back flush the sand filter every 5 hours, and flush the screen/disc filter at the end of each day's operation. Do not allow the pressure difference across the sand filter to be more than 0.3 kg/cm2. Do not allow the pressure difference across the screen filter to be more than 0.2kg/cm^2.

Periodic or Fortnightly Maintenance

Repeat 1 to 6 operations and take corrective actions.

Please take out the element of the screen/disc filter and clean it thoroughly. Never use a brush to rub the screen surface. Open the lid of the sand (media) filter manhole, allow the water to come out through the manhole, and stir the sand thoroughly by moving the hand in between filter mushrooms (candles) without disturbing their position to thoroughly separate accumulated foreign material with media (sand) for recharging its filtering capacity.

Chemical Treatment

Clogging or plugging of emitters will be due to precipitation and accumulation of certain dissolved salts like carbonates, iron, calcium and manganese salts. The clogging is also due to microorganisms and the related iron and sulphur slimes due to algae and bacteria. The clogging can be cleared by chemical treatment of water. Chemical treatment commonly used in drip irrigation systems includes the addition of chloride and/or acid to the water supply. The commercial Grade of Acid recommended for Acid Treatment are:

1. Hydrochloric Acid: HCl – 35 per cent

2. Nitric Acid: HNO_3 – 33 per cent
3. Sulfuric Acid: H_2SO_4 - 65 per cent
4. Ortho Phosphoric Acid: H_3PO_4 – 85 per cent

In most cases, HCl can be used, which is highly effective and the cheapest of all, except for crops which cannot tolerate Chloride. In that case, HNO_3 or H_2SO_4 could be used. H_2SO_4 is not recommended where Calcium and Magnesium salts are above 500ppm. Acid treatment is always done before chlorination as chlorination is effective under ph range of 6.5-to-8.5. Do not use orthophosphoric acid (H_3PO_4) if iron is present, as iron would precipitate with phosphoric acid. Always use goggles and surgical/rubber hand gloves and never pour water in acid, but always add acid into the water as a safety precaution before handling acid. Always use plastic containers for acids. Select the most suitable acid and proceed as follows.

Determining the Acid Quantity to be Injected

Take a 10-liter bucket and gradually start adding acid in small portions, measuring the accepted pH. Once you reach a pH of 2.0, calculate the amount of acid required to receive this value in your system by multiplying the acid quantity by 100 and injecting this amount per 1 m^3 of the system discharge.

Example

Acid needed for receiving pH (2.0) in the 10 liter bucket = 12 cc

- ☆ 12 cc X 100 = 1200 cc = 1.2 liters
- ☆ Inject 1.2 liters of acid per 1 m^3 of the system discharge
- ☆ System discharge (of the treated section) = 10 m^3/h
- ☆ System discharge during the 15 minute treatment= 2.5 m^3
- ☆ Acid required = 1.2 liter X 2.5 = 3.0 litter
- ☆ Maximum Injector discharge = 80 l/h
- ☆ Total solution volume required (¼ of 80 liters) = 20 liters
- ☆ 20 liters of solution = 3 liters of acid + 17 liters of water
- ☆ Injection time = 15 minutes (20 liters injected with a 80 l/h injector)

1. After the injection of acid, allow acidified water to react with precipitated salts for about a minimum of 4 - 6 hours (It is desirable to prolong the period for 24 hours). Then, open the ends of the laterals and sub-main flush valve. Start on the pump and allow all the water to flow out. Measure the discharges of marked drippers. Flush Main, sub-main and laterals. If there is no significant improvement, repeat the treatment as above for all the sections.
2. At the end of acid treatment, wash the equipment and vessels with clean water, whip, and dry, removing the acid residue. If clogging is observed due to algae or other causes, carry out chlorination treatment.

3. Run the system for half an hour longer than the normal irrigation schedule so that an extra quantity of acid is removed from the root zone.

A higher concentration of iron and/or manganese oxidation by aeration allows iron to precipitate faster. Use a settlement tank by enabling the water to fall freely into the settlement tank for increased aeration and then pump the water for your system. Chlorination along with aeration can enhance rate of oxidation. Please note that manganese impurities react slowly with chlorine hence they coagulate after the main filters. In such cases, either allow some additional reaction and precipitation time or use plot filters as secondary fine filters to avoid drip per clogging, which cannot be cleaned by any chemical means.

Chlorination Treatment

Chlorine is a biocide that kills microorganisms, bacteria, algae, *etc.* Chlorine injection will reduce clogging and help keep the irrigation lines clean. It is recommended as an intermittent or ongoing preventive treatment in systems that use water containing a high concentration of organic materials. The most commonly used material is sodium hypochlorite 10-12 per cent.

Example:

- ☆ Chlorine concentration required at injection point = 10 PPM
- ☆ System discharge (of the treated section) = 10 m3/h
- ☆ Chlorine quantity required: 10ppm*10 m3/hr/10 per cent/10=1.0 liter
- ☆ Injector discharge = 80 l/h
- ☆ Solution preparation: Mix the1.0 liter with79 liter of water.

This volume will be injected now in 1.0 hr at 10ppm of chlorine

Precautions for Chlorination

Chlorine is toxic to human and animal. Do not have direct contact with skin, eyes, nose and mouth with any Chlorine substance or Cl_2 gas; as it is poisonous for human and animal. Wear goggles, hand gloves, safety shoes *etc.* during chlorination treatment.

Chapter 13

Micropropagation for Production of Disease Free Elite Fruit Plants

Krishan Kumar, Subhash Chander, Harsimran Singh and P.K. Arora

Punjab Agricultural University, Dr J.C. Bakhshi Regional Research Station, Abohar, Punjab

The present day fruit crops are the results of intensive domestications for traits like high yield and edible fruit quality. The transitions in perpetuation, from seeds (sexual) to vegetative mean (asexual) played a key role in this voyage. The vegetative propagation means of plant production like cuttings, layering or the grafting ensured fixing of these forms and in turn brought uniformity of the progeny plants. The technique of grafting further facilitated the cultivation of fruit plants across diverse set of edaphic conditions. However, vegetative propagation also contributes towards disease transmission and accumulation in the progenies. It is noted that incidence of various diseases, especially virus mediated diseases can cause 30 to 100 per cent yield losses or induce plant destruction. The vegetative means or conventional propagation are unable to meet the demand of requisite plant types in crops like date palm and papaya. Under these events, micropropagation can play a significant role in mass multiplication of true to type plants of a candidate genotype. Direct organogenesis (plants regeneration from pre-formed meristems), somatic embryogenesis and meristem culture/shoot tip grafting are the biotechnological techniques being used for clonal plant propagation. The first two techniques are being used for mass multiplication of healthy plants in several fruit crops like banana, strawberry, papaya and date palm *etc.* The meristem culture (or its deviation shoot tip grafting) and somatic embryogenesis are vital in cleansing the planting material. These techniques can produce healthy plants even from a diseased (especially viruses) source plant. The disease filtering techniques are now being used for producing disease free stocks of citrus varieties and are of great significance in facilitating exchange of disease free germplasm globally. The use of bio-reactors for optimized protocols not only would cut the cost but would also can aid in their commercialization.

Introduction

Fruits have immense importance in good human health. They are rich source of dietary fiber, vitamins, minerals and antioxidants. The regular intake of fruits, especially of those rich in anthocyanin, prevents the onset of cardiovascular diseases (Habauzit *et al.,* 2013).

The modern day fruit crops have resulted from domestication process for yield and edible fruit quality traits. This selection, gradually enabled cultivation of delicious and high yielding varieties across different fruit crops. The transitions in mode of propagation played a key role in emergence of the present day fruit forms or cultivars. In wild, the fruit crops used to be multiplied through seeds. These seed propagation led to the development of forest gardens or home gardens. The seed propagation was taken over by vegetative propagation methods like cuttings, layering *etc.* Vegetative propagation helped in preserving uniformity, reducing juvenility, and also suppressed the traits associated with juvenility like excessive thorniness (Goldschmidt, 2013). The next innovation was grafting, which allowed to take advantages from two different components (variety and rootstock) in a single plant. This breakthrough allowed use of differential characteristics from two different partners in a single plant. However, the vegetative propagation also had a significant bearing associated with plant health. In vegetative propagation, if the mother plant is diseased or infected with a disease, the disease passes to the progeny/offspring. There are many examples of disease transmission through vegetative propagation. Citrus tristeza virus, porosis, greening in citrus; bunchy top, infectious chlorosis, banana streak virus, moko wilt in banana; leaf roll, little leaf, stem pitting, fan leaf, yellow mosaic, vein bending, necrosis, corky bark, Pierce's disease, bacterial canker in grapes; mosaic, chlorotic leaf spot, star crack in apple; pineapple wilt virus; plum line pattern virus, plum and peach mosaic, peach X disease, peach root knot nematode of stone fruits are examples of diseases, which are transmitted and spread through the use of infected planting material (scions or rootstocks) and propagating tools for nursery production.

The yield loss due to these different diseases have been found significant. Alone, the viruses cause significant losses. CTV infection caused destruction of 50-100 million citrus trees world-wide (Moreno *et al.,* 2008). Similarly, the BBTV infection limited the area of cultivation of a unique flavoured variety Virupakshi from 18,000 to meagre 2,000 ha in Tamil Nadu. It led to the loss of production to the extent of $50 million (Selvarajan and Balasubramanian, 2013). Banana streak virus (BSV) was reported to cause yield losses of 48 per cent in cv. Poovan in India (Thangavelu *et al.*, 2000). The grapevine fan leaf virus induces a significant decline in yield and cause a degradation in quality (Andret-Link *et al.*, 2004). Similarly, PRSV among the viral diseases causes huge losses to yield amounting to 70 per cent in Australia (Dillon, 2005). In India, its incidence varies from 50 to 100 per cent. Sharma *et al.* (2006) reported that the PRSV infection caused about 41.12 per cent yield loss in papaya between flowering to fruit set and a 34 per cent yield loss in its infection post fruit set. Similarly in citrus, various viruses like citrus ringspot virus, citrus mosaic virus, citrus tristeza virus and even greening is associated with citrus decline; most of these are bud transmissible (Ahlawat, 1997).

The viruses move into different plants through plant vascular system. Extreme plant apex lacks the connection with plant vascular system. The *de novo* developed somatic embryos also arise from a non-vascular tissues. Due to this, the plants produced from somatic embryos/meristem can help produce plants free of viruses. The micropropagation is the clonal perpetuation of plants on artificial media under aseptic conditions. Most of the fruit crops are vegetatively propagated. In some other crops like date palm, vegetative or conventional propagation doesn't able to produce the required amount of plants. Under these circumstances, micropropagation is of great significance. The micropropagation is of great value in fruit plant production under the following conditions:

1. For rapid multiplication of clonal plants
2. For overcoming seasonal constraints
3. For multiplication of desirable types plants
4. For production of disease free plantlets

Factors Necessitating the Use of Micropropagation

Limited vegetative propagules in certain crops: There are crops where vegetative propagation cannot meet the required plant demand for establishing new orchards. In date palm, the vegetative mean of propagation is an offshoots, which emerges from the main stem at the ground level. These offshoots develop roots in soil contact. The offshoot produced plants are uniform and homogeneous for raising a commercial orchard. However, the offshoot production capacity of mother tree is limited to vegetative phase of the tree. A tree can produce a maximum of 20-30 offshoots in its entire life cycle (Fki *et al.*, 2011). Further, the offshoot production potential is genotype dependent and establishment success of grown up offshoot derived plants is also less. The commercial date palm varieties are dioecious in nature *i.e.* male and female flowers are borne on different plants. The seed propagation lead to production of male and female plants in 1: 1 proportion. The tissue culture based direct organogenesis or somatic embryogenesis can enable mass production of this valuable palm.

For Production of a Specific Plant Type

The commercial varieties of papaya and date palm are polygamous in nature. The commercially important papaya varieties represent dioecious (male and female flowers borne on different plants) and gynodioecious (Generates female and hermaphrodite plants) or pistillate sex forms. Traditionally, seed is the commercial mean of propagation. But, propagation of dioecious and gynodioecious through seeds lead to emergence of different sex plants. For instance, dioecious varieties gives rise to male and female plants while seeds of hermaphrodite plants produce hermaphrodite and female plants in ratio of 2:1. The female plants depending upon the pollinizer type (hermaphrodite or male) produce pistillate: hermaphrodite or pistillate: male plants in 1:1 ratio (Ming *et al.,* 2007). For establishing the stand of desired hermaphrodite/pistillate plants, seeding of 3-4 seeds or 3-4 plants is done per spot or hole and at flowering emergence, the extra unwanted plants (male/ female) are removed, which adds cost of production.

Similarly, date palm, *Phoenix dactylifera,* a crop of hot arid areas is dioecious in nature. Due to this, the seed propagation results in heterogeneous plant stand of male and female plants. The vegetative propagule structure called offshoots cannot produce the desired number of plants. In citrus, rootstock has a significant bearing on the performance of scion varieties. The rootstocks have a variable degree of embryony. Rough lemon, Rangpur lime and Carrizo citrange are the commonly used citrus rootstocks in India. Rough lemon has high degree of polyembryony while in the latter two rootstocks, the degree of polyembryony is less (Bodade *et al.,* 2022). In such circumstances, micropropagation can produce the uniform stock and can significantly improve the tree performance.

For Production of Disease Free Plantlets

Vegetative propagation from infected source can lead to transmission of the disease to the new plantlets. Continuous vegetative propagation leads to the accumulation of these diseases into the planting material. The accumulated disease load lowers productivity and economic life span of the fruit trees. Many a viruses in fruit crops like citrus are bud transmitted. Tissue culture techniques like meristem tip culture or shoot tip grafting can cleanse the planting material free of the viruses or other diseases. This has been extensively used in recovering clean material from infected tissues in citrus, stone fruits and kiwi fruit.

Rapid Multiplication of Plants

The conventional propagation techniques allow production of plants at a small scale and cannot realize the desired plant demand. The adoption of micropropagation techniques can ensure year-round multiplication of plants and thus can overcome the seasonal specificity of plant response.

Techniques of Micropropagation

The micropropagation is the clonal propagation of plants by maintaining their genetic stability relative to mother plant. The micropropagation is performed through three tissue culture techniques:

1. Direct organogenesis
2. Somatic embryogenesis
3. Meristem tip culture/shoot tip grafting

Direct Organogenesis

In direct organogenesis, the shoots are directly induced from the explant. The plantlet regeneration in direct organogenesis passes through four different phases: explant establishment/shoot initiation, shoot proliferation, rooting and hardening (Figure 13.1). The explant (initial propagule used for explant establishment) may be from field or lab grown plant. When the explant is directly procured from field plants/trees, the explants may carry outside microbes, which cause contamination of cultures and prevent explant establishment. In explants collected from field trees, disinfection is required. Different surface sterilants like sodium hypochlorite (0.5%), calcium hypochlorite (4-8%) and mercuric chloride ($HgCl_2$) (0.1%) have been used for disinfecting the explants (Hutchinson and Zimmerman, 1987). To prevent fungal contamination, a pre-sterilization with systemic or contact fungicides has also been attempted in several fruit crops like grapes, pomegranate, *jamun etc.* (Kumar *et al.,* 2006; Kumar *et al.,* 2019; Kamboj *et al.,* 2023). After sterilization, the explants need to be thoroughly washed with sterile water to prevent carry over of the sterilant to the cultured explant. The examples of *in vitro* propagation using different explants, medium composition are detailed in Table 13.1.

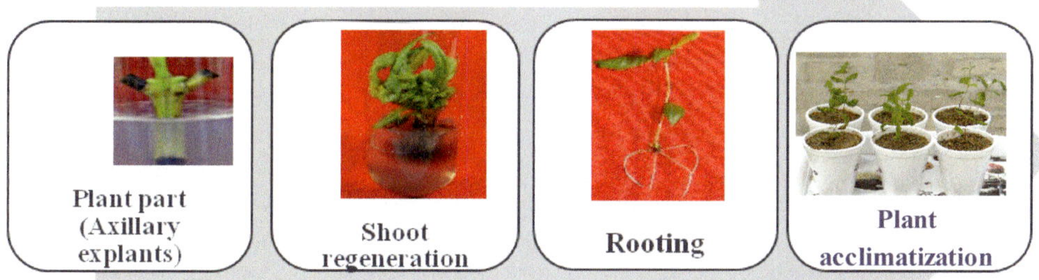

Figure 13.1: Illustration of Direct Organogenesis from Axillary Explants of Pomegranate cv. Bhagwa.

Explant

The type of explant has a significant influence on production of variation free fruit plants. Usually, the explants with pre-formed meristems produce genetically stable plants than highly differentiated explants like stem, leaves or roots (Duncan, 1997). Even among the explants with preformed meristems, the nodal explants (with axillary meristems) has shown superiority over shoot tips (explants with terminal meristems) for establishment or regeneration response in crops like grapes (Singh *et al.*, 2002; Kumar *et al.*, 2008), guava (Amin and Jaiswal 1987), *jamun* (Kamboj *et al.*, 2023). In already differentiated explants, the shoot meristems have to be *de novo* formed. In that case, the degree of juvenility increases the chances of shoot buds formation. In grapes, the youngest leaves wrapped around the meristem showed higher regeneration competence than the older ones (Kumar *et al.*, 2006), indicating their closeness to the meristem in developmental pattern.

The state of the explant, culture month and location have a considerable impact on culture initiation response from field trees in fruit crops. In grapes under Ludhiana (Punjab) conditions, best culture establishment was recorded in April month (Kumar *et al.*, 2006). In pomegranate and *jamun*, the best explant establishment response under arid conditions of Punjab was recorded during May-June (Kumar *et al.*, 2019; Kamboj *et al.*, 2023), which was linked to lower per cent contamination attributable to retarding effect of higher temperature of these months.

In crops like papaya, where seedlings of the mother plant serve as the source of explant; the identification of the plant sex before stating micropropagation is essential.

The physiological state of the explant also influences the explant establishment. One of the biggest challenge in explant establishment or the cause for the recalcitrance of fruit crops to *in vitro* propagation is the explant mortality induced by browning (Bhat *et al.*, 2022). Browning is caused by exudation of phenols from the cut portion of explants. The phenols are converted to quinones by enzymes called polyphenol oxidase (PPO), which is toxic to the tissue. The use of various anti-browning measures like pre-treatment with antioxidants (El-Agamy *et al.*, 2009), addition of activated charcoal (Kumar *et al.*, 2019), transfer of explants to fresh medium at regular interval (Murkute *et al.*, 2003) can reduce the problem of browning. In grapes, the higher establishment success during April month was also linked to lower polyphenol content during this period. The period of November to March is considered suitable for culture initiation in date palm (Sidky, 2017).

Medium: The nutrient medium provide macronutrients, micronutrients and vitamins. These are required for any morphogenetic response from the cultured explant. More than 60 different culture media have been devised. However, the suitability of the medium is species specific. Out of the three different basal media (MS, WPM and B5 medium), WPM was found superior for guava cultures (Meghwal *et al.*, 2010). Raju and Divya (2020) compared two different basal media (MS and

WPM) with different growth regulator for morphogenetic culture response in *S. densiflorum*, a tropical forest tree. The MS medium showed better performance than WPM medium. In finger lime, MS medium was better out of three different basal media- MS, C2D, and DKW media (Mahmoud *et al.,* 2020).

Growth Regulators and morphogenetic response: Growth regulators composition of the medium evokes a particular response in the cultured explants. The morphogenetic response under direct organogenesis includes shoot induction, proliferation and rooting. The auxin cytokinin ratio determines the type of morphogenetic response (Skoog and Miller, 1957). The first morphogenetic response under *in vitro* direct organogenesis is the formation of shoots. The high cytokinin to auxin ration promotes shoot regeneration, which is probably related to *de novo* formation of new shoot buds and release of bud dormancy from axillary meristems (Muller and Leyser, 2011). Various cytokinin's [Thidiazuron (TDZ), benzyl amino purine or benzyl adenine (BAP/BA), kinetin] alone or in combinations have been used for shoot bud induction in different fruit crops. The optimum concentration of a cytokinin is species dependent, which is determined by culturing the explants to variable levels and various combinations. In grapes cv. Perlette, the best shoot initiation response was recorded with a combination of BAP and kinetin (1.0 mg l^{-1} each) while for Flame seedless, the combination of BAP (1.0 mg l^{-1}) and kinetin (0.5 mg l^{-1}) induced better response (Kumar *et al.,* 2006). Even the cytokinin's have been complemented with auxins to evoke shoot regeneration in crops like jamun, pomegranate and date palm (Sidky, 2017; Kumar *et al.,* 2019; Kamboj *et al.,* 2023). The low level of auxins with cytokinin's exert a synergistic effect possibly by helping in activation of differentiated or induced meristems (Muller and Leyser, 2011).

The next response in direct organogenesis is shoot proliferation. During this stage, the original explant after harvest of the regenerated shoots or regenerated shoots itself are cultured on the medium with decreased cytokinin concentrations. In *Citrus limon* nodal explants, the axillary shoots emerged on MS medium amended with 9.0 μM BAP (Rathore *et al.,* 2007). The researchers achieved repeated shoot harvest for six cycles by culturing the initial explants on ½ MS medium + 0.22 μM BAP. The decreased cytokinin concentrations helps maintain the integrity of the shoots while the elevated cytokinin's adversely affect shoot morphology (Kieber, 2002). In some of the studies, like in pomegranate (Kumar *et al.,* 2019), citrus (Mahmoud *et al.,* 2020), jamun (Kamboj *et al.,* 2023), the shoot proliferation is achieved by adding silver nitrate. The optimum silver nitrate concentration for shoot proliferation has been found variable among different crops *viz.*, pomegranate (1.5 mg l^{-1}), jamun (3.0 mg l^{-1}), and date palm (10.0 mg l^{-1}) (Kumar *et al.,* 2019; Kamboj *et al.,* 2023; Sidky, 2017).

Rooting is the next stage, which is essential to make the *in vitro* plants self-reliant. Variations of the strength of basal nutrient medium, supplementations of auxins (IBA, IAA and NAA) alone or in combinations at low concentration (< 2.0 mg l^{-1}) and activated charcoal have been used for *in vitro* root formation in fruit crops

(Table 13.1). The rooting can be induced *in vitro* or *ex vitro*. For *ex vitro* rooting, the *in vitro* derived micro-shoots are given short exposure to high concentration of the auxins (200 to 400 mg l^{-1}). The *ex vitro* mode of rooting reduces the time of plantlet production and the roots have strong lateral network (Sharma *et al.*, 2017).

Table 13.1: Examples of Direct Organogenesis Mode of Micropropagation in Fruit Crops

Fruit Crop	*Variety*	*Explant*	*Medium and Response*	*Reference*
Date palm	-	5 cm size shoot tip prepared from 7-8 Kg offshoot (3-5 year old offshoot)*	**Initiation (six subcultures at 4 week intervals):** Modified MS medium+1 mg l^{-1} NAA + 1 mg l^{-1} NOA + 2.5 mg l^{-1} BA + 2.5 mg l^{-1} 2ip **Shoot multiplication:** **Shoot bud differentiation:** Modified MS medium+ 0.5 mg l^{-1} BA + 0.5 mg l^{-1} 2iP **Shoot bud proliferation:** ½ MS + 10 mg l^{-1} silver nitrate **Shoot elongation:** ½ MS + 0.5 mg l^{-1} GA_3 + 1 mg l^{-1} NAA **Rooting:** : ½ MS + 0.1 mg l^{-1} NAA + 4 mg l^{-1} paclobutrazol **Pre-acclimatization:** ½ MS + PEG-8000 MW	Sidky, 2017
	Maktoom	Inflorescence explant	**Shoot bud initiation:** MS +2.0 mg l^{-1} NAA, 4 mg l^{-1} BAP + 40 g l^{-1} sucrose **Shoot multiplication:** 3.0 mg l^{-1} 2-ip + 2.0 mg l^{-1} BAP **Shoot elongation:** 0.5 mg l^{-1} BAP + 0.5 mg l^{-1} 2ip + 1 mg l^{-1} GA_3 **Rooting:** ½ strength MS medium + 1 mg l^{-1} NAA	Khierallah *et al.*, 2017
	Medjool	Shoot tips	**Shoot bud induction:** MS + 1 mg l^{-1} TDZ + 1 mg l^{-1} BA + 1 mg l^{-1} NAA **Shoot multiplication:** MS + 3 mg l^{-1} TDZ + 2 mg l^{-1} BA **Shoot elongation:** MS + 3 mg l^{-1} GA_3 **Rooting:** MS + 1.5 mg l^{-1} NAA	Hussein and Jawad, 2021

Fruit Crop	*Variety*	*Explant*	*Medium and Response*	*Reference*
	Medjool, Barhee and Selmi	Inflorescence explant	**Organogenesis:** MS + 2 mg l^{-1} TDZ +1 mg l^{-1} BA **Shoot bud multiplication:** MS + 0.5 mg l^{-1} TDZ +1 mg l^{-1} BA **Rooting:** MS +1.0 mg l^{-1} IAA + 0.5 mgl^{-1} NAA + 0.5 mg l^{-1} IBA	Taha *et al.*, 2021
Papaya	Rainbow	Nodal segments	**Shoot initiation or multiplication:** MS + 3.0 mg l^{-1} BA **Shoot elongation:** MS + 0.25 mg l^{-1} BA + 0.3 mg l^{-1} GA_3 **Root initiation:** MS + 1.0 mg l^{-1} IBA	Caple and Cheah, 2016
	Surya	Shoot tips	**Shoot initiation or multiplication:** MS + 2.0 mg l^{-1} BAP + 1.0 mg l^{-1} NAA **Root initiation:** MS + 3.0 mg l^{-1} NAA	Ryavalad *et al.*, 2019
	Red Lady	Shoot tips	**Shoot Induction:** MS +0.5 mg l^{-1} BAP + 0.1 mg l^{-1} NAA **Shoot growth:** MS +160 mg l^{-1} Adenine sulphate **Rooting:** ½ MS + 1.0 mg l^{-1} IBA	Patel *et al.*, 2013
	Laie Gold	Shoot tips	**Shoot Induction:** MS +0.2 mg l^{-1} BAP + 0.2 mg l^{-1} NAA **Shoot proliferation:** MS +0.05 mg l^{-1} BAP + 0.02 mg l^{-1} NAA **Rooting:** MS + 4.0 mg l^{-1} IBA	Fitch *et al.*, 2022
	Pusa Dwarf	Shoot tips	**Shoot initiation:** MS + 100 ppm glycine, 5 ppm casein hydrolysate and 0.05 per cent charcoal **Shoot development:** MS + 3 mg l^{-1} BAP + 0.5 mg l^{-1} NAA **Root initiation:** ½ MS + 3 mg l^{-1} IBA	Kumari *et al.*, 2023
Passiflora edulis Sims. f. *flavicarpa* Deg	-	Nodal segments	**Shoot induction:** MS + 2.0 mg l^{-1} BAP **Shoot proliferation:** MS + 0.5 mg l^{-1} BAP + 0.5 mg l^{-1} kinetin **In vitro rooting:** ½ MS + 2.0 mg l^{-1} IBA ***Ex vitro* rooting:** 200-300 mg l^{-1} IBA	Shekhawat *et al.*, 2015

Fruit Crop	*Variety*	*Explant*	*Medium and Response*	*Reference*
Citrus	Rangpur lime	Nodal segments	**Shoot induction:** MS + 2.0 mg l^{-1} BAP **In vitro rooting:** MS + 2.0 mg l^{-1} IBA + 0.1 mg l^{-1} NAA + 1000 mg l^{-1} activated charcol	Bodade *et al.*, 2022
Grapes	Perlette and Flame seedless	Nodal segments	**Shoot induction:** MS + BAP (1 mg l^{-1}) + kinetin (0.5 – 1.0 mg l^{-1})	Kumar *et al.*, 2006
Jamun	Konkan Bahadoli	Nodal segments	**Shoot induction:** MS + 2.0 mg l^{-1} BAP + 0.25 mg l^{-1} NAA **Shoot proliferation:** MS + 2.0 mg l^{-1} BAP + 3.0 mg l^{-1} $AgNO_3$ **Rooting:** MS + 0.5 mg l^{-1} NAA	Kamboj *et al.*, 2023
	Seedling	Nodal segments	**Shoot induction:** MS + 10 µM BA **Rooting:** ½ MS + 5 µM NAA	Naaz *et al.*, 2014
Syzygium densi-florum	-	Nodal segments	**Shoot induction:** WPM+ 1.5 mg l^{-1} BAP + 1.5 mg l^{-1}IBA **Rooting:** ½ WPM + 0.5 mg l^{-1}IBA	Raju and Divya, 2020
Pome-granate	Bhagwa	Nodal segments	**Shoot induction:** MS + 1 mg l^{-1} BAP + 0.25 mg l^{-1} NAA **Shoot proliferation:** MS + 0.5 mg l^{-1} BAP + 0.5 mg l^{-1} NAA + 1.5 mg l^{-1} $AgNO_3$ + 40 mg l^{-1} adenine sulphate **Rooting:** ½ MS + 1.0 mg l^{-1} NAA	Kumar *et al.*, 2019

* Shoot tips dipped in antioxidant solution.

Somatic Embryogenesis

Somatic embryogenesis is the process by which an embryo is produced from a somatic cell (von Arnold *et al.*, 2002). This structure lacks any vascular connection from the original tissue (von Arnold *et al.*, 2002). In terms of developmental pattern, somatic and zygotic embryos follow similar pathway and typically pass through globular, heart, torpedo and cotyledonary stages in dicot plants (Mordhorst *et al.*, 1997). The somatic embryo just like zygotic embryo is bipolar in nature, possess both shoot (plumule) and root (radicle) structures. This precludes the requirement of separate rooting step in somatic embryos. The somatic embryogenesis has a higher rate of plant multiplication and the plants derived are of single cell origin. The single cell origin makes it attractive even for improvement through genetic engineering. The process of embryogenesis is under the gene regulation. To date, several genes related with somatic embryogenesis have been revealed; notable among them are Citrin, a globulin protein (Koltunow *et al.,* 1996), two SERK genes, CitSERK1 and CitSERK1-like (Shimada *et al.,* 2005; Ge *et al.,* 2010), CsHpt1 gene encoding for a phospho-transmitter protein (Maul *et al.,* 2006), and 11 candidate genes spanning over a 80 Kb region (Wang *et al.,* 2017). In fruit crops, there are numerous reports on somatic embryogenesis from citrus (Kumar *et al.,* 2018), however, its direct significance for propagation lies in fruit crops like date palm. The process of somatic embryogenesis passes through multiple steps: embryogenic callus initiation; formation and maturation of somatic embryos; conversion or germination of somatic embryos. The somatic embryos may also be directly formed. The process of somatic embryogenesis is influenced by the factors of genotype, explants and growth regulator composition. In date palm, actively growing variety exhibited early embryogenic induction than the slow growing (El Fatih and Hind Badereldin, 2010). Besides genotype, the influence of explant is critical in the process of somatic embryogenesis. The ovular origin explants (ovule and nucellus) are reported to be better than the non-ovular type explants for somatic embryogenesis in citrus (Fiore *et al.,* 2002, Savita *et al.,* 2015). However, the choice of explant for somatic embryogenesis varies per fruit crop. In papaya, immature zygotic embryos has been the choicest explant for somatic embryogenesis (Mishra *et al.,* 2007) while in monocot fruit crop like date palm, the zygotic embryos showed lower embryogenic competence (Fki *et al.,* 2011). In date palm, the actively growing tissues like shoot tip originating from offfshoot (Hassan and Taha, 2012), inflorescence (Hassan *et al.,* 2021; Solangi *et al.,* 2023) and leaf explant from field trees, while bud explants from *in vitro* cultures (Mazri *et al.,* 2017) have been used for inducing somatic embryos.

Initial growth regulators compositions is vital for embryogenic induction. In most of the cases, auxins alone or in combination with low cytokinins induce somatic embryos. Among the auxins, 2,4-D has been used with concentration varying per crop. For papaya cv. Sahi, 1.5 mg l^{-1} 2,4-D induced embryogeneic callus while for date palm the concentration of the growth regulator varied from 1 mg l^{-1} to 100 mg l^{-1} (Table 13.2). However, higher concentration may cause varitions in the regnerants (Fki *et al.,* 2011; Al-khayri and Naik, 2017). Picloram is the second most frequent auxin used for inducing embryogenic cultures (Table 13.2).

Post induction, next step is maturation and development of the somatic embryos. This step requires reduction in auxins concentration or addition of cytokinins and giberellin synthesis inhibitors in the medium as demonstrated in date palm and papaya. In date palm, the use of anti ethylene compound significantly improved the percent embryo formation (Al-Khayri and Al-Bahrany, 2001). One major problem during embryo maturation is presence of embryos in different stages; use of ABA and poly-ethylene glycol is known to induce embryo synchronization (Al-Khayri and Al-Bahrany, 2012).

Once the development of somatic embryo is complete; it is can germinates even on basal MS medium. The use of GA_3 similar to seed germination improves germination of somatic embryos.

Table 13.2: Examples of Somatic Embryogenesis for Propagation in Fruit Crops

Fruit Crop	*Variety/ Cultivar*	*Explant*	*Medium and Response*	*Reference*
Date palm	Medjool	Spikelets from 10-15 cm long inflorescence	**Initiation:** MS+ 2.0 mg l^{-1} picloram + 0.25mg l^{-1} BA + 0.25 mg l^{-1} 2iP **Maturation:** MS+ 0.1 mg l^{-1} picloram + 0.1mg l^{-1} ABA or + 0.1 mg l^{-1} 2iPMultiplication: ¾ MS + 0.25 mg l^{-1} BA +/- AC **Rooting:** MS+ 0.2 mg l^{-1} NAA, 0.1 mg l^{-1}IBA, 0.1 mg l^{-1} PBZ, 30 g l^{-1} sucrose,2.0 mg l^{-1} glycine, 5.0 mg l^{-1} thiamine HCl + 1.0 mg l^{-1} biotin **Acclimatization :** Liquid MS medium	Hassan *et al.*, 2021
	Samany and Berta-moda	Spikelets from 15 cm long spathe	**Initiation:** MS + 0.5 mg l^{-1} 2,4-D, 3.0 mg l^{-1} 2-isopentenyl adenine (2iP) **Embryo formation:** MS + 0.05 mg l^{-1} 2,4-D + 3.0 mg l^{-1} 2-ip+ 3 g l^{-1} activated charcoal + 200 mg l^{-1} glutamine **Maturation:** MS+ 0.1 mg l^{-1} picloram + 0.1mg l^{-1} ABA or + 0.1 mg l^{-1} 2iPMultiplication: ¾ MS + 0.25 mg l^{-1} BA +/- AC **Embryo germination:** 0.1 mg l^{-1} BA, 0.1 mg l^{-1} NAA **Acclimatization:** on potting mixture of peatmoss, river sand, hill sand (4:1:1 v/v)	Solangi *et al.*, 2023
	Najda	Bud explant from in vitro grown seeedlins	**Embryo initiation:** MS + 45.0 μM 2,4-D + 4.5 μM kinetin or 4.5 μM 6-2iP **Embryo germination:** Basal MS medium	Mazri *et al.*, 2017

Fruit Crop	Variety/ Cultivar	Explant	Medium and Response	Reference
	Zaghlol (soft), Amry (semidry) and Malakaby	Shoot tips	**Callus initiation:** MS + 10 mg l^{-1} 2,4-D+3 mg l^{-1} 2ip+2 g l^{-1} activated charcoal+100 mg l^{-1} glutamine+5 mg l^{-1} thiamine HCl +1 mg l^{-1} biotin + 0.15 M sucrose **Somatic embryo formation:** ½ MS + 0.1 mg l^{-1} NAA+ 0.05 mg l^{-1} BA+ 1.0 g l^{-1} AC	Hassan *et al.*, 2012
	'Mishrig Wad Khateeb' and 'Mishrig Wad Laggai'	Shoot tip explant	**Embryogenic callus induction:** MS + 100 mg l^{-1} 2, 4-D + 3.0 mg l^{-1} 2-ip **Germination:** MS + 0.1 mg l^{-1} NAA	El Fatih and Hind Badereldin, 2010
	Esta-maran	Leaf explant	**Somatic embryo formation:** MS + 5 mg l^{-1} 2,4-D + 10 mg l^{-1} BAP + 5 mg l^{-1} TDZ	Baharan and Mohammadi, 2018
Papaya	Shahi	Young leaves from *in vitro* raised shoots	**Embryogenic callus induction:** MS + 1.5 mg l^{-1} 2, 4-D **Somatic embryo differentiation:** MS + mg l^{-1} Kin + 0.5 mg l^{-1} NAA + 200 mg l^{-1} glutamine **Embryo germination:** MS + 1.0 mg l^{-1} GA_3 + 10 per cent coconut water	Roy *et al.*, 2016

Meristem Tip Culture/Shoot Tip Grafting

Meristem is the extreme tip of 0.1 to 0.5 mm size, which is utilized for the production of virus free plantlets (Figure 13.3). Shoot tip grafting is employed where, the meristem is extremely small (citrus and its excision may cause injury or meristem tip culture is not successful. The shoot tips are grafted on to the stock under *in vitro* conditions. The apical shoot apex lacks vascular connection. This characteristic feature keeps these structures free of viruses or the viral titer is on lower side in this zone.

Virus free bud-woods are the pre-requisite for any bud-wood certification program. Viral diseases cause a great reduction in yield in fruit crops. In strawberry, viral infections can cause huge yield reductions (30-80%) annually. In fig, fig mosaic emaravirus is the major threatening disease. Across world, different viruses have been associated with this disease and spread by insects like aphids, mites and mealy bugs (Jamous *et al.,* 2020). Since, most of the viruses are transmitted either by budding/grafting of the infected planting material or through insect transmission. So, maintenance of the virus free planting material can help in overcoming the problem to some extent. Thus, potential of the meristem tip culture and shoot tip grafting techniques need to be exploited for virus elimination. The technique has been used in various fruit crops for virus elimination and is given in tabulated from (Table 13.3).

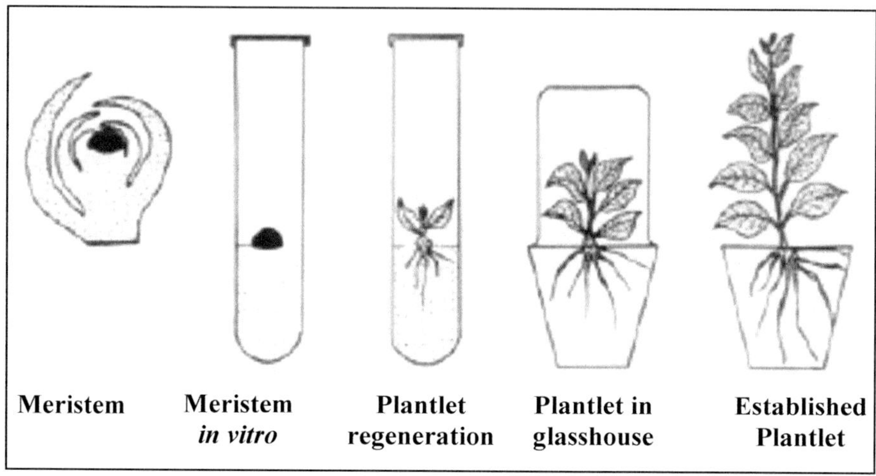

Figure 13.3: Schematic View of Meristematic Based Plantlet Regeneration.

The chances of getting a virus free plant in fruit crops is dependent upon meristem size. In fig (*Ficus carica*), the meristem size (0.5 mm and 1.5 mm) had influence on the recovery of plants from two major viruses [Fig cryptic virus (FCV), Fig milde-mottle-associated virus (FMMaV)]. Across five fig varieties, the average virus free plants ranged from 60.9 per cent for 1.5 mm meristem size to 86.7 per cent for 0.5 mm size (Bayoudh *et al.,* 2017). In apple, the meristem size (0.1 to 1.0 mm) influenced the production of virus free plants (Bhardwaj *et al.,* 1998). The plants regenerated from larger than 0.2 mm meristem were found carrying apple mosaic virus. Similarly, Vivek and Modgil (2018) found that the plants regenerated from 0.3-0.6 mm meristem size were free of four viruses *i.e.* ACLSV, ApMV, ASGV and ASPV whereas the plants from 0.5-0.6 mm meristem carried contamination of ACLSV. In such events, the combination of thermal treatments/therapy with meristem culture seems a reliable technique for obtaining virus free plantlets. Hu *et al.* (2017) tried the combinations of thermotherapy and meristem culture in apple. The regenerated plants were found to be free from the viruses. Against the recovery of virus free plants, the size of meristem has positive relationship with explant survival success and regeneration (Vivek and Modgil, 2018). In apple cv. 'Oregon Spur', the morphogenetic response of different size meristems was investigated.

Table 13.3: Meristem Culture and Shoot Tip Grafting for Getting Virus Free Plantlets in Fruit Crops

Fruit Crop	*Explant*	*Medium*	*Response*	*Virus or Disease Sanitation Success*	*Reference (s)*
Strawberrycv. Oso Grande	Meristem	MS + 0.5 mg l^{-1} BA + 0.1 mg l^{-1} IBA	Shoot induction and proliferation	-	Rattanpal *et al.*, 2011
		½ MS + 1.0 mg l^{-1} IBA	Rooting		
F. chiloensis accessions Contulmo and Purén	Meristem	MS + PVP-300 mg l^{-1}	Culture regeneration	-	Quiroz *et al.*, 2017
		MS + 0.5 mg l^{-1} BAP	Shoot multiplication		
		MS medium	Shoot development and rooting		
Fragaria ananasa cvs.	Meristem (0.3-0.5mm)	MS + 0.5 mg l^{-1} kinetin	Shoot development	-	Naing *et al.*, 2019
		MS + 0.75 mg l^{-1} IBA	Root induction		
Fragaria × ananassa var. Aroma	Meristem (0.3-0.5mm)	MS + 1 mg l^{-1} zeatin	Shoot development and root induction		Neri *et al.*, 2022
Apple cv. Oregon Spur II	Meristem (0.3-0.6 mm), source tree exposed to 37-40°C for 4 weeks	MS + 1 mg l^{-1} BA + 0.05 mg l^{-1} IBA + 0.1 mg l^{-1} GA_3MS + 0.5 mg l^{-1}BA + 0.08 mg l^{-1} IBA+ 0.5 mg l^{-1} GA3	-Explant establishment-Shoot proliferation	Virus free plants	Vivek and Modgil, 2018
Papaya *cv.* Hermaphrodite plants	Meristem	Semi-solid/liquid double layer MS medium + 0.15 mg l^{-1} CPPU	-Shoot induction and development	-	Gatambia *et al.*, 2015
Papaya *cv.* MSXJ	Meristem (0.3 mm)	MS + 0.4439 µM BAP, 0.5708 µM IAA + 3 per cent sucrose	87 per cent survival and 22 per cent regeneration	100 per cent virus free plants	González-Lázaro *et al.*, 2019

Fruit Crop	*Explant*	*Medium*	*Response*	*Virus or Disease Sanitation Success*	*Reference (s)*
Fig. cvs. Alkuden and Bursa Siyahi	Meristem	MS + 0.2 mg l^{-1} GA_3 + 0.5 mg l^{-1} BAMS + 0.2 mg l^{-1} GA_3 + 0.5 mg l^{-1} BA	-Explant establishment-Shoot proliferation	-	Comlekcioglu *et al.*, 2007
Fig. cvs. Zidi (ZDI), Soltani (SNI), Bither Abiadh (BA) and caprifig Assafri (ASF)	Meristem (0.5 mm)	MS + 0.2 mg l^{-1} BAP + 0.1 mg l^{-1} NAA + 0.1 mg l^{-1} GA_3MS + 0.5 mg l^{-1} BAP + 0.1 mg l^{-1} NAA	-Explant establishment-Shoot multiplication	86.7 per cent virus free plants	Bayoudh *et al.*, 2017
Shoot tip grafting					
Passion fruit	Shoot tip (2 mm)	MS + 0.5 mg l^{-1}BA	Shoot regeneration	-	Prammanee *et al.*, 2011
Prunus persica var. *nectarina* Max, cv. 'Arm King'	Shoot tip (1.3 to 2.0 mm) from thermos-treated source trees at 35°C for 3 weeks	WPM + 8.0 µM BAPWPM + 8.0 µM BAP	-38 per cent Explant regeneration-Shoot multiplication	86 per cent plants free of plum pox virus (PPV); 81 per cent free of prunus necrotic ringspot virus (PNRSV)	Manganaris *et al.*, 2003
21 Citrus genotypes	Scion meristem -0.2 mm, Troyer citrange as rootstock	MS + White's vitamins + 7.5 per cent sucrose	-Meristem regeneration	76 per cent plants free of citrus tristeza virus (CTV)87 per cent plants free of greening	Vijayakumari *et al.*, 2006
Citrus reticulata cv. Darjeeling manadarin	Scion meristem (0.2 mm) grafted on Sour orange/ Rangpur lime/ Rough lemon	MS medium	-		Mukhopadhyay *et al.*, 1997

Fruit Crop	*Explant*	*Medium*	*Response*	*Virus or Disease Sanitation Success*	*Reference (s)*
Citrus reticulata cv. Khasi manadarin	Scion meristem (0.5 mm) grafted on Champra Maounthabi rootstock	Liquid MT media + 0.5 mg l^{-1} BAP +6 per cent sucrose	51.6 per cent explant regeneration	Virus free plants	Sanabam *et al.*, 2015
Citrus reticulata cv. Nagpur manadarin	Shoot tip and decapitated rootstock pre-treated with 1 mg l^{-1}kinetin or 10 mg l^{-1}2,4-D	Liquid MS + 7.5 per cent sucrose, 0.2 mg l^{-1} thiamine HCI + 1 mg l^{-1} pyridoxine HCl + 1 mg l^{-1} nicotinic acid.	66 per cent graft success	-	Vijayakumari and Singh, 2000
Kinnow mandarin	Shoot tip on Rough lemon rootstock	-	20.83 per cent success	Plats free from: *citrus ringspot virus* (ICRSV), *Citrus tristeza virus* (CTV), *Citrus Yellow Mosaic badnavirus* (CYMBV), *Citrus yellow vein clearing Virus* (CYVCV)	Hoa *et al.*, 2004
Almond cvs. Marcona, Fournat, Tuono and Ferradduel	Meristem with 2-4 leaf primordia	-	50 per cent graft success	Virus free plants	Rizqui *et al.*, 2001

The meristem below 0.3 mm size didn't form shoot instead turned into callus; the shoots were formed only in meristems with > 0.3 mm size, while in 0.7 mm size, the regeneration/explant establishment was found doubled (Vivek and Modgil, 2018).

There are also crops, where meristem culture does not yield any success. In such crops, *in vitro* grafting of the meristem/shoot tip onto the rootstock is practiced (Chaturvedi and Sharma, 1988). The technique is termed as *in vitro* shoot tip grafting. This has been attempted in few crops with most successful deployment examples come from citrus. In citrus, this was first used by Murashige *et al.* (1972) to produce plants free of some viruses. This was later improvised by Navarro *et al.* (1975) for establishing virus free nursery stock. This is now being used as a standard practice to filter out the planting material free of virus, viroid and bacterial diseases (Juárez *et al.,* 2015; Vijyakumari *et al.,* 2006). Because of these advantages, the meristem/shoot tip grafted plants are considered a great component for bud wood certification and as a safe mean of germplasm exchange. The shoot tip grafting success is influenced by many factors including rootstocks (Murashige *et al.,* 1972; Vijyakumari *et al.,* 2006), their age, medium and culture conditions (Navarro *et al.,* 1975; Vijyakumari *et al.,* 2006). In citrus, the adventitious shoots emerge from the cut portion of the rootstock. In view of this, the use of a trifoliate rootstock like trifoliate orange or its hybrids is advantageous. The trifoliate leaves serves as a marker for identification of unsuccessful grafts (Navarro *et al.,* 1975). However, the grafting success is dependent upon the combinations of rootstock and scion. In micrografting of Khasi mandarin over four different rootstocks ['Kachai' lime (*Citrus jambhiri* Lush), Champra maounthabi (*Citrus jambhiri* Lush), Solom (*Citrus* sp.) and Phouheiree (*Citrus* sp.)], the maximum 40 per cent success was recorded with Champra maounthabi rootstock, a strain of rough lemon (Sanabam *et al.,* 2015). Lahoty *et al.* (2013) found higher shoot tip grafting success of Nagpur mandarin with Carrizo rootstock (53.33%) than rough lemon (43.33%) and Rangpur lime rootstock (40.00%). For Kinnow mandarin, the better success was obtained with rough lemon than trifoliate orange (Hoa *et al.,* 2004). However, grafting success of Kinnow with Carrizo citrange was better than rough lemon and four other rootstocks (Chand *et al.,* 2016). Like rootstock, the micrografting success also gets influenced by the scion genotype. Vijayakumari *et al.* (2006) performed micro-grafting in 21 exotic citrus genotypes. The maximum success was recorded in 'Ruby' Blood orange (80.0%) followed by Dhakula Local mandarin and Caffin clementine (66.7% each). The age of the rootstock at the time of grafting significantly influence the success rate of the technique. With Troyer orange as rootstock, best grafting success was recorded with two weeks old seedlings. The 1-week old rootstock seedlings produced excessive callus while grafts on 3 or 4 weeks old seedling turned dead (Navarro *et al.,* 1975), indicating a decrease of dedifferentiation potential with increase of age. Mukhopadhyay *et al.* (1997) found the right age of the rootstock for obtaining better micro-graft success varieties with the rootstock used. For Darjeeling mandarin, 21 days of sour orange seedlings, and 14 days for Rangpur lime or rough lemon were found to be the optimum age. Lahoty *et al.* (2013) showed 13 days for Carrizo, 14 days for

Rangpur lime and 16 days for rough lemon as the optimum age for getting higher shoot tip grafting success in Nagpur mandarin.

In citrus, the micrografting success was improved with alterations in concentration of sucrose or treatments of growth regulators and incubation conditions. The increase of sucrose from 3 to 6 per cent significantly improved the micro-grafting success of Khasi mandarin (Sabanam *et al.*, 2015). Similarly, Vijayakumari and Singh (2000) noted a significantly enhanced graft success (2-fold increase over control) with a pre-dipping growth regulators treatment (1.0 mg l^{-1} kinetin or 10 mg l^{-1} 2,4-D) of meristem and the decapitated rootstock. The maximum micrografting success of Darjeeling mandarin with Sour orange rootstock was obtained when the micrograft's were incubated at 25-30 °C under 4 days darkness, for 15 days 16 h day^{-1} low light (750 W) and for 39 days 16 h day^{-1} high light (1500-2000 W) exposure (Mukhopadhyay *et al.*, 1997).

The homogeneity of the meristem culture derived plants was tested using RAPD markers; all the plants were found genetically stable (Naing *et al.*, 2019).

Table 13.4: Comparative Features of Three Micropropagation Techniques

Particulars	*Organogenesis*	*Somatic Embryogenesis*	*Meristem/ Shoot Tip Culture*
Applications	Mass multiplication of the candidate genotypes	Rapid multiplication Production of virus free plants Mean of improvement through cell based genetics improvement techniques Amenable to artificial seed production	Production of disease (virus) free plants For exchange of disease free germplasm
Root induction	A separate step	Simultaneous	A Separate step
Somaclonal variation	Likely if shoots are induced from already differentiated tissues	Likely during indirect somatic emryogensis	Less chances
Examples	Most of the fruit crops	Most of the fruit crops	Strawberry, Pome fruits, banana, grapes, citrus *etc.*

Technological Advancements

Most of the tissue culture techniques require a skilled labour and are cost intensive. Another cost consuming item is gelling agent (Nagori *et al.*, 2004). The above two components amounts to most of the cost. The high cost is one of the drawback in way of routine use of micropropagation in plant production. Since, the plants are produced in closed vessels, the optimum gases requirements are not met and high humidity leads to abnormality in the cultures. It has been come to fore that liquid media induce better growth. In recent, bioreactors based propagation method has been developed to overcome the stated limitations of conventional micropropagation techniques. Bio reactors are closed vessels in which cultures are provided optimum physical and chemical environment, which not only increase their multiplication rate but also shorten the time for plantlet production, and makes the system commercial (Mishra *et al.,* 2024). There are two types of bioreactors: continuous immersion system (CIS) and temporary immersion system (TIS). The TIS provides pathogen free environment for *in vitro* cultures. A standard TIS bioreactor contains two plastic or glass jars/vessels; one jar containing liquid medium and the other contains the cultures. The use of the optimized micropropagation protocols through bio-reactors will enable commercialization of micropropagation. Both direct organogenesis (using different explants like nodal explants, shoot tip and shoots or inflorescences) and somatic embryogenesis based automated system has been practiced in many fruit crops. The bioreactors based propagation has been performed in banana, citrus, date palm, strawberry and pome fruits (apple, pear) and in nut crops like almond, pistachio and chestnut (Mishra *et al.,* 2024).

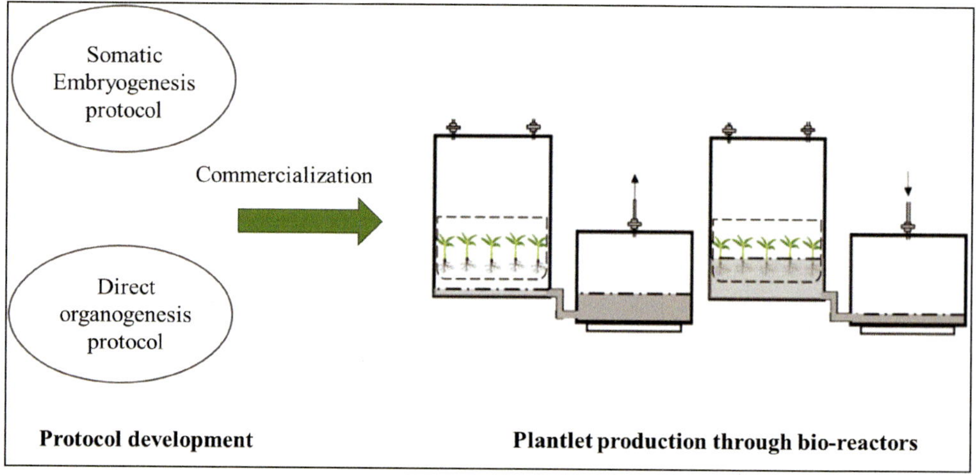

Figure 13.3: Scheme for Optimization and Commercialization of Micropropagation.

Conclusion

Tissue culture techniques like somatic embryogenesis, direct organogenesis and meristem tip culture/shoot tip grafting has significance for production of healthy and disease free homogenous plants. Somatic embryogenesis can produce rooted plantlets in a single step. This mode hastendency to produce somaclonal variation under the influence of growth regulators or the age of the culture. The technique of direct organogenesis is a boon for providing the plant material needs of desired sex form in fruit crops like date palm and papaya. The virus load which deteriorates the plant performance. The virus and disease cleansing techniques of meristem tip culture/shoot tip grafting can restore the actual yield potential of a variety. The plants produced through these techniques are an ideal material for disease free germplasm exchange and also work as a foundation stock for production of virus free plantlets. To commercialize the use of micropropagation, the optimized variety specific protocols need to be linked with bioreactor based technology.

REFERENCES

Ahlawat YS, 1997. Viruses, greening bacterium and viroids associated with citrus (*Citrus* species) decline in India. *Indian J. Agri. Sci.* **67:** 51-57.

Al-Khayri JM and Al-Bahrany AM, 2001. Silver nitrate and 2-isopentyladenine promote somatic embryogenesis in date palm (*Phoenix dactylifera* L.). *Scientia Hortic.* **89:** 291-98.

Al-Khayri JM and Al-Bahrany AM, 2012. Effect of abscisic acid and polyethylene glycol on the synchronization of somatic embryo development in date palm (*Phoenix dactylifera* L.). *Biotechnol.* **11:** 318-25.

Al-Khayri JM and Naik PM, 2017. Date palm micropropagation: Advances and applications. *Ciência e Agrotecnologia* **41:** 347-58.

Amin MN and Jaiswal VS, 1987. Rapid clonal propagation of guava through *in vitro* shoot proliferation on nodal explants of mature trees. *Plant Cell Tiss. Org. Cult.* **9:** 235-43.

Andret-Link P, Schmitt-Keichinger C, Demangeat G, Komar V and Fuchs M, 2004. Grapevine fanleaf virus: still a major threat to the grapevine industry. *J. Plant Pathol.* **86:** 183195.

Baharan E and Mohammadi P, 2018. Induction of direct somatic embryogenesis and callogenesis in date palm (*Phoenix dactylifera* L.) using leaf explants. *BioTechnologia* **99:** 197-203.

Bayoudh C, Elair M, Labidi R, Majdoub A, Mahfoudhi N and Mars M, 2017. Efficacy of tissue culture in virus elimination from caprifig and female fig varieties (*Ficus carica* L.). *The Plant Pathol. J.* **33:** 288-95.

Bhardwaj SV, Rai SJ, Thakur PD and Handa A, 1998. Meristem tip culture and heat therapy for production of apple mosaic virus free plants in India. *Acta Hortic.* **472:** 135-40.

Bhat SN, Khalil A, Nazir N, Mir MA, Khan I, Mubashir SS, Dar MS, Wani SH and Hossain MA, 2022. *In vitro* prevention of browning in persian walnut (*Juglans regia* L.) cv. Sulaiman. *Int. J. Plant Biol.* **13:** 330-42.

Bodade S, Gahukar S and Akhare A, 2022. *In-vitro* propagation of Rangpur lime (*Citrus limonia*). *The Pharma Innovation J.* **10:** 2978-80.

Caple AD and Cheah KT, 2016. Micropropagation of hermaphrodite *Carica* papaya L. 'Rainbow' seedlings via axillary bud pathway. *Biotechnology* **12:** 1-5.

Chand L, Sharma S and Kajla S, 2016. Effect of rootstock and age of seedling on success of *in vitro* shoot tip grafting in Kinnow mandarin. *Indian J. Hort.* **73:** 8-12.

Chaturvedi HC and Sharma AK, 1988. Citrus tissue culture. **In:** *Proc. National Seminar Plant Tissue Culture*, ICAR New Delhi, pp. 36-46.

Comlekcioglu S, Kuden AB, Kacar YA and Kamberoglu MA, 2007. Meristem culture of two fig cultivars in Turkey. *Fruits* **62:** 125-31.

Dillon S, 2005. Characterization, genetic mapping and development of marker selection strategies for the resistance to the Papaya Ring Spot Virus Type P (PRSV-P) in highland papaya. Ph.D. Thesis, School of Biomolecular and Biomedical Science, Griffith University, Brisbane, Australia, p. 271.

Duncan RR, 1997. Tissue culture-induced variation and crop improvement. *Adv. Agron.* **58:** 201-40.

El-Agamy SZ, Mostafa RAA, Shaaban MM and El-Mahdy MT, 2009. *In vitro* propagation of Manfalouty and Nab El-gamal pomegranate cultivars. *Res. J. Agric. Biol. Sci.* **5:** 1169-75.

El Fatih MM and Hind Badereldin H, 2010. Cultivar differences of date palm (*Phoenix dactylifera* L.) in somatic embryogenesis micropropagation. *Acta Hortic.* **882:** 193-98.

Fiore S, De Pasquale F, Carimi F and Sajeva M, 2002. Effect of 2,4-D and 4-CPPU on somatic embryogenesis from stigma and style transverse thin cell layers of citrus. *Plant Cell Tiss. Org. Cult.* **68:** 57-63.

Fki L, Masmoudi R, Kriaa W, Mahjoub A, Sghaier B, Mzid R, Rival A and Drira N, 2011. Date palm micropropagation via somatic embryogenesis. *Date palm biotechnology*, 47-68.

Ge XX, Fan GE, Chai LJ and Guo WW, 2010. Cloning, molecular characterization and expression analysis of a SOMATIC EMBRYOGENESIS RECEPTOR-LIKE KINASE gene (CitSERK1-like) in Valencia sweet orange. *Acta Physiol. Plant.* **32:** 1197-1207.

Gatambia EK, Kihurani AW, Rimberia FK and Waiganjo MM, 2015. *In vitro* meristem culture for rapid regeneration of papaya plantlets in liquid media. *Ann. Res. Rev. Biol.* **9:** 1-7.

Goldschmidt EE, 2013. The evolution of fruit tree productivity: a review. *Economic Bot.* **67:** 51-62.

González-Lázaro M, Hernández-Rosas F, Gómez-Merino FC, Pastelín-Solano MC, Bulbarela-Marini JE, Solano-Rodríguez LA and Castro OC, 2019. PCR diagnosis and *in vitro* sanitation of the papaya MSXJ hybrid with ringspot virus disease. *Chil. J. Agr. Res.* **79:** 376-84.

Habauzit V, Milenkovic D and Morand C, 2013. Vascular protective effects of fruit polyphenols. **In:** *Polyphenols in Human Health and Disease*. R. Watson, V. Preedy and S. Zibadi Eds. 1st Ed, Elsevier Inc., London, pp. 875-93.

Hassan MM and Taha RA, 2012. Callogenesis, somatic embryogenesis and regeneration of date palm *Phoenix dactylifera* L. cultivars affected by carbohydrate sources. *Intl. J. Agri. Res.* **7:** 231-42.

Hassan MM, Allam MA, El Din IS, Malhat MH and Taha RA, 2021. High-frequency direct somatic embryogenesis and plantlet regeneration from date palm immature inflorescences using picloram. *J. Genet. Eng. Biotechnol.* **19:** 33.

Hoa NV, Ahlawat YS and Pant RP, 2004. Production of virus-free Kinnow mandarin and Mosambi sweet orange nucleus planting material through shoot tip grafting. *Indian Phytopath.* **57:** 482-87.

Hu GJ, Dong YF, Zhang ZP, Fan XD, Ren F and Li ZN, 2017. Efficacy of virus elimination from apple by thermotherapy coupled with *in vivo* shoot tip grafting and *in vitro* meristem culture. *J. Phytopathol.* **165:** 701-06.

Hussein NH and Jawad HA, 2021. Micropropagation of date palm var. Medjool by adventitious shoot induction. *Int. J. Agricult. Stat. Sci.* **17:** 1879-83.

Hutchinson JF and Zimmerman RH, 1987. Tissue culture of temperate fruit and nut trees. **In:** Horticultural Reviews. J. Janick (Ed.). https: //doi. org/10.1002/9781118060827.ch8.

Jamous RM, Zaitoun SYA, Mallah OB, Shtaya M, Elbeaino T and Ali-Shtayeh MS, 2020. Detection and phylogenetic analysis of viruses linked with fig mosaic disease in seventeen fig cultivars in Palestine. *The Plant Pathol. J.* **36:** 267-79.

Juárez J, Aleza P and Navarro L, 2015. Applications of citrus shoot-tip grafting *in vitro*. *Acta Hortic.* **1065:** 635-42.

Kamboj H, Kumar K, Brar JS, Kumar A and Arora PK, 2023. Optimization of *in vitro* propagation of jamun (*Syzygium cumini* Skeels) variety Konkan Bahadoli. *Indian J. Hort.* **80:** 245-50.

Kieber JJ, 2002. Cytokinins. *The arabidopsis book* **1:** e0063.

Khierallah HSM, Bader SM and Al-Khafaji MA, 2017. NAA-induced direct organogenesis from female immature inflorescence explants of date palm. *Methods Mol. Biol.* **1637:** 17-25.

Koltunow AM, Hidaka T and Robinson SP, 1996. Polyembryony in Citrus: Accumulation of seed storage proteins in seeds and in embryos cultured *in vitro*. *Plant Physiol.* **110:** 599-609.

Kumar K, Kaur H, Gill MIS and Rattanpal HS, 2006. Studies on *in vitro* shoot regeneration in grape cvs. Perlette and Flame Seedless. *Indian J. Hort.* **63:** 347-51.

Kumar K, Gill MIS, Sangwan A and Gosal SS, 2008. *In vitro* shoot regeneration in nematode tolerant grape rootstock 1613C. *Indian J. Hort.* **65:** 255-57.

Kumar K, Gill MIS and Gosal SS, 2018. Somatic embryogenesis, *in vitro* selection and plantlet regeneration for citrus improvement. **In:** Biotechnologies of Crop Improvement, Gosal, S. and Wani, S. (Ed.). Volume 1. Springer, Cham.

Kumar K, Arora PK, Brar JS, Bhatia D and Kumar A, 2019. Influence of explant collection period, anti-browning strategy and growth regulators composition on *in vitro* propagation of Bhagwa pomegranate. *Indian J. Hort.* **76:** 273-78.

Kumari R, Kundu M, Mir H, 2023. *In vitro* regeneration technique of papaya (*Carica papaya* l.) cv. Pusa Dwarf through shoot tip culture. *Natl. Acad. Sci. Lett.* **46:** 1-5.

Lahoty P, Singh J, Bhatanagar P, Rajpurohit D and Jain SK, 2013. Shoot tip grafting in Nagpur mandarin (*Citrus reticulata* Blanco). *Plant Arc.* **13:** 173-75.

Mahmoud LM, Grosser JW and Dutt M, 2020. Silver compounds regulate leaf drop and improve *in vitro* regeneration from mature tissues of Australian finger lime (*Citrus australasica*). *Plant Cell Tiss. Org.* **141:** 455-64.

Manganaris GA, Economou AS, Boubourakas IN and Katis NI, 2003. Elimination of PPV and PNRSV through thermotherapy and meristem-tip culture in nectarine. *Plant Cell Rep.* **22:** 195-200.

Maul P, Bausher M, McCollum G, Mozoruk J and Niedz R, 2006. CsHPt1, a putative histidine-containing phosphotransmitter protein induced during early somatic embryogenesis in Valencia sweet orange. *Plant Science* **170:** 44-53.

Mazri MA, Belkoura I, Meziani R, Mokhless B and Nour S, 2017. Somatic embryogenesis from bud and leaf explants of date palm (*Phoenix dactylifera* L.) cv. Najda. *3 Biotech* **7:** 58.

Meghwal PR, Sharma HC and Singh SK, 2010. Micropropagation studies on guava. *Indian J. Hort.* **67:** 55-58.

Ming R, Yu Q and Moore PH, 2007. Sex determination in papaya. **In:** Seminars in cell and developmental biology, Vol. 18, No. 3, pp. 401-408. Academic Press.

Mishra M, Shukla N and Chandra R, 2007. Micropropagation of papaya (*Carica papaya* L.). **In:** Protocols for Micropropagation of Woody Trees and Fruits. Jain,. S.M. and Häggman H. (Ed.). Springer, Dordrecht.

Mishra M, Rajan S and Damodaran T, 2024. New paradigm shifts in micropropagation of fruit crops through bioreactors-a review. *Indian J. Hort.* **81:** 1-10.

Mordhorst AP, Toonen MA, de Vries SC and Meinke D, 1997. Plant embryogenesis. *Crit. Rev. Plant Sci.* **16:** 535-76.

Moreno P, Ambros S, Albiach-Marti MR, Guerri J and Pena L, 2008. Citrus tristeza virus: a pathogen that changed the course of the citrus industry. *Mol. Pl. Pathol.* **9:** 251-68

Mukhopadhyay S, Raj J, Sharma BC, Gurung A, Sengupta RK and Nath PS, 1997. Micropropagation of Darjeeling orange (*Citrus reticulata* Blanco) by shoot tip grafting. *J. Hort. Sci.* **72:** 493-99.

Muller D and Leyser O, 2011. Auxin, cytokinin and the control of shoot branching. *Ann. Bot.* **107:** 1203-12.

Murashige T, Bitters WP, Rangan TS, Nauer EM, Roistacher CN and Holliday BP, 1972. A technique of shoot apex grafting and its utilization towards recoveringvirus-free citrus clones. *HortScience* **7:** 118-19.

Murkute AA, Patil S and Mayakumari, 2003. Exudation and browning in tissue culture of pomegranate. *Agric. Sci. Dig.* **23:** 29-31.

Nagori R and Purohit SD, 2004. *In vitro* plantlet regeneration in *Annona squamosa* through direct shoot bud differentiation on hypocotyl segments. *Scientia Hortic.* **99:** 89-98.

Naing AH, Kim SH, Chung MY, Park SK and Kim CK, 2019. *In vitro* propagation method for production of morphologically and genetically stable plants of different strawberry cultivars. *Plant Methods* **15:** 36.

Navarro L, Roistacher CN and Murashige T, 1975. Improvement of shoot tip grafting *in vitro* for virus-free citrus. *J. Am. Soc. Hort. Sci.* **100:** 471-79.

Neri JC, Meléndez-Mori JB, Tejada-Alvarado JJ, Vilca-Valqui NC, Huaman-Huaman E, Oliva M, and Goñas M, 2022. An optimized protocol for micropropagation and acclimatization of strawberry (*Fragaria × ananassa* Duch.) variety 'Aroma'. *Agronomy* **12:** 968.

Patel JR, Patel RM, Patel SR and Virendra Singh VS, 2013. *In vitro* micro propagation of papaya var. Red Lady.*Intl. J. Plant Protection* **6:** 449-56.

Prammanee S, Thumjamras S, Chiemsombat P and Pipattanawong N, 2011. Efficient shoot regeneration from direct apical meristem tissue to produce virus-free purple passion fruit plants. *Crop Protect.* **30**: 1425-29.

Quiroz KA, Berríos M, Carrasco B, Retamales JB, Caligari PDS and García-Gonzáles R, 2017. Meristem culture and subsequent micropropagation of Chilean strawberry (*Fragaria chiloensis* (L.) Duch.). *Biol. Res.* **50:** 20.

Raju R and Divya C, 2020. Micropropagation of *Syzygium densiflorum* Wall. ex Wight and arn.: An endemic and endangered semi-evergreen tree species of the Western Ghats, India. *Trees Forests People* **10:** 1-9.

Rathore JS, Rathore MS, Singh M, Singh RP and Shekhawat NS, 2007. Micropropagation of mature tree of *Citrus limon*. *Indian J. Biotech.* **6:** 239-44.

Rattanpal HS, Gill MIS and Sangwan AK, 2011. Micropropagation of strawberry through meristem culture. *Acta Hortic.* **890:** 149-53.

Rizqi A, Zemzami M and Spiegel S, 2001. Recovery of virus-free almond plants by improved *in vitro* shoot-tip grafting. *Acta Hortic.* **550:** 447-54.

Roy PK, Roy SK, Hakim L and Mamun ANK, 2016. Somatic embryogenesis and plant regeneration of papaya (*Carica papaya* L. cv. Shahi). *Nucl. Sci. App.* **25:** 23-26.

Ryavalad S, Malabasari TA, Shantappa T, Uppar DS, Biradar BD and Mantur SM, 2019. Micropropagation studies in papaya (*Carica papaya* L.) cv. 'Surya'. *Int. J. Curr. Microbiol. App. Sci.* **84:** 2362-67.

Sanabam R, Handique PJ and Huidrom S, 2015. Microshoot tip grafting as tool for production of disease free foundation block of *Citrus reticulata* Blanco Khasi mandarin. *Acta Hortic.* **1065:** 643-48.

Savita Pati PK, Virk GS and Nagpal A 2015. An efficient somatic embryogenesis protocol for Citrus jambhiri and assessment of clonal fidelity of plantlets using RAPD markers. *J. Plant Growth Regul.* **34:** 309-19.

Selvarajan R and Balasubramanian V, 2013. Natural occurrence of banana bunchy top virus in Ensete superbum in India. *Virus Dis.* **24:** 97-98.

Sharma SK, Chavan VM, Tomer SPS, Darade RS, Dhale MG and Kadam UM, 2006. Minimizing losses in papaya due to papayaringspot virus-a case study of IARI, regional station, Pune. Abstract-OP 9/20. **In:** Annual Convention of IVS and InternationalSymposium on Management of Vector-Borne Viruses, held at Patancheru, Hyderabad, India.

Sharma U, Kataria V and Shekhawat NS, 2017. *In vitro* propagation, *ex vitro* rooting and leaf micromorphology of *Bauhinia racemosa* Lam.: a leguminous tree with medicinal values. *Physiol Mol Biol Plants* **23:** 969-77.

Shekhawat MS, Manokari M and Ravindran CP, 2015. An improved micropropagation protocol by *ex vitro* rooting of *Passiflora edulis* Sims. f. *flavicarpa* Deg. through nodal segment culture. *Scientifica* **1:** 578676.

Shimada T, Hirabayashi T, Endo T, Fujii H, Kita M and Omura M, 2005. Isolation and characterization of the somatic embryogenesis receptor-like kinase gene homologue (CitSERK1) from *Citrus unshiu* Marc. *Scientia Hortic.* **103:** 233-38.

Singh SK, Khawale RN and Singh SP, 2002. Effect of season, type of explant and pretreatments to minimize polyphenolics exudation on in vitro culture establishment in grape. *Indian J. Hort.* **59:** 233-38.

Sidky R, 2017. Optimized direct organogenesis from shoot tip explants of date palm. *Methods Mol Biol.* **1637:** 37-45.

Skoog F and Miller CO, 1957. Chemical regulation of growth and organ formation in plant tissue cultures *in vitro*. *Symp. Soc. Exp. Biol.* **11:** 118-31.

Solangi N, Abul-Soad AA, Jatoi MA, Mirani AA and Markhand GS, 2023. Micropropagation of elite date palm (*Phoenix dactylifera* L.) cultivars Samany and Bertamoda through immature inflorescence explants. *J. Saudi Society Agri. Sciences* **22:** 430-38.

Taha RA, Allam MA, Hassan SAM, Bakr BM and Hassan MM, 2021. Thidiazuron-induced direct organogenesis from immature inflorescence of three date palm cultivars. *J. Genetic Engineering Biotech.* **19:** 14.

Thangavelu R, Selvarajan R and Singh HP, 2000. Status of banana streak virus and banana bract mosaic diseases in India. **In:** *Proc. Banana Improvement, Production and Utilization.* H.P. Singh and K.L. Chadha (Ed.). NRC-Banana, Trichy, pp. 364-77.

Vijayakumari N, Ghosh DK, Das AK, Awtar S and Shyam S, 2006. Elimination of citrus tristeza virus and greening pathogens from exotic germplasm through *in vitro* shoot tip grafting in citrus. *Indian J. Agri. Sci.* **76:** 209-10.

Vijayakumari N and Singh S, 2000. Shoot tip grafting with growth regulators for virus elimination of Nagpur Mandarin (*Citrus reticulata* Blanco). *Indian J. Agri. Sci.* **70:** 396-97.

Vivek M and Modgil M, 2018. Elimination of viruses through thermotherapy and meristem culture in apple cultivar 'Oregon Spur-II'. *Virus disease* **29:** 75-82.

von Arnold S, Sabala I, Bozhkov P, Dyachok J and Filonova L, 2002. Developmental pathways of somatic embryogenesis. *Plant Cell Tiss. Org. Cult.* **69:** 233-49.

Wang X, Xu Y, Zhang S, Cao L, Huang Y, Cheng J and Xu Q, 2017. Genomic analyses of primitive, wild and cultivated citrus provide insights into asexual reproduction. *Nature Genetics* **49:** 765-72.

Chapter 14

Machinery for Nursery and Orchard Management

Ramesh Kumar Sahni and V.P. Chaudhary

ICAR–Central Institute of Agricultural Engineering, Bhopal, Madhya Pradesh

The increasing demand for efficient and sustainable horticultural practices has led to the widespread adoption of machinery in nursery and orchard management. In nurseries, equipment such as soil mixers, seed drills, and transplanters streamline plant production, ensuring uniformity and reducing labour requirements. In orchards, specialized machinery, including tractors, ploughs, sprayers, and mechanical harvesters, enhances productivity while mitigating the risks associated with manual labour. The mechanization of these horticultural systems not only increases operational efficiency but also contributes to sustainable practices. Modern machinery is designed to minimize environmental impact, optimize resource usage, and integrate data-driven management techniques. For example, precision irrigation systems and GPS-enabled machinery allow for better resource allocation and improved plant health monitoring. This chapter explores the technological advancements in nursery and orchard machinery, their role in promoting sustainability, and their future potential. Furthermore, it discusses the role of mechanization in horticulture by boosting productivity, reducing manual labour, and fostering eco-friendly practices, making it an important for modern nursery and orchard management.

Introduction

India is the largest producer of fruits and the second largest producer of vegetables in the world. India's production of horticultural crops was about 342.3 million tonnes (MT) from an area of 28.1 million ha during 2021-22. The production of vegetables, flowers, plantation crops, spices and honey were 107.2, 204.8, 3.1, 15.5, 108.1 and 0.13 million tonnes, respectively during the same year (Anonymous. 2023, Table 14.1). However, the productivity of horticultural crops is quite low as compared to other countries. One of the approaches for increase in productivity is mechanization of farm operations involved in horticultural crops including production and post harvesting operations. Most of the farm operations are still carried out manually leading to drudgery and delay in the operations (Tiwari *et al.,* 2019b). Many of the developed technologies have still to be adopted on a large scale (Singh and Sahni, 2019). The industry has a major role to play in the manufacture and promotion of these technologies in collaboration with extension agencies.

Table 14.1: Area and Production of Horticultural Crops in India (2021-22)

Crop	*Area ('000 ha)*	*Production ('000 MT)*
Fruits	7049	107242
Vegetables	11348	204835
Flowers	283	3128
Plantation Crops	4239	15488
Spices	4487	10814
Honey	-	133
Total	28075	342329

Source: Agriculture Statistics at a Glance 2022.

The advancement of agricultural practices has significantly transformed the way we cultivate crops, manage farms, and optimize productivity (Tiwari *et al.,* 2017). One of the crucial sectors that have witnessed substantial technological intervention is horticulture, specifically in the management of nurseries and orchards. These horticultural systems require precision, efficiency, and sustainability to ensure high-quality plant growth, fruit production, and the overall health of the ecosystem (Tiwari *et al.,* 2019a).

Nursery and orchard management involve a diverse range of activities, from soil preparation, planting, and irrigation to pruning, pest control, and harvesting. Traditionally, many of these tasks were labour-intensive and time-consuming. However, with the evolution of agricultural machinery, there has been a paradigm shift towards mechanization and automation, leading to improved efficiency and scalability in these operations (Sahni *et al.,* 2018c). Whether it is the use of tractors for soil preparation or specialized equipment for pruning and harvesting, the integration of machinery has become indispensable in modern horticulture.

Nurseries, where young plants are grown before being transplanted to the field or orchard, require meticulous care and management. The machinery used in nurseries ranges from soil mixers, seed drills, and transplanting machines to irrigation systems and plant protection equipment. These machines not only ensure uniformity in plant growth but also reduce the time and labour involved in nursery operations. For instance, automated seed drills can handle thousands of seeds per hour with precision, ensuring optimal spacing and depth, which are critical factors for seed germination and plant health.

Similarly, orchards, which involve the cultivation of fruit-bearing trees, demand specialized equipment for tasks like planting, irrigation, pruning, pest management, and harvesting. The terrain and scale of operations in orchards often require machinery that is adaptable to varying conditions. Tractors with specialized attachments, tree sprayers, and mechanical harvesters are some of the key machines used in orchard management. These machines not only enhance productivity but also reduce the risks associated with manual labour, such as injuries during pruning or harvesting. Modern machines are increasingly being designed to minimize environmental impact by reducing fuel consumption, optimizing water usage, and limiting the application of chemical inputs. For instance, precision irrigation systems, which use sensors to determine soil moisture levels, ensure that water is applied only, when necessary, thus conserving water resources and preventing over-irrigation (Kumar *et al.,* 2018 a,b). Moreover, the use of machinery in these operations facilitates better data collection and management. Many modern machines are equipped with GPS and data sensors that allow farmers to monitor various parameters such as soil conditions, plant health, and pest infestations in real-time (Sahni *et al.,* 2018a). This data-driven approach enables more informed decision-making, leading to better management practices and increased productivity (Rathnayake *et al.,* 2022; Mahapatra *et al.,* 2024; Sahni *et al.,* 2024).

The role of machinery in nursery and orchard management is pivotal for improving efficiency, reducing labour, and promoting sustainability. As the demand for high-quality produce continues to grow, the integration of advanced machinery will play an essential role in meeting these needs while ensuring the long-term health of horticultural systems. This chapter discusses various types of machinery employed in nursery and orchard management starting from field preparation to seeding/transplanting to harvesting. Furthermore, mechanization status of horticultural crops and steps for accelerating mechanization in horticultural crops has been discussed in detail.

Mechanization Status of Horticultural Crops

The Indian agriculture has undergone a sea change as regards to mechanization of production operations in cereal crops. Farm mechanization inputs are essential in increasing agricultural production and productivity through timeliness in operations, efficient application of inputs, conserving soil and water resources; and reducing losses, pollution and drudgery of the task involved apart from increase in cropping intensity. There has been a rapid transformation from use of manual tools and bullock farming to utilization of improved equipment and power farming, using tractors and self-propelled equipment, during the last two decades, which has resulted in increased cropping intensity, efficient utilization of farm inputs, crop productivity and production.

Because of diverse nature of horticultural crops, the mechanization requirements of different horticultural crops are widely varied and crop specific. Horticultural crop cultivation so far continues to be excessively labour intensive, and country is lagging with respect to horticultural mechanization. Most of critical operations like nursery raising, transplanting, earthing up, weeding and interculture, pruning, picking/harvesting, *etc.* are generally carried out manually and are labour intensive and involve lot of drudgery. Operations like coconut and oil palm harvesting involve climbing of tall tree, which requires especially skilled manpower and endanger human life. Scarcity of labourers during peak season, high wages, low work rate, *etc.* often delay the critical farm operations causing loss of yield and increasing the cost of cultivation. Untrained, careless workers reduce the quality of operations. Mechanization of horticultural crops has not been duly addressed yet. There is an urgent need for mechanization of farm operations for horticultural production in India for timeliness of farm operations, efficient utilization of farm inputs and reduce drudgery of farm workers. Mechanization has to be considered with multidisciplinary approach based on agro-climatic region and specific crop requirements. The need to introduce mechanization is imperative to enable farmers to overcome the constraints of horticultural crop production system.

Mechanization of Nursery and Orchard Crops

Mechanization of nursery and orchard crops enhances efficiency in ploughing, planting, pruning, plant protection, and harvesting. It involves using machines like seed drills, sprayers, and tree shakers, reducing labour costs and time. Mechanization improves productivity, ensures uniformity in operations, and helps maintain crop health, leading to better yields and profitability (Sahni *et al.*, 2018b).

Land Preparation

Majority of fruit and plantation crops are perennial in nature. Establishment of orchards/vineyards and plantations is one time operation and land can be prepared using conventional land development and tillage equipment. The seedbed preparation/tillage operations for cultivation of different horticulture crops are mostly mechanized and equipment for different power sources are commercially available. Ridger, raised bed former, laser levellers (Figure 14.1) are getting popular and being adopted at farmer's level. These need to be promoted among the farmers based on their specific requirements. Levelling of field is mainly done by using leveller's or drapers. In case the fields are levelled the surface compaction is achieved by using 2 to 3 m wide plank pulled by tractor. Laser guided land leveller can be adopted for precision land levelling and grading. Besides savings in irrigation water (20-25%) and avoiding waterlogging, the laser levelled field supports the crops for uniform growth and yield. The well levelled field also helps in precise operation of sowing/planting/transplanting equipment resulting in uniformity in placement of seeds/plants.

Figure 14.1: Tractor Operated MB Plough, Rotovator and Laser Guided Land Leveller.

Some crops are planted in pits. Tractor and power tiller operated post-hole diggers are commercially available for digging of pits up to 30 cm diameter. For tillage of inter-row spaces of pits, conventional power tiller or tractor operated tillage equipment like MB plough, disc harrow, tyne cultivator, rotovator, *etc.* can be used to control weeds, increase aeration and conserve moisture. Tractor operated rotovators have advantage of preparing seedbed of uniform depth with more uniform size of soil aggregates as compared to tine or disc type tools and it also helps in mixing of plant residues in soil and prepare levelled seedbed in a single pass of implement. The field capacity of tractor operated tillage equipment varies from 0.2 to 0.3 ha/h for ploughs and 0.4 to 0.5 ha/h for duck foot or shovel cultivator, disc harrows. Improved implements give higher work output than traditional implements and need to be promoted among the farmers growing horticulture crops based on their specific requirements.

Majority of the vegetable crops are seasonal in nature. Seed-bed preparation for raising vegetable crops involves farm operations like loosening of soil, clod breaking, smoothening, *etc.* to provide the optimum environment for plant growth. Loosening of soil is done to achieve a desirable granular soil structure for seedbed and to allow rapid infiltration and good retention of moisture, to provide adequate air exchange capacity within the soil and to minimize the resistance to root penetration and shoot growth. Smoothening of seed bed is required for proper operation of planting/transplanting machinery, better distribution of irrigation water and quick disposal of excess rainwater.

The indigenous animal drawn implements like local plough and blade harrow (Bakhar) are widely used by small farmers. These have advantage of low cost, versatility, ease of repairs and maintenance and easy availability. However, these implements operate at shallow depth and repeated operations are required to obtain satisfactory seedbed resulting in low work output. Improved designs of manually operated, animal drawn, power tiller and tractor operated implements/ equipment for seedbed/nursery beds preparation are commercially available and have been adopted by the farmers to large extent. These include mould board plough (fixed and reversible type), disc plough, tine cultivator (9-13 tines), ridger ploughs, sweep cultivators, disc harrows, rotovator, leveling planks, bed formers, *etc.* Animal drawn multipurpose tool carriers (Figure 14.2) or tool frames with changeable tools like MB plough, sweeps, ridger, bund former, *etc.* for farm operations like tillage, bed forming, sowing, weeding *etc.* are alternate systems to utilize the animal's draft power more efficiently. Many designs have been developed at different research institutes. Tool frame/carriers with different attachments give up to 50 per cent higher output as compared to traditional implements.

Figure 14.2: Animal Drawn Multipurpose Tool Frame for Ploughing and Seed Sowing Operation.

Plant Propagation and Sowing/Transplanting

Most of the fruit plants are raised by vegetative propagation. The seedlings required for fruit crops are grown in pots or plastic bags in the field. For raising the seedlings in field, the seed beds can be prepared using manual, animal drawn, or tractor operated implements as per requirement. The fruit seeds and stones are sown manually for raising nursery/seedlings for grafting or direct planting. Rootstock for plant is prepared by planting the stem cutting or by sowing of seeds or stones depending upon the fruit crop. The root stock is converted into desired variety of plant by budding and grafting techniques. Hand tools like budding and grafting knife, pruning knife, and secateurs are used for the purpose. Plant replacers come handy in uprooting the plant from nursery/seedbed along with sufficient soil without damage to the roots. Hand tools like transplanting trowel and hand hoes are also used for the purpose.

Transplanting of fruit and plantation crops is carried out manually in duly prepared pits. Digging of pits or trenches and then placement of seedlings are presently done manually. There is lot of drudgery involved in pit digging operations. Tractor and engine operated post hole diggers are commercially available and can be successfully adopted for digging of pits for fruit and plantation crops like mango, citrus crops, pomegranate *etc.* (Figure 14.3). In addition to pit digging basin preparation around orchard trees for manuring and fertilizer application can also be mechanized by adopting power tillers.

Figure 14.3: Tractor Operated Post Hole Digger, Excavator, Subsoiler.

Cultivation of vegetable crops is carried out both by direct seeding and transplanting. Traditional methods for direct seeding of vegetable crops include broadcasting of seeds, manual dropping of seeds in furrow opened by manual tools or a country plough and dropping seeds through a bamboo/metal funnel attached to a country plough (Figure 14.4). Manual seed dropping results in non-uniform distribution of seeds, uneven inter-row and intra-row distribution of seeds, resulting in bunching and gaps in the field. Farmers use high seed rates to get a higher plant population and bring the plant population to desired level by thinning. Manual dibbling of seeds is also practiced by the farmers for crops with wider plant to plant spacing.

Figure 14.4: Manual and Animal Drawn Planters.

Different designs of manual, animal drawn, power tiller and tractor operated seed drills, seed-cum-fertilizer drills and planters with inclined plate, cup-feed or pneumatic type seed metering systems have been developed for sowing of directly seeded vegetable crops like pea, okra, beans, *etc.* and can be successfully adopted for sowing of medium and bold size seeds (Figure 14.5). Animal drawn and tractor operated-seed-cum fertilizer drills are commercially available and are mostly provided with either agitator with opening or fluted roller type seed metering mechanisms. The output of tractor operated machines varies from 0.4 to 0.6 ha/h depending on row to row spacing. Small farmers who cannot purchase the tractor operated equipment can take the benefit of these equipment on custom hiring basis. Pneumatic planters work well for small seed vegetables like lettuce, sugar beet, cabbage, turnips as well as large seed crops (Kumar *et al.,* 2024a).

Mechanized sowing/planting of small seed vegetable crops like onion, carrot, *etc.* which are cultivated at narrow row to row spacing and require very close and precise plant to plant spacing is still a challenge and R&D efforts are continuing for development of such equipment. For direct sowing of such small, seeded vegetable crops like onion, garlic, carrot, *etc.* mechanical/pneumatic planting systems need to be adopted or developed (Kumar *et al.,* 2020).

Figure 14.5: Tractor Operated Seed-cum-Fertilizer Drill, Inclined Plate Planter and Potato Planter.

For transplanting nursery seedlings are traditionally grown on soil beds in open field. Nursery preparation of vegetable crops is predominantly of bare root type. This method involves low costs and skills however, results in considerable loss of seeds on account of poor germination and subsequent plant establishment in the field. Development of improved varieties and technologies have helped in enhanced production and improved productivity in various crops. Major portion of the area particularly, under vegetable cultivation is now sown with costly hybrid seeds having higher yields and quality and it is necessary to achieve maximum germination with healthy plants. However, of late cup type nursery produced in rectangular trays having cells and using either soil or soil-less media has come up commercially.

Transplanting of vegetable crops is mostly carried out manually all over the country. Worldwide the operation is mechanized with the use of tractor operated and self-propelled semi-automatic and automatic vegetable transplanters. In fully automatic transplanters feeding of seedlings, metering and placement are mechanized. However, semi-automatic type transplanters are more popular. In India, tractor operated two/four row vegetable transplanters suitable for bare root seedlings. About 0.8 to 1.0 ha can be covered in a day with these transplanters leading to timeliness of operations (Figure 14.6).

Figure 14.6: Tractor Operated Vegetable Transplanters.

Plastic mulching has been found to be useful for moisture conservation, reduction of weeds and increase in yield for vegetable and fruit crops. However, the application of plastic mulch is mostly done manually which is very tedious and lot of drudgery is involved. Tractor operated plastic mulch laying equipment have been developed (Figure 14.7). There is a need for popularization of this technology.

Figure 14.7: Tractor Operated Plastic Mulch Laying Machine and Planting Machine.

Pruning

The main purposes of pruning operation are removing unwanted twigs or branches of trees/vines and developing a proper shape or form and regulate the bearing habit and quality of fruit with the minimum possible interference with the natural growth habit of the variety. Several improved manual tools have been developed and are commercially available. Hand tools like pruning knives, secateurs, pruning saws, tree pruners, loppers, *etc.* (Figure 14.8) are used. Tools with telescopic handles are used for pruning of the twigs/branches, which are not in reach of human hand. The hydraulic and power operated pruning secateurs are used in developed countries for ease of operation, coverage of more area and lessen the human drudgery. However, their adoption is still low as many farmers are not aware of them. Power chain saws are commercially available, but they are still not widely adopted. Pruning in medium and tall trees is still a problem.

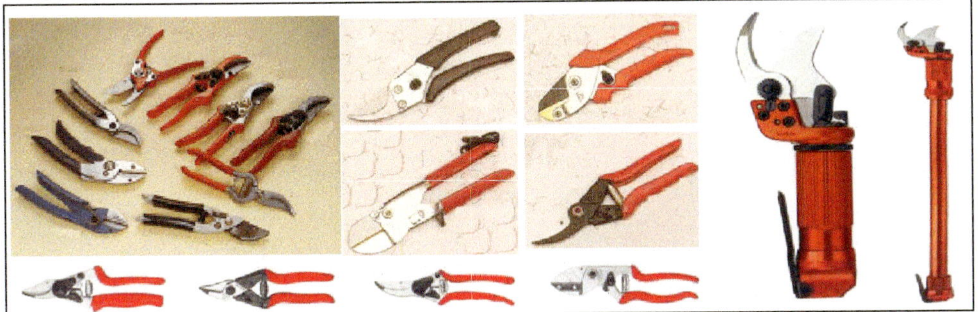

Figure 14.8: Manual Pruning Tools.

High-capacity power operated mechanized pruning equipment are widely used in developed countries. However, selective pruning of twigs or branches of the trees/vines, as practiced in India, is not possible by these machines. However, handheld battery operated, hydraulic and pneumatically operated pruning tools can be promoted among the farmers to increase their work capacity and reduce human drudgery in pruning operation. For crops like guava, citrus, *etc.* where pruning operation is required for developing proper shape/geometry of plants mechanized power operated pruning equipment can be introduced.

Weeding and Interculture

Weeding and Interculture operation is essential for the basin areas of fruit trees (Sahni *et al.*, 2020). Hand tools or small power tillers with rotary tillers are used. Mowers/bush cutters are available which can be used to cut or remove the grass/weeds from the space which is not under the fruit tee basins. For weeding and interculture in the basin and cutting/uprooting the weeds in the space not occupied by plant basins specially designed tractor mounted offset mowers and offset cultivators are available in developed countries. Some progressive vineyard and orchard farmers have procured these foreign designs of offset weeding and Interculture equipment. There is urgent need for development of offset design of weeding and Interculture tools for use in orchards, vineyards and plantation crops (Figure 14.9).

Weeding and interculture is important and labour-intensive operation in vegetable cultivation. Main objectives of weed control are to improve soil conditions by reducing the evaporation from the soil surface, improve infiltration of rain or surface water, reduce run-off to maintain ridges or beds on which the crop is grown and to reduce competition of weeds with the crop for light, nutrients and water. Losses in crop yield in India due to weed growth are reported from 10 to 75 per cent for various crops.

Figure 14.9: Tractor Operated Weeding-cum-Interculture and Earthing Up Equipment.

At present weeding and interculture in vegetables crops is carried out manually by traditional tools like *Khurpi, kudali, spade, etc.* and to a limited extent by bullock operated weeding tools. These traditional small weeding tools are efficient in removing the weeds but are very slow in operation and involve drudgery. Several improved designs of manual weeders have been developed, which are ergonomically designed for maximum operator's comfort.

Improved manually operated long handle weeders and wheel hoes such as three tined hoe (Grubber), V-blade hand hoe, Karjat hoe, *etc.*, are operated in push or push-pull or pull mode in standing posture of the operator. The soil working tool consists of one or more blades of different shape (Straight, Convex, V- shape, sweep, serrated, *etc.*) and size mounted on tines which in turn are fixed on a socket for fitting to the handle. The weeders are designed to work under favourable soil moisture conditions and give high work output at early stages of crop growth.

Wheel hoe type weeders (Figure 14.10) work on push action of operator and are usually provided with one or two wheels or a weeding roll/drum. Wheels or weeding roll support weight of weeder, control depth of operation and add to stability of the weeder. The weeders having peg or star type weeding drum also mix the cut and uprooted weeds, in the soil, and form soil mulch, which helps in conserving soil moisture. The field capacity of this type of weeders vary from 0.1 to 0.15 ha/man-day. These need to be adopted for vegetable and other horticultural row crops.

(a) (b) (c) (d)

Figure 14.10: Manually Operated Weeders: (a) Three-tined hoe; (b) Twin wheel hand hoe; (c) Peg type dry land weeder; (d) Wheel hand hoe.

The traditional animal drawn weeders (blade hoes) are mostly used for weeding between two rows and mostly single row equipment. Animal drawn weeding and inter-cultivation tools are suitable for row spacing above 30 cm and when the crop is sown in straight lines. Accurate row spacing and straight rows are essential for effective weeding without crop damage. They can cover an area of one hectare in 10 to 20 hours depending upon row spacing and number of rows operated at a time and soil crop conditions. The three-time cultivator or Triphali, Akola hoe, Bardoli hoe and animal drawn sweeps of different designs are improved designs of animal drawn weeders (Figure 14.11). Animal drawn tools reduce the cost of operation and time as compared to manual weeding.

Few of small engine operated power weeders, both sweep and rotary type, have been developed for inter-row cultivation. These equipment can be very useful when large area is to be covered in short period. Since row spacing of most vegetable crops is 45 cm or more, there is good scope for adoption of power weeders. Most of the technologies for weeding and interculture have been commercialized and these need to be adopted for horticultural crops.

Figure 14.11: Animal Drawn and Power Operated Weeders.

The tractor operated implements can also be used for inter cultivation but require wider row spacing and leaving of space at the headlands for allowing the tractor to operate and turn before entering into the rows. The work rate for various weeding implements varies due to variation in crop growth, row and plant spacing, weed intensity, soil conditions and other factors.

Plant Protection

Pest control and chemical application is important operation in orchard management (Bhalekar *et al.*, 2024). Different types of manually and power operated sprayers and dusters are available, which include hand sprayers and dusters, foot operated sprayers, power operated mist blowers and dusters, controlled spectrum droplet sprayers and granule applicators (Figure 14.12). Special stretcher type tree sprayers and dusters are used which are moved manually around the basin of the plant. Aero blast sprayers have also been introduced in the country, which can be used in the regions where plants are planted in rows. Plant protection equipment like fixed spray systems (Figure 14.13), fogging machines and mist blowers are available and should be used with proper safety gadgets to avoid chemical exposure to the operator's body (Sahni *et al.*, 2022a,b, 2023).

In recent years farmers have also started using airblast and electrostatic sprayers, especially in vineyards and orchards (Sahni *et al.*, 2022a). Electrostatic sprayers produce charged spray particles for more effective deposition on the plant leaves. Many designs of electrostatic sprayers have been imported by the farmers for vineyards and orchards (Figure 14.6). Small and medium farmers who cannot afford to own this expensive equipment hire these machines on area coverage basis. However, imported equipment are very expensive and there is need to develop more efficient and cost effective indigenous electrostatic sprayers.

Figure 14.12: Tractor Operated Duster, Sprayer and Electrostatic Sprayer.

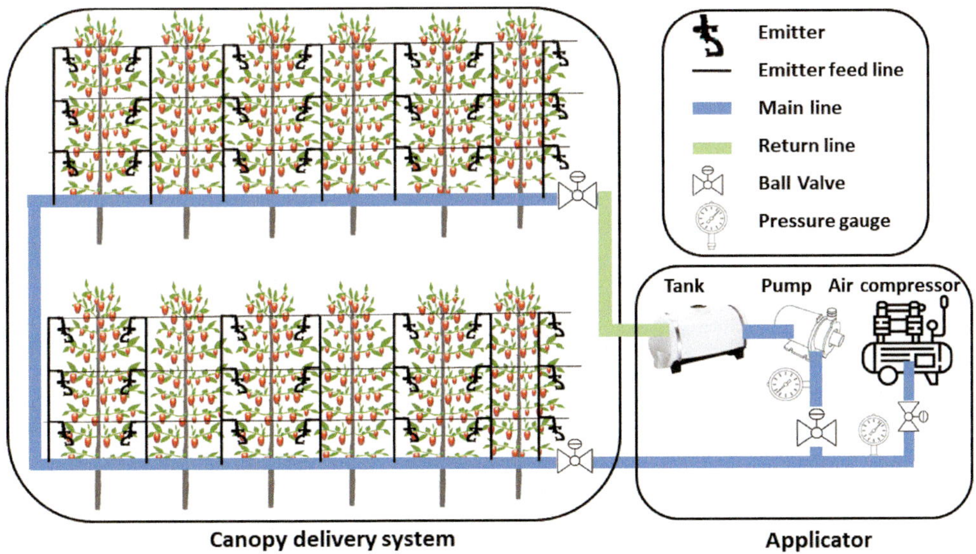

Figure 14.13: The General Schematic of a Fixed Spray System for Vineyards and Orchard Crops.

Timely and efficient application of chemicals is vegetable crops is most important. The application may be on plants and soil in the form of spray, dust or mist. Dusting, the simpler method of applying chemical, and usually requires simple equipment but it is less efficient than spraying, because of the low retention of the dust. Spraying is done for different purposes such as application of herbicides, fungicides, insecticides, and micro-nutrients. Different designs of spraying equipment have been developed and commercially available for different types of applications and field and crop conditions (Figure 14.14). Conventional designs of sprayers such as knapsack sprayers, foot sprayers, power knapsack sprayers, and tractor-operated sprayers, *etc.* are widely used in vegetable and fruit crops. The present trend is to apply concentrated pesticides by means of low and ultra-low volume sprayers. This has been possible through the development of better formulations and special nozzles.

Figure 14.14: Manually Operated and Animal Drawn Sprayers.

Effective application of chemicals is still a challenge. The repeated use of chemicals results in accumulation of larger amount of pesticide residues, much higher than the permissible limits as per international standards. Some estimates have suggested that up to 80 per cent of the total pesticide applied to the plants may eventually reach the soil.

Recently drone technology has been used in nursery and orchard management to monitor crop health, assess plant growth, and detect stress factors such as pests, diseases, or nutrient deficiencies (Anand *et al.,* 2023). They enable precise irrigation, fertilization, and pesticide application by identifying problem areas, reducing waste and labour (Sahni *et al.,* 2023; Kumar *et al.,* 2024). In orchards, drones can map terrain and create 3D models to optimize planting patterns and pruning strategies (Dixit *et al.,* 2024). Their aerial vantage point also aids in fruit yield estimation and tracking the ripening process, improving harvest timing. This technology enhances productivity, sustainability, and resource management in modern nurseries and orchards.

Harvesting Equipment

Harvesting of fruits is most labour-intensive operation and contributes major cost in fruits production system. Failure to harvest at right stage of maturity means loss of quality and loss of market price. Since most of the produce is for table purposes, where properly matured, fresh, free of blemishes and mechanical injury products are needed, mechanical options are not many. As a result, currently more than 80 per cent of the produce is harvested manually using mechanically aided hand tools.

Commonly used manual tools/devices for fruit harvesting are secateurs or knives, and handheld or pole mounted picking shears (Figure 14.15).

Figure 14.15: Hand Tools Used for Harvesting of different Types of Fruits.

Harvesting bags with shoulder or waist slings can be used for fruits with firm skins, like citrus and avocados. For hand picking mobile ladders/platforms are also used. There are tree climbers developed for harvesting fruits from coconuts/ palms. The contents of the bag are emptied through the bottom into a field container without tipping the bag. Smooth plastic buckets/containers without sharp edges are suitable for harvesting fruits that are easily crushed, such as tomatoes. The main problems with hand harvesting are labour management and picker skill. Primary advantages of hand harvest include:

- Manual picking avoids mechanical damage and maintains quality
- Selection of fruits is based on visual observation of the picker regarding maturity and appearance, damage *etc.* and multiple harvests are possible
- Minimum capital investment. Cost of labour is lower than in developed countries and sufficient availability of workers.

Picking of fruits from tall trees are also carried out manually by skilled climbers. Manual aids for harvesting/picking of fruits are available. The fruit harvesters are mainly twist and pull type having a small basket for catching the fruit. For all crops, hydraulically operated elevated platforms have been developed and are available (Figure 14.16). With these lifts/platforms picker can reach up to a height of 6.5 meter for picking of fruits from tall trees.

Figure 14.16: Self-Propelled Hydraulic Platform for Fruit Harvesting and Climbing Device for Coconut Harvesting.

Harvesting of most of vegetable crops is done manually all over the country. Harvesting of vegetable crops include plucking, picking, uprooting, digging, *etc.*, depending on type of vegetable. Some of the vegetables like pea, beans, okra, *etc.* selected as green fruits are plucked from the plants as and when these are ready and multiple rounds of harvest are done. In such crops mechanization is not possible due to selective and multiple rounds of harvesting.

For harvesting of root crops/rhizomes digging with country plough or blade harrow at proper moisture level and manual pulling and gathering of crop are traditional method, practiced by the farmers. Animal drawn harvesters and tractor operated harvesters/diggers/digger shakers, windrowers have been developed for crops and rhizomes like potato, onion, garlic, carrot, ginger, turmeric, *etc.* (Figure 14.17). The digging units consist of V-shaped or straight blade and lifter rods are attached behind the share. These lifter rods are spaced to allow the clods and residual material to drop while operating the implement. However, their adoption is limited to certain areas of the country. Efforts are continuing for development of combine harvesters for crops like potato, onion, *etc.* However, higher crop losses are associated with the combine harvesting of root crops.

Figure 14.17: Tractor Operated Root Crop Harvesters.

Some Steps Required for Accelerating Mechanization of Horticulture Crops

- The commercially available seedbed preparation/tillage equipment for cultivation of horticulture crops (fruits and vegetables) need to be popularized among the farmers based on operational requirements. Tractor operated sub-soiler/chisel ploughs, rotovators, and laser guided land leveler are some of the new equipment which needs to be promoted among farmers.
- Planters with mechanical/pneumatic seed metering systems for small seeds of vegetable crops like onion, carrot, spinach *etc.* and other crops (bare or pelleted) need to be developed.
- Equipment for mechanized nursery raising system, which presently is highly labour intensive, need to be developed and popularized among the vegetable growing farmers.
- Vegetable transplanter developed for bare root type nursery needs to be popularized. Semi-automatic and automatic transplanters for plug tray type nursery need to be developed.
- Tractor operated plastic mulch laying machines need to be promoted for mechanized plastic mulch laying for moisture conservation, reduction of weeds and increase in yield for vegetable and fruit crops.
- Improved ergonomically designed manually operated weeding and interculture tools need to promoted for higher efficiency and maximum operator comfort.

- ✰ Appropriate equipment for pest control (spraying and dusting) should be selected for judicious and effective use of chemicals. Use of protective clothing for the operator need to be enforced during use of spraying equipment.
- ✰ Manual tools for pruning, budding and grafting in fruit crops have been developed and are commercially manufactured. Hydraulic, pneumatic and battery-operated pruners need to be introduced. Pruning in medium and tall trees is still problem and more efforts are required to mechanize the operation.
- ✰ Harvesting of most of vegetable crops is manual due to selective and multistage harvesting/picking. Use of manual harvesting tools/aids may be encouraged for increasing the working efficiency of labour. For harvesting of root crops/rhizomes adoption of mechanical harvesting need to be encouraged.
- ✰ Mechanical/hydraulic ladders, lift platforms, and self-propelled hydraulic platforms have been developed in the country. These equipment may be introduced in orchards and plantation crops for harvesting and other crop management operations.
- ✰ Entrepreneurs need to be identified and encouraged for custom hiring of package of equipment for horticulture mechanization.

There are number of ICAR institutes and State Agricultural Universities which are engaged in the research work on different horticultural crops. Most of them are crop specific and located in the areas having highest concentration of horticultural crops dealt with. A collective mechanization effort is required to adapt available horticultural tools/equipment developed at different research institutes or imported from aboard and develop new tools and machinery in order to bridge the identified mechanization gaps.

Conclusions

Using proper machinery in nursery and orchard management brings many benefits. Machines help save time, reduce labour, and improve the efficiency of unit operations like ploughing, digging pits, planting, pruning, spraying, weeding, and harvesting. For nurseries, equipment like planters and irrigation systems ensure plants grow evenly and stay healthy. In orchards, tractors, sprayers, and mechanical harvesters make it easier to care for large areas of fruit trees, reducing the need for manual work. While these machines can be expensive, they often pay off by boosting productivity and improving crop quality. They also help manage resources better, such as water and fertilizers, reducing waste. However, it's important to train workers properly to use the machines safely and maintain them regularly. In the future, as technology continues to advance, machines will become even more essential in nursery and orchard management, helping farmers grow healthier plants and produce more food efficiently.

REFERENCES

Anand R, Sahni RK, Kumar SP, Thorat DS and Kumar AK, 2023. Advancement in agricultural practices with use of drones in the context of precision farming. *Global Journal of Engineering Science*, **11**(2), 1-7.

Anonymous, 2023. Agricultural statistics at a glance, 2022. Department of Agriculture, Cooperation and Farmers Welfare, Ministry of Agriculture and Farmers Welfare, New Delhi.

Bhalekar DG, Sahni RK, Schrader MJ and Khot LR, 2024. Pneumatic spray delivery–based fixed spray system configuration optimization for efficient agrochemical application in modern vineyards. *Pest Management Science*, **80**(8), 4044-4054. https: //doi.org/10.1002/ps.8111.

Dixit PS, Sahni RK and Kumar P, 2024. Application of unmanned aerial vehicle (UAV) in smart agriculture. Advances in Agriculture Sciences Volume I (pp. 68-76). Bhumi Publishing. https: //www.bhumipublishing.com/wp-content/uploads/2024/03/Advances-in-Agriculture-Science-Volume-I.pdf

Kumar M, Gaikwad BB and Sahni RK, 2020. Mechanization of garlic (*Allium sativum*) cultivation in India: An overview. *Pantnagar Journal of Research*, **18**(1), 76-83. https: //www.gbpuat.res.in/uploads/archive/18.1.14.pdf

Kumar M, Sahni RK and Agarwal KN, 2024a. Development of a mechanically assisted pneumatic metering mechanism for high-speed planting of soybean. *Journal of Scientific and Industrial Research*, **83**(1), 84-91. https: //doi.org/10.56042/jsir.v83i1.2863.

Kumar M, Sahni RK, Waghaye AM and Nayak AK, 2018. Precision irrigation technologies: A present need to save for future. Biotech Articles. https: //biotecharticles.com/Agriculture-Article/Precision-Irrigation-Technologies-A-Present-Need-to-Save-for-Future-4382.html

Kumar M, Sahni RK, Waghaye AM, Nayak AK and Kumar D, 2018. Automated irrigation system for rice: A review. *The Andhra Agricultural Journal*, **65**: 324-329.

Kumar SP, Jat D, Sahni RK, Jyoti B, Kumar M, Subeesh A, Parmar BS and Mehta CR, 2024. Measurement of droplets characteristics of UAV based spraying system using imaging techniques and prediction by GWO-ANN model. Measurement, 234, 114759. https: //doi.org/10.1016/j.measurement.2023.114759.

Mahapatra J, Tiwari PS, Singh KP, Nandede BM, Sahni RK, Pagare V, Singh J, Shrinivasa DJ and Mandal S, 2024. An adjustable pneumatic planter with reduced source vibration for better precision in field seeding. *Sensors*, **24**(11), 3399.

Rathnayake AP, Sahni RK, Khot LR and Hoheisel GA, 2022. Intelligent sprayer spray rates optimization to efficiently apply chemicals in modern apple orchards. *Journal of ASABE*, **65**(6), 1411–1420. https: //doi.org/10.13031/ja.14654.

Sahni RK and Khot LR, 2023. Fixed spray systems for perennial specialty crops. In Zhang, Q. (Ed.), Encyclopedia of Smart Agriculture Technologies. Springer. https: //doi.org/10.1007/978-3-030-89123-7195-1

Sahni RK, Kumar D, Tiwari PS, Kumar V, Kumar SP, Chandel NS, 2018a. DGPS based on-the-go soil nutrient mapping system: A Review, *The Andhra Agric J* **65**(spl): 1–6.

Sahni RK, Kumar SP, Thorat D, Rajwade Y, Jyoti B, Ranjan J and Anand R, 2023. Drone spraying system for efficient agrochemical application in precision agriculture. In Chouhan, S. S., Singh, U. P., and Jain, S. (Eds.), *Applications of Computer Vision and Drone Technology in Agriculture 4.0*. Springer. https: // doi.org/10.1007/978-981-99-8684-2_13

Sahni RK, Kumar V, Kumar SP, Chandel NS and Tiwari PS, 2018b. Precision agriculture technologies. *Biotech Articles*. https: //biotecharticles.com/ Agriculture-Article/Precision-Agriculture-Technologies-4383.html

Sahni RK, Kumari S, Kumar M, Kumar M and Kumar A, 2020. Status of litchi cultivation in India. *International Journal of Current Microbiology and Applied Sciences,* **9**(4): 1827-1840.

Sahni RK, Patel A, Kumar M, Thorat DS and Kumar V, 2018c. Status of farm mechanization in Indian agriculture. Biotech Articles. https: //biotecharticles. com/Agriculture-Article/Status-of-Farm-Mechanization-in-Indian-Agriculture-4468.html

Sahni RK, Ranjan R, Hoheisel GA, Khot LR, Beers EH and Grieshop MJ, 2022a. Pneumatic spray delivery–based solid set canopy delivery system for oblique banded leaf roller and codling moth control in a high–density modern apple orchard. *Pest Management Science,* **78**(11): 4793-4801.

Sahni RK, Ranjan R, Khot LR, Hoheisel GA and Grieshop MJ, 2022b. Fixed spray delivery systems for efficient crop input applications in deciduous crops. *Acta Horticulture,* **1346**: 527-536.

Sahni RK, Schrader MJ, Rathnayake AP, Khot LR, Hoheisel GA and Zhu H, 2024. Evaluation of suitable base spray rate estimation methods for precision chemical applications in vineyards with different training systems. *American Journal of Enology and Viticulture, 75*(1), 0750009. https: //doi.org/10.5344/ ajev.2024.22064

Singh RS and Sahni RK, 2019. Transformation of Indian agriculture through mechanization. *Economic Affairs,* **64**(2): 297-303.

Tiwari PS, Gurung TR, Sahni RK and Kumar V, 2017. Agricultural mechanization trends in SAARC region. *Gurung, TR, Kabir, W., and Bokhtiar, SM* (*eds.*). Mechanization for Sustainable Agricultural Intensification in SAARC Region. SAARC Agriculture Centre, Dhaka, Bangladesh, 1-40.

Tiwari PS, Sahni RK, Kumar SP, Kumar V and Chandel NS, 2019a. Precision agriculture applications in horticulture. *Pantnagar Journal of Research,* **17**(1): 1-10. https: //www.gbpuat.res.in/uploads/archive/Vol.17(1).1.pdf

Tiwari PS, Singh KK, Sahni RK and Kumar V, 2019b. Farm mechanization–trends and policy for its promotion in India. *The Indian Journal of Agricultural Sciences,* **89**(10): 1555-1562.

Chapter 15

Mechanized Grafting in Horticultural Crops

Anant Bahadur and Swati Sharma

Division of Vegetable Production, ICAR–Indian Institute of Vegetable Research, Varanasi, Uttar Pradesh

Grafting is an asexual propagation technique of horticultural crops wherein tissues of the same family plants are joined so as to develop a new plant together. It is a fusion of plant parts so that vascular continuity is established between them, and the resulting genetically composite organism functions as a single plant. The upper part of the combined plant is called scion while the lower part is called rootstock. The rootstocks used for a particular scion are usually close relatives or wild selections, mostly within the genera of the scion crop, however the natural grafts between different families have also been reported (Warschefsky *et al.*, 2016). The rootstock used generally shows tolerance for biotic (pathogens), abiotic stresses (environmental stress) and plant vigour, whereas the scion (upper part) possesses qualitative and quantitative horticultural traits.

Grafting technology has been adopted for fruit propagation for the past 7000 years (Mudge *et al.*, 2009). Currently, grafting is still employed as the most common propagation method for 80 per cent of the most popular fruit and nut trees (Warschefsky *et al.*, 2016). It has been in wide use for the commercial propagation of important fruit crops like grapes, citrus and apples. The requirement for grafting technology was primarily due to the need for overcoming obstacles of biotic and abiotic stresses, quick multiplication of desirable scion and utilizing unique advantages of specific rootstocks. It has been mentioned in old Chinese, Korean and Japanese writings in different forms. The historical records of grafting date back to 1560 BC (Mudge *et al.,* 2009). In vegetables, grafting began in later parts of 1920s in Japan when watermelon was grafted on *Cucurbita moschata* to control *Fusarium* wilt diseases (Ashita *et al.,* 1927; Sakata *et al.,* 2007). Large-scale commercial cultivation of grafted solanaceous and cucurbitaceous vegetables can be traced back to late 1950s to early 1960s in Japan and Korea. In European countries, vegetable grafting was commercially started in early 1990s while commercial production of grafted seedlings was initiated in first decade of 21st century in USA.

Discovery of polythene film, its application in seedling productions and simultaneous intensive research for identification of new fruit rootstocks in 1960s paved path for commercial production of grafted seedlings, particularly in solanaceous and cucurbitaceous vegetables. Consequently, percentage of grafted vegetables in commercial production rose significantly in Japan (59%) and Korea (81%) (Lee *et al.,* 1999). In India, grafting in vegetable crops was initiated in later part of first decade of 21st century at VNR Nursery Pvt. Ltd., Raipur, and thereafter at ICAR–IIHR, Bangalore, ICAR–IIVR, Varanasi, CSKHPKV, Palampur and ICAR–CAZRI, Jodhpur, *etc.*

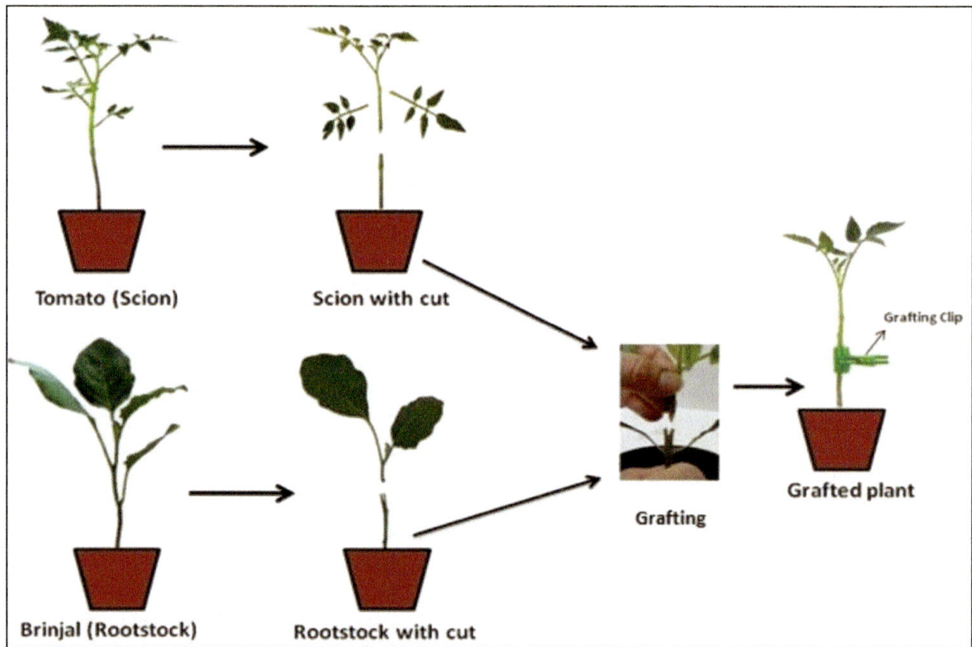

Figure 15.1: Depiction of Manual Grafting of Tomato on Brinjal Rootstock.

Diving into history of fruit grafting, the apple variety "Paradise" was adopted as rootstock for its strong dwarfing effects and rapid rooting and establishment. Likewise, to overcome the disaster to the European grapes due to the pest

Phylloxera, use of resistant American rootstocks was practised. Several grafting technologies are used worldwide based on ease of doing, success rate, species and season. Some of them are namely, wedge, chip, splice, saddle, bark, whip and tongue, side-veneer, hole insertion, approach and inflorescence grafting. The overwhelming importance of grafting in horticulture, particularly fruit propagation emphasizes the need to make the process rapid and economic for easy and bulk production of grafts. However, the process, especially automated graft production needs intricate management to manage prevention of graft failure and ensure strong graft union with maintenance of distinct phenotypes. In addition, virus free grafting material and time of grafting effect the graft success considerably in woody fruit species. For example, whip and tongue grafting has been found to have higher success at the end of dormancy period in apple, while t-budding has been found more appropriate during active growth phase. The selection of grafting method and the required tissue is important to guide the designing and development of an appropriate robotics technique for mechanized graft production. The self-healing and growth rate of grafted plants produced by automation can be further enhanced by effective modification of environment *i.e.* temperature, light quality and relative humidity. Notably, it was also observed that inter-stocks and interstock length affected the plant size of peach, which can be utilized to modulate tree density and canopy management. Rootstock effects on growth, yield and fruit quality have been reported.

An effective graft process is initiated with cutting and separation of scion from mother plant, followed by accurate alignment of plant tissues. The self-healing process is initiated by the plant cells to begin tissue adhesion through cell wall modifications, expansion, division and differentiation. Vascular cambium regeneration is important for success of grafts. The development of automated robotic graft production can also incorporate hormones application to enhance success rate. Earlier studies have reported that the blocking of auxin pathway in cambium is detrimental to graft attachment and development of new vascular reconnection for effective transport of organic compounds, nutrients and water. The concentration of auxin and cytokinin is known to peak after grafting process at the union. Ethylene hastens the formation of a successful graft union. Strong attachment, upregulation of growth hormones, down regulation of peroxidases, vascular (phloem and xylem) reconnection development, enhanced expression of specific genes associated with grafting indicate graft success and compatibility. In woody species, grafting is often marred by higher phenolic concentrations, limited xylem and phloem development, necrosis at graft joint, rooting, bulging or stunted/ abnormal development.

Mechanization is important for smooth and quick farm operations for different operations like training-pruning, harvesting, grafting *etc.* The advent of artificial intelligence, robotics and mechanization has also initiated concerns among growers regarding precision, accuracy and effects on trees. Mechanized grafting will enable development of integrated automated farm operations for orchard management.

Grafting robots have also been developed for facilitating propagation of horticultural crops. Previously, USDA has prepared operation simulator for semi and completely automated systems for mechanized grafting. Recently, the interventions of high-tech robotic systems in horticulture for future ready orchards and the importance of automated grafting has been catching momentum. Sensors, human-robot combined approaches, and advancement in mechanization can improve the efficacy of orchard operations. The grafting machine generally uses a fixed-angle cutting technique, which could be modified for suitability to variable angle for improved precision and adaptability in automation. Model can be developed by using regression models and polynomial functions can be used for fitting/understanding dynamic growth process of intersection. There is need to develop effective design for application of adaptive cutting for automation in grafting. Use of integrated simulation models and analytical hierarchy process techniques can be done for designing grafting operations for vegetable seedlings. Some advantages of grafting technology have been mentioned below.

Table 15.1: Advantages of Grafting Technology in Horticultural Crops

Fruit Plants	*Vegetable Crops*
☆ **Ease in propagation:** Grafting is most effective and cost-efficient technique where scion is difficult to multiply by other vegetative method. ☆ **Precocious:** Grafting technology may reduce the juvenile phase of plant and induce early fruiting. For example, juvenile phase of mangosteen may prolong to 15 years, while grafting with mature scion can result in fruiting in two years. ☆ **Dwarfing:** In some fruit plants (*e.g.*, apple), grafting is used to induce dwarfing to enable high density planting. ☆ **Hardiness:** Grafting helps to induce tolerance or resistance to cold, salinity, drought, waterlogging, *etc.* ☆ **Reduced time for hybrid breeding program:** Developing hybrids takes 10 or more years, whereas grafting can shorten time required for fruiting and breeding drastically.	☆ **Disease and nematode tolerance:** Several soil-borne diseases, *e.g.*, bacterial wilt, *Fusarium* wilt, *Verticillium* wilt, collar rot, gummy stem blight and root knot nematodes may be manged by grafting over tolerant rootstocks. ☆ **Tolerance to abiotic stresses:** Effect of drought, waterlogging, salinity, low temperature and heavy metals can be alleviated through grafting. ☆ **Increased plant vigour and yield:** Vigorous rootstocks can be used to enhance yield due to improved nutrient and water uptake. ☆ **Improved produce quality:** In many solanaceous and cucurbitaceous vegetables, fruit quality such as, lycopene, β-carotene, ascorbic acid, TSS, *etc.* have been improved due to grafting on suitable rootstocks. ☆ **Extended harvest period:** Grafting often enhances cropping period, *e.g.*, tomato or brinjal grafted on *Solanaum torvum* showed an extended harvest period.

Fruit Plants	*Vegetable Crops*
☆ **Multiple fruit cultivars:** Multiple cultivars of fruits (*e.g.*, apple) are sometimes grafted on a single tree, providing more varieties, and also serving as pollinizer. "Kaleem Ullah Khan" known as Mango man from Lucknow has succeeded in grafting more than 300 varieties of mango on a single tree. ☆ **Plant sturdiness:** Scions are grafted over well-established rootstocks, thereby giving sturdiness. ☆ **Resistance to pest and diseases:** Several insects and diseases, particularly soil-borne can be managed by using tolerant rootstocks. ☆ **Virus indexing:** Presence or absence of virus in the suspect plant can be confirmed by grafting scion to highly susceptible rootstock to display prominent symptoms (*e.g.*, Tristeza virus in citrus). ☆ **Availability of pollens:** A good polliniser grafted over self-incompatible fruit tree (*e.g.*, apple) may serve pollen need at blossom time. ☆ **Repair/Rejuvenation of old/ Damaged tree:** Bridge grafting/ inarching can be used for rejuvenation of old fruit plants. ☆ **Genetic consistency:** In commercial fruit industry, consistency is maintained by grafting a scion with desired fruit traits onto a hardy rootstock.	☆ **Harvesting dual crop and developing fancy plant for exhibition:** Through grafting technology, two or more vegetables can be taken from single plant, *e.g.*, Pomato (Potato+Tomato), Brimato (Brinjal+Tomato). Similarly, flowering plant Petunia can be grafted over brinjal rootstocks for extended flowering period and higher flower production. ☆ **Higher income for nursery entrepreneurs:** Grafted seedlings of solanaceous and cucurbitaceous vegetables can be sold at higher price (INR 10-15/seedling) to get more income (2-3 times). For example, several nurserymen in Chhattisgarh have been selling more than 5 million grafted seedlings of brinjal, tomato and chilli per annum with average cost of about INR 11 per seedling.

Now-a-days, most of the fruit plants are produced by bud grafting, popularly known as budding. A bud of a desired productive fruit tree is cloned or grafted into a pencil-sized stem of a seedling of the same or related species. Bud eventually develops as the scion, while the trunk and roots below the budding area are considered as rootstock. The bud or scion exhibits more vigour when grafted on a rootstock seedling. For example, Santa Rosa plum, Elberta peach, Bing cherry and Gala apple trees developed from clonal cuttings result in less vigorous growth than the buds of these trees grafted on suitable rootstocks. Many a times, grafted trees increase production, yield better quality fruits, and exhibit tolerance to disease, pest, nematode and cold. In fruit plants, interspecific grafting is frequently adopted in *Prunus* or stone fruits. Almond (*Prunus amygdalus*) buds may be grafted on peach (*Prunus persica*), plum (*Prunus salicina*) or apricot (*Prunus armeniaca*) seedlings. Similarly, quince (*Cydonia oblonga*) buds may be grafted on pear (*Pyrus communis*) or loquat (*Eriobotrya japonica*) trees.

Table 15.2: Characteristics of some Selected Rootstocks Useful in Overcoming Biotic and Abiotic Stresses in Fruit Crops by Grafting

Crop	*Rootstock*	*Characteristics*
Apple	M 7, M 26, MM 106, MM 111, G 11, G 30, G 41, Alnarp 2, *Malus baccata*	Cold hardiness
	Northern Spy, *Malus seiboldii*, MM 102, MM 105. MM 106, MM 108, MM 111, CG 5179, CG 6210, CG 7707	Woolly aphid resistance
	M 27, MM 111, Ottawa 3, B 9	*Phytophthora* resistance
	CG 5179, CG 6210, CG 7707, MM 106	Fire blight resistance
	M 26, M 27, Ottawa 3, P2, B 9	Dwarfing
	M26, MM 106	Salinity tolerance
	EMLA 7, EMLA 26, EMLA 27,EMLA.106	Virus free, moderate to highly productive, wide adoptability
	Bemali, MI 793	Resistance to woolly apple aphid
	Geneva 11	Fire blight resistance, tolerant to crown and root rot
	Geneva 16	Highly fire blight resistant, tolerant to crown rot and immune to scab
	Geneva 41	Fire blight resistant, very winter hardy
	Geneva 202	Fire blight resistant, highly resistant to wooly aphid
	Geneva 210, Geneva 214, Geneva 222, Geneva 890, Geneva 814, Geneva 969	Resistant to woolly apple aphid, highly resistant to fire blight, tolerant to replant disease complex and crown and root rot
Pear	*Pyrus calleryana*	Resistant to fire blight
	Pyrus betulaefolia	Winter hardy, vigorous, moderately tolerant to fire blight and salinity
	Pyrus ussurensis	Resistant to fire blight, pear psylla and cold hardy
	Quince C, Quince A, Pyro dwarf, Pyro 233, O x F97, OH x F51, Oregon 211	Dwarfing

Crop	Rootstock	Characteristics
	Pyrus betulaefolia and *Cydonia oblonga*	Higher nutrient uptake
	OHF series (OHF 59, 87, 333)	Cold hardy (OHF 87- tolerant to fire blight)
	QR 193/16	Induces dwarfing and large fruit size
	Quince C	Early bearing but tart and astringent flavour
	Oregon 211 and Oregon 249, 260, 261 and 264	High temperature tolerance
	Oregon 211 x 249, Oregon 249, 260, 261 and 264	Drought and high soil moisture tolerance
	Oregon series	Low soil pH tolerance, nematode resistance
	Oregon 211 x 249	Resistance to powdery mildew
	Oregon series x OH x F series	Resistance to crown gall and root rot
	Oregon 249, 260, 261 and 264	Resistance to canker/gummosis
	Ct. S 212, Ct. S 214	Vigorous and tolerant to high soil pH
Peach	Nemaguard, Nemared, Floraguard, Guardian and Okinawa	Tolerance to nematodes
	Western Sand Cherry, Almond, Nanking Cherry, Tetra, Rubira, P.S. A5, Everica, and Junior	Moderately dwarf rootstocks
	GF 677	Drought tolerance, Salt tolerance, Tolerance to replant disease, Tolerance to calcareous soil
	St. Julian Hybrid No. 2	Semi dwarf, Tolerance to salt and bacterial canker
	Marianna GF 8/1	Precocious, Resistant to water logging, viruses and root knot nematode
	Damas GF 1869	Semi-dwarf, Resistant to high soil pH, water logging and bacterial canker
	Everica 99, VVA-1, Kuban 86, and VSV-1	Cold hardy
	Sirio, Julier, Paramount, Jasper, Cadaman, Montizo, Adarcias, Adesoto 101, peach x almond hybrids (Monegro, Paramount, Garnem), and peach × *P. davidiana* (Barrier 1, PeDa and Cadman)	Drought tolerant

Crop	*Rootstock*	*Characteristics*
Plum	Myrobalan (*P. cerasifera*)	Root asphyxia, cold hardiness, drought resistance
	Hybrid myrobalan 29C (almond x peach) and L2 cherry	Resistance to plum pox virus
	Mr. S. 2/5	Tolerant to waterlogging condition and alkalinity
	ISG 1/5	Tolerant to lime induced chlorosis
	GF 677	Salt tolerance
	Black Damas	Resistance to bacterial canker
	Brompton	Resistance to cold temperature
	Prune GF43	Resistance to root and collar rot, water logging
	Pixy	Tolerant to bacterial canker
	Pixy, Maridon and Ferlenain	Dwarfing rootstocks suitable for HDP
	St. Julien A	Resistance to low temperature
	Marianna	Resistant to rosette virus
	Marianna 2624	Resistance to waterlogging, oak root rot, prune brown line, root knot nematodes
	Marianna 4001	Tolerance to drought, bacterial canker
	Marianna GF-8-1	Tolerance to waterlogging, calcareous soil, resistance to root knot nematodes and virus
Apricot	Mariana GF 8-1, Greengage CD-4 and Damas 1869	Improved yield
	Pollizo prune (*Prunus institia* L.)	Tolerance to flooding condition
	GF305 (peach) and Real Fino (apricot)	Resistance to Apple Chlorotic Leaf Spot Virus (ACLSV)
Cherry	Mazzard (wild sweet cherry)	Vigorous rootstock
	P. mahaleb	Vigorous rootstock, better adapted to calcareous soils and drought condition
	Edabriz	Dwarfing and precocious
	Krymsk 5	Cold hardy, semi-dwarfing
	Colt (*P. avium* F 299/2 x *P. pseudocerasus*)	Semi-vigorous, resistance to *Phytophthora* root rot, bacterial canker and virus
	Stockton Morello	Semi-dwarf and higher yield
	Edabriz and Weiroot	Dwarfing rootstocks
	F12/1	Resistant to bacterial canker

Crop	*Rootstock*	*Characteristics*
Almond	Nemaguard	Immune to root knot nematode
	Empyrean 1	Resistance to root knot nematode, tolerance to sodium and chloride
	Viking	Resistance to root knot nematode, tolerance to ring nematode, high pH and bacterial canker
	Atlas	Tolerance to high pH and root knot nematode
	Almen 88 and Almen 201	Resistance to root knot nematode
Walnut	Paradox	Resistance to root lesion nematodes and crown gall, tolerance to salinity and heavy wet soil
	Valch	Vigorous, high disease tolerance
	VX211	Tolerance to *Phytophthora* and nematodes
	RX1	Excellent tolerance to replant disease complex
Mango	Olour, Bappakai, Gomera 1, 13/3, Kurukkan, Nekkare, ML-2, Gomera 1 and 3, *M. zeylanica*, *M. odorata*, *M. foetida*, 13-1	Salt tolerance
	Arauca, Latra and Mylepalian, Vellaicolumban, Brodie	Dwarfing
	Carabao, Pico Manga d Agua	Resistance to wilt
Guava	*Psidium cattleianum*, *P. mole*	Tolerance to wilt
	Psidium friedrichsthalianum	Resistance to wilt and nematodes, Dwarfing
	P. guajava x *P. guinense*	Tolerance to nematodes
	P. molle x *P. guajava*	Resistance to wilt
	P. cattleianum	Vigorous, high yield
	P. pumilum	Dwarfing, higher TSS and total sugars
	P. cujavillis	Resistance to wilt, high ascorbic acid content
	Pusa Srijan (Aneuploid no. 82)	Dwarfing, high yield
	P. friedrichsthalianum, *P. cattleianum* var. lucidum	Resistance to root knot nematode
	P. guineense	Resistance to root knot nematode
	Crioula	Salinity tolerance

Crop	*Rootstock*	*Characteristics*
Grape	Riparia/Gloire	High resistance to *Phylloxera*, moderate resistance to root knot nematode
	Saint George	High resistance to *Phylloxera*, high tolerance to drought, moderate Tolerance to salt
	3309 Couderc	High resistance to *Phylloxera*
	101-14	High resistance to *Phylloxera*, moderate tolerance to root knot nematode
	Oppenheim, 5 BB Kober, 5C Teleki, Swarzmann, *Vitis labrusca*	High resistance to *Phylloxera*
	1103 Paulsen	High resistance to *Phylloxera* and drought
	RS-3, RS 9, Kingfisher	High resistance to root knot nematode
	GRN-1, GRN-2, GRN-3 and GRN-4	Very high resistance to *Phylloxera* and root knot nematode
	Dimple, Jonelle, GU8	Nematode and salinity tolerance
	GU8 *P. longipes* and *P. arayan*	Root knot nematode resistant
	Ramsey and Dogridge	Root knot nematode resistant
	UCD GRN1 and VR 039-16	Resistant to root lesion nematode
	Harmony and Freedom	Resistant to root knot nematodes
	Ramsey and 1103 Paulsen	Tolerance of *Phylloxera* and nematodes
	Salt creek and Dogridge	Tolerance to salt
	Beta and 3309 C	Cold hardiness
	A15 and A17	Tolerance to alkalinity
	Hybrids of *V. berlandieri* x *V. riparia*	Tolerance to drought
	196-17, CH-1 and CH-2	Tolerance to salinity
	99-R, 110-R, 1103-P, B-2-56, Dogridge, Salt Creek, Ramsey, Riparia Glorie	Salt tolerance
	Borner (*Vitis cinerea*)	Drought tolerance
	Dogridge and 1613	Nematodes and phylloxera root louse resistance
Citrus	Rangpur lime	Suitable for south Indian condition, tolerance to Tristeza, salt, drought; moderate resistance to root rot
	Marmalade orange	Moderate tolerance to root rot and citrus nematode; resistance to Tristeza, salt and drought
	Rough lemon	Tristeza and exocortis resistance, improved fruit size in Coorg mandarin

Crop	*Rootstock*	*Characteristics*
	Cleopatra mandarin	Resistance to almost all viruses and exocortis, moderate tolerance to salt
	Sour orange	Root rot resistance, moderate tolerance to drought, susceptible to Tristeza; thin peel, excellent quality fruits in sweet orange and grapefruit
	Sweet lime	Immune to collar rot, moderate tolerance to nematode
	Trifoliate orange	Hardy to low temperature, resistance to root rot, nematode and Tristeza
	Troyer and Carrizo citrange	Moderate tolerance to root rot and Tristeza, attractive bright orange colour and early bearing in kinnow
	Sweet orange	Moderate tolerance to Tristeza and resistance to exocortis
	Karna Khatta	Suitable for heavy soil, moderate tolerance to nematode
	Jatti khatti	Suitable for arid and semi-arid climates for kinnow
	Nasnaran	Moderate tolerance to citrus nematode; tolerance to Tristeza and resistance to exocortis
	US 802, US 942	Greening tolerance
	F and A 418	Dwarfing
	US 852	Resistance to Tristeza virus, citrus nematode and *Phytophthora*
	Citrus macrophylla	Salinity resistance
	Pearl, Mosambi x Kinnow and Mosambi x Nagpur	Resistance against citrus psylla
	Star Ruby and Ruby Red	Resistance against citrus psylla
	F and A 418, Troyer Citrange x Cleopatra mandarin	Dwarfing rootstock
Avocado	G755C13-1	Salinity tolerance
	Duke 7 and G6	Resistance to *P. cinnamomic*
Loquat	Anger	Salinity tolerance

Source: Darshan and Solanki, 2022; Dhurve *et al.,* 2023; Nimbolkar *et al.,* 2016; Santhi *et al.,* 2020; Sharma *et al.,* 2020).

Grafting technology has several advantages, however its large-scale adoption is hampered by a number of factors, such as graft incompatibility due to failure of graft union formation, which largely depends on environmental conditions or grafting skill. It is time consuming and labour-intensive practice, if carried out manually. Therefore, efforts were made to mechanise grafting technique in the last five decades. Mechanised grafting is more efficient over manual grafting, because of high output and greater graft success rate.

Development of grafting machines or robots are an important innovation for commercial production of grafted vegetables and fruits seedlings. Rapid expansion of vegetable nursery industry and introduction of commercial plug seedling production has stimulated development of an efficient grafting machine to reduce expenses needed to produce grafted seedlings manually. The initiatives for development of efficient and reasonably priced grafting robots for vegetables were taken by Japan and Korea. With advancement in grafting mechanisation, dependency on manual grafting has reduced drastically, particularly in developed countries, such as South Korea, Japan, China, Spain and France. Many manual works required in grafting can be eliminated by employing grafting machine or robots. Mechanised grafting can also eliminate the mismatch of vascular tissues, allowing grafted seedlings to recover and grow faster. Large-scale production of grafted seedlings (more than 10 million seedlings per year) in shorter time can be aided by mechanisation and automation in grafting technology. Grafting robot can easily recognise the size and feature of the seedlings to be grafted. The robotic machine carries seedlings, sliced in appropriate size and manner, and attach two parts of the plants with help of elastic grafting clips. Robotic grafting machine can perform grafting at the rate of 600-1200 seedlings per hour with more than 95 per cent success rate, whereas a grafting specialist human can graft maximum up to 150 seedlings per hour. It has been estimated that a semi-automatic grafting machine can graft 600-800 plants per hour, which is comparable to 5-6 skilled workers engaged for cucumber grafting, and 2-3 skilled workers for tomato.

Mechanization in Vegetable Grafting

In vegetable crops, research on mechanisation in grafting was started in 1980s by Japan followed by Korea in the early 1990s. In China, research on vegetable grafting robots began in 1993, whereas in European countries like, Netherlands, Italy, Spain, *etc.* some achievements in development of grafting robots were made in the beginning of 21st century (Yan *et al.,* 102). Due to the increasing demand for grafted seedlings and simultaneously increasing labour cost, researchers switched over for mechanisation in grafting to lower production costs. Grafting machine/ robot can improve grafted seedling productivity with more precision and success rate.

In mechanical grafting, mostly splice grafting (SG), tube grafting (TG), cleft grafting (CG), tongue approach grafting (TAG) and hole insertion grafting (HIG) have been followed. Splice or side grafting is mainly adopted in solanaceous and cucurbitaceous vegetables, wherein top of the rootstock and bottom of the scion is cut into slanted surfaces (45° angle), and then two slanted cut surfaces are perfectly joined to prepare grafted seedlings. Based on SG, ISEKI company of Japan launched the GR800 series of vegetable grafting robot in 1994. It was semi-automatic robot with the capacity to prepare about 800 grafted plants/h with survival rate of 95 per cent. Another company in Japan, *i.e.*, Japan Yanmar developed AG1000 series automatic grafting robot in 1994 for solanaceous vegetables. It consists of an automated provision for transportation, clamping, cutting and docking of rootstocks and scions accomplished by three different conveyor belts, and six seedlings can be grafted at the same time, however the quality of grafted seedlings were highly affected by plant height and stem diameter. A semi-automatic grafting robot EMP 300 with silicone clips was developed in 2010 by Conic Systems Company in Spain. In this system, clamping and cutting and docking operations were arranged symmetrically which ensured consistency of cutting angle, and accuracy of incision docking. However, it showed lower efficiency (400-600 plants/h) and small scale seedling production suitability. Further in 2011, ISEKI company launched GRF800-U, an automatic grafting robot with artificial seedling feeding operation.

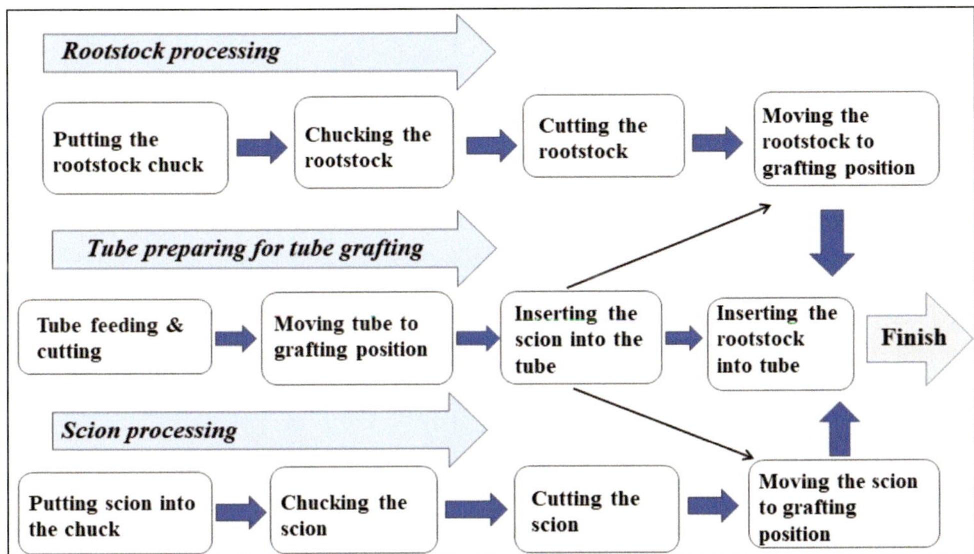

Figure 15.2: Flow Chart for Mechanised Grafting Process in Vegetable Crops.

China Agricultural University launched single-arm automatic vegetable grafting robot- 2JSZ-600 in 1998, which could perform multiple operations such as picking, cutting, joining, fixing and rowing of seedlings with grafting speed of 600 plants/h. Later on in 2009, for higher grafting speed (854 plants/h), they launched the 2JSZ-600B double-arms bi-directional vegetable grafting robot, wherein the rootstock and scion seedlings enters from the left and right directions for cutting mechanism at the same time. In 2011, an automatic grafting robot- TJ-800 was developed in China. It has double-station feeding seedlings, and equipped with clamping hands and cutting devices for the rootstock and scion, respectively. The grafting efficiency was improved to 800 plants/h with survival rate of 95 per cent. In 2015, China launched a solanaceous vegetable grafting robot which achieved rapid operation by horizontal seedling supply and linear sliding cutting. Its average production efficiency was 512 plants/h, and survival rate of grafted seedlings was 99.5 per cent. Based on the assembly line, South China Agricultural University developed a semi-automatic grafting robot for solanaceous seedlings in 2020. It is suitable for performing six operations (clamping, cutting, gathering seedlings, docking, clip sending and seedling taking) into four working stations with grafting efficiency of 1000 plants/h. In 2021, Hefei Jiafute Robot Technology Co., Ltd. developed a semi-automatic vegetable grafting robot (JFT-A1500T), and its maximum grafting speed was 1500 plants/h.

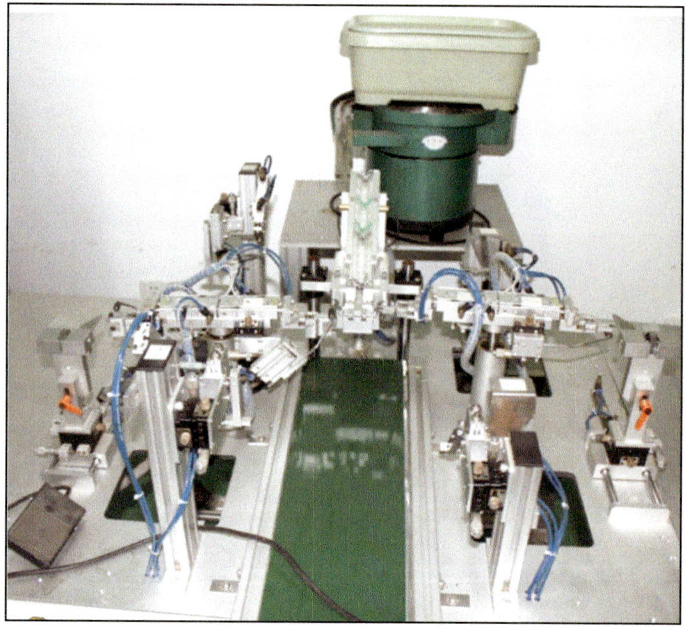

Figure 15.3: Semi-Automatic Vegetable Grafting Machine.

Grafting robots based on tube grafting were developed in many parts of the world. In this grafting technique, scions and rootstocks with similar diameters are selected to cut, and they are joined by some special plastic casings or tubes to support both rootstock and scion. For TG, first automatic grafting robot- MGM600 was developed in 1992 by Japanese Mitsubishi Group. Later on in 2003, Japanese Yanmar company developed T600 semi-automatic vegetable grafting robot, which can automatically perform cutting, docking, fixing, casing, and other operations of the seedling fixing with only one worker. In Netherlands, ISO Group company developed the Graft1000 automatic grafting robot in 2007 with image recognition system and conveying system of seedlings for matching and grafting of rootstocks and scions. Further in 2010, the firm developed a Graft1200 automatic grafting robot. These two robots were highly automatic but more expensive. Therefore, in 2014 the company again developed a Graft1100 semi-automatic grafting robot with 1000 plants/h productivity and a 98 per cent grafting survival rate.

HIG and CG based semi-automatic melon grafting robot 2JC-350 was developed by China in 2005. The grafting machine performs operations, such as the clamping of the rootstock, cutting of rootstock, making hole in rootstock, clamping of scion, cutting of scion, and the docking between scion and rootstock. Subsequently, 2JC-450, 2JC-500 and 2JC-600 grafting robot series were developed, and the grafting rate gradually increased to 600 plants/h. In 2010, 2JC-1000A grafting robot with two robotic arms was designed to graft five seedlings at a time with 1000 plants/h grafting speed. In 2020, China designed a visually guided robot for watermelon grafting based on HIG. The robot uses machine vision to recognize and locate the rootstock and scion then accurately punch the hole to insert the watermelon scion into the hole.

Table 15.3: Characteristics and Comparison of Vegetable Grafting Robots (Adopted from Yan *et al.*, 102)

Year	*Model*	*Automation*	*Firm*	*Grafting method*	*Vegetable crop*	*Productivity (grafts/h)*	*Survival Rate (%)*
1992	MGM600	Semi	Mitsubishi, Japan	Tube	Solanum	600	95
1994	GR800	Semi	ISEKI, Japan	Splice	Melons and Solanum	800	95
1994	AG1000	Full	Yanmar, Japan	Splice	Solanum	1000	97
2003	T600	Semi	Yanmar, Japan	Tube	Melons	600	98
2004	Needle connected Robot	Full	Ideal System, Korea	Pin	Solanum	1200	95
2007	Graft1000	Full	ISO Group, Netherlands	Tube	Solanum	1000	99
2009	2JSZ-600B	Full	CAU, China	Splice	Melons and Solanum	854	95
2009	2JC-450	Semi	Northeast Agricultural University, China	HIG	Melons	450	90

Year	*Model*	*Automation*	*Firm*	*Grafting method*	*Vegetable crop*	*Productivity (grafts/h)*	*Survival Rate (%)*
2010	EMP300	Semi	Conic System, Spain	Splice	Solanum	400-600	98
2010	Graft1200	Fully	ISO Group, Netherlands	Tube	Melons and Solanum	1050	99
2010	2JC-1000A	Fully	SCAU and NAIEETRC, China	HIG	Melons	1000	>90
2011	GRF800-U	Fully	ISEKI, Japan	Splice	Melons	800	95
2013	AFGR-800CS	Fully	Helper Robotech, Korea	TAG	Melons and Solanum	800	95
2014	Graft1100	Semi	ISO Group, Netherlands	Tube	Melons and Solanum	1000	98
2018	JS-6	Fully	QUA and SZAE, China	Cleft	Solanum	720	96
2021	JFT-A1500T	Semi	Hefei Jiafute Robot Tech., China	Splice	Melons and Solanum	1500	98
2022	HAU-22	Fully	HAU, China	Splice	Melons	2134	67

Applications of Grafting in Vegetable Crops

Biotic Stress Tolerance

Grafting technology in vegetable crops originated in later part of 1920s with primary intent to manage soilborne pathogens. The earliest report on grafting was from Japan wherein *Cucurbita moschata* was used to control *Fusarium* wilt in watermelon production (Sakata *et al.,* 85). Grafting technology usage for disease management has rapidly expanded to other solanaceous and cucurbitaceous vegetables. By 1990s, the use of grafting in vegetable production systems has expanded to manage a broad range of pathogens including fungi, oomycetes, bacteria, nematodes and viruses. Intraspecific or hybrid rootstocks including wild accessions have been found good source of resistance, particularly for soil-borne pathogens. These rootstocks have fewer negative effects on crop productivity, fruit quality or graft compatibility, but often carry resistance gene against one or two major pathogens. The major diseases that can be managed through grafting technology includes, bacterial wilt, *Fusarium* wilt, *Verticillium* wilt, Gummy stem blight, *Macrophomina*, *Sclerotium* root rot, tomato spotted wilt virus, *Rhizoctonia*, root knot nematodes, *etc.* The list of common rootstocks used in solanaceous and cucurbitaceous vegetables and their purpose has been given in Table 15.4.

Table 15.4: Important Rootstocks and their Impact on Solanaceous and Cucurbitaceous Vegetables (Adopted from Bahadur *et al.*, 8,13,14)

Scion	*Rootstock*	*Purpose*
Tomato	*Solanum lycopersicum*, *S. pimpinellifolium*, *S. habrochaites*, *S. melongena*, *S. torvum*, Interspecific hybrid (*Solanum lycopersicum* x *S. habrochaites*), *S. cheesmaniae*	Tolerance to bacterial wilt, *Fusarium* wilt, root knot nematode, *Verticillium* wilt, corky root, flooding, drought, extended fruiting period, higher yield, fruit quality
Brinjal	*Solanum torvum*, *S. integrifolium*, *S. melongena*, *S. sisymbriifolium*	Plant vigour, tolerance to bacterial wilt, *Verticillium* wilt, *Fusarium* wilt, low temperature and nematodes
Watermelon	*Lagenaria siceraria*, *Cucurbita moschata*, Interspecific hybrid squash (*C. maxima* x *C. moschata*), *Benincasa hispida*	Vigorous root system, tolerance to *Fusarium* wilt, physiological disorders, low and high temperature, *B. hispida* for multiple disease tolerance
Muskmelon	*Cucurbita moschata*, *C. pepo*, interspecific hybrid squash (*C. maxima* x *C. moschata*), *Cucumis melo*, *Cucumis metuliferus* (AH cucumber)	Vigorous root system, Tolerance to *Fusarium* wilt, low and high temperature, *Phytophthora* diseases, high soil moisture, nematode and fruit quality
Cucumber	*Cucurbita moschata*, *C. ficifolia* (Fig-leaf gourd), *C. maxima*, *Sicyos angulatus* (Bur cucumber), AH cucumber	Vigorous root system, tolerance to *Fusarium* wilt, low temperature, *Phytophthora*, high soil moisture and nematode

Abiotic Stress Tolerance

Abiotic stresses, such as those caused by water (drought, flooding/hypoxia), temperature (heat, cold chilling/frost), chemicals (mineral deficiency/excess, pollutants heavy metals/pesticides, gaseous toxins), radiation (UV, ionising radiation), mechanical (wind, soil erosion, submergence), *etc.*, are the reason for about 50 per cent decrease in agricultural output. Among these abiotic stresses, high temperature (40%), salinity (20%), drought (17%), low temperature (15%) are primary stressors (Ashraf *et al.*, 7). The scope of grafting technology in last three decades broadened with salt tolerance receiving the most focus. Grafting in vegetables besides enhancing plant vigour, harvesting period (Lee *et al.*, 2010), yield and fruit quality (Huang *et al.*, 2009; Khah *et al.*, 2006; Riga *et al.*, 2016; Rouphael *et al.*, 2010; Tsaballa *et al.*, 2013, 2021), has also been utilized for tolerance to low and high temperatures (Ma *et al.*, 2023; Schwarz *et al.*, 2010), alleviates salinity and heavy metal stress (Arao *et al.*, 2008; Bayoumi *et al.*, 2021; Estan *et al.*, 2005; Fu *et al.*, 2022; Penella *et al.*, 2016;http://journal.frontiersin.org/article/10.3389/fpls.2016.01457/full - B106 Rouphael *et al.*, 2008; Sanwal *et al.*, 2022), increases drought and flooding stress tolerance (Bahadur *et al.*, 2016, 2015; Bhatt *et al.*, 2015), improves water use efficiency (Cantero-Navarro *et al.*, 2016), nutrient uptake and utilization efficiency (Coskun, 27; Lee *et al.*, 2010; Padilla *et al.*, 2023; Rouphael *et al.*, 2008). The cultivated area of grafted solanaceaeous and cucurbitaceaeous vegetables have increased tremendously in recent years because the objective of grafting has been greatly expanded (Lee *et al.*, 2010).

Water Deficit or Drought Stress

Water deficit or drought reduces plant water and turgor potential to the point wherein plants struggle to carry out basic physiological processes. Water deficit condition causes the concentration of solutes in plant cells to rise, which ultimately lowers the water potential and interferes with membranes and cell functions including photosynthesis. The loss due to drought depends on the frequency, severity, and length of drought spells. Vegetables consist of more than 90 per cent water, and drought stress at any stage greatly influences vegetable productivity and quality. Soil moisture deficit at critical growth stages such as active growth, flowering and fruit enlargement greatly reduces vegetable production and quality (Bahadur *et al.*, 2023). Under drought conditions, grafting high-yielding vegetable cultivars onto rootstocks that can lessen the impact of water stress on the shoot is an effective strategy to reduce losses and increase water use efficiency (WUE) (Coskun, 27; Padilla *et al.*, 2023). Inter-specific grafting in cucumber significantly improved plant height, biomass, leaf area and chlorophyll content index (CCI) with remarkable reduction in lipid peroxidation content in leaves under drought stress condition (Coskun, 2023). Padilla *et al.* (2023) reported that severe water stress condition (48 h induced by polyethylene glycols, PEG) in pepper can be mitigated by use of tolerant rootstock 'Niber' which caused higher water retention and WUE because of stomatal closure and synthesis of defensive phytohormones such as abscisic acid (ABA), jasmonic acid (JA) and salicylic acid (SA). Grafting of mini watermelons onto a commercial rootstock, PS 1313 (*Cucurbita maxima* × *Cucurbita moschata*) recorded 60 per cent higher marketable yield when grown under deficit irrigation conditions (Rouphael *et al.*, 2008). The increased marketable production from grafting was mostly due to improved water and nutrient intake, as seen by greater N, K, and Mg concentrations in the leaves, as well as higher CO_2 assimilation. Tomato plants grafted onto Beaufort and Multifort (*Solanum lycopersicum* x *S. habrochaites*) registered an average increase of 30 per cent in marketable fruit yield relative to non-grafted plants (Djidonou *et al.*, 2013). One of the plant strategies to avoid drought stress effect is a larger and more vigorous root system of the rootstock that can absorb water and nutrients much more efficiently than non-grafted plants (Lee *et al.*, 2010). It has been observed that cherry tomato as rootstock mitigated the adverse effect of moderate and severe drought stress in tomato, and enhanced plant growth and yield mainly because of vigorous root system, improved uptake of P and K with reduction in electrolyte leakage (Sadeghi *et al.,* 2023). Reduction in chlorophyll degradation, over production of reactive oxygen species (ROS) in different plant tissues, and induction of super-oxide-dismutase (SOD) activity are the first line of cellular defence under drought stress conditions (Ahmad *et al.,* 2009). It has been reported that drought tolerant tomato rootstock Zarina shows higher SOD and catalase (CAT) activities, and less ROS concentration in leaf tissues of scion under water stress conditions (Sanchez-Rodriguez *et al.,* 2012). Besides, induced activities of ascorbate, glutathione reductase (GR), ascorbate peroxidase (APX), ascorbic acid (AsA) and lower production of hydrogen-peroxides (H_2O_2) and malondialdehyde (MDA) are also imparting drought tolerance in plants (Bahadur *et al.,* 2011). In one of the recent studies in cucumber, grafting mediated drought tolerance was resulted due to regulation of stress responsive genes as well as modulations of antioxidant enzyme activities (Shehata *et al.,* 2022).

Waterlogging or Flooding Stress

Waterlogging conditions cause oxygen starvation, which arises from the slow diffusion of gases in water and from oxygen consumption by micro-organisms and plant roots. Flooding problem may be solved by grafting susceptible plants onto tolerant rootstocks (Bahadur *et al.,* 2016, 2015; Bhatt *et al.,* 2015). More adventitious roots and parenchyma formation have been associated to waterlogging tolerance in grafted plants (Bahadur *et al.,* 2015). It has been reported that 4-6 days of waterlogging/flooding tolerance in tomato can be achieved by using brinjal as rootstock (Bahadur *et al.,* 2016, 2015, 2023, 2024c; Bhatt *et al.,* 2015). It has also been reported that grafting improved flooding tolerance in bitter melon when grafted on sponge gourd (*Luffa cylindrica*) (Liao and Lin, 1996). In cucumber, waterlogging causes reduction of chlorophyll content in leaves, however, grafting onto squash rootstocks retained chlorophyll in leaves (Kato *et al.,* 2001). It has been noticed that when cucumber was grafted into 'Ferro' and 'Cobalt' (*Cucurbita maxima* × *Cucurbita moschata*) rootstocks, the harmful effect of severe flooding stress (15 days) can be alleviated due to maintenance of higher leaf water potential, efficient uptake of K and improved photosynthesis rate (Haghighi and Khosravi, 43). Waterlogging study conducted by Peng *et al.* (2020) revealed that grafting can improve the waterlogging tolerance in bitter melon by increasing the activity of anaerobic respiration enzymes (pyruvate dehydrogenase, alcohol dehydrogenase and lactate dehydrogenase) in the taproots and adventitious roots, besides expanding the aerenchyma of adventitious roots. Similarly, Bahadur *et al.* (2015) observed no leaf chlorosis and wilting in plant, less reduction in CCI and chlorophyll fluorescence yield (Fv/Fm), when tomato hybrid Arka Rakshak was grafted over brinjal rootstocks IC-354557 and IC-111056 and exposed to 72-96 h waterlogging situation; whereas, the non-grafted plants showed 39.6-41 per cent reduction in Fv/Fm and 41-100 per cent reduction in CCI after 96 h exposure to waterlogging condition. These findings were also corroborated by Bhatt *et al.* (2015), who reported that tomato grafted over brinjal rootstock 'Arka Neelkanth, recovered from 6 days of submergence condition during reproductive stage. Higher gas exchange, leaf water potential and chlorophyll fluorescence yield was observed in grafted plants. Grafting study in World Vegetable Centre, Taiwan concluded that wild species of chilli (*C. baccatum, C. frutescens* and *C. chacoense*) have ability to tolerate flooding situation in sweet pepper under hot-humid condition with improved growth and yield (Palad and Wu, 2008).

Salinity and Alkalinity Stress

Excess soluble salts in soil, notably in arid and semi-arid zones reduce vegetable yield in many parts of the world. Salinity has affected approximately 20-33 per cent of agricultural land across the world. Higher salinity levels caused by high accumulation of Na^+ and Cl^- ions often produce a decrease in water potential in the root media, resulting in a water deficit in plants, as well as ion-toxicity and nutritional imbalance, particularly of K^+, Ca_2^+, and Mg_2^+ by interfering with their uptake and/or transport to the shoots. Grafting aids improvement in plant growth and yield of fruit vegetables under saline conditions. Grafted plants have shown better performance under saline conditions in several cucurbitaceous vegetables, such as watermelon, musk melon and cucumber. Grafting the watermelon cv. Fantasy onto Strongtosa (*Cucurbita maxima* × *C. moschata*) showed less decrease in shoot weight and leaf area with increased salinity tolerance as compared to un-grafted plants (Goreta *et al.,* 2008). Furthermore, in watermelon (cv. Crimson Tide), higher plant growth was observed when grafted on squash (*Cucurbita maxima*) and bottle gourd (*Lagenaria siceraria*) rootstocks under saline conditions (Yetisir and Uygur, 2010). Huang *et al.* (2010) showed that the cucumber (cv. Jinchun No. 2) performance on three graft combinations *viz.,* self-grafted, grafted on fig-leaf gourd (*Cucurbita ficifolia*) and bottle gourd responded differently as to EC increases from 1.9 to 5.7 and 9.8 dSm^{-1} compared to the self-grafted plants. Cucumber plants grafted onto fig-leaf gourd had higher scion dry weight at 5.7 and 9.8 dSm^{-1}. These rootstocks could significantly alleviate scion growth reduction, maintain higher soluble sugar and manganese (Mn) contents, higher SOD and POD activities, but lower electrolyte leakage and MDA at 5.7 dSm^{-1}. Salinity study in tomato conducted by Sanwal *et al.* (2022) enhanced fruit yield of tomato (cv. Kashi Aman) by 20.3, 49.1, 24.41 and 55.9 per cent, respectively on brinjal rootstocks IC-354557 and IC 111056, over un-grafted tomato with saline water irrigation (EC 6 and 9 dS m^{-1}). They found that tolerant rootstocks mediated the partitioning of toxic saline ions in the scions by promoting higher Na^+ accumulation (24%) in the older and lower leaves than in the active younger leaves (14%) of the grafted plants. Grafted plants grown under saline conditions often exhibited better growth and yield, higher photosynthesis and leaf water content, greater root-to-shoot ratio, higher accumulation of compatible osmolytes, abscisic acid and polyamines in leaves, greater antioxidant capacity in leaves, and lower accumulation of Na^+ and/or Cl^- in shoots than un-grafted or self-grafted plants (Colla *et al.,* 2010a, 2010b). In tomato, salt stress decreased both vegetative growth and fruit yield of both grafted and non-grafted plants, however, at moderate salinity level (3.76 dS m^{-1}), no significant decrease in fruit yield was observed in grafted tomato plants over

Unifort rootstock (Al-Harbi *et al.,* 2017). Tolerance of grafted plants under salt stress condition is also dependent on the root architecture. A vigorous root system is more efficient in increasing salt tolerance. The better root system of rootstock such as, length, density, root hairs and surface area play a crucial role in ion and water uptake, thus determining salt tolerance of grafted plants (Singh *et al.,* 2020). Study conducted by Penella *et al.* (2016) revealed that salt stress (80 mM NaCl for 14 days) induced stunted growth in sweet pepper plants, while no alterations were observed in grafted sweet pepper on A25 rootstock. This may be attributed to enhanced root system resulting in maintenance of shoot and root growth with no oxidative stress under salinity conditions. Mohsenian *et al.* (2012) evaluated the impacts of five rootstocks (cultivated brinjal, datura, *S. luteum*, tobacco and tomato) under alkaline circumstances (0, 5, and 10 mM $NaHCO_3$). Plants grafted onto datura rootstock showed higher TSS, photosynthetic pigments, maximum quantum yield of PS II (Fv/Fm), and improved performance index values. Grafting study carried out by Bayoumi *et al.* (2021) revealed that grafting of cucumber plants on hybrid of *Cucurbita maxima* x *C. moschata* may mitigate adverse effects caused by salinity and heat.

Chilling and Sub-Optimal Temperature

Most chilling-sensitive fruit and vegetables, including melons, capsicum, brinjal, cucumber, tomatoes, *etc.* require a temperature range of 8 to 12°C to thrive. Many tropical or sub-tropical vegetable crops suffer from physiological abnormalities below this temperature, which can lead to permanent malfunction, cell death, and plant death depending on the severity and length of exposure. Studies related to the effects of low temperature on root systems were mostly focused on water uptake, causing chilling-sensitive species to suffer from water stress. The most common and quickly visible symptom of chilling injury is wilting of leaves. Cold-tolerant rootstocks may overcome restriction to water absorption at chilling temperatures by an increase of root hydraulic conductance, decreased induction of cell wall suberin layers, lipid peroxidation and closure of stomata (Bloom *et al.,* 2004). In tomato, rootstocks accession LA 1777 (*Solanum habrochaites* syn. *S. hirsutum,*), KNVF (the interspecific hybrid of *S. lycopersicum* × *S. habrochaites,*), and chill-tolerant lines from backcrossed progeny of *S. habrochaites* LA 1778 × *S. lycopersicum* cv. T5 are able to alleviate low root-temperature stress for different scion cultivars of tomato (Bloom *et al.,* 18; Venema *et al.,* 2008). Fig-leaf gourd (*Cucurbita ficifolia*) and bur cucumber (*Sicos angulatus*) are used as potential rootstocks for cucumber under sub-optimal temperature (Li and Yu, 2007; Zhou *et al.,* 2007). Low temperature stress disrupted osmotic regulation system, sucrose and N metabolism in bitter gourd; however, use of pumpkin and ridge gourd rootstocks alleviated low temperature (8°C) stress by maintaining osmotic regulation, gas exchange and improved sucrose and nitrogen metabolism in bitter gourd (Liang *et al.,* 2028). It has been reported that pumpkin rootstock alleviated chilling stress in cucumber seedlings by changes in physiological traits, mobile mRNA profiles, transcriptomic and metabolomic changes in above and underground tissues of the graft combinations (Liu *et al.,* 2022). Ropokis *et al.* (2019) reported that low temperature reduces pepper yield by 33-50 per cent depending upon varieties. However, grafting pepper (cv. Sammy) on 'Robusto, and 'Terrano' rootstocks increased total fruit yield by 39 and 34 per cent, respectively compared to self-grafted control. The uptake of K, Ca, Mg, N, Mn and WUE were also enhanced in grafted plants under low temperature.

High Temperature/Supra-Optimal Temperature

Vegetable production can also be constrained by high temperature during the hot-wet and hot-dry season in the lowland tropics (Palada and Wu, 70). The high temperature leads to growth reduction, a drop-in photosynthetic rate and an increase in respiration, assimilate partitioning towards the fruits, osmotic and oxidative damage, reduced water and ion uptake/movement, and cellular dehydration. Supra-optimal root-zone temperatures severely reduce root elongation and increase average root diameter (Qin *et al.,* 2007). Use of brinjal as rootstock for grafted tomato may confer a certain degree of resistance against thermal stress since brinjal is better adapted to hot arid climate and has a better tolerance against supra-optimal soil temperature (Rivero *et al.,* 2003a). In tomato, reported that heat tolerant tomato cultivar Summerset and brinjal rootstock 'Black Beauty' showed significantly higher values of chlorophyll fluorescence, greater leaf area, leaf weight, number of pollen grains and lower values of electrolyte leakage than non-grafted plants at 37/27°C (Day/Night) temperature exposure (Abdelmageed and Gruda, 1). Similarly, Lee *et al.* (2023) demonstrated potential use of 'Maxifort' rootstocks (*S. lycopersicum* x *S. habrochaites*) in enhancing thermotolerance in tomato during vegetative stage by increase in antioxidant enzyme activities, ascorbate peroxidase, superoxide dismutase, catalase, and decreased guaiacol peroxidase under heat stress (38/30°C, day/night for 14 days followed by 7 days of recovery at 25/18°C). Rivero *et al.* (2003b) revealed that grafting tomato on a heat tolerant rootstock (cv. RX-335) resulted in decreased H_2O_2 concentration as evidenced by lower oxidative stress at high temperature. H_2O_2 was most likely catalysed by antioxidant enzymes. High temperatures also promote phenolic buildup in tomato plants by accelerating biosynthesis and preventing oxidation. Similarly, tomato grafted on brinjal reduced electrolyte leakage under supra-optimal temperature stress, indicating less membrane damage and a higher ability to retain solutes, ascorbic acid and water with lower proline content (Abdelmageed and Gruda, 2009).

Heavy Metals and Pesticides

Non-nutrient heavy metals such as cadmium, arsenic, lead, and mercury have been introduced to agricultural ecosystems through industry, reclaimed wastewater, and soil amendments. These sources may result in excessively high levels of NO_3, SO_4, $H_2PO_4^-$, K^+, Ca_2^+, Mg_2^+, and metallic micronutrients (*e.g.,* Cu) in soil. Grafting fruit vegetables onto appropriate rootstocks may reduce the entry of heavy metals, excessive trace mineral buildup, and pesticide accumulation in aerial portions, hence limiting their negative impact on crop performance and human health. Grafting cucumber (cv. Akito) on commercial rootstock Shintoza (*Cucurbita maxima* × *Cucurbita moschata*) restricted uptake and translocation of Cu to shoot, thereby mitigating adverse effects of excessive Cu supply on plant biomass and fruit yield (Rouphael *et al.*, 2008). The leaf Cu concentration in grafted plants treated with Cu rich nutrient solution was reduced by 97 and 211 per cent, respectively over ungrafted plants. Savvas *et al.* (2009) found that transport of Cu to leaves and roots of tomato cv. Belladona was restricted when the plants were grafted on He-Man rootstock (*Solanum lycopersicum* × *Solanum habrochaites*). The effects of grafting on heavy metal and trace mineral concentrations in musk melon fruit was assessed (Edelstein *et al.*, 2007). Lower quantities of B, Zn, Sr, Mn, Cu, Ti, Cr, Ni, and Cd were found in fruit from grafted than from ungrafted musk melons, which were grown on the commercial Cucurbita rootstock TZ-148 and irrigated with water of sub-standard quality. Similarly, muskmelon plant exposed to different B concentrations (0.1 to 10 mgL^{-1}) showed significantly lower tissue B concentrations in plants grafted on commercial rootstock TZ-148 (*C. maxima* × *C. moschata*) than in self-rooted plants (Edelstein *et al.*, 2007). Arao *et al.* (2008) found that grafting of brinjal on *S. torvum* reduced fruit Cd concentration by 63-75 per cent in Cd-polluted soil compared to grafting on *Solanum melongena* and *Solanum integrifolium*. Persistent organic pollutants such as, aldrin, dieldrin and endrin are highly toxic, and their high bioaccumulation and persistence in the environment is a serious concern. Otani and Seike (2007) investigated influence of three *Cucurbita* spp. rootstocks on dieldrin content in grafted cucumber fruit grown in contaminated soils. Dieldrin content dropped by 50-70 per cent in cucumber fruits grafted on Yuyuikki-black (*Cucurbita mixta*) compared to those grafted on Shintosagou (*C. maxima* x *C. moschata*) and by 30-50 per cent in cucumber fruits grafted on Hikaripower-gold (*C. moschata*).

Plant Growth and Yield

In fruit vegetables, an improvement in plant growth and yield was observed in several graft combinations. Khah *et al.* (2006) noticed 11-32.5 per cent higher fruit yield in Big Red tomato grafted onto He-man and Primavera (hybrid tomatoes) than ungrafted in greenhouse and open-field conditions. Yetisir and Sari (2003) evaluated improvement in growth and yield of watermelon grafted on 10 different cucurbitaceous rootstocks. They noticed that grafted watermelon exhibited 42-180 per cent higher dry weight and 58-100 per cent larger leaf area, whereas watermelon on *Lagenaria* type rootstocks produced 27-106 per cent higher yield, while in *Cucurbita* type rootstocks 127–240 per cent reduction in yield over ungrafted control was noticed. Two squash interspecific hybrid rootstocks (Shintoza and Tetsukabuto) increased yield and fruit size of watermelon by 90 and 26 per cent, respectively compared to ungrafted plants (Miguel *et al.*, 2004). Martinez-Rodriguez *et al.* (2008) observed around 40 per cent improvement in tomato yield on grafting Moneymaker on either Radja or Pera compared to self-grafted at 50 mM NaCl. Estan *et al.* (2005) noticed 80 per cent increase in yield using same rootstock with different scion (Jaguar) at same salt concentration. Similarly, brinjal cv. Suqiqie grafted on its wild relative- *Solanum torvum* showed improved plant growth under saline stress (Liu *et al.*, 2007). Grafting has resulted in better performance in several cucurbitaceous vegetables, such as watermelon, muskmelon and cucumber. Grafting watermelon cv. Fantasy onto Strongtosa rootstock (*Cucurbita maxima* × *C. moschata*) showed lesser decrease in shoot weight and leaf area and increased salinity tolerance over ungrafted plants (Goreta *et al.*, 2008). Similarly, two muskmelon cultivars grafted on three hybrids of squash (*Cucurbita maxima* × *C. moschata*) exhibited higher fruit yield compared to ungrafted plants under saline level of 4.6 dS m^{-1} (Huang *et al.*, 2009). Similarly, higher fruit weight (22%) was obtained in watermelon cv. Crispy grafted on different interspecific hybrid, under plastic tunnel (Alan *et al.*, 2007). Crino *et al.* (2007) reported that melon plants grafted on *Cucurbita* interspecific hybrid rootstocks, especially on 'RS 841', produced fruits with 24 per cent increased weight than ungrafted plants, and plants grafted on *Cucumis melo* rootstocks. Traka-Mavrona *et al.* (2000) used a variety of *Cucurbita* spp. as rootstocks for winter melons (*Cucumis melo*). They observed that fruit yield and other qualitative and descriptive traits of grafted plants were unaffected by any rootstock, with exception of taste and texture, which showed significant decline in some stock/scion combinations. About 27 per cent higher marketable yield was obtained on grafting cucumber on pumpkin rootstock compared to non-grafted plants (Seong *et al.*, 2003). Similarly, in watermelon, seedlings grafted on bottle gourd exhibited 27-106 per cent more yield over control, whereas rootstock *Cucurbita spp.* reduced fruit yield by 127-240 per cent (Yetisir and Sari, 2003; Yetisir *et al.*, 2003) than vigorous rootstock. Likewise, Chawda (2021) noticed significant improvement in growth and yield of muskmelon and cucumber on using Summerfit (inter-specific hybrid of snap melon x acidulous melon) as a rootstock. Furthermore, Cansev and Ozgur (2010) observed that grafting of hybrid cucumber on P360 (*Cucurbita maxima* × *Cucurbita moschata*) and Arican-97 (*Cucurbita maxima* Duch.) rootstocks induced earliness and higher total yield by 53-120 and 87-209 per cent, respectively in cv. Marathon F1 and by 20-100 and 54-154 per cent, respectively in cv. Assos F1, compared to control. Maximum number of fruits and 200 per cent higher fruit yield was recorded in watermelon grafted on bottle gourd rootstock over ungrafted seedy watermelons (Khankahdani *et al.*, 2012). In contrast to these findings, Farhadi *et al.* (2016) reported a significant reduction in hybrid cucumber yield with various commercial *Cucurbita* interspecific hybrid rootstocks, *e.g.*, 44 per cent on Ghalyani, 73 per cent on 913 and 35 per cent on 64-19 rootstocks.

Fruit Quality Traits

Grafting technique affects the quality of fruits and vegetables, including physical properties, flavour and contents of health-related compounds mainly vitamin C, carotenoids and minerals (Rouphael *et al.*, 81). Reports differ on whether grafting effects are beneficial or detrimental to quality attributes, therefore the rootstock/scion combination must be carefully chosen for best fruit quality. It has been observed that grafting in watermelon induce poor fruit quality as exhibited by reduced soluble solids content, increased number of yellowish bands in the flesh, insipid taste, poor texture, higher fiber and decreased firmness (Lee and Oda, 2003). However, some other reports indicate positive effects of grafting on watermelon quality, including increased fruit firmness, TSS and lycopene content (Davis and Perkins-Veazie, 2005). Brinjal grafted on *Solanum torvum* not only showed increased marketable yield (53%), but also higher proteins (22.9%), K (7.2%), Fe (20%) and Zn (2.4%), particularly when sprayed with iodine at 100 or 300 mg L^{-1} (Consentino *et al.*, 2022a, 2022b). Sabatino *et al.* (2023) reported that grafting tomato on cherry tomato rootstock followed by spraying with selenium (2.0 and 4.0 μmol SeL^{-1}) enhanced fruit firmness, soluble solid, polyphenol, total carotenoids, ascorbic acid and lycopene content in fruits. Yetisir *et al.* (2003) reported that fruit quality (TSS, firmness, rind thickness and fruit shape) of grafted watermelon was greatly affected by grafting, but results were dependent on the rootstock used. However, Miguel *et al.* (2004) found no significant difference in soluble solids concentration of watermelon fruits grafted on *C. maxima* x *C. moschata* hybrid rootstock compared to control.

Grafting is known to influence mineral and vitamin contents in fruit vegetables. In some studies, grafting did not have any effect on vitamin C contet (Barrett *et al.*, 2012), while in others it reduced the amount of vitamin C in tomatoes (Turhan *et al.*, 2011; Nicoletto *et al.*, 2013). Furthermore, some studies reported no differences in lycopene contents of tomatoes from grafted and ungrafted plants (Nicoletto *et al.*, 2013), while others reported reductions of 6.7 per cent (Nicoletto *et al.*, 2013), 14 per cent (Krumbein and Schwarz, 2013) and 23.5 per cent (Helyes *et al.*, 2009), relative to ungrafted plants. Matsuzoe *et al.* (1996) concluded that fruit quality of Momotaro tomato was not affected by grafting over wild *Solanum spp.*, *Solanum sisymbriifolium*, *Solanum torvum* and *Solanum toxicarium*. In contrary, Vinkovic *et al.* (2011) reported that tomato (cv. Tamaris) grafted on Efialto, He-Man and Maxifort rootstocks resulted in significant decrease of vitamin C, total phenolics and antioxidant content in fruits. Turhan *et al.* (2011) revealed that grafting reduced total sugar and vitamin C content in fruits, whereas no significant difference in lycopene and pH content was observed. Gisbert-Mullor *et al.* (2020) observed that rootstock and scion combination play an important role in improving fruit yield and marketable quality of peppers. They observed significant enhancement in lycopene content and volatile compounds in both red and green fruits grafted on Niber rootstock. Proietti *et al.* (2008) reported that dehydro-ascorbate and total vitamin C contents in grafted mini-watermelon plants were higher by 13 and 7 per cent, respectively than ungrafted plants. They also reported that watermelon grafted on hybrid rootstock PS 1313 (*C. maxima* × *C. moschata*) enhanced lycopene concentration by 40 per cent. Similarly, Davis *et al.* (31,32) reported that grafting in watermelon increased lycopene and total carotenoids by 20 per cent. According to Di Gioia *et al.* (2010), vitamin C content in tomato fruit grafted onto Beaufort F1 and Maxifort F1 dropped by 14-20 per cent when compared to ungrafted treatment. Fernandez-Garcia *et al.* (39) reported that tomato fruits from grafted plants showed 2-fold higher lycopene concentration. Riga *et al.* (2016) observed

48 per cent increase in lycopene content in grafted tomatoes cv. Brigeor than in non-grafted plants, whereas self-grafted fruit registered 39 per cent increase in lycopene content. They also found that Alligator rootstock resulted in a reduction of 32 per cent amino acid and 16.5 per cent protein content, while Primed induced an increase of 19 per cent in total essential amino acids and 17 per cent in protein content in tomato fruits. Recent study on brimato conducted by Bahadur *et al.* (2024c) revealed that using brinjal rootstock IC 111056 significantly enhanced ascorbic acid and β-carotene content in tomato fruits.

Flavour in vegetables is constituted by acids, sugars, volatiles and other compounds individually or in combination which elicit specific sensory responses. Most reports on grafting suggest that changes in fruit qualities of scion are controlled by rootstock through controlled uptake, synthesis, and translocation of water, minerals, and plant hormones (Lee and Oda, 2003; Rouphael *et al.,* 2008). Lee *et al.* (1999) reported significant reduction in flavour compounds of cucumber fruits grafted onto Heukjong rootstock (*C. ficifolia*). Significant reduction in soluble solids and fructose concentration were also observed in cucumber fruits grown over different rootstocks. Proietti *et al.* (2008) demonstrated that mini-watermelon grafted on commercial hybrid rootstock PS 1313 (*C. maxima* × *C. moschata*) exhibited higher titratable acidity (TA) and TSS/TA ratio. The juice pH and acidity of musk melon grafted on pumpkin inter-specific hybrid P 360 was reported to be lower than ungrafted fruit (Colla *et al.,* 2006); whereas Crino *et al.* (2007) reported that TA of grafted melon was similar to ungrafted fruit. Furthermore, Pogonyi *et al.* (2005) reported that fruit quality parameters of tomato cv. Lemance were lower in total soluble solids, carbohydrate and organic acids concentrations when grafted on Beaufort (*S. lycopersicum* × *S. habrochaites*) rootstock than fruit obtained from ungrafted.

Mechanized/robot-aided grafting can be used to increase production rates of successful grafts (Lee *et al.,* 2010). The mechanized production of grafts can be checked for their strength using manual bending or extensometer. Formerly, mechanized grafting system have been developed for solanaceous vegetables with an hourly capacity of about 2250 and graft success of about 100 per cent. Thus, mechanized grafting may prove to be useful for giving impetus to quality production of successful grafts. This will also rationalize the excess cost of automated production, enhancing availability of grafts while reducing the requirement of skilled labour. It will also provide increased opportunities to promote rootstock breeding and grafting research. Exploration of rootstock and suitable graft combinations is required to overcome challenges associated with several viruses of orange and grape. Higher throughput through mechanized grafting may prove useful for further research on graft incompatibility in fruits crops like grape, pear, quince, apricot, cherry, passion fruit, kiwi fruit and litchi. In future, the mechanized graft production should also allow for provision to include inter-gaft to prevent graft incompatibility by double working. This will prove useful for a viable fruit grafting industry in the light of numerous new emerging agronomic, microbial and climatic challenges. The science and practise of grafting both stand to benefit immensely with the development of a robust low cost automated grafting system for woody fruit crop species.

REFERENCES

Abdelmageed AHA and Gruda N, 2009. Influence of grafting on growth, development and some physiological parameters of tomatoes under controlled heat stress conditions. *Eur. J. Hortic. Sci.* **74**(1): 6-20.

Ahmad S, Ahmad R, Ashraf MY, Ashraf M and Waraich EA, 2009. Sunflower (*Helianthus annuus* L.) response to drought stress at germination and seedling growth stages. *Pak. J. Bot.* **41**(2): 647-54.

Alan O, Ozdemir N and Gunen Y, 2007. Effect of grafting on watermelon plant growth, yield and quality. *J. Agro.* **6**(2): 362-5.

Al-Harbi A, Hejazi A and Al-Omran A, 2017. Responses of grafted tomato (*Solanum lycopersicon* L.) to abiotic stresses in Saudi Arabia. *Saudi J. Biol. Sci.* **24**(6): 1274-80.

Arao T, Takeda H and Nishihara E, 2008. Reduction of cadmium translocation from roots to shoots in eggplant (*Solanum melongena*) by grafting onto *Solanum torvum* rootstock. *Soil Sci. Plant Nutr.* **54**(4): 555-9.

Ashita E, 1927. Grafting of watermelons. *Korea Agric. Newslett.* **1**: 9.

Ashraf MA, Iqbal M, Rasheed R, Hussain I Riaz M and Arif MS, 2018. Environmental stress and secondary metabolites in plants: an overview. *Plant Meta. Reg. Env. Stress.* DOI: 10.1016/B978-0-12-812689-9.00008-X.

Bahadur A and Kumar R, 2024a. Grafting in tomato for improving abiotic stress tolerance, yield and quality traits. *Veg. Sci.* **51**(SI): 22-33.

Bahadur A, Chatterjee A, Kumar R, Singh M and Naik PS, 2011. Physiological and biochemical basis of drought tolerance in vegetables. *Veg. Sci.* **38**(1): 1-16.

Bahadur A, Jangid KK, Singh A K, Singh U, Rai KK, Singh MK, Rai N, Singh PM, Rai AB and Singh B, 2016. Tomato genotypes grafted on eggplant: Physiological and biochemical tolerance under waterlogged condition. *Veg. Sci.* **43**(2): 208-15.

Bahadur A, Rai N, Kumar R, Tiwari SK, Singh AK, Rai AK, Singh U, Patel PK, Tiwari V, Rai AB and Singh M, 2015. Grafting tomato on eggplant as a potential tool to improve waterlogging tolerance in hybrid tomato. *Veg. Sci.* **42**(2): 82-7.

Bahadur A, Rajeev K, Krishna H and Behera TK, 2023. Abiotic stress in vegetable crops: challenges and strategies. *J. Biotechnol. Biores.* **5**(1): 640-648.

Bahadur A, Singh AK, Krishna H, Kumar R, Patel VK and Behera TK, 2024b. Evaluation of brinjal and tomato scions for enhancing yield and quality attributes in grafted brimato plants. *Indian J. Hortic.* **81**(3): 92-8.

Bahadur A, Singh PM, Rai N, Singh AK, Singh AK, Karkute SG and Behera TK, 2024c. Grafting in vegetables to improve abiotic stress tolerance, yield and quality. *J. Horti. Sci. Biotech.* **99**(4): 385-403.

Barrett CE., Zhao X, Sims CA, Brecht JK., Dreyer EQ and Gao Z, 2012. Fruit composition and sensory attributes of oranic heirloom tomatoes as affected by grafting. *Hort Technol.* **22**(6): 804-9.

Bayoumi, Abd-Alkarim E, El-Ramady H, El-Aidy F, Hamed ES, Taha N, Prohens J and Rakha M, 2021. Grafting improves fruit yield of cucumber plants grown under combined heat and soil salinity stresses. *Horticul.* **7**: 61.

Bhatt RM, Upreti KK, Divya MH, Bhat S, Pavithra CB and Sadashiva AT, 2015. Interspecific grafting to enhance physiological resilience to flooding stress in tomato (*Solanum lycopersicum* L.). *Sci. Hortic.* **182**: 8-17.

Bloom AJ, Zwieniecki MA, Passioura JB, Randall LB, Holbrook NM and Clair DA, 2004. Water relations under root chilling in a sensitive and tolerant tomato species. *Plant, Cell Env.* **27**(8): 971-9.

Cansev A and Ozgur M, 2010. Grafting cucumber seedlings on *Cucurbita* spp.: comparison of different grafting methods, scions and their performance. *J. Food Agri. Env.* **8**(3/4): 804-9.

Cantero-Navarro E, Romero-Aranda R, Fernandez-Munoz R, Martinez-Andujar C, Perez-Alfocea F and Albacete A, 2016. Improving agronomic water use efficiency in tomato by rootstock-mediated hormonal regulation of leaf biomass. *Plant Sci.* **251**: 90-100.

Chawda V, 2021. Development of suitable rootstock and standardization of appropriate grafting technology for dry and humid areas of India. *Acta Hortic.* **1302**: 45-8.

Colla G., Roupahel Y, Cardarelli M and Rea E, 2006. Effect of salinity on yield, fruit quality, leaf gas exchange, and mineral composition of grafted watermelon plants. *HortSci.* **41**(3): 622-7.

Colla G, Rouphael Y and Cardarelli M, 2010a. The effectiveness of grafting to improve alkalinity tolerance in watermelon. *Env. Exp. Bot.* **68**(3): 283-91.

Colla G, Rouphael Y, Leonardi C and Bie Z, 2010b. Role of grafting in vegetable crops grown under saline conditions. *Sci. Hortic.* **127**(2): 147-55.

Consentino BB, Rouphael Y, Ntatsi G, Pasquale CD, Iapichino G, D'Anna F, Bella S L and Sabatino L, 2022a. Agronomic performance and fruit quality in greenhouse grown eggplant are interactively modulated by iodine dosage and grafting. *Sci. Hortic.* **295**: 110891.

Consentino BB, Sabatino L, Vultaggio L, Rotino GL, La Placa GG, D'Anna F, Leto C, Iacuzzi N and De PC, 2022b. Grafting eggplant onto underutilized *Solanum* species and bio-stimulatory action of *Azospirillum brasilense* modulate growth, yield and nutritional functional traits. *Hortic.* **8**(8): 722.

Coskun OF, 2023. The effect of grafting on morphological, physiological and molecular changes induced by drought stress in cucumber. *Sustain.* **15**: 875.

Crino P, Bianco CL, Rouphael Y, Colla G, Saccardo F and Paratore A, 2007. Evaluation of rootstock resistance to Fusarium wilt and gummy stem blight and effect on yield and quality of a grafted 'Inodorus' melon. *HortSci.* **42**(3): 521-5.

Darshan D and Solanki SPS, 2022. Rootstock problem and their influence in temperate fruit crops. *Planta* **4**: 912-32.

Davis AR and Perkins-Veazie P, 2005. Rootstock effects on plant vigor and watermelon fruit quality. *Rep.-Cuc. Gen. Coop.* **28**: 39.

Davis AR, Perkins-Veazie P, Hassell R, Levi A, King SR and Zhang, X, 2008a. Grafting effects on vegetable quality. *HortSci.* **43**(6): 1670-2.

Davis AR, Perkins-Veazie P, Sakata Y, Lopez-Galarza S, Maroto JV, Lee SG and Lee J M, 2008b. Cucurbit grafting. *Crit. Rev. Plant Sci.* **27**(1): 50-74.

Dhurve L, Mathew D, Kumar AK, Joseph AV and Mehara H, 2023. Rootstocks: importance in fruit crop improvement. *Intl. J. Environ. Clim. Change.* **13**(11): 4479-90.

Di Gioia BF, Serio F, Buttaro D, Ayala O and Santamaria P, 2010. Influence of rootstock on vegetative growth, fruit yield and quality in Cuore di Bue an heirloom tomato. *J. Horti. Sci. Biotechnol.* **85**: 477-82.

Djidonou D, Zhao X, Simonne EH, Koch KE and Erickson JE, 2013. Yield, water and nitrogen use efficiency in field-grown, grafted tomatoes. *HortSci.* **48**(4): 485-92.

Edelstein M, Ben-Hur M and Plaut Z, 2007. Grafted melons irrigated with fresh or effluent water tolerate excess boron. *J. Amer. Soc. Horti. Sci.* **132**(4): 484-91.

Estan MT, Martinez-Rodriguez MM, Perez-Alfocea F, Flowers TJ and Bolarin MC, 2005. Grafting raises the salt tolerance of tomato through limiting the transport of sodium and chloride to the shoot. *J. Exp. Bot.* **56**(412): 703-12.

Farhadi A, Aroeii H, Nemati H, Salehi R and Giuffrida F, 2016. The effectiveness of different rootstocks for improving yield and growth of cucumber cultivated hydroponically in a greenhouse. *Hortic.* **2**(1): 1-7.

Fernandez-Garcia N, Martinez V, Cerda A and Carvajal M, 2004. Fruit quality of grafted tomato plants grown under saline conditions. *J. Horti. Sci. Biotechnol.* **79**(6): 995-1001.

Fu S, Chen J, Wu X, Gao H and Lu G, 2022. Comprehensive evaluation of low temperature and salt tolerance in grafted and rootstock seedlings combined with yield and quality of grafted tomato. *Hortic.* **8**: 595.

Gisbert-Mullor R, Ceccanti C, Padilla YG, Lopez-Galarza S, Angeles C, Conte G and Guidi L, 2020. Effect of grafting on the production, physico-chemical characteristics and nutritional quality of fruit from pepper landraces. *Antioxi.* **9**: 501.

Goreta S, Bucevic-Popovic V, Selak GV, Pavela-Vrancic M and Perica S, 2008. Vegetative growth, superoxide dismutase activity and ion concentration of salt-stressed watermelon as influenced by rootstock. The *J. Agri. Sci.* **146**(6): 695- 704.

Haghighi M and Khosravi S, 2022. Effects of grafting on cucumber growth under flooding stress during 15 days in vegetative stage. *J. Agric. Sci. Technol.* **24**(4): 873-83.

Helyes L, Lugasi A, Pogonyi A and Pek Z, 2009. Effect of variety and grafting on lycopene content of tomato (*Lycopersicon lycopersicum* L. Karsten) fruit. *Acta Alimen.* **38**(1): 27-34.

Huang Y, Bie Z, He S, Hua B, Zhen A and Liu Z, 2010. Improving cucumber tolerance to major nutrients induced salinity by grafting onto *Cucurbita ficifolia*. *Env. Exp. Bot.* **69**(1): 32-38.

Huang Y, Tang R, Cao Q and Bie Z, 2009. Improving the fruit yield and quality of cucumber by grafting onto the salt tolerant rootstock under NaCl stress. *Sci. Hortic.* **122**(1): 26-31.

Kato C, Ohshima N, Kamada H and Satoh S, 2001. Enhancement of the inhibitory activity for greening in xylem sap of squash root with waterlogging. *Plant Physiol. Biochem.* **39**(6): 513-9.

Khah EM, Kakava E, Mavromatis A, Chachalis D and Goulas C, 2006. Effect of grafting on growth and yield of tomato (*Lycopersicon esculentum* Mill.) in greenhouse and open-field. *J. Appl. Hortic.* **8**(1): 3-7.

Khankahdani HH, Zakeri E, Saeedi G and Shakerdargah G, 2012. Evaluation of different rootstocks and grafting techniques on graft union percent, yield and yield components of watermelon cv. 'Crimson Sweet'. *World Appl. Sci. J.* **18**(5): 645-51.

Krumbein A and Schwarz D, 2013. Grafting: A possibility to enhance health-promoting and flavour compounds in tomato fruits of shaded plants. *Sci. Hortic.* **149**: 97-107. https: //doi.org/10.1016/j.scienta.2012.09.003

Lee C, Joshua TH, Qin K and Daniel IL, 2023. Physio-biochemical responses of grafted tomatoes differing in thermotolerance to heat stress and recovery. *Sci. Hortic.* **308**: 111546.

Lee JM and Oda M, 2003. Grafting of herbaceous vegetable and ornamental crops. *Horti. Rev.* **28**: 61-124.

Lee JM, Bang HJ and Ham HS, 1999. Quality of cucumber fruit as affected by rootstock. *Acta Hortic.* **483**: 117-23.

Lee JM, Kubota C, Tsao SJ, Bie Z, Echevarria PH, Morra L and Oda M, 2010. Current status of vegetable grafting: Diffusion, grafting techniques, automation. *Sci. Hortic.* **127**(2): 93-105.

Li T and Yu X, 2007. Effect of Cu^{2+}, Zn^{2+}, and Mn^{2+} on SOD activity of cucumber leaves extraction after low temperature stress. *Acta Hortic. Sinica* **34**(4): 895-900.

Liang L, Lian H, Li H, Dong Y, Tang W, Zhang R, Peng X, Li X and Tang Y, 2023. Interspecific rootstocks improve the low-temperature resistance of bitter gourd through sucrose and nitrogen metabolism regulation. *Acta Physiol. Plant.* **45**(8): 97.

Liao CT and Lin CH, 1996. Photosynthetic responses of grafted bitter melon seedlings to flood stress. *Env. Exp. Bot.* **36**(2): 167-172.

Liu W, Wang Q, Zhang R, Liu M, Wang C, Liu Z, Xiang C, Lu X., Zhang X, Li X, Wang T, Gao L and Zhang W, 2022. Rootstock–scion exchanging mRNAs participate in the pathways of amino acid and fatty acid metabolism in cucumber under early chilling stress. *Hortic. Res.* **9**: uhac031.

Liu ZL, Zhu YL, Wei GP, Yang LF, Zhang GW and Hu CM, 2007. Metabolism of ascorbic acid and glutathione in leaves of grafted eggplant seedlings under NaCl stress. *Acta Bot. Boreal-Occiden. Sin.* **27**: 1795-800.

Ma Q, Niu C, Wang C, Chen C, Li Y and Wei M, 2023. Effects of differentially expressed microRNAs induced by rootstocks and silicon on improving chilling tolerance of cucumber seedlings (*Cucumis sativus* L.). *BMC Geno.* **24:** 250.

Martinez-Rodriguez MM, Estan MT, Moyano E, Garcia-Abellan JO, Flores FB. Campos JF, Al-Azzawi M J, Flowers TJ and Bolarin MC, 2008. The effectiveness of grafting to improve salt tolerance in tomato when an 'Excluder' genotype is used as scion. *Env. Exp. Bot.* **63**: 392-401.

Matsuzoe N, Aida H, Hanada K, Ali M, Okubo H and Fujieda K, 1996. Fruit quality of tomato plants grafted on *Solanum* rootstocks. *J. Jap. Soc. Horti. Sci.* **65**(1): 73-80.

Miguel A, Maroto JV, San Bautista A, Baixauli C, Cebolla V, Pascual B, Lopez S and Guardiola JL, 2004. The grafting of triploid watermelon is an advantageous alternative to soil fumigation by methyl bromide for control of Fusarium wilt. *Sci. Hortic.* **103**(1): 9-17.

Mohsenian Y, Roosta HR, Karimi HR and Esmaeilizade M, 2012. Investigation of the ameliorating effects of eggplant, datura, orange nightshade, local Iranian tobacco, and field tomato as rootstocks on alkali stress in tomato plants. *Photosyn.* **50**: 411-21.

Mudge K, Janick J, Scofield S and Goldschmidt EE, 2009. A history of grafting. *Hortic. Rev.* **35**: 437-93.

Nicoletto C, Tosini F and Sambo P, 2013. Effect of grafting and ripening conditions on some qualitative traits of 'Cuore di Bue' tomato fruits. *J. Sci. Food Agri.* **93**: 1397-1403.

Nimbolkar PK, Awachare C, Reddy YTN, Chander S and Hussain F, 2016. Role of rootstocks in fruit production–a review. *J. Agric. Engg. Food Tech.* **3**(3): 183-8.

Otani T and Seike N, 2007. Rootstock control of fruit dieldrin concentration in grafted cucumber (*Cucumis sativus*). *J. Pest. Sci.* **32**(3): 235-42.

Padilla YG, Gisbert-Mullor R, Lopez-Galarza S, Albacete A, Martinez-Melgarejo P A and Calatayud A, 2023. Short-term water stress responses of grafted pepper plants are associated with changes in the hormonal balance. *Fron. Plant Sci.* **14**: 1170021.

Palada MC and Wu DL, 2008. Evaluation of chili rootstocks for grafted sweet pepper production during the hot-wet and hot-dry seasons in Taiwan. *Acta Horti.* **767**: 167-174.

Penella C, Landi M, Guidi L, Nebauer SG, Pellegrini E, San Bautista A, Remorini D, Nali C, Lopez-Galarza S and Calatayud A 2016. Salt-tolerant rootstock increases yield of pepper under salinity through maintenance of photosynthetic performance and sinks strength. *J. Plant Physiol.* **193**: 1-11.

Peng YQ, Zhu J, Li WJ, Gao W, Shen RY and Meng LJ, 2020. Effects of grafting on root growth, anaerobic respiration enzyme activity and aerenchyma of bitter melon under waterlogging stress. *Sci. Hortic.* **261**: 108977.

Pogonyi A, Pek Z, Helyes L and Lugasi A, 2005. Effect of grafting on the tomato's yield, quality and main fruit components in spring forcing. *Acta Alim.* **34**(4): 453-62.

Proietti S, Rouphael Y, Colla G, Cardarelli M, De Agazio M, Zacchini M, Rea E, Moscatello S and Battistelli A, 2008. Fruit quality of mini–watermelon as affected by grafting and irrigation regimes. *J. Sci. Food Agri.* **88**(6): 1107-14.

Qin L, He J, Lee SK and Dodd IC, 2007. An assessment of the role of ethylene in mediating lettuce (*Lactuca sativa*) root growth at high temperatures. *J. Exp. Bot.* **58**(11): 3017-24.

Riga P, Benedicto L, Garcia-Flores L, Villano D, Medina S and Gil-Izquierdo A, 2016. Rootstock effect on serotonin and nutritional quality of tomatoes produced under low temperature and light conditions. *J. Food Comp. Anal.* **46**: 50-9.

Rivero RM, Ruiz J M and Romero L, 2003a. Can grafting in tomato plants strengthen resistance to thermal stress? *J. Sci. Food Agri.* **83**: 1315-9.

Rivero RM, Ruiz JM, Sanchez E and Romero L, 2003b. Does grafting provide tomato plants an advantage against H_2O_2 production under conditions of thermal shock? *Physiol. Plan.* **117**: 44-50.

Ropokis A, Ntatsi G, Kittas C, Katsoulas N and Savvas D, 2019. Effects of temperature and grafting on yield, nutrient uptake, and water use efficiency of a hydroponic sweet pepper crop. *Agro.* **9**: 110.

Rouphael Y, Cardarelli M, Colla G and Rea E, 2008. Yield, mineral composition, water relations, and water use efficiency of grafted mini-watermelon plants under deficit irrigation. *HortSci.* **43**(3): 730-6.

Rouphael Y, Schwarz D, Krumbein A and Colla G, 2010. Impact of grafting on product quality of fruit vegetables. *Sci. Hortic.* **127**(2): 172-9.

Sabatino L, Bella S L, Ntatsi G, Iapichino G, D'Anna F, Pasquale CD, Consentino BB and Rouphael Y, 2021. Selenium biofortification and grafting modulate plant performance and functional features of cherry tomato grown in a soilless system. *Sci. Hortic.* **285**: 110095.

Sadeghi Z, Shamshiri MH, Soroush F and Karimi HR, 2023. Evaluation of growth, yield and elements uptake of grafted tomatoes on several solanaceous rootstocks exposed to deficit irrigation. *J. Plant Nutr.* **46**(17): 4326-39.

Sakata Y, Ohara T and Sugiyama M, 2007. The history and present state of the grafting of cucurbitaceous vegetables in Japan. *Acta Hortic.* **731**: 159-70.

Sakata Y, Ohara T and Sugiyama M, 2008. The history of melon and cucumber grafting in Japan. *Acta Hortic.* **767**: 217-28.

Sanchez-Rodriguez E, del Mar Rubio-Wilhelmi M, Blasco B, Leyva R, Romero L, and Ruiz JM, 2012. Antioxidant response resides in the shoot in reciprocal grafts of drought-tolerant and drought-sensitive cultivars in tomato under water stress. *Plant Sci.* **188**: 89-96.

Santhi VP, Kumar N, Vasugi C, Parthiban S and Masilamani P, 2020. Role of rootstocks to mitigate biotic and abiotic stresses in tropical and subtropical fruit crops: A review. *Intl. J. Chem. Stud.* **8**(5): 499-510.

Sanwal SK, Mann A, Kumar A, Kesh H, Kaur G, Rai AK, Kumar R, Sharma PC, Kumar A, Bahadur A and Singh B, 2022. Salt tolerant eggplant rootstocks modulate sodium partitioning in tomato scion and improve performance under saline conditions. *Agri.* **12**(2): 183.

Savvas D, Papastavrou D, Ntatsi G, Ropokis A, Olympios C, Hartmann H and Schwarz D, 2009. Interactive effects of grafting and manganese supply on growth, yield, and nutrient uptake by tomato. *Hortic. Sci.* **44**(7): 1978-82.

Schwarz D, Rouphael Y, Colla G and Venema JH, 2010. Grafting as a tool to improve tolerance of vegetables to abiotic stresses: Thermal stress, water stress and organic pollutants. *Sci. Hortic.* **127**(2): 162-71.

Seong K, Moon J H, Lee SG, Kang YG, Kim K Y and Seo HD, 2003. Growth, lateral shoot development, and fruit yield of white of white-spined cucumber (*Cucumis sativus* cv. Baekseong-3) as affected by grafting methods. *J. Kor. Soc. Horti. Sci.* **44**(4): 478-82.

Sharma JB, Chauhan N, Rana K and Bakshi M, 2020. Evaluation of rootstocks for temperate fruit crops-A review. *Intl. J. Curr. Microbiol. Appl. Sci.* **9**(11): 3533-9.

Shehata SA, Omar HS, Elfaidy AG, El-Sayed SS, Abuarab ME and Abdeldaym EA, 2022. Grafting enhances drought tolerance by regulating stress-responsive gene expression and antioxidant enzyme activities in cucumbers. *BMC Plant Biol.* **22**(1): 1-17.

Singh H, Kumar P, Kumar A, Kyriacou MC, Colla G and Rouphael Y, 2020. Grafting tomato as a tool to improve salt tolerance. *Agro.* **10**: 263.

Traka-Mavrona E, Koutsika-Sotiriou M and Pritsa T, 2000. Response of squash (*Cucurbita spp.*) as rootstock for melon (*Cucumis melo* L.). *Sci. Hortic.* **83**(3-4): 353-62.

Tsaballa A, Athanasiadis C, Pasentsis K, Ganopoulos I, Nianiou-Obeidat I and Tsaftaris A, 2013. Molecular studies of inheritable grafting induced changes in pepper (*Capsicum annuum*) fruit shape. *Sci. Hortic.* **149**: 2-8.

Tsaballa A, Xanthopoulou A, Madesis P, Tsaftaris A and Nianiou-Obeidat I, 2021. Vegetable grafting from a molecular point of view: the involvement of epigenetics in rootstock-scion interactions. *Fron. Plant Sci.* **11**: 621999.

Turhan A, Ozmen N, Serbeci MS and Seniz V, 2011. Effects of grafting on different rootstocks on tomato fruit yield and quality. *Horti. Sci.* **38**(4): 142-9.

Venema JH, Dijk BE, Bax JM, van Hasselt PR and Elzenga JTM, 2008. Grafting tomato (*Solanum lycopersicum*) onto the rootstock of a high-altitude accession of *Solanum habrochaites* improves suboptimal-temperature tolerance. *Env. Exp. Bot.* **63**(1-3): 359-67.

Vinkovic VI, Samobor V, Bojic M, Medic-Saric M, Vukobratovic M, Erhatic R, Horvat D and Matotan Z, 2011. The effect of grafting on the antioxidant properties of tomato (*Solanum lycopersicum* L.). *Span. J. Agri. Res.* **9**: 844-51.

Warschefsky EJ, Klein LL, Frank MH, Chitwood DH, Londo JP, von Wettberg EJ, and Miller AJ, 2016. Rootstocks: diversity, domestication, and impacts on shoot phenotypes. *Trends Plant Sci.* **21**(5): 418-37.

Yan G, Feng M, Lin W, Huang Y, Tong R and Cheng Y, 2022. Review and prospect for vegetable grafting robot and relevant key technologies. *Agri.* **12**: 1578.

Yetisir H and Sari N 2003. Effect of different rootstock on plant growth, yield and quality of watermelon. *Aus. J. Exp. Agri.* **43**(10): 1269-74.

Yetisir H and Uygur V, 2010. Responses of grafted watermelon onto different gourd species to salinity stress. *J. Plant Nutr.* **33**(3): 315-27.

Yetisir H, Sari N and Yucel S, 2003. Rootstock resistance to *Fusarium* wilt and effect on watermelon fruit yield and quality. *Phytopar.* **31**(2): 163-9.

Zhou S, Wei S, Boone B and Levy S, 2007. Microarray analysis of genes affected by salt stress in tomato. *Afr. J. Env. Sci. Technol.* **1**(2): 14-26.

Chapter 16

Nursery Accreditation: Its Importance, National Status and Nursery Quarantine

B.N.S. Moorty

Former Director, Indian Institute of Horticulture Research; Former Horticulture Commissioner, GOI, New Delhi

Accreditation plays a critical role in ensuring the highest standards of quality and reliability in fruit nursery operations. This chapter explores the multifaceted importance of accreditation, encompassing genetic purity, compliance with industry standards, traceability, customer confidence, market access, and continuous improvement initiatives. Horticulture production is a cornerstone of agricultural development, playing a crucial role in ensuring food security, nutritional diversity and economic growth in the country. It encompasses the cultivation of fruits, vegetables, flowers, and ornamental plants, contributing significantly to the agricultural GDP and providing livelihoods to millions of farmers and workers (Dhillon and Gill, 2004).

Importance of Horticulture Production

Horticulture production is vital for several reasons:

- **Nutritional Security:** Horticultural crops are rich in vitamins, minerals and antioxidants, essential for a balanced diet and overall health (Aruoma, 2003).

- **Economic benefits:** Horticulture generates higher income per unit area compared to other crops, thus boosting farmers' income and contributing to rural development (Singh and Kalloo, 2000).
- **Employment Generation:** The sector provides substantial employment opportunities in farming, processing, packaging, and distribution.
- **Export potential:** Horticultural products, particularly fruits, vegetables, and flowers, have significant export potential, contributing to foreign exchange earnings (Pretty, 2008).
- **Sustainable agriculture:** Horticulture promotes sustainable agricultural practices through efficient use of resources, crop diversification, and reduced chemical inputs.

India is looking for bringing in an initiative for supply of quality planting material for the benefit of farmers in a project mode. The project aims at improving target crop farmers' access to disease-free planting materials. The project focuses on increasing the propagation of disease-free planting materials of target crops through monitoring, and expanding disease-free planting material sales. In the long-term, this will contribute to improving yields and quality of horticulture crops in India.

Quality Planting Material

Quality Planting Material (QPM) is an essential input in horticulture ecosystem for maximizing revenues, improving adaptability to adverse environmental conditions, and meeting quality requirements of markets. The quality of planting material is determined by origin, the authenticity of variety and stock, vegetation development, and health state in terms of inherent pathogens. Most of the horticultural plants, particularly the fruit trees, are perennial in nature. And hence, any mistake done in selection of plants for orchard establishment would prove very costly, both monetarily and spatially. Most of the horticultural crops are propagated vegetative for which nursery units are necessary. At present the planting material requirement is met by nurseries established under State Department of Horticulture/Agriculture, the State Agriculture Universities (SAUs) and ICAR Institutes. However, these nurseries in public domain provide only 30-40 per cent of the demand for planting material. The major part of the demand is met by the unregulated private nurseries, which lacks modern infrastructure such as green house, mist chamber, efficient nursery tools and gadget, implements and machinery. Establishment of a network of well-equipped nursery to ensure the availability of good quality, disease free, certified planting material of desired high yielding variety will have a tremendous impact on production, productivity and quality of the fruits produced. The Ministry of Agriculture and Farmers Welfare, Government of India has recognized the inadequate availability of genuine quality planting material of important fruit crops for orchard establishment leading to frustration of farming community and non-availability of produce for domestic and export markets in sufficient quantity and quality. It is estimated that only a limited quantity of planting material is produced by private and government nurseries and a voluntary system in place for accreditation of them by Notational Horticulture Board on the basis of Nursery Registration Act approved in 1954, to produce planting materials these nurseries. However, the current guidelines for nursery accreditation does not check that produced planting materials are disease-free. Nurseries are rated on a scale of no star (not eligible) to three stars (good) based on visual inspection only and focuses on the presence of facility infrastructure. Further, as plant diseases of various types causes direct and indirect losses in terms of reduced quality and quantity of crop produce, increased cost of production, threat to animal health and environment, limiting the type of crops/varieties grown, loss of natural resources it is necessary to check the propagules are disease free to begin with since, success in disease management, as in most walks of life, depends on having right tools and proper timely implementation of the same. One such tool is the starting of the orchards with disease free planting material and the clean plant certification program aims in this direction.

Why Certification?

As considerable increase in trade and travel in the recent past due to globalization has resulted in a significant rise in the number of plant pathogens affecting plant health due to the movement of plant materials. The impact of pathogens on vegetative growth and fruit production and quality is well known. In view of this, certification of planting material has received considerable attention worldwide including in India. It is the process of incorporating value addition encompassing true-to-type and plant health features to the planting material, particularly in clonally propagated fruit crops. It has been defined as the production of uniform, healthy, disease-free planting material raised through vegetative methods with an overarching goal of improving the physiological and phytosanitary quality of plants (APAARI, 2019). Certification not only ensures the production of quality planting material but also promotes profitable agricultural diversification with increased farmer income. Realizing the importance of quality planting materials, a brain storming session was organized by the National Academy of Agricultural Sciences (NAAS) in 2022. Subsequently, Government of India announced clean plant programme (CPP) in collaboration with Asian Development Bank (ADB) in February 2023 to provide disease free planting materials to farmers. Participation in such programs not only benefits growers by enhancing the marketability of their products but also provides consumers with assurance regarding the quality, sustainability, and safety of the crops they purchase. Further, it prevent the spread of regulated pests outside of regulated areas. Would help in certify nurseries for domestic movement of plants. Implement and document best management practices related to pest management for production of nursery stock and plant traceability for production of nursery stock in the country.

The Process

Nursery Certification involves a formal recognition or certification process for nurseries, which are establishments that propagate and cultivate plants, especially for sale or transplantation. Accreditation ensures that nurseries adhere to specific standards and guidelines set by relevant authorities. Plant Health Certification standards vary by region and are often established and regulated by national or international plant protection organizations. It's important for growers and nurseries to be familiar with and adhere to the specific Plant Health Certification standards applicable to their region. Local agricultural authorities and plant protection organizations can provide detailed information and guidance on these standards. A Crop-Specific Certification Program is designed to ensure that a particular crop is produced in accordance with specific standards and practices to meet quality, safety, and sustainability criteria. The specifics can depend on the type of plants, the intended use of the planting material, and the potential risks associated with the spread of diseases. However, some common elements found in Plant Health Certification standards include:

- ✰ **Disease-free status:** Certification typically requires that the planting material is free from specific quarantine pests and diseases. This is often determined through inspections, testing, and other phytosanitary measures.
- ✰ **Traceability:** Certification programs often emphasize the need for traceability, ensuring that the origin of the planting material can be identified and tracked. This is crucial for managing disease outbreaks and preventing their spread.
- ✰ **Phytosanitary inspections:** Rigorous inspections are conducted to assess the health and quality of the planting material. Inspectors look for signs of diseases, pests, and other factors that could compromise plant health.
- ✰ **Testing for pathogens:** Laboratory testing may be required to detect the presence of specific pathogens. This can include testing for bacteria, viruses, fungi, and other harmful organisms.
- ✰ **Quarantine measures:** Certification programs often involve quarantine measures for incoming planting material. This allows for a period of observation and testing to ensure that the material is free from diseases before it is released for general use.
- ✰ **Sanitation protocols:** Certification standards may specify sanitation protocols for the production, handling, and transportation of planting material. This includes guidelines for cleaning equipment and facilities to prevent contamination.
- ✰ **Phytosanitary certificates:** Once the planting material meets the required standards, a phytosanitary certificate is issued. This document confirms that the material has been inspected, tested, and found to comply with plant health regulations.

- **Compliance with international standards:** Some certification programs align with international standards set by organizations like the International Plant Protection Convention (IPPC). This ensures consistency and facilitates international trade.
- **Record keeping:** Comprehensive record-keeping requirements are often part of certification standards. This includes documentation of all activities related to the production, testing, and movement of planting material.
- **Training and accreditation:** Individuals involved in the production and inspection of planting material may need to undergo training and accreditation to ensure they have the necessary knowledge and skills to comply with certification standards.

Certification: Its Prevalence

Among biotic agents, viruses and viroids are the major constraints affecting the cultivation of important fruit crops. Despite being simple pathogens with a small genome, these are the most difficult pathogens to detect and manage. Although few viruses/viroids have been recorded so far, yet the possibility of occurrence of others reported elsewhere cannot be ruled out. Hence there is a need to generate region-wise viruses/viroids profile for different crops using next generation sequencing to know the unknown, leading to the identification of targeted pathogens. Recent studies on virome profiling of apple and grape has shown the presence of viruses/ viroids not previously recorded.

Importance of Accreditation

Accreditation serves as a pivotal mechanism for maintaining genetic purity, adherence to industry standards, fostering customer confidence, facilitating market access, and driving continuous improvement in fruit nursery operations. By upholding rigorous standards and practices, accredited nurseries contribute to the integrity and sustainability of the fruit production industry (Chadha and Mishra, 2002).

Accreditation Status of Nurseries by the National Horticulture Board

According to the National Horticulture Board, a total of 692 nurseries have been accredited. Among these, 11 nurseries have been recognized as three-star nurseries, 67 as two-star nurseries, and 614 as one-star nurseries.

Overview of Fruit Nursery Operations

Fruit nurseries are vital components of the agricultural supply chain, responsible for propagating high-quality planting materials essential for fruit production. Accredited nurseries prioritize genetic purity, comply with industry standards, maintain traceability, instill customer confidence and strive for continuous improvement to meet the evolving needs of growers and consumers alike.

Genetic Purity and Propagation

Accredited fruit nurseries prioritize genetic purity through meticulous selection of parent plants and controlled pollination techniques. By preventing genetic contamination and divergence, these nurseries ensure that propagated plants retain desired characteristics and traits consistent with their intended use (Hancock and Siefker, 1982).

Significance of Genetic Purity

Genetic purity is paramount in fruit nursery operations to preserve the integrity and quality of plant materials. Accredited nurseries employ rigorous selection criteria and propagation methods to maintain genetic consistency and enhance the overall performance of fruit crops.

Broad Points for Successful NCP-CPP

- Effective networking of clean plant centres (CPCs), accredited test laboratories (ATLs), state agricultural universities (SAUs), and registered nurseries.
- Establishment of crop-based CPCs with state-of-the-art facilities and competent human resources. Besides, facilities at ATLs, SAUs, and registered nurseries need to be upgraded and equipped.
- Development of region-wise virus profile for identified crops to identify target pathogens.
- Preparedness and continuous refinement for diagnostics and elimination protocols for targeted viruses and development of suitable SOPs.
- Establishment of foundation stock of clean source plants. Proper labeling with a unique identification number (Bar/QR code).
- Revisiting of the accreditation system of nurseries by NHB to ensure availability and supply of virus-free and true-to-type planting materials.
- Proper indexing of exotic planting material under PEQ setup.
- Competence building of human resources available at CPCs, ATLs, SAUs, and nurseries.
- Community approach development and impact assessment.

The Clean Plant Certification Standards are applicable to all fruit crops which are eligible for certification, and with field and propagules standards for the individual crops, shall constitute the Minimum clean plant certification standards. The purpose of clean plant certification is to maintain and make available to the farmers, through certification, high quality propagating materials of crops and varieties so grown and distributed as to ensure genetic fidelity and is also designed to achieve prescribed standards. Certification shall be conducted by the Certification Agency designated by National Horticulture Board, Government of India. Certified Clean plant Nursery is an agency/organisation who produces and/or distributes certified propagules in accordance with the procedures and standards of the clean plant certification program.

Structure of NCP-CPP

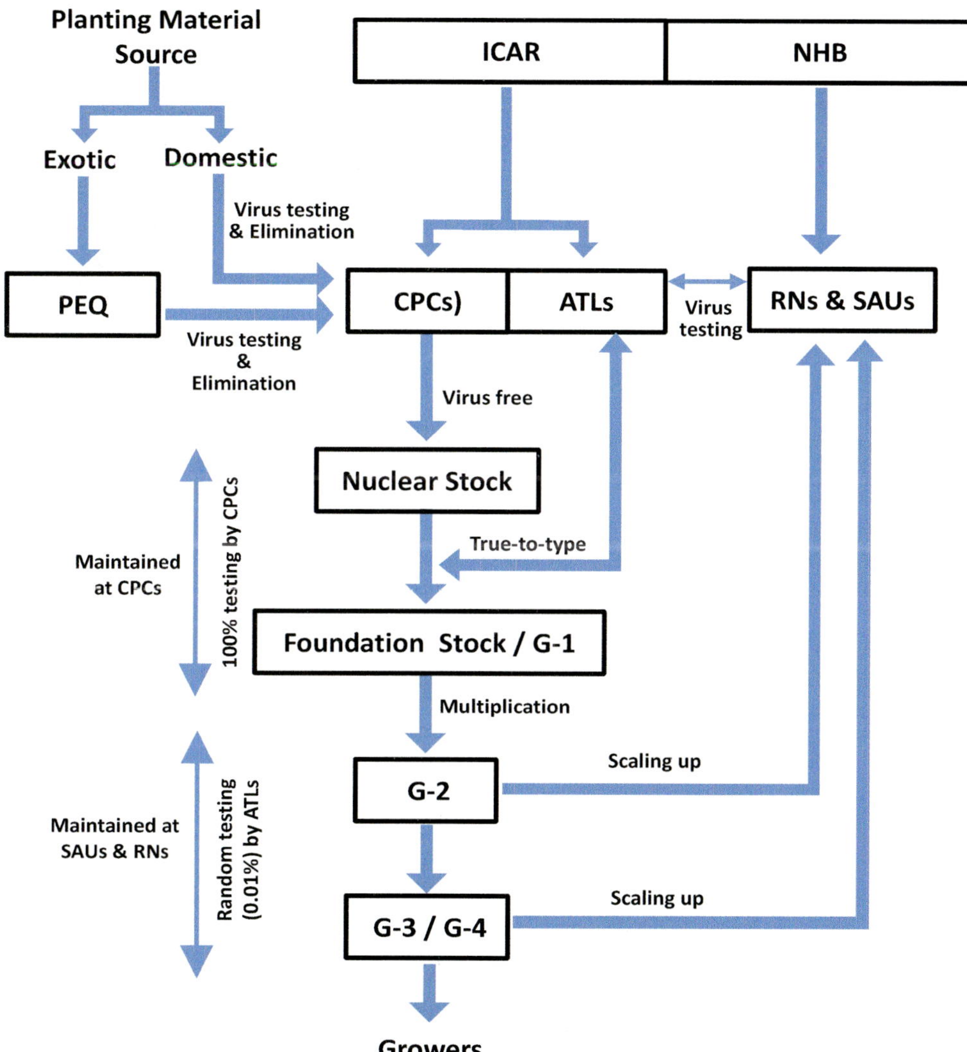

ATLs: Accredited Test Laboratories; CPCs: Clean Plant Centers; ICAR: Indian Council of Agricultural Research; NHB: National Horticulture Board; PEQ: Post Entry Quarantine; RNs: Registered Nurseries; and SAUs: State Agricultural Universities.

Initial planting materials tested free from viruses/viroids at CPCs would constitute nuclear stock and if required will also be tested for true-to-type at ATL. Such planting materials, usually five to ten of each accession depending upon plant species, would be designated as Foundation Stock (G-1) and will be kept in a controlled set-up free from insect vectors. The planting materials from the G-1 block will then be propagated in the field under net-house conditions and referred to as the G-2 materials. The clean planting material from the G-2 block will be taken by registered nurseries or SAUs for further multiplication in G-3 block and subsequently in G-4 block depending upon the quantum of materials required. Planting materials from G-3 or G-4 will be provided to growers with registration-cum-certification NHB label with a unique barcode/QR code containing about initial planting material source, quality attributes and genetic trueness.

Planting materials at all generation levels (G1-G4) will have to be maintained in an insect-free environment. ATLs will be responsible for developing and updating rapid and reliable diagnostic protocols for the targeted viruses/viroids. Subsequently, crop-specific standard operating procedures (SOPs) will be developed. Initially, it is proposed that entire planting materials (100%) of the foundation block (G1) will be indexed once or twice a year depending upon vector transmissibility in the area. Further, the plants of generation levels G2-G4 will be subjected to indexing through random stratified sampling. Plants tested positive or found showing symptoms at any time would be immediately removed to check the virus spread.

Schematic Diagram of National Clean Plant Program for Clonally Propagated Crops

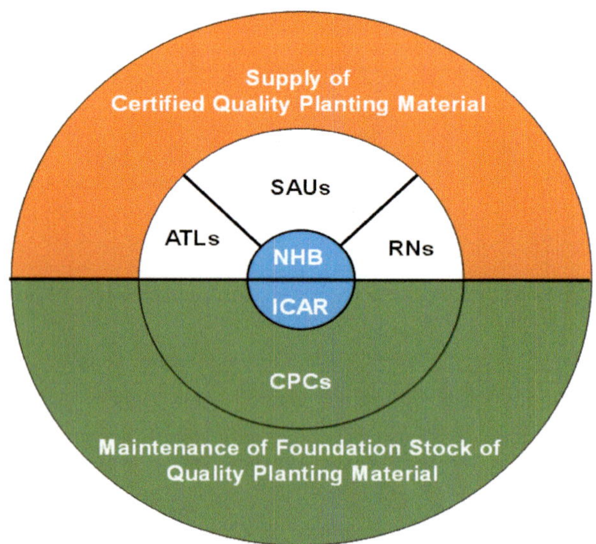

ATLs: Accredited Test Labs; CPCs: Clean Plant Centers; ICAR: Indian Council of Agricultural Research; NHB: National Horticulture Board; RNs: Registered Nurseries; and SAUs: State Agricultural Universities.

Structure of NCP-TCP

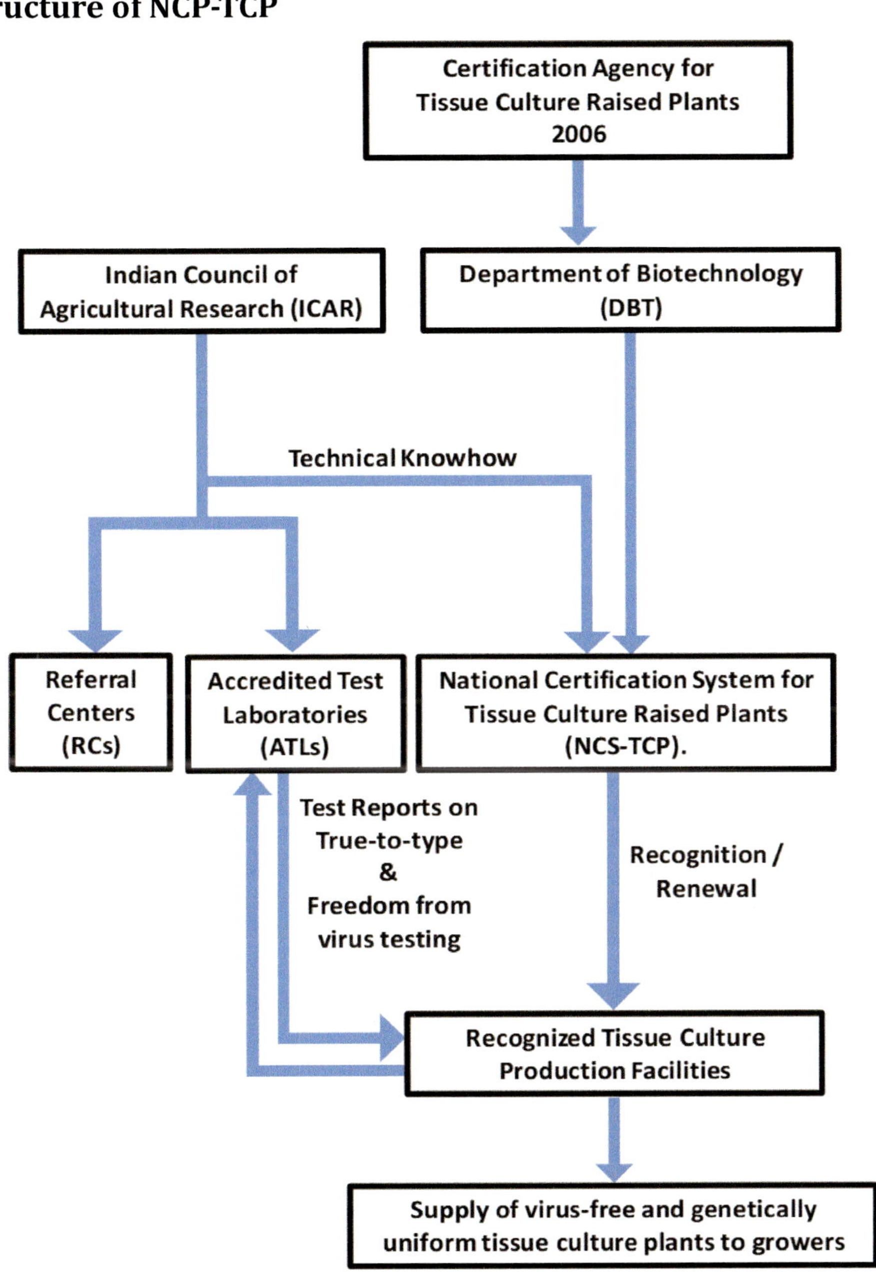

NCP-TCP National Clean Plant Program for Tissue Culture Raised Plants.

Scope: The scope of this program is to establish a network of quality nurseries across the country for the purpose of propagation and distribution of genuine quality disease-free planting materials of specified fruit crops.

The recognition shall be accorded to the nurseries for:

a. Production of quality planting material of one or more specified crops by evaluating the good Nursery Management Practices adopted by them.
b. The mother plants for such propagules production should have been obtained by G1 centers, established for the supply of genuine disease free planting material.
c. Nursery premise only where sale of specified quality planting material of recognized source are being carried out by creating necessary infrastructure facilities and proper record keeping.

   Note: If different premises are being used for nursery, recognition of all these are required to be obtained separately and individually.
d. Recognition of nursery with NHB shall be period specific.

Sampling Strategy for Certification

Number of samples to be tested will vary based on the batch size. The sampling details are given below:

Batch Size (number of plants/propagules)	*Number of Plants to be Sampled (sample size)*
Up to 1000	1 per cent plants subject to a minimum of 10
1001 to 10000	0.5 per cent of plants subject to a minimum of 10
10001 to 10000	0.1 per cent of plants subject to minimum of 50
10000 and above	0.05 per cent subject to minimum of 75

Genetic Fidelity Testing

For genetic fidelity testing each sample should contain minimum 10g of tissue.

Certification process step by step:

1. Application submission by nursery - The fruit crop nursery seeking NHB Recognition under CPP shall apply in Form- I
2. Payment of fees by the nursery
3. Preparation of Clean Plants Nursery document by the nursery
4. Implementation of the Clean Plants Nursery document at the nursery
5. Evaluation and approval of:
   - ✰ the application,
   - ✰ the initial facility evaluation
   - ✰ the Clean Plants Nursery document
6. Audit – Initial Facility Evaluation by approved auditors
7. Signed Compliance Agreement between the nursery and the NHB

8. Certification is granted by NHB to the nursery
9. An external audit must be completed the year following the initial facility evaluation
10. Maintenance – annual internal audits to be submitted to NHB, external audits (determined randomly over a three year period), compliance with corrective action requests within the agreed time line,
11. Issuance of annual certification status by NHB – electronic stamp with nursery specific certification number and a list of pest specific modules, for which the nursery is certified.

(Suspension of Certification, if required, is done by the NHB.)

Application

Each such application shall be accompanied by following details:

a. Layout of nursery showing location of infrastructure components and land utilization plan.
b. Details of technically qualified staff in the nursery.
c. Major farm machineries and equipment's in use.
d. Operational manual of procedures followed for selection and maintenance of mother plants, nursery production techniques and management of planting materials.
e. Details regarding source of mother plants used for propagation of each crop
f. Processing fee in the form of a Demand Draft in favour of Department of Agriculture and Cooperation is also to be enclosed along with the application.

The nurseries should also maintain separate registers for production (stock), sale of seedlings and attendance of staff. Duly filled applications with processing fee in the form of a demand draft should be submitted. Each application will be considered by NHB based on criteria specified for recognition of nursery. The recognition of nursery by NHB shall be considered product wise as required.

On receipt of the application for recognition, inspection and assessment will be done by the Technical Committee appointed by NHB shall be conveyed to the concerned nursery. In case of recognition, the period of validity of recognition with other terms and conditions shall be indicated. In case of any deficiency, the nursery shall be given time frame for compliance; failing which, the application shall be rejected. In case of requirement, additional assessment visits may be undertaken.

Assessment Committee

NHB shall nominate one Assessment Committee with a Consultant as head, which may have representatives from:

i. State Agricultural University in the State concerned
ii. Apex Horticulture Institute, NRC for Regional Station of lCAR
iii. Nursery Men Association
iv. NHB
v. State Directorate of Horticulture/Agriculture

Assessment Committee shall conduct pre-assessment, final assessment and periodic/surveillance of the nurseries. The committee shall submit report and their recommendations to NHB.

Assessment Criteria

The criteria have been aligned with infrastructural requirement of Model Nursery and Product specific technical requirement and adoption of Good Nursery Practices for propagation of good quality planting materials as detailed below:

a) Product specific criteria to asses for capability and competence of nursery to follow technical programme for specific crop as laid down in guidelines
b) Prescribed Nursery Management Practices and Adoption of Model layout plan. The assessment will be based on the continuous evaluation of source of parent material, propagation in disease free condition by adoption of technically prescribed method, adoption of Good Nursery Management Practices, reliable record keeping and training of staff. Each parameter will be critically examined by assessment team as per laid down criteria.

Validity Period of Recognition

In case of approval, initially, recognition shall be granted for a period of three years. The effective date of recognition shall be considered from the date of issuance of certificate.

Surveillance/Monitoring

Surveillance visit shall be carried out at least once in a year or as and when required depending on the performance of the nursery. The assessment team will conduct surveillance visit during validity of recognition.

Issue of Recognition Certificate

In case, Technical Committee satisfies that the nursery conforms to the requirements of this recognition, it will recommend for recognition of the nursery. The recognition shall bear an identification number. The Recognized Nursery shall have to sign a Memorandum of Understanding (MoU) with NHB for terms and conditions of recognition. Any change in the location, layout, design or capacity of the nursery shall be intimated toNHB. The date of validity of the nursery recognition shall be specified on the recognition certificate. The recognition certificate once issued shall continue to be in force till the date specified in the certificate unless suspended or cancelled at earlier date by Competent Authority. The issuing Authority may institute surprise checks through its officers in order to ensure that the nursery has maintained the standard as required for issue of the recognition certificate.

Renewal of Recognition

The Nursery seeking renewal of recognition shall apply in prescribed form. The renewal of recognition shall be done based on the satisfactory performance reported as per the surveillance/periodic inspection carried out during the validity period and fresh assessment if felt necessary.

Refusal/Cancellation of Nursery Recognition

Issue of recognition certificate may be refused or, if issued, may be cancelled or suspended.

a. If the nursery does not conform/fail to perform as per requirements of this scheme.
b. If there are adverse reports from the farmers/users or any other complaints made to NHB by any other entity and upon enquiry duly conducted it is established that the Nursery has breached any of the conditions of recognition.
c. On expiry of the recognition date specified in the Recognition Certificate the recognition ceases to be valid unless renewed.

Appeal Against Refusal/Cancellation of Recognition

Appeal against refusal/cancellation of recognition should reach to the Project Director, CPP within 30 days from the receipt of such order. In case of cancellation of recognition. The appeal shall accompany the original certificate. Project Director, CPP will consider the application on merit. and in case, found necessary, order re-assessment of the nursery.

Promotion of Recognized Nursery

NHB shall host list of recognized nurseries in its website and promote use of its planting materials through schemes implemented by it.

Procedure for Accreditation of Nurseries

1. Application for accreditation in prescribed form with Fee
2. Appraisal of application by the Accreditation Agency based on prescribed criteria
3. Monitoring of performance of nurseries (withdrawal of certificate on breach of conditions)
4. Inspection of nursery site by an expert(s) nominated by NHB
5. Report on the status in the prescribed format and deficiencies, if any

6. Reappraisal, if required, after recommendations are implemented
7. Certification of accreditation issued in prescribed format
8. Renewal of accreditation

Procedure for Certification of Quality Planting Material

1. Selection of mother plant/Generation of passport data
2. Confirmation of species/Clonal identity (DNA barcoding/fingerprinting)
3. Certification of mother plant/Virus indexing
4. Maintenance in seed/Clonal orchard
5. Sexual/clonal multiplication in Accredited Nursery
6. Assessment of planting material for nursery quality parameters
7. Issue of Certificate

Clean Plants Certification Stamp

The nurseries which have successfully completed certification process would be provided with an electronic version of the Clean Plants Stamp which is known as the Certification Seal showing details of certification details including the identity number and validity period would be issued to the nursery. The nurseries have to use the stamp electronically to "stamp" invoice documents. The Clean Plants Certification Stamp must be used only as authorized.

Standards of Planting Material Production

The vegetative propagation of fruit crops makes them vulnerable to transmission of several diseases and pests through the plant material. Thus, importance of testing of material in the process of its preparation at various stages needs due attentions. The targets of the enhancing fruit. production in the coming years will be achieved only through production and distribution of healthy, genuine and high quality planting material of commercial/improved varieties of fruit crops in sufficient quantities. The maintenance of purity is easy in vegetatively propagated fruit crops as compared to seed propagated ones, still it requires a close monitoring at different stages in the nursery to avoid mixing with other varieties. Similarly, adequate measures are taken in the preparation of plant material to produce disease and pest free plant material.

Progeny Trees/Mother Plants

The bud sticks/graft wood should always be taken from healthy and true to type progeny trees of commercial/new varieties, which are free from viruses, disease and pest occurrence. A nurseryman should have progeny trees of all the promising cultivars of fruits that can be grown in that particular area.

Criteria for Selection of Mother Plants

Mother plants of the variety should be genetically true to type.

a. The plants should be healthy and free from any diseases; pest infestations and physiological disorder.
b. The plants should have known pedigree records regarding bearing potential, fruit quality and problems, if any.
c. The plants should be precocious and prolific bearer.
d. They should be sourced from G1 CPP centers

Criteria for Selection of Rootstocks

a. Dwarfing/semi-dwarfing in nature
b. Compatibility with the known commercial variety
c. Resistance/tolerance to biotic (diseases and pests) and abiotic stresses.
d. Rootstock should have well developed root system.
e. It should be easy to propagate rootstock by vegetative means

Preventing Genetic Contamination

Accredited nurseries implement strict protocols to prevent genetic contamination from external sources. By isolating parent plants, monitoring pollination processes and implementing quality control measures, these nurseries minimize the risk of unintended genetic changes and ensure the reliability of propagated plant materials.

Compliance with Industry Standards

Accreditation mandates compliance with industry standards governing various aspects of nursery operations, including propagation methods, nursery management practices, pest and disease control strategies and sustainable cultivation practices.

- **Nursery operation standards:** Accredited nurseries adhere to established standards for nursery operations, encompassing all stages of plant production from propagation to distribution. By following best practices and industry guidelines, these nurseries ensure the ethical and responsible management of plant materials and minimize adverse environmental impacts.
- **Sustainable cultivation practices:** Sustainable cultivation practices are integral to accreditation requirements, emphasizing resource conservation, biodiversity preservation, and ecosystem resilience. Accredited nurseries implement strategies such as integrated pest management, water conservation measures, and soil health management to promote environmental sustainability and long-term viability.

- **Environmental impact mitigation:** Accredited nurseries mitigate environmental impacts associated with fruit production through responsible land use, waste management, and pollution prevention measures. By adopting sustainable practices and minimizing their ecological footprint, these nurseries contribute to the preservation of natural resources and the protection of ecosystems.
- **Traceability and documentation:** Accredited fruit nurseries maintain robust systems for traceability and documentation of plant stock, ensuring transparency and accountability throughout the supply chain (Smith and Roosen, 2006).
- **Importance of traceability:** Traceability is essential for tracking the origin and movement of plant materials, identifying potential sources of contamination, and maintaining the integrity of products supplied to customers. Accredited nurseries implement traceability systems to provide accurate information about the parentage, propagation techniques, growth conditions, and treatments applied to plants, ensuring transparency and accountability.
- **Documentation in accredited nurseries:** Accredited nurseries maintain detailed records of all aspects of nursery operations, including plant propagation, cultivation practices, pest and disease management, and quality control measures. Comprehensive documentation enables nurseries to demonstrate compliance with accreditation requirements, facilitate regulatory inspections, and respond effectively to customer inquiries or concerns.
- **Transparency and accountability:** Transparency and accountability are fundamental principles of accreditation, fostering trust and confidence among customers, stakeholders, and regulatory authorities. Accredited nurseries proactively disclose information about their practices, policies, and performance, enabling stakeholders to make informed decisions and hold nurseries accountable for their actions.
- **Customer confidence and market access:** Accreditation instills confidence among customers, including commercial growers, farmers and orchardists, who rely on fruit nurseries for high-quality planting materials. The assurance of accreditation mitigates risks associated with crop failure or yield losses attributable to substandard plant material, thereby fostering trust and long-term relationships within the industry.
- **Building customer trust:** Customer trust is essential for the success and sustainability of fruit nursery operations. Accredited nurseries demonstrate their commitment to quality, reliability, and professionalism, earning the trust and loyalty of customers through consistent performance and adherence to accreditation standards.

- **Market advantage through accreditation:** Accreditation provides nurseries with a competitive advantage in the marketplace by enabling access to lucrative markets with stringent phytosanitary regulations and quality standards. Nurseries accredited to international standards are better positioned to meet regulatory requirements, expand their market reach, and enhance competitiveness on a global scale (Garcia Martinez and Poole, 2004).
- **Global market reach:** Accreditation facilitates market access and trade opportunities for accredited nurseries, enabling them to compete in domestic and international markets. By demonstrating compliance with recognized standards and certifications, accredited nurseries gain access to premium markets and premium prices, enhancing their profitability and sustainability in the long term.
- **Continuous improvement:** Accreditation fosters a culture of continuous improvement within fruit nurseries, driving innovation, efficiency, and excellence in all aspects of operations (Deming, 1986).
- **Culture of innovation:** Innovation is essential for maintaining competitiveness and relevance in the dynamic fruit production industry. Accredited nurseries embrace innovation by investing in research, technology, and best practices to enhance productivity, quality, and sustainability throughout the supply chain.
- **Monitoring and evaluation:** Accredited nurseries continuously monitor and evaluate their performance against accreditation standards, identifying areas for improvement and implementing corrective actions as necessary. Regular assessments, audits, and feedback mechanisms enable nurseries to track progress, measure outcomes, and drive continuous improvement initiatives.

Adaptation to Emerging Trends

The fruit production industry is constantly evolving, driven by changing consumer preferences, market dynamics, and technological advancements. Accredited nurseries remain agile and adaptable, embracing emerging trends such as organic farming, digital agriculture, and precision horticulture to meet evolving customer needs and market demands.

- **Recognition of spice nurseries:** The significance of spice nurseries within the broader agricultural landscape underscores the need for specialized recognition criteria and standards tailored to their unique requirements.
- **Importance of spice nurseries:** Spices represent a valuable segment of agricultural production, contributing to culinary, medicinal, and cultural practices worldwide. Spice nurseries serve as the foundation for spice

cultivation, providing growers with high-quality planting materials essential for the production of diverse spice varieties.

- **Model nursery components:** Recognizing the diversity and complexity of spice cultivation, model spice nurseries incorporate specialized infrastructure, propagation techniques, and management practices optimized for different spice species. These model nurseries serve as benchmarks for excellence, showcasing best-in-class methodologies and technologies for spice propagation and cultivation.
- **Establishment and recognition criteria:** Establishing recognition criteria for spice nurseries involves assessing factors such as genetic diversity, propagation techniques, pest and disease management strategies, and adherence to quality standards. Recognized spice nurseries must demonstrate proficiency in spice propagation, cultivation, and management, meeting predefined benchmarks for quality, reliability, and sustainability.
- **Scheme scope and application procedure:** The scope of accreditation schemes encompasses a wide range of nursery operations, including fruit, spice, and horticultural nurseries. Understanding the scheme's objectives and application procedures is essential for nurseries seeking accreditation and recognition within their respective sectors.

Objectives of the Scheme

Accreditation schemes aim to uphold industry standards, promote best practices, and ensure the integrity and reliability of nursery operations. By establishing clear objectives and performance criteria, these schemes facilitate continuous improvement and drive excellence across the nursery sector.

- **Recognition criteria and application process:** The recognition criteria for accreditation schemes outline the requirements and benchmarks that nurseries must meet to qualify for accreditation. Nurseries seeking accreditation must undergo a rigorous application process, which typically involves submission of documentation, on-site inspections, and assessment of compliance with accreditation standards.
- **Assessment and oversight:** Accreditation schemes involve ongoing assessment and oversight to maintain the integrity and credibility of accredited nurseries. Regulatory bodies or accreditation agencies oversee the accreditation process, conduct periodic inspections, and monitor compliance with accreditation standards to ensure continued adherence to quality and performance requirements.

Horticulture Nursery Recognition Scheme

The Horticulture Nursery Recognition Scheme addresses the unique needs and challenges of horticultural nurseries, encompassing a diverse range of plant species, cultivation practices, and market dynamics.

- **Bridging regulatory gaps:** The Horticulture Nursery Recognition Scheme fills regulatory gaps by establishing standardized criteria and performance benchmarks for horticultural nurseries. By providing clear guidelines and expectations, the scheme promotes consistency, transparency, and accountability across the horticulture sector.
- **Scheme scope and objectives:** The scope of the scheme extends to all aspects of horticultural nursery operations, including plant propagation, cultivation, and distribution. The primary objectives of the scheme are to ensure the quality, reliability, and sustainability of horticultural nursery products, thereby enhancing consumer confidence and industry competitiveness.
- **Recognition criteria and application process:** Nurseries seeking recognition under the scheme must satisfy predefined criteria related to genetic quality, production practices, environmental stewardship, and business ethics. The application process involves the submission of comprehensive documentation, on-site inspections, and assessment of compliance with accreditation standards by qualified evaluators.
- **Assessment standards and review process:** Accreditation standards and review processes ensure the consistency, integrity, and relevance of accreditation schemes, driving continuous improvement and adaptation to evolving industry needs.
- **Technical evaluation:** Technical evaluation criteria encompass a wide range of factors, including genetic purity, propagation techniques, cultivation practices, pest and disease management, environmental sustainability, and business ethics. Evaluators assess nurseries based on predefined performance indicators, benchmarks, and best practices relevant to their respective sectors.
- **Certification system:** Certification under accreditation schemes provides formal recognition of nurseries' compliance with industry standards and best practices. Certified nurseries demonstrate their commitment to quality, reliability, and professionalism, distinguishing themselves as leaders within the nursery sector and gaining credibility and trust among customers and stakeholders.
- **Renewal and monitoring:** Accredited nurseries undergo periodic renewal and monitoring to maintain their certification status and ensure continued adherence to accreditation standards. Renewal processes typically involve reevaluation of nurseries' performance, documentation

review, and on-site inspections to verify ongoing compliance with accreditation requirements.

- **Coordination and oversight:** Effective coordination and oversight mechanisms are essential for the success and integrity of accreditation schemes, involving collaboration among stakeholders, regulatory bodies, accreditation agencies, and industry associations.
- **Stakeholder involvement:** Stakeholder engagement fosters transparency, accountability, and buy-in from all parties involved in the accreditation process. Collaboration among nurseries, government agencies, industry associations, consumers, and academia promotes dialogue, knowledge sharing, and consensus-building, strengthening the credibility and effectiveness of accreditation schemes.
- **Recognition duration and monitoring:** The duration of accreditation recognition varies depending on the scheme and regulatory requirements, with accredited nurseries typically subject to periodic monitoring and evaluation to maintain their certification status. Ongoing monitoring ensures that accredited nurseries continue to uphold the highest standards of quality, reliability, and sustainability over time.
- **Appeal mechanism:** Accreditation schemes incorporate appeal mechanisms to address grievances, disputes, or challenges related to accreditation decisions. Nurseries have the right to appeal adverse findings or decisions through formal channels, such as appeals committees or independent arbitration, ensuring procedural fairness and accountability within the accreditation process.

Understanding Nursery Quarantine Practices

Nursery quarantine practices are essential for preventing the introduction and spread of pests, diseases, and invasive species within the nursery environment, safeguarding plant health and minimizing economic losses.

Introduction to Nursery Quarantine

Nursery quarantine practices play a pivotal role in safeguarding the health and integrity of horticultural crops. By implementing rigorous quarantine measures, nurseries can prevent the introduction and spread of pests, diseases, and invasive species, thereby promoting healthy and resilient crop production systems.

Quarantine Procedures

Quarantine procedures typically involve a series of steps to evaluate and manage the health status of incoming plant material. This may include visual inspections, laboratory testing, and phytosanitary certifications to identify and address potential pests, diseases, and pathogens. Quarantine protocols are tailored to the specific requirements of each nursery and may vary based on regional regulations and industry standards (Heuser, 2003).

- **Quarantine facility design:** Designing an effective quarantine facility is crucial for preventing the escape of pests and diseases into the surrounding environment. Quarantine facilities should feature controlled access points, secure containment structures, and appropriate ventilation systems to minimize the risk of contamination. Additionally, quarantine facilities may incorporate quarantine cages or isolation areas for high-risk plant material.
- **Monitoring and surveillance:** Regular monitoring and surveillance are essential components of nursery quarantine practices. Nurseries should implement systematic monitoring programs to detect any signs of pests, diseases, or abnormalities in quarantined plant material. Early detection allows for prompt intervention and mitigation measures to prevent the spread of potential threats.
- **Quarantine period:** The duration of the quarantine period may vary depending on factors such as the origin of the plant material, its intended use, and the presence of any specific pests or diseases of concern. Quarantine periods typically range from a few days to several weeks, during which time the health status of the plant material is carefully monitored before it is released for production or distribution.

Conclusion

Nursery quarantine practices are indispensable for safeguarding the health and quality of horticultural crops. By adhering to stringent quarantine protocols, nurseries can mitigate the risk of introducing and disseminating pests, diseases, and invasive species, thus preserving the long-term viability and sustainability of crop production systems. Robust quarantine measures underscore the significance of proactive risk management and biosecurity in nursery operations.

Accreditation serves as a cornerstone in maintaining excellence and fostering continual enhancement in fruit nursery operations. Accredited nurseries prioritize genetic integrity, adherence to industry benchmarks, traceability, instilling customer trust, gaining market access, and driving ongoing improvement initiatives. These accredited establishments contribute significantly to the integrity, sustainability, and competitiveness of the fruit production sector.

As stewards of plant health and custodians of agricultural biodiversity, accredited nurseries play a pivotal role in ensuring the availability of top-tier planting materials crucial for global food security and economic prosperity. Their dedication to upholding rigorous standards fortifies the foundation of the agricultural landscape and supports the resilience of food supply chains.

Glossary/Definition of Terms

Accreditation Panel (AP): A team of technical experts with knowledge about audit techniques and criteria for accreditation of laboratory facilities for virus diagnosis/genetic fidelity testing of specified fruit crops and recognition/ renewal of recognition of nurseries.

Accredited Test Laboratory (ATL) : An institute/organization granted accreditation by NHB, Govt. of India for virus diagnosis/genetic fidelity testing of specified fruit plants. under CPP.

Appellate Authority : An administrative Department/Ministry to consider the appeal in decision taken by the NHB or Accredited Test Laboratory.

Appeal Panel : A group of experts nominated by the NHB to review the decision taken by the Accredited Test Laboratory.

Appendix: An appendix is a section that provides additional information or supplementary material. This could include charts, graphs, tables, or other data that supports the main content but is not essential to understanding it. For the purposes of this Certification Standard, the explanatory documents following the Standard are called 'the Appendix' or 'Appendices'.

Attachment: An attachment may be additional documents or exhibits that are appended to the main document to provide further details, specifications, or terms. For the purposes of this Certification Standard, 'Attachments' are used by the nursery to provide the corroborating or additional information.

Biosecurity: *Measures taken to prevent the introduction and/or to minimize the risk of establishment and spread of a specific pest.*

Certification Status : Status of a nursery that has undergone all components of a Certification Program and has been audited to verify that all components are properly implemented.

Certification Identification Number: The unique number assigned by the NHB to a Clean Plants certified nursery.

Certification Stamp: The stamp or seal issued time to time by NHB providing the nursery's certification identification number as well as any additional pest specific modules for use by the nursery on their own documents.

Certification Manager: Nursery manager that is responsible for all aspects of the Clean Plants program.

NHB Directives: Policy documents that outline specific requirements for the import, export and domestic movement of plants and plant products.

Clean Plants: The propagules produced/maintained by any nursery based certification program.

Clean Plants Standard: Document that outlines all the required components of the Clean Plants program.

NHB: National Horticulture Board, the organization with the responsibility for accreditation of auditors and certification of nurseries under Clean Plant Program.

CNCI Accredited Auditor : Auditors who have been trained and accredited by NHB for purposes of conducting Clean Plants audits.

Adjudication Committee: Committee which reviews Clean Plants Nursery Manuals, initial facility audit reports and audit reports, when critical non-conformances occur.

Control Measures: Methods of pest control to achieve the desired result as per the certification standards.

Corrective Action Plan: The plan, created by the nursery and agreed to by the auditor, to correct non-conformances identified in Corrective Action Requests.

Crop Protection Manager: A member of the Clean Plants certified nursery management team or a contractor who is responsible and accountable for the overall implementation of the Pest Management Plan.

Growing Cycle: The period in which plants grow from a propagative state (*e.g.* cuttings, tissue culture plantlets, pre-finished plants, *etc.*) to a commercially acceptable size for either retail or wholesale trade..

Growing Season: Period or periods of the year when plants actively grow in an area, place of production, or production site.

Isolation Area: An area of the nursery set aside to hold plants that may pose a pest risk to other plants on the nursery.

Mother Plants: : Mother plants are the plants maintained in the nursery as the source for supply of scion and rootstocks. As the success of any nursery depends on the health and vigour of its mother/progeny plants, it is therefore necessary to obtain genuine mother plants to produce healthy and vigorous saplings. The mother plants must necessarily be genetically pure and superior in quality. Further, not only selection of mother plants necessary, but also proper care and maintenance of these plants is essential to obtain healthy growth. Only when vegetative growth is permitted and maximum bud sticks are produced. Mother plants are kept healthy by regular testing of the plant material for viruses and other organisms.

Non-Conformance: Minor, major, and critical non-fulfillment of requirements of the Clean Plants Certification Program.

Non-Conforming Plant Material: Plant material that is found to not conform with the requirements of the Clean Plants Certification Program.

Net house: : Insect proof structure stretched over mother plants to avoid insect attack and viral contamination.

Non-Regulated Pests: A pest that is not under official control. Often referred to as "other pests.".

Nursery: Nursery is the managed site, having a standard design, layout, infrastructure and other facilities, and manned by trained manpower, designed to produce seedlings, cuttings, layers, composite plantlets *etc.* grown under favorable conditions until they are ready for final planting. Also it is an Institute/organization seeking accreditation/recognition.

Nursery Stock: Any propagule for planting, including greenhouse, containerized and field grown plants.

Pest: Anything that is injurious or potentially injurious, whether directly or indirectly, to plants or to products or by-products of plants and includes any plant prescribed as a pest .

Pest Management Plan (PMP): A written description of procedures or processes designed to control a pest population, to exclude it from entering the system, or when present to either to eliminate it or to suppress it to a level that meets phytosanitary requirements.

Pest Specific Module: A certification program designed to control the spread of a specific pest which is integrated as part of a nursery's overall Clean Plants Certification Program.

Phytosanitary Management System (PMS): A systems approach that is used to direct and control an organization with regard to consistently meeting all the requirements of the Clean Plants program.

Phytosanitary Management Practices: A series of production practices designed to ensure a nursery meets all phytosanitary regulations.

Phytosanitary Risk: The possibility of introducing a potentially harmful pest.

Poly house: : A structure created with the help of polyethylene sheets with humidity and temperature control mechanism primarily used for keeping freshly grafted/budded plants so as to get early union and high success rate.

Quality Planting Material: Quality Planting Material is 'the production of uniform, healthy, disease-free planting material raised through seed or vegetative methods with an overall goal to raise the physiological and phytosanitary quality of the plant available to stakeholders.

Quarantine Pest: A pest of potential economic importance to the area endangered thereby and not yet present there, or present but not widely distributed and being officially controlled.

Record: An authenticated document (electronic or print), product or sample statement, which confirms that a procedure (or part of the procedure) has been carried out.

Referral Laboratory (RL): An official laboratory for the performance of confirmatory tests in the event of dispute of test results.

Regulated Area: An area into which, within which and/or from which plants, plant products and other regulated articles are subjected to phytosanitary regulations or procedures in order to prevent the introduction and/or spread of quarantine pests or to limit the economic impact of regulated non-quarantine pests.

Regulated Pest: A quarantine pest or a regulated non-quarantine pest.

Regulated Non- Quarantine Pest: A non-quarantine pest whose presence in plants for planting affects the intended use of those plants with an economically unacceptable impact and which is therefore regulated .

Segregate: Keep separate, identifiable as a separate unit.

Shade net: : Simple structure stretched for providing partial shade to mother plants and for keeping young plants for hardening.

Standard Operating Procedures (SOPs): Standard Operating Procedures (SOPs) are sets of written instructions that document the routine or repetitive activity followed by an organization. The development and use of SOPs are an integral part of a successful quality system as it provides individuals with the information to perform a job properly.

Standard Protocol: A standardized/validated test procedure established either nationally/internationally.

Surveillance Audit: An audit of an entire facility to ensure all phytosanitary processes are in place as specified in the Clean Plants Certification System. Surveillance audits should also include field inspections to verify freedom from regulated pests.

Suspension: Nursery can no longer distribute the propagules as being Clean Plants certified and loses its privilege of using the stamp/logo.

Traceability: The ability to follow an item or group of items – animals, plants, food products or ingredients – from one point in the supply chain to another.

Wind break: A group of fast growing fruit and forest plants raised along the boundaries of the field. Such grown up trees can minimize the wind velocity and reduce.

Terminologies Used for Planting Materials

Generation Levels	*Other Terminologies Used*	*Location where Raised*	*Features*
G-1	Foundation, Nuclear, Elite, Pre-elite, Pre-pre basic, Pre-selection, Mother plants	Clean Plant Centres	Original mother plants tested negative for all targeted pathogens, Source for further propagation (100 per cent Pass in testing)
G-2	Elite, Foundation, pre-basic plants	Maintained by registered nurseries	Propagated from G-1

Generation Levels	*Other Terminologies Used*	*Location where Raised*	*Features*
G-3	Basic, Elite, Increase Block plants	Certified nurseries	Propagated from G-2
G-4	Certified Plants	Certified nurseries	Propagated from G-2/G-3 and given to consumers

References

Alchanatis V and Cohen Y, 2010. A review of the accreditation and certification systems for nurseries. *HortScience*, **45**: 734-740.

Aruoma OI, 2003. Methodological considerations for characterizing potential antioxidant actions of bioactive components in plant foods. *Mutation Research/ Fundamental and Molecular Mechanisms of Mutagenesis*, 523-524.

Chadha KL and Mishra R, 2002. Accreditation standards for horticultural nurseries. *Journal of Horticultural Science and Biotechnology*, **77**: 407-414.

Deming WE, 1986. Out of the crisis. *MIT Press*, Cambridge, MA. ISBN: 978-0-262-54115-2

Dhillon WS and Gill PPS, 2004. Role of horticulture in the nutritional and economic security of India. *Journal of Food Science and Technology*, **41:** 605-608.

Garcia Martinez M and Poole ND, 2004. The development of private fresh produce safety standards: implications for developing Mediterranean exporting countries. *Food Policy*, **29**: 229-255.

Hancock JF and Siefker JH, 1982. Levels of inbreeding in highbush blueberry cultivars. *Hort. Science*, **17** : 363-365.

Heuser CW, 2003. The role of plant quarantine in protecting agriculture and the environment. *Hort Technology* **13:** 542-548.

Pretty J, 2008. Agricultural sustainability: concepts, principles and evidence. *Philosophical Transactions of the Royal Society B: Biological Sciences*, **363**: 447-465.

Singh HP and Kalloo G, 2000. Role of horticulture in the livelihood security of farm households in India. *Indian Journal of Agricultural Economics*, **55**: 155-168.

Smith RL and Roosen J, 2006. The role of traceability in the agricultural supply chain: from theory to practice. *Agricultural Economics*, **35**: 165-173.

Chapter 17

Intelligent Technologies in Orchard Management: Applications and Challenges

Sita Rani and O.P. Gupta*

Department of Computer Science and Engineering, Guru Nanak Dev Engineering College, Ludhiana–141006, Punjab
**Associate Director IT, Post Graduate Studies Complex, Punjab Agricultural University, Ludhiana–141004, Punjab*

Orchard management incorporates smart technologies such as IoT sensors, AI, and drones to enhance precision agriculture. These technologies enable real-time monitoring of soil conditions, crop health, and pest activity, allowing for optimized irrigation, targeted pest control, and efficient harvesting. This integration improves productivity, resource efficiency, and sustainability in orchard operations. In this chapter, the authors explore the integration of advanced technologies in modern orchard management. It discusses the use of IoT devices, AI, and data analytics to optimize orchard operations, including precision agriculture techniques for monitoring soil health, crop growth, and pest management. The chapter highlights applications such as automated irrigation systems, drone-based crop surveillance, and robotic harvesting, which enhance productivity and resource efficiency. It also addresses the challenges faced in implementing these technologies, including high costs, the need for technical expertise, and data privacy concerns. The chapter emphasizes the importance of developing robust, user-friendly systems and fostering collaboration between technology providers and farmers. Through these discussions, the chapter provides a comprehensive overview of the potential benefits and obstacles in adopting intelligent technologies for sustainable and efficient orchard management.

Introduction

Orchard management involves cultivating fruit trees through careful selection, planting, pruning, and maintenance. Key practices include pest and disease control, irrigation, fertilization, and pollination management. The goal of orchard management is to maximize the health, growth, and yield of fruit trees through a variety of techniques (Zhang *et al.,* 2021). The first step in the process is to carefully choose tree varieties that are appropriate for the local climate, soil, and market needs. Establishing a healthy orchard requires careful site preparation, which includes soil testing and amendment. Planting strategies, such as arrangement and spacing, are designed to optimize airflow and sunlight exposure to lower the risk of disease (Montanaro *et al.,* 2017).

Pruning is an essential step in shaping trees to allow lighter and air to reach them, improving the quality of the fruit and lowering the risk of disease (Ali, 2023). Trees and fruit are protected while using fewer chemicals when pest and disease management, which frequently involves integrated pest management (IPM) techniques, is implemented. Tree Vigor and fruit quality require regular irrigation, especially in arid climates. Drip irrigation systems increase water efficiency.

Thinning techniques help control fruit load, enhancing the size and quality of the remaining fruits, while fertilization supplies vital nutrients that are specific to the orchard's needs. Bee use is one of the pollination strategies that is essential for fruit set (Lei *et al.,* 2023). For the purpose of tracking tree health, growth, and yield–which allows for timely interventions and continuous improvement–continuous monitoring and record-keeping are crucial. There is a growing emphasis on sustainable practices to lessen their impact on the environment and increase the longevity of orchards.

Modern orchard management relies heavily on intelligent technologies to increase productivity, sustainability, and efficiency (Wang, *et al.,* 2020). Drones and other precision agriculture instruments, like sensors and sensors give precise information about weather patterns, tree health, and soil conditions, allowing for focused interventions (Bresilla, 2019; Shuai *et al.,* 2019). This data-driven strategy minimizes waste and its negative effects on the environment by improving irrigation, using less water, and applying fertilizers more wisely. Reliance on chemical pesticides is decreased through early detection and accurate treatment made possible by automated systems for monitoring pests and diseases. Furthermore, labour-intensive operations like pruning, thinning, and harvesting are made easier by robotics and automation, which lowers expenses and addresses the labor shortage. In order to predict crop yields and determine the best times to harvest, decision-making processes are further refined by advanced analytics and machine learning algorithms. In general, intelligent technologies improve crop quality, yields, and environmental stewardship by increasing orchard management's sustainability and efficiency.

The objective of this chapter is to examine the incorporation of cutting-edge technologies in orchard management, emphasizing the advantages and disadvantages that come with them. Examining different intelligent technologies like automation, machine learning, drones, and sensors as well as their applications in precision agriculture, pest and disease monitoring, irrigation control, and labor optimization are all included in the scope of the study. The aims encompass evaluating the advantages of these technologies in augmenting efficacy, durability, and crop excellence, alongside tackling obstacles like elevated implementation expenses, intricate technicalities, and the requirement for specialized competencies. In order to integrate intelligent technologies in orchard management and eventually contribute to more productive and sustainable agricultural practices, the chapter attempts to give a thorough overview while also providing insights into current trends, best practices, and future directions in this area.

Intelligent Technologies and Orchard Management

Intelligent technologies refer to advanced tools and systems that use data analytics, automation, artificial intelligence, and machine learning to enhance efficiency and decision-making in various fields. They gather, process, and analyze data to optimize processes, improve accuracy, and reduce human intervention. By incorporating cutting-edge instruments like drones, sensors, automation, and machine learning, intelligent technologies are revolutionizing orchard management (Jin *et al.,* 2020; Verbiest *et al.,* 2021; Zhang *et al.,* 2021). Precision agriculture is made possible by these technologies, which gather and analyze extensive data on environmental factors, tree health, and soil. By optimizing irrigation, fertilization, and pest management, this information lowers resource consumption and improves crop quality. Drones can monitor tree health from the air and identify problems early; automated systems can handle labour shortages by streamlining repetitive tasks like pruning and harvesting. Data is analysed by machine learning algorithms to forecast yields, improve harvest schedules, and aid in decision-making. Although there are certain drawbacks, such as high expenses and technical complexity, these technologies have a big impact on sustainability, productivity, and orchard management in general.

Historical Background of Intelligent Technologies in Orchard Management

In the last few decades, there has been a significant evolution in the application of intelligent technologies to orchard management. At first, mechanization was the main focus of advancements, leading to the widespread use of machines like mechanical sprayers and harvesters. A big step forward was made possible by the 1990s introduction of Geographic Information Systems (GIS) and Global

Positioning Systems (GPS), which allowed for accurate mapping and orchard monitoring. The Internet of Things (IoT) and data analytics have completely changed orchard management in the twenty-first century. Real-time information on pest activity, weather, and soil moisture is now available thanks to sensors and drones. Sophisticated software systems that help with decision-making, fertilization, irrigation, and pest control optimization incorporate this data. Predicting crop yields and early disease detection are becoming more and more dependent on machine learning and artificial intelligence (AI). In order to further increase efficiency and lower labor costs, autonomous machinery and robots are being developed to automate tasks like pruning, harvesting, and sorting. All of these technologies work together to improve yield quality and quantity while reducing environmental impact, leading to more sustainable and effective orchard management.

Current Trends and Developments

The current trends in intelligent technologies for managing orchards center on data-driven decision-making, automation, and precision agriculture (Figure 17.1). The following are some significant developments:

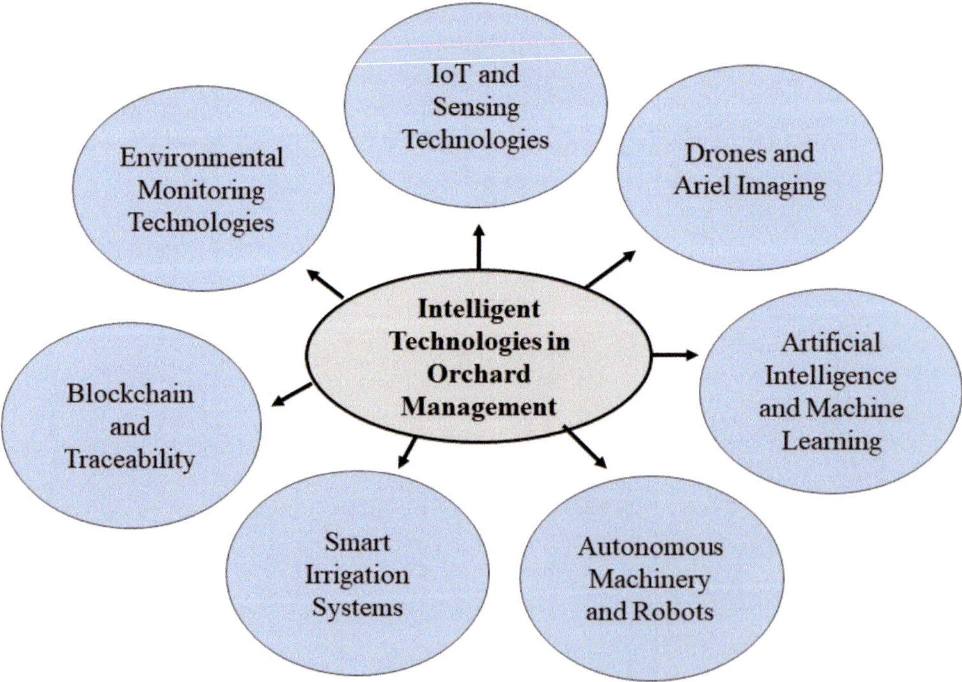

Figure 17.1: Intelligent Technologies in Orchard Management.

- ***Advanced sensors and IoT integration:*** More and more sophisticated sensors are being installed in modern orchards to keep an eye on insect activity, temperature, humidity, nutrient levels, and soil moisture. The Internet of Things (IoT) connects these sensors, enabling real-time data collection and analysis (Varandas *et al.,* 2020). By maximizing fertilization, irrigation, and pest control, this integration contributes to improved resource management and a smaller environmental impact.

- ☆ ***Drones and ariel imaging***: Aerial mapping and imaging are being done with drones fitted with thermal and multispectral cameras. By assisting in the early detection of disease outbreaks, water stress, and crop health problems, these technologies allow for more focused interventions. Drones can also be used to swiftly and effectively cover large areas when spraying fertilizer and pesticides (Zhang *et al.,* 2021).
- ☆ ***Artificial intelligence and machine learning:*** Data gathered from sensors and drones is increasingly being analysed using AI and machine learning algorithms. These technologies optimize harvest schedules, forecast crop yields, and identify illnesses. In addition to improving decision-making, AI models can spot patterns in data that human operators might not notice right away (Jin *et al.,* 2020).
- ☆ ***Autonomous machinery and robots:*** Orchard management is changing due to the development of autonomous machinery, such as robotic harvesters and self-driving tractors. These devices can precisely complete tasks like pruning, thinning, harvesting, and sorting, which lowers labour costs and boosts productivity (Abeyrathna and Ahamed 2023).
- ☆ ***Smart irrigation systems:*** Weather forecasts and real-time data are used by smart irrigation systems to maximize water usage. By ensuring that crops receive the proper amount at the right time and helping to conserve water, these systems enhance crop quality and yield (Xie *et al.,* 2022).
- ☆ ***Blockchain and traceability:*** Investigations into blockchain technology are being conducted to enhance supply chain transparency and traceability. It improves food safety and quality assurance by enabling secure and verifiable tracking of produce from the orchard to the consumer.
- ☆ ***Sustainability and environmental monitoring:*** A greater focus is being placed on sustainability, and orchard practices' environmental impact is being monitored and mitigated through the use of technologies. This covers monitoring carbon footprints, cutting back on waste, and utilizing renewable energy sources.

Applications of Intelligent Technologies

Precision agriculture is improved by intelligent technologies in orchard management, which include yield prediction, automated irrigation, and pest detection. They monitor crop health, optimize resource use, and boost overall productivity by using sensors, drones, and AI algorithms. Some of the prominent applications of intelligent technologies in orchards management are as follows (Jung *et al.*, 2021; Shaikh *et al.*, 2022; Talaviya *et al.*, 2020):

Precision Agriculture

Intelligent technologies contribute to better orchard management through automated irrigation, yield prediction, and pest detection in precision agriculture. By employing sensors, drones, and AI algorithms, they can monitor crop health, maximize resource utilization, and increase overall productivity. IoT sensors keep an eye on weather patterns and soil moisture levels, which makes it possible to precisely manage water, cut down on waste, and improve plant health. Drones and imaging systems with AI capabilities detect disease symptoms and pest infestations early on, allowing for targeted treatments and reducing the need for chemicals. Planning and marketing strategies are aided by machine learning algorithms, which forecast yields and track fruit growth by analysing data from sensors and historical records. Accurate application of pesticides and nutrients is ensured by GPS-guided machinery, reducing expenses and environmental impact. Pruning and thinning techniques are informed by the assessment of canopy health and density by remote sensing technologies. These tools facilitate better decision-making, increase productivity, and support environmentally friendly orchard management techniques.

Automation and Robotics

Automation and robotics are transforming orchard management through the implementation of intelligent technologies. These advancements enhance efficiency, reduce labour costs, and optimize fruit production. Key applications include:

- **Automated Harvesting:** Robotic harvesters equipped with computer vision and AI algorithms can identify ripe fruits, pick them delicately, and sort them, minimizing damage and increasing harvesting speed.
- **Pruning and Thinning:** Robotic systems use advanced sensors and machine learning to perform precise pruning and thinning, optimizing tree health and fruit quality.
- **Weed Control:** Autonomous robots equipped with vision systems and AI can detect and eliminate weeds, reducing the need for chemical herbicides and promoting sustainable practices.
- **Autonomous Transport:** Automated vehicles transport harvested fruits and other materials within the orchard, streamlining logistics and reducing manual labor.
- **Data Collection and Analysis:** Robots equipped with sensors and cameras collect data on tree health, soil conditions, and fruit development, enabling real-time monitoring and decision-making.

These technologies enhance productivity, quality, and sustainability in orchard management.

Remote Sensing and GIS

With accurate and timely data provided by intelligent technologies in GIS and remote sensing, orchard management is being revolutionized. These technologies track various aspects of orchard conditions, such as soil moisture, plant health, and pest infestations, using sensors, drones, and satellite imagery. These data are integrated by GIS systems, which provide comprehensive maps and analytical tools to support decision-making. For example, farmers are able to plan effective harvesting routes, identify early symptoms of disease, and optimize irrigation schedules. These technologies support profitable and sustainable orchard management by improving accuracy and using less resources, which improves crop yields and quality.

Data Analytics and IoT

Real-time insights and automation from intelligent technologies in data analytics and IoT are revolutionizing orchard management. Large volumes of data are gathered from the orchard by Internet of Things devices like soil sensors, weather stations, and plant health monitors. After processing this data, data analytics tools find patterns and trends that help with decision-making. Predictive analytics, for instance, can forecast the best times to harvest, and information from Internet of Things (IoT) devices can be used to optimize fertilization and irrigation schedules based on current soil and weather conditions. With the help of these technologies, orchard operations can operate more sustainably and efficiently thanks to improved crop quality, labor cost savings, and precise resource management.

Case Studies

Here are some case studies highlighting the successful implementation of intelligent technologies in orchards worldwide:

- ☆ **Apple orchards in Washington State, USA**: Washington State apple orchards have successfully used IoT devices and data analytics. Smart irrigation systems equipped with soil moisture sensors and weather forecasting tools have optimized water use, reduced water consumption and improving fruit quality. These systems adjust irrigation schedules based on real-time data, ensuring that trees receive the right amount of water, enhancing both yield and resource efficiency (Silwal *et al.,* 2017).
- ☆ **Almond orchards in California, USA**: California almond growers have adopted remote sensing and GIS technologies to address water scarcity. By utilizing satellite imagery and aerial drones, farmers can monitor tree health and soil moisture levels. This data allows for precise irrigation management, reducing water usage by up to 20 per cent. The technology also helps in detecting pest infestations early, enabling timely interventions and minimizing crop losses (Gillispie 2021).
- ☆ **Kiwifruit orchards in New Zealand**: Kiwifruit growers in New Zealand have implemented precision agriculture technologies, including

satellite-based GIS mapping and data analytics. These tools provide detailed insights into soil conditions, vine health, and climate patterns. By analyzing this data, farmers can optimize fertilizer application and manage diseases more effectively, leading to higher-quality fruit and improved yields (Mills *et al.,* 2019).

- **Citrus orchards in Spain**: Spanish citrus orchards have integrated drone technology and multispectral imaging to monitor tree health. This technology helps detect diseases like citrus greening early, allowing farmers to take preventive measures. By mapping the health status of each tree, growers can target interventions precisely, improving overall orchard productivity and reducing losses.
- **Olive groves in Italy**: In Italy, olive groves have benefited from the use of GIS and remote sensing technologies. By analyzing soil types, elevation, and climate data, farmers have optimized the selection of olive varieties suited to specific microclimates. This has resulted in better yields and higher-quality olive oil production, demonstrating the value of intelligent technologies in improving agricultural outcomes (Anastasiou *et al.,* 2023).

These case studies illustrate how the adoption of intelligent technologies in orchard management can lead to significant improvements in resource efficiency, crop quality, and overall productivity.

Table 17.1: Comparative Analysis of Various Intelligent Technologies in Orchard Management

Technology	*Application*	*Benefits*	*Challenges*
IoT Sensors	Soil moisture, temperature, humidity	Real-time data, precise irrigation, reduced water usage	High initial cost, requires network connectivity
Drones	Aerial imaging, crop health monitoring	High-resolution imagery, early disease detection	Limited flight time, weather-dependent
GIS and Remote Sensing	Mapping, soil and climate analysis	Spatial data integration, optimized planting strategies	Requires specialized software and skills
Data Analytics	Yield prediction, disease forecasting	Data-driven decision making, trend analysis	Data privacy concerns, need for large datasets
Smart Irrigation	Automated watering based on data	Water conservation, reduced labor costs	Installation complexity, maintenance requirements

Technology	*Application*	*Benefits*	*Challenges*
Satellite Imagery	Large-scale monitoring, seasonal analysis	Broad coverage, multi-temporal analysis	Lower resolution than drones, delayed data
Machine Learning	Pest and disease prediction, yield forecasting	Improved accuracy in predictions, adaptive learning	Requires significant data, complexity in model training
Robotics	Automated harvesting, pruning	Labor savings, consistent quality	High cost, technical maintenance

Each technology offers distinct advantages and faces unique challenges. Combining these technologies can enhance overall orchard management, optimizing resource use and improving crop quality and yield.

Major Challenges and Limitations

Deploying smart technologies in orchard management can offer numerous benefits, such as increased efficiency, better yield prediction, and improved resource management. However, there are also several challenges and limitations that need to be addressed (Figure 17.2) (Rehman, *et al.,* 2022; Symeonaki *et al.,* 2019), discussed below:

High Initial Costs

- **Technology Investment**: The initial cost of purchasing and installing smart technologies, such as sensors, drones, and automation systems, can be substantial. This can be a significant barrier, especially for smaller or family-owned orchards.
- **Maintenance and Upgrades**: Ongoing costs for maintenance, software updates, and potential hardware upgrades can also add to the financial burden.

Data Management and Integration

- **Data Overload**: The deployment of various sensors and data collection devices can lead to an overwhelming amount of data. Effective data management systems are needed to store, process, and analyse this information.
- **Integration Challenges**: Integrating data from different sources (*e.g.,* weather stations, soil sensors, drones) into a cohesive system can be complex. Compatibility issues between different technologies and platforms can also pose challenges.

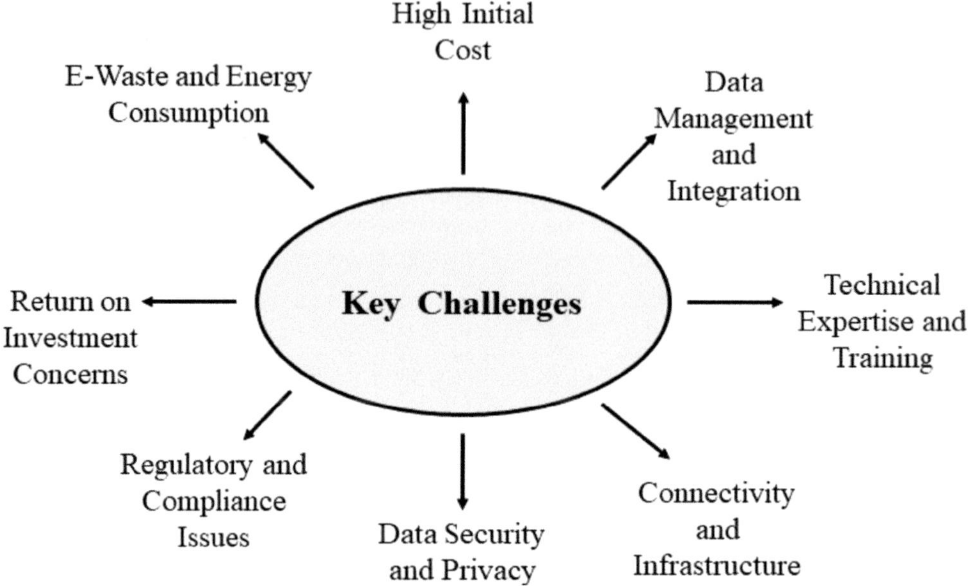

Figure 17.2: Key Challenges in Deploying Smart Technologies for Orchard Management.

Technical Expertise and Training

- **Skill Requirements**: Implementing and maintaining smart technologies require technical skills that many orchard managers and workers may not possess. This necessitates training or hiring skilled personnel, which can be costly and time-consuming.
- **User Adoption**: Resistance to adopting new technologies can be a barrier, particularly among workers or managers who are accustomed to traditional methods.

Connectivity and Infrastructure

- **Internet Connectivity**: Many orchards are located in rural or remote areas with limited or unreliable internet access. This can hinder the real-time data transmission and control capabilities of smart systems.
- **Power Supply**: Ensuring a stable power supply for sensors, automated systems, and data centers can be challenging, especially in off-grid locations.

Data Security and Privacy

- **Cybersecurity risks**: The increased reliance on digital systems exposes orchards to cybersecurity risks, including data breaches and hacking.

- ☆ **Privacy concerns**: Handling large amounts of data, particularly when it involves personal information or proprietary business data, raises concerns about privacy and data protection.

Environmental and Climatic Factors

- ☆ **Weather sensitivity**: Many smart technologies rely on weather data and predictions. Unexpected weather events or climate variability can affect the accuracy and effectiveness of these systems.
- ☆ **Adaptability**: Technologies need to be adaptable to different environmental conditions and crop types, which can vary significantly between regions.

Regulatory and Compliance Issues

- ☆ **Regulation compliance**: There may be regulations governing the use of drones, sensors, or automated equipment in agriculture. Compliance with these regulations can be complex and vary by region.
- ☆ **Data regulations**: Ensuring compliance with data protection laws, especially in regions with strict regulations like the GDPR in Europe, can be challenging.

Return on Investment (ROI) Concerns

- ☆ **Uncertain ROI**: The benefits of smart technologies may not be immediately apparent, making it difficult for orchard managers to justify the investment. It may take several seasons to see a clear return on investment.
- ☆ **Yield Variability**: Factors beyond the control of technology, such as pest infestations or market fluctuations, can impact the expected ROI.

Environmental Impact

- ☆ **E-waste**: The use of electronic devices and sensors can contribute to electronic waste if not properly managed at the end of their lifecycle.
- ☆ **Energy consumption**: The energy consumption of smart systems, particularly in large-scale operations, can be significant and may not always align with sustainable practices.

Addressing these challenges requires a strategic approach that includes proper planning, investment in training, and consideration of both the technological and human factors involved in orchard management.

Future Prospects and Trends

The future of orchard management is increasingly intertwined with the deployment of smart technologies, driven by the need for higher efficiency, sustainability, and productivity. Here are some key prospects and trends expected to shape the field in the coming years (Ayaz *et al.*, 2019; Watteyne *et al.*, 2016 and Zhang *et al.*, 2023):

Increased Adoption of Precision Agriculture

- **Variable rate technology (VRT)**: Technologies that allow precise application of inputs (such as water, fertilizers, and pesticides) based on real-time data and crop needs will become more prevalent. This can lead to significant cost savings and environmental benefits.
- **Site-specific management**: Use of GPS and GIS technologies to manage specific areas within orchards differently, optimizing resource use and improving yields.

Advancements in IoT and Sensor Technologies

- **Enhanced sensing capabilities**: Continued improvements in sensor technologies will provide more accurate and comprehensive data on soil moisture, temperature, humidity, and other critical factors.
- **Real-time monitoring and alerts**: Integration of IoT devices with cloud platforms for real-time data analysis and alert systems will enable faster responses to issues such as pest infestations or irrigation needs.

Automation and Robotics

- **Automated harvesting**: Robotics will play an increasingly significant role in harvesting, reducing labour costs and minimizing crop damage.
- **Drones and UAVs**: Drones equipped with advanced cameras and sensors will be used for tasks such as crop monitoring, spraying, and health assessment.

Artificial Intelligence and Machine Learning

- **Predictive analytics**: AI-driven models will enhance forecasting abilities for crop yields, disease outbreaks, and market trends, allowing better planning and decision-making.
- **Decision support systems**: AI-powered decision support systems will provide recommendations on planting, pruning, and harvesting based on data analytics, weather patterns, and historical data.

Blockchain and Supply Chain Transparency

- **Traceability**: Blockchain technology will offer greater transparency and traceability in the supply chain, from farm to market. This will enhance food safety, quality assurance, and consumer trust.

- ☆ **Smart contracts**: Implementation of smart contracts for transactions and agreements in the agricultural sector could streamline operations and reduce administrative burdens.

Sustainable Practices and Environmental Monitoring

- ☆ **Sustainable irrigation**: Smart irrigation systems that use real-time data to optimize water usage will become more common, promoting water conservation.
- ☆ **Climate resilience**: Technologies that help orchards adapt to changing climate conditions, such as weather-resistant crop varieties and adaptive management practices, will gain importance.

Enhanced Data Analytics and Integration

- ☆ **Big data utilization**: The aggregation and analysis of large datasets from various sources will enable more comprehensive insights into orchard management.
- ☆ **Interoperability and integration**: Efforts to standardize data formats and enhance interoperability between different systems and devices will improve the efficiency and usability of smart technologies.

Personalized and Precision Marketing

- ☆ **Consumer data integration**: Integration of consumer preferences and trends into production planning will allow for more targeted marketing and product development.
- ☆ **Direct-to-Consumer Models**: Smart technologies will facilitate direct sales channels, such as e-commerce and farm-to-table initiatives, enhancing profitability and customer engagement.

Development of Ecosystem Services

- ☆ **Biodiversity monitoring**: Smart technologies will be used to monitor and promote biodiversity within orchards, supporting ecosystem services like pollination and natural pest control.
- ☆ **Carbon footprint reduction**: Data-driven strategies will help orchards reduce their carbon footprint, contributing to sustainability goals.

Education and Training

- ☆ **Digital literacy**: Increasing emphasis on digital literacy and training for orchard managers and workers will be crucial to effectively utilize smart technologies.
- ☆ **Collaboration and Knowledge Sharing**: Platforms for sharing best practices, research, and innovations will support the broader adoption of smart technologies in orchard management.

These trends and prospects indicate a future where technology plays a pivotal role in transforming orchard management, making it more efficient, sustainable, and responsive to market demands and environmental challenges.

Conclusion

In this Chapter, the authors emphasize how intelligent technologies can revolutionize orchard productivity, efficiency, and sustainability. It highlights the important developments in automation, data analytics, and precision agriculture that are transforming conventional methods. The chapter does, however, also address the difficulties, including the high upfront costs, the complexity of data management, and the requirement for technical know-how. It highlights how crucial infrastructure development, training, and strategic planning are to maximizing the advantages of these technologies. In the end, the chapter makes the argument that even though intelligent technologies offer exciting new possibilities, their effective integration into orchard management will depend on careful assessment of the difficulties they raise.

REFERENCES

Abeyrathna RM, Rasika D and Tofael Ahamed, 2023. Autonomous Robots in Orchard Management: Present Status and Future Trends. IoT and AI in Agriculture: *Self-sufficiency in Food Production to Achieve Society 5.0 and SDG's Globally:* 335-366.

Ali Mujahid, 2023. Pruning Management in *Psidium guajava* L. Orchards–An Overview. *Gesunde Pflanzen* **75**(5): 1441-1447.

Anastasiou Evangelos, Athanasios T Balafoutis and Spyros Fountas, 2023. Trends in remote sensing technologies in olive cultivation. *Smart Agricultural Technology* **3**: 100103.

Ayaz Muhammad *et al.,* 2019. Internet-of-Things (IoT)-based smart agriculture: Toward making the fields talk. *IEEE access* **7**: 129551-129583.

Bresilla Kushtrim, 2019. Sensors, Robotics and Artificial Intelligence in Precision Orchard Management (POM). Doctoral Thesis, Alma Mater Studorum, University od Bolonga, Italy.

Gillispie George, 2021. Internet of Things (IoT) Controlled Irrigation for Almond Orchardists in California: A Multiple Case Study, Capella University.

Jin Yufang *et al.,* 2020. Advancing agricultural production with machine learning analytics: yield determinants for California's almond orchards. *Frontiers in Plant Science* **11**: 290.

Jung Jinha *et al.,* 2021. The potential of remote sensing and artificial intelligence as tools to improve the resilience of agriculture production systems. *Current Opinion in Biotechnology* **70**: 15-22.

Lei Xiaohui *et al.,* 2023. Technologies and Equipment of Mechanized Blossom Thinning in Orchards: A Review. *Agronomy* **13**(11): 2753.

Mills Linda *et al.,* 2019. Prediction of kiwifruit orchard characteristics from satellite images. *Precision Agriculture* **20**: 911-925.

Montanaro Giuseppe *et al.,* 2017. Orchard management, soil organic carbon and ecosystem services in Mediterranean fruit tree crops. *Scientia Horticulturae* **217**: 92-101.

Rehman Amjad *et al.,* 2022. A revisit of internet of things technologies for monitoring and control strategies in smart agriculture. *Agronomy* **12**(1): 127.

Shaikh Tawseef Ayoub, Tabasum Rasool and Faisal Rasheed Lone, 2022. Towards leveraging the role of machine learning and artificial intelligence in precision agriculture and smart farming. *Computers and Electronics in Agriculture* **198**: 107119.

Shuai Yuan *et al.,* 2019. Design and Implementation of a Kind of Intelligent Orchard System. *Agricultural Engineering* **9**(4): 20-25.

Silwal Abhisesh *et al.,* 2017. Design, integration, and field evaluation of a robotic apple harvester. *Journal of Field Robotics* **34**(6): 1140-1159.

Symeonaki Eleni G, Konstantinos G Arvanitis and Dimitrios D Piromalis, 2019. Current trends and challenges in the deployment of IoT technologies for climate smart facility agriculture. *International Journal of Sustainable Agricultural Management and Informatics* **5**(2-3): 181-200.

Talaviya Tanha *et al.,* 2020. Implementation of artificial intelligence in agriculture for optimisation of *irrigation and application of pesticides and herbicides. Artificial Intelligence in Agriculture* **4**: 58-73.

Varandas Leonor *et al.,* 2020. Low-cost IoT remote sensor mesh for large-scale orchard monitorization. *Journal of Sensor and Actuator Networks* **9**(3): 44.

Verbiest Rafaël *et al.,* 2021. Automation and robotics in the cultivation of pome fruit: Where do we stand today? *Journal of Field Robotics* **38**(4): 513-531.

Wang Zuliang *et al.,* 2020. Application of internet of things technology in intelligent orchard. Journal of Physics: Conference Series, 2020. Vol. 1533, pp. 042055. IOP Publishing.

Watteyne, Thomas *et al.,* 2016. Peach: Predicting frost events in peach orchards using iot technology.

Xie Jiaxing *et al.,* 2022. Smart fuzzy irrigation system for litchi orchards. *Computers and Electronics in Agriculture* **201**: 107287.

Zhang Chenglong *et al.,* 2021. Orchard management with small unmanned aerial vehicles: A survey of sensing and analysis approaches. *Precision agriculture* **22**(6): 2007-2052.

Zhang Moran *et al.,* 2023. Smart horticulture as an emerging interdisciplinary field combining novel solutions: past development, current challenges, and future perspectives. *Horticultural Plant Journal* **10**(6): 1257-1273.

Chapter 18

Use of Drones in Orchard Management

N.R. Prasanna Kumar, M.C. Keerthi, Hemant Kumar*, E. Vidya Madhuri*, J.S. Rupali*, K. Logeswaran and Tusar Kanti Behera

Division of Crop Protection, ICAR–Indian Institute Horticultural Research, Bengaluru–560089, Karnataka
**Division of Entomology, ICAR–Indian Agricultural Research Institute, New Delhi–110012*

The agricultural sector is the primary engine of India's economy. Despite technological advancements, the sector is facing numerous challenges. The use of age-old farming practices by large majority of Indian farmers still results in crop failure due to abiotic and biotic stresses. Therefore, even with the continuous efforts of farmers, the quality and quantity of agricultural products may need to be improved. Farmers can prevent crop failure by taking remedial actions if they can detect both abiotic and biotic stresses in advance. Conventional agriculture systems typically use heavy amounts and indiscriminate application of fertilizers, pesticides, and other agrochemicals. Crops do not properly utilize the higher doses of fertilizers and pesticides; thus, unutilized substances act as sources of environmental pollution. Managing these resources sustainably can be achieved through precision agriculture. Precision farming has become significantly impacted by drones, or unmanned aerial vehicles (UAVs), which have become indispensable in recent years. The timely and detailed crop health and pest data drones benefit orchards. The potential of drones, or unmanned aerial vehicles (UAVs), is to revolutionize current agriculture through their ability to improve sustainability, efficiency, and precision in orchard management. UAVs have a vast array of applications, including aerial imaging for early detection of disease, agricultural stress assessment, and tree health monitoring. The use of UAVs can help in solving the problems of labour shortages in agriculture lower operating expenses. Advanced sensors equipped with drones have brought about a revolution in fruit crop monitoring. The use of chemicals and their impact on the environment is reduced by their precise spraying of fertilizers and insecticides. Drones aid in yield estimation, canopy management, and post-harvest monitoring. This chapter focuses on drones' applications, advantages, challenges, and future prospects in orchard management. Furthermore, the challenges like payload limits, weather sensitivity, and regulations that are addressed through research and technological advancements are also discussed in this chapter.

Introduction

In recent years, the agricultural industry has experienced a profound technological revolution, with unmanned aerial vehicles (UAVs), commonly known as drones, emerging as indispensable tools in precision farming (Bajpeyi *et al.,* 2019). Orchards, in particular, have reaped the benefits of this transformation, as drones provide timely and detailed information about crop health, pest infestations, and overall orchard management. The continuous integration of cutting-edge technology, including sophisticated sensors, Internet of Things (IoT) devices, innovative algorithms, and contemporary gear, has reshaped agriculture. The development of UAVs has empowered them to monitor the long-distance crop fields and detect plant illnesses at an early stage, ushering in a new era of orchard management (Abbas *et al.,* 2023).

Farmers can now monitor and manage fruit crops with drones equipped with advanced sensors and imaging capabilities. Early detection of diseases, pests, and other stress factors that could impact crop health can be achieved through aerial surveillance that enables high-resolution imaging of orchards and vineyards. Precision and targeted interventions can be achieved by collecting data collected by drones, which optimizes resource use and enhances overall crop productivity (Anushi *et al.,* 2023). Furthermore, drones aid in estimating yield, managing canopy, and monitoring post-harvest. While the technology offers significant advantages, challenges such as payload limitations, weather sensitivity, and regulatory compliance must be addressed (Toscano *et al.,* 2024). Drones will become indispensable tools for sustainable and efficient fruit crop management, and ongoing research and technological advancements are expected to overcome these limitations. Sustainable management of these natural resources can be achieved through precision agriculture (Rejeb *et al.,* 2022).

Precision agriculture (PA), a method of farm management that leverages information technology (IT) to ensure crops and soil receive the right amount of nutrients for optimal health and productivity, is promoting sustainable agriculture and horticulture. This method utilizes remote sensing, artificial intelligence (AI), sensors, handheld Global Positioning Systems (GPS), UAVs, and manual field sampling. The recent integration of UAVs or drones has further modernized farming (Anand *et al.,* 2023). These UAVs, equipped with sensors, provide real-time information on plant, nutrient, and water status, leading to more efficient and precise management decisions in the orchard (Aiswarya*et al.*, 2024).

Realizing the benefits of drones in enhancing agricultural efficiency, the orchard growers of oil palm, chestnut, olive, peach, lychee, citrus, and mango increasingly use UAVs (Ok and Ozdarici-Ok, 2018). Further, the Government of India has introduced the Kisan Drone scheme, to accomplish Atmanirbha Bharat in the agricultural and related industries (Mahesh *et al.,* 2023). As technology progresses, grower operations increasingly use drones, particularly in nut orchards. By efficiently collecting valuable data for comprehensive crop analysis and decision-making, they provide new opportunities while taking on labour-intensive tasks like pest spraying (Singh and Singh, 2023). This chapter examines the various applications of drones in orchard management, emphasizing their benefits, struggles, and future possibilities.

Components of UAVs

The design of drones, or UAVs, determines whether propellers, wings, or a combination of both guide them. The two types of onboard propellers that drones equipped with propellers have are meant to provide direction and thrust. All other components are held in the chassis, constituting the quadcopter's main body (Mohsan*et al.,*2023). The flight controller ensures that the electronic speed controller's power supply is maintained, that the drone's orientation is detected, that the motors are operated, and that the drone is airborne. To create direction and thrust commands, the transmitter communicates signals from the controller to the drone, and the receiver gets them and transmits them to the Flight Controller PCB (Muhamad *et al.,* 2024). Images are captured, and remote signals are sensed by sensors and cameras installed on the chassis. Flight control, stability, remote communication, and tasking sensors are the primary sensors for operating and controlling the UAV. Imaging, sensing, tracking, and measuring sensors are secondary and may or may not be necessary for their function (Gaddikeri *et al.,* 2023).

Uses of Drones in Orchards

Aerial Imaging and Mapping

The way we perceive and understand our crop ecosystem has been revolutionized by advances in drone technology, which facilitate aerial imaging and mapping(Shah *et al.,* 2024). The drones equipped with high-resolution cameras and sophisticated sensors enable the capture of detailed aerial imagery over large areas with remarkable precision. Aerial imaging can comprehensively overview landscapes, infrastructure, and natural resources in urban planning, environmental monitoring, and agriculture (Tahar *et al.*, 2012). Accurate maps can only be created by gathering accurate data through aerial mapping, assessing changes in land use, and identifying potential areas of concern like environmental degradation or urban expansion. Disaster response has found this technology particularly valuable, as it can aid in assessing damage and planning relief efforts through rapid and detailed mapping (Rejesus*et al.*, 1999). Our ability to comprehend complex spatial relationships is enhanced by aerial imaging and mapping, empowering decision-makers in various sectors to make wise decisions using real-time, high-quality geospatial information. The advancement of drone technology will lead to the expansion of aerial imaging and mapping applications, which will contribute to more effective and sustainable resource management practices (Wang *et al.*,2022).

Mapping can be significantly improved using Unmanned Aerial Vehicles (UAVs) that produce highly accurate data. There are numerous UAV models and designs available in the market. UAVs capture images at various flight altitudes, from ground level up to 60m depending upon the capacity and type of the UAV and purpose. However, in many countries, aviation regulations often limit the visibility and altitude of UAVs. UAVs can acquire images without a pilot, just as manned flights do, enabling quick deployment and high-resolution imaging in confined areas (Whitley *et al.*, 2000). UAVs can be utilized in various applications and surveillance missions, including aerial tracking. An operator can operate fixed-wing UAVs through radio-controlled or autonomous operations. The accuracy of autonomous flight control can be affected by the impact of wind direction on the turbulence of UAVs. However, large study areas can be covered without human vision limitations through autonomous flight missions, but regional aviation laws must be considered (Molin *et al.*, 2007). During a mission, manual and autonomous flights require identical levels of control, including monitoring the flight path and examining any anomalies in the UAV's instrumentation from beginning to end. The safety of a UAV can be significantly affected by the operator's proficiency during flight missions (Aggelopoulou *et al.*, 2006).

UAVs and Cameras Used for Image Acquisition in Orchard Monitoring

Modern orchard monitoring now uses UAVs as the preferred platform (Zhang *et al.*, 2019). In this field, it is crucial to use UAVs, which are also known as unmanned fixed-wing devices or multicopters. The popularity of multicopters is particularly high, which are divided into quadcopters (four motors), hexacopters (six motors), and octocopters (eight motors). While octocopters may have redundancy and increased stability, their larger size and increased velocity make them more expensive and require expert handling, otherwise can lead to property damage or accidents. However, with the right safety measures in place, such as the use of redundant propellers in hexacopters to improve stability and reliability under hardware loss (Ju *et al.,* 39), the risks are significantly mitigated. The strengths and weaknesses of each platform are unique. Fixed-wing UAVs, for instance, have a significantly greater endurance and sometimes higher payload capacity, but they are not flexible enough, but due to their requirement for a minimum speed to avoid stalling and operating at higher altitudes, they are unable to be flexible. In large areas, fixed-wing UAVs have traditionally been employed to perform photogrammetry, monitoring, spraying, and data acquisition (Pederi and Cheporniuk, 2015). Multicopters can hover and approach objects closely, but their battery life is typically limited to 20 to 30 minutes (within tens of centimetres when safety measures are deactivated). The main tool in small and medium-scale precision agriculture is multicopters due to their relatively low cost. It's challenging to categorize multicopters comprehensively, but the number of motors and commercial models (mainly DJI or Parrot variants) or custom-made ones for specific research or application projects are the main factors that determine their classification. The drones employed for crop monitoring, particularly in orchards, are medium or small-sized, capable of gathering image or sensor data. Spraying, picking, or planting are all possible uses for larger drones, but they are still rare. Orchard monitoring has become popular with electric multi-rotor drones due to their short travel distances and modern batteries that provide enough autonomy (Popescu *et al.,* 2023).

The Phantom 3 is a popular DJI quadcopter model (Menshchikov*et al.*, 2021), Mavic 2 Pro (Nguyen *et al.,* 2022), and Inspire 2 (Hani, 2020) models are in use. Other models include the GYRO-500X4 quadcopter (Zortea*et al.*, 2018) and the microdrone MD4-1000 (Torres-Sánchez *et al.,* 2018). Hexacopters like the Tarot 960 are used in orchard applications (Nevalainen *et al.,* 2017), whereas, larger payload capacities and increased stability are achieved through the use of octocopters (Horstrand*et al.*, 2019; Deng *et al.,* 2020). Quadcopters are the most commonly used in orchard monitoring, but hexacopters are becoming more popular because of their redundant propellers, which improve stability and reliability under hardware loss (Ju *et al.,* 2022).

GPS modules, a standard feature in most commercial UAVs for outdoor localization, are reliable and precise. While GPS issues may arise in urban areas with tall buildings, the 'canyon effect,' in relatively smooth environments like orchards, the GPS, and avoidance sensors perform admirably, with position errors typically within 1 meter (Nevalainen *et al.,* 2017). To further enhance accuracy, an RTK module can be added, improving planimetry to 2 cm and altimetry to 3 cm (Torres-Sanchez*et al.*, 2018). Physical targets for GPS correction, like AeroPoints, have largely been replaced by RTK modules. These modules accurately position visible elements and use them as references in drone images (Johansen *et al.,* 2018). A comprehensive overview of research on the use of UAVs with various cameras to monitor orchards is presented in Table 18.1.

Table 18.1: Use of UAVs with Various Cameras to Monitor Orchards (Popescu *et al.*, 64)

UAV/Type	*Camera/Type*	*Reference*
DJI Mavic 2 Pro (quadcopter, DJI Corporation)	Hasselblad L1D-20c, 20MP/ RGB	Barmpoutis *et al.*, 2023
DJI Mavic 3 (DJI Corporation)	Included	Hani, 2020
Phantom 3 Professional (quadcopter, DJI Corporation)	RGB, Multispectral 5 channels, 12 MP	Horton *et al.*, 2017
Phantom 4, 4 PRO, 4 RTK (quadcopter, DJI Corporation)	Included, RGB, Multispectral 5 channels, 12 MP	Lobo Torres *et al.*, 2020
DJI Matrice 100 (quadcopter, DJI Corporation)	Logitech C310 webcam, MicaSenseRedEdge-M/ multispectral	Hulens*et al.*, 2017
DJI Matrice 210 (quadcopter/ possible RTK, DJI Corporation)	Two cameras/RGB - 48 MP (Sony Alpha 7) and multispectral 4 channels (Parrot Sequoia)	Ampatzidis*et al.*, 2024
HSE-EVO (quadcopter, ITALDRON)	MicaSenseRedEdge-M/ multispectral	Adamo *et al.*, 2021
DJI Inspire 1 (quadcopter, DJI Corporation)	Included, RGB	Hu and Li, 2020
DJI Inspire 2 (quadcopter, DJI Corporation)	Included, RGB	Mu *et al.*, 2018
DJI Matrice 600 (hexacopter/ possible RTK, DJI Corporation)	Zenmuse, SpecimFX10, Multispectral 5 channels, Reson Pika L 2.4 hyperspectral, MicaSenseRedEdge-M/ multispectral	Abdulridha *et al.*, 2019
OktoXL6S12 (octocopter, Mikrokopter)	Alpha 7R, Sony/RGB	Schiefer *et al.*, 2020
eBee Sense Fly (fixed-wing, Mikrokopter GmbH)	Parrot SEQUOIA, Multispectral 4 channels, senseFlyS.O.D.A.	Duarte *et al.*, 2020
Trimble UX5 (fixed-wing, Applanix)	RGB and multiple bands	Adhikari *et al.*, 2021

The use of UAVs allows for observing many cameras and their applications. UAVs can carry payloads that contain various types of image or video sensors, including RGB cameras, multispectral cameras, hyperspectral cameras, thermal sensors, and SAR sensors commonly utilized in orchard monitoring applications (Table 18.1). Small UAVs with built-in video/photo cameras are widely employed in crop monitoring applications, as they do not permit the attachment of additional cameras. Larger UAVs can accommodate different cameras based on specific requirements. The types and numbers of payloads, including thermal cameras, vary significantly despite the limited number and types of UAVs (Mesas-Carrascosa *et al.,* 2018; Padua *et al.,* 2020), video cameras (Torres-Sanchez *et al.,* 2018), multispectral cameras (Horton *et al.,* 2017) and even spectrometers like Ocean Optics (Nevalainen *et al.,* 2017). Due to cost reductions, energy requirements, and weight reductions, UAV applications are starting to incorporate cameras with integrated machine-learning features.

Characteristics of UAV Trajectories in Orchard Monitoring

Navigating complex 3D trajectories both above and within the orchard canopy is a unique challenge when monitoring an orchard with UAVs. This overview aims to examine the critical parameters and considerations that go into designing effective UAV trajectories for orchard monitoring applications.

Key Parameters in UAV Trajectories: Total Trajectory Time and Ground Velocity

Effective orchard monitoring often requires effortless coverage of significant distances. Studies have shown that the ground velocity of the UAV can vary between 2.5 m/s and 5 m/s depending on its capabilities and mission objectives (Mu*et al.,* 2018; Cheng *et al.,* 2024; Dong *et al.,* 2020). Torres-Sanchez *et al.* (2015) reported covering 6.2 km in 1.5 hours for almond orchards, highlighting the practical difficulties that arise during large-scale monitoring missions.

Flight Altitude

The precision of UAVs in orchard monitoring is evident in their ability to vary flight altitudes based on mission goals. For top-down observations like photogrammetry, heights range from 30 meters to 120 meters (Mu *et al.,* 56; Mesas-Carrascosa *et al.,* 2018; Dong *et al.,* 2020). For detailed inspections near treetops or within orchard rows, UAVs can confidently operate at altitudes of 1 m to 5 m.

Spacing of Orchard Rows and Trees

Effective trajectory planning requires a clear understanding of orchard layout (Figure 18.1). Inter-tree spacing of 4 meters and inter-row spacing of 5 meters are typical for apple trees (Cheng *et al.,* 2024). These dimensions influence the path and movement capabilities of the UAV when monitoring operations.

Challenges in UAV Navigation

Maneuvering Complexity and Battery Management

To navigate orchards with UAVs, it is essential to address the complexities of maneuvering, particularly for fixed-wing UAVs that are limited in agility. Despite the increased flexibility of rotary drones, strategic battery swapping and efficient battery management strategies are necessary to ensure continuous operations in large orchards, even though flight durations have increased.

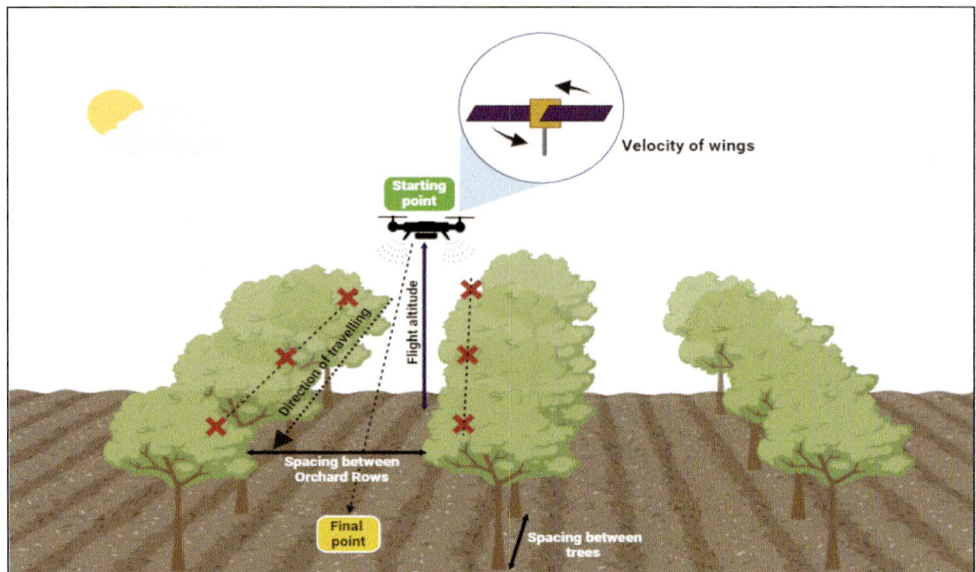

Figure 18.1: Role of Trajectories in UAVs
"X" mark indicates the collide pathway which makes the UAVs into collision.

Impact of Environmental Factors

Environmental conditions, particularly wind patterns, significantly impact UAVs' stability and energy efficiency. To maintain stable paths and reduce energy usage, flights are frequently scheduled during periods of low wind, which is crucial for long-term monitoring missions.

Trajectory Design and Planning

Autonomous Navigation and Obstacle Avoidance

Safe UAV operations in dynamic orchard environments can be ensured by integrating autonomous navigation and obstacle avoidance systems in pre-computed trajectory designs (Horton *et al.,* 33). This approach minimizes collision risks and improves data collection efficiency during photogrammetry missions.

Regulatory Frameworks and Future Directions

Role of Regulatory Frameworks

UAV operations in agriculture are shaped by the development of regulatory frameworks, such as Europe's U-space Regulation(European Union, 2019). Safe airspace management and operational standards are established through these frameworks, which ensure sustainable and compliant UAV use for orchard monitoring. The balance between technological advances and environmental constraints is necessary for effective trajectory planning for UAVs in orchard monitoring. Enhancing monitoring practices and supporting sustainable and efficient agricultural operations requires continuous innovations in UAV technology and regulatory frameworks.

Applications of Drones in Orchard Management

Throughout the growing season, fruit orchards necessitate customized management for every site or individual tree, including flowering, fruit development, ripening, harvesting, and tree dormancy. Growers used to rely heavily on their professional experience to make decisions based on visual inspections of fruit and trees' colour, shape, size, and other characteristics. This method frequently results in unreliable observations and necessitates skilled workers (Zhang *et al.,* 2021). Pesticides applications in orchard are particularly challenging to manage pests and diseases because of their taller heights, which require more labour and are less efficient. UAVs and machine learning techniques can effectively replace manual inspection of fruit ripening and maturity, which enhances decision-making (Figure 18.2). UAVs can watch for irrigation and nutrient changes in the field, which can help to lessen water and nutrient stress in trees. These tasks require a lot of work, are expensive, inefficient, and take time. Managing orchards more effectively can be achieved by integrating ICT with human efforts (Zhang *et al.,* 80).

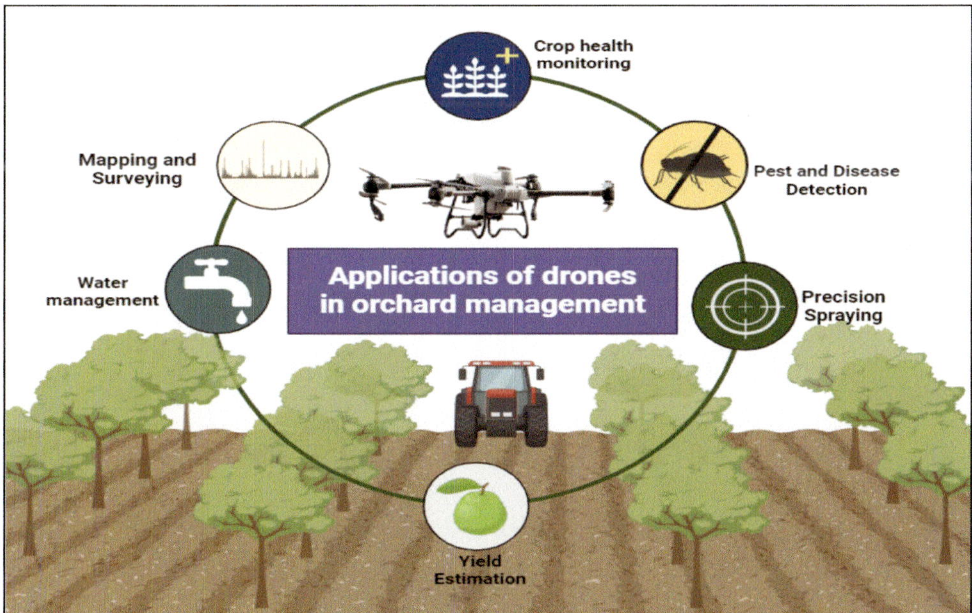

Figure 18.2: Applications of Drones in Orchard Management.

Crop Health Monitoring

Effective orchard management throughout the growing season requires attention to individual trees and specific sites. Traditionally, farmers made decisions by visually inspecting fruits for colour, shape, size, and other characteristics, relying heavily on personal experience, often leading to inaccuracies (Venkatesh *et al.*, 2022). The advent of drones offers a transformative solution for monitoring crop conditions, enabling timely and precise actions. With high-resolution cameras and multispectral or hyperspectral sensors, drones can capture detailed images of orchards, allowing close-up inspection and high spatial resolution. Their flexibility enables frequent revisits, ensuring high temporal-resolution imagery (Ballester *et al.*,2019). Farmers can detect early signs of stress, nutrient deficiencies, or disease outbreaks using drone imagery, allowing for prompt and appropriate interventions to prevent yield loss. Regular aerial surveys help monitor vegetation vigor and inform irrigation, fertilization, and pest control decisions. This technology eliminates manual crop inspection, making it particularly useful in remote or difficult-to-access areas such as mountainous regions. It also facilitates efficient monitoring of tall crops and trees, which are challenging to scout physically (Dutta and Goswami, 2024). Basic RGB or multispectral sensors on drones can identify signs of wilting or colour changes, providing critical information for assessing and monitoring overall crop development (FNG, 2023). The effectiveness of crop monitoring hinges on selecting appropriate sensors tailored to applications such as disease detection, nutrient assessment, and water status identification (Figure 18.3). Ongoing research and development have led to the creation of drones designed explicitly for crop monitoring, enhancing their utility in modern agriculture (Abdul *et al.*, 2023).

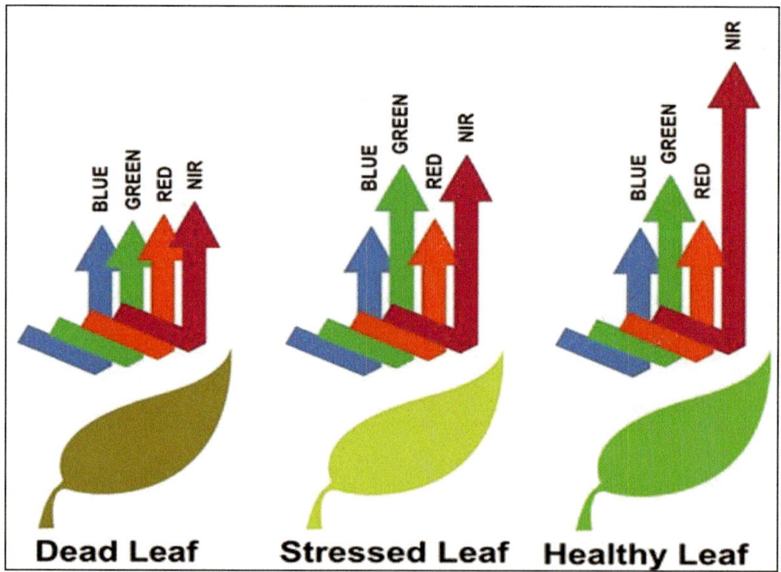

Figure 18.3: The Basic Principle of NDVI Relies on the Fact That, due the spongy layers found on their backsides, leaves reflect a lot of light in the near infrared, in stark contrast with most non-plant object. When the plant becomes dehydrated or stressed, the spongy layer collapses and the leaves reflect less NIR light, but the same amount in the visible range. Thus, mathematically combining these two signals can help differentiate plant from non-plant and healthy plant from sickly plant (Image courtesy Agribotrix.com)

Pest and Disease Detection

During cropping season, higher yield loss is caused by pests and diseases, which are the most devastating. Farmers can manage yield loss by proactively detecting pests or diseases before they spread. Crop disease diagnosis is achieved through visual observation of symptoms, microscopy, and isolation of pathogen cultures, a method that is often tedious, time-consuming, and cumbersome (Kalischuk *et al.,* 2019; Wang *et al.,* 2018). In this scenario, due to their versatility in revisiting the field and their ability to capture high-resolution imagery, drones are becoming more popular than other airborne imagery; the plant can be seen at a much closer distance (Maes and Steppe, 2019). Using drones in orchard management has one of the most significant advantages, as they can detect pests and diseases early. Temperature differences or spectral signatures linked to pest infestations or fungal infections can be identified using thermal imaging sensors and specialized software. Targeted interventions can be implemented through this early detection, reducing the need for broad-spectrum pesticides and decreasing crop losses (Picanco Filho *et al.,* 2024).

Disease and pest surveillance in fruit crops has been made possible by drones equipped with advanced sensors and imaging technology. Farmers can gain detailed insights into crop health by using their aerial capabilities to efficiently and thoroughly monitor extensive orchards and vineyards. By capturing high-quality images and thermal data with these unmanned aerial vehicles, it is possible to detect potential disease outbreaks and pest infestations that may not be visible from the ground (Jain *et al.,* 2023). Timely interventions, including targeted pesticide application or disease management strategies, can be facilitated by conducting a quick and precise survey of the entire crop area. Drones are particularly effective in discovering patterns of disease spread or pest movement due to their ability to cover large areas relatively quickly. Using specialized software, drone data is processed to create detailed maps highlighting areas of concern (Lawrence *et al.,* 2024).

Farmers can improve crop yields and the overall health of fruit crops by integrating drone technology into disease and pest surveillance and taking proactive measures. With the advancement of drone technology, it is anticipated that it will play a more significant role in improving the efficiency and effectiveness of disease and pest management in fruit agriculture. Despite detecting yellowing plants in a field of green with the help of spectral imaging technology, pathogens causing crops to wither and die can often go unnoticed (Aliloo *et al.*, 2024). Drones are used by Schale at Virginia Tech to detect pathogens that have not yet settled in visible spaces. Crop diseases are highly destructive and can be caused by bacteria, viruses, or fungal agents. Infrared cameras on drones can inspect the interior state of plants, giving a clear picture of their condition. Farmers can take action, such as removing the affected plant to prevent its spread to nearby plants, when they detect an infection early on. Identifying and diagnosing plant diseases requires image-based tools when human assessments are not appropriate, unreliable, or unavailable, especially considering the increased coverage of UAVs. Hyperspectral and thermal images have been examined, but RGB and multispectral images dominate the data collection market (Ameer *et al.,* 2019). High-resolution aerial imaging can be used by drone technology to detect citrus greening disease, also known as Huanglongbing (HLB). Multi-band imaging sensors are used by drones equipped with them to capture images at adjusted altitudes to achieve the required resolution. Drone-based sensors were compared with aircraft-based sensors that had less spatial resolution. Regression analysis was used to analyze data from seven vegetation indices (VIs) and six spectral bands ranging from 530 to 900 nm. According to the findings, high-resolution aerial sensing is reliable for recognizing citrus trees infected with HLB (Garcia-Ruiz *et al.,* 2013; Deng *et al.,* 2020).

The threat of deadly diseases like *Verticillium*Wilt (VW) in olive trees seriously threatens fruit industries worldwide. Water flow is severely impacted and is significantly hindered by VW infecting the vascular system. Using UAVs, multispectral, hyperspectral, and thermal images can help detect Volkswagen as the most limiting disease of olive trees and assess its severity levels early. The investigation was carried out to investigate multiple vegetation indexes, such as chlorophyll a+b, blue/green/red/B/G/R indices, NDVI, and CWSI (Montes-Osuna and Mercado-Blanco, 55). Further insights were provided by investigations carried out at the leaf and tree-crown levels. Early detection was achieved through CWSI, visible ratios B/BG/BR, and fluorescence, as demonstrated by the results; the R/G ratio is the best method to assess severity levels, whereas the photochemical reflectance index (PRI), chlorophyll, carotenoid indices, and the photochemical reflectance index (PRI) are the best (Calderon *et al.,* 2019).

Precision Spraying

Traditional pesticide application methods often cause inefficient chemical use and environmental concerns. Drones with precision spraying systems and reliance on real-time data and geospatial mapping can accurately deliver pesticides or fertilizers to particular areas within the orchard (Figure 18.4). This targeted approach can reduce chemical usage while minimizing labour costs and environmental impact. The spraying of chemicals, such as fertilizers and pesticides, can be done by drones based on the spatial variability of crops and fields. Adjustments can be made to the chemical amount to be sprayed based on the crop conditions or the severity of the insect pest attack. Pest management could be made more accessible by integrating drones with spray nozzles. However, spraying drones is currently restricted due to a lack of understanding of the drift they produce compared to conventional spraying. Both spraying platforms in the South Iberian Peninsula were subjected to trials to assess their sedimented drift. The drone sprayer's results indicated that it required significantly less distance to deposit 90 per cent of the sprayed volume than a conventional mist blower (Luis *et al.,* 2023).

Drone spraying has advantages over conventional methods when landscapes prevent using ground sprayers. Drones can be used to spray small and irregular

Figure 18.4: Spray Drone.

fields with ease and efficiency. Drones significantly reduce the risk of pesticide exposure/contamination for applicators. Large fields and tall trees, especially in orchards, would be accurately sprayed using drones (Figure 18.5).

Figure 18.5: Drone Spraying in Pomegrante and Mango Orchard at ICAR–IIHR.

To perform efficient aerial pesticide spraying, drones that are easy to operate are necessary. Using drones in spray platforms has benefited growers in Israel who grow almonds, walnuts, pistachios, wine and cannery grapes, citrus and avocados. Deploying beneficial insects and releasing pheromones is done to prevent pests from mating (FGN, 2023). The main focus of drone applications in agriculture is pesticide application. However, before commercializing drones in agriculture, it is crucial to optimize drone operational parameters such as spray volume, droplet size, spread of droplets, droplet drift, penetrability flight speed, flight height, nozzle type, payload, drone type, and insecticidal efficiency (Lou *et al.*,2018; Lv*et al.*, 2019).

Mapping and Surveying

Farmers can obtain detailed information about the topography, drainage patterns, and overall layout of orchards by creating high-resolution maps and 3D models of orchards by drones. This data is crucial for maximizing planting densities, constructing drainage systems, and evaluating the efficiency of land management strategies. Modern agriculture relies heavily on drones with various controllers and sensors, such as control systems and LIDAR sensors. The ability to map agricultural fields in real-time efficiently provides detailed spatial and temporal information over large areas (Mekala *et al.*, 2023). Drones can be beneficial in surveying orchards and monitoring soil conditions across different types and nutrient levels. Drones are utilized in regions like India to prevent environmental threats like unlawful stubble burning and provide oversight and surveillance capabilities. Moreover, drones permit accurate 3-D mapping for detailed soil analysis, including properties, moisture content, and erosion. This is essential for maximizing seed planting patterns and managing irrigation and nitrogen. One of the significant challenges in modern farming is addressed by their ability to monitor crops across vast fields. (Kumar *et al.*, 2022). The reliability of UAV-derived algorithms in mapping fruit crops' geometric features is impressive. They can achieve a root mean square error of 0.39m for mapping 3-D almond tree volume and growth, as demonstrated by a dataset gathered over various phenological stages over two years (Torres-Sanchez *et al.*, 2015). Another study shows that canopy area, tree heights, and crown volumes can be calculated accurately with 97 per cent quantification accuracy. This level of precision is a *testament to the* potential of UAV technology in agriculture.

It's time-consuming to generate DSMs for 3-D reconstruction in UAV photogrammetry. The optimization of processing time was achieved by testing multiple DSMs with different forward laps. Using a 100m flying altitude and a 95 per cent forward lap led to the best outcomes, which included 95 per cent accuracy in estimating tree volume and an 85 per cent time-saving performance (Torres-Sanchez *et al.*, 2018). The accuracy of 2-D delineation must be improved to achieve effective 3-D reconstruction. As evidenced by comparisons with eight DSMs, a new approach using dense photogrammetric DSMs achieved a good balance between accuracy and recall (Ok *et al.*, 2018). Multispectral UAV images can assess pruning effects by analyzing fruit crop structural properties like crown perimeter, width, height, and area. After pruning, significant changes can be seen in these properties. Different flying heights measure Fruit tree structures differently: higher altitudes decrease crown perimeter and height measurements, but they only affect crown width and Plant Projective Cover (PPC) (Johansen *et al.*, 2018). Consumer-grade

UAVs can accurately measure crown width and crown projection area (CPA) using Digital Surface Models (DSM) and help map orchard management dynamics. UAV-based methods can replace field measurements and match the accuracy of manual delineation (Mu *et al.,* 2018). An F1-score of 92.5 per cent was obtained by integrating orientation symmetry and local maximum cues into a novel approach for DSM application in fruit crop detection (Ok *et al.,* 2022). Tree growth can be influenced by various pruning strategies, including traditional, adapted, and mechanical. A year later, using UAV technology and object-based image analysis (OBIA) to quantify pruning effects showed that adapted pruning has an advantage for tree height when intensity is less than 10 per cent. Tree height, crown volume, and projected canopy area (PCA) were the parameters that had quantitative data (Jimenez-Brenes *et al.,* 2017).

Water Management

Remote sensing is used to monitor soil moisture, the use of Drone technology has led to a significant transformation in water management in agriculture. Assessing and optimizing water usage in fields is possible with drones equipped with advanced sensors and imaging. Insights into soil moisture, irrigation efficiency, and crop health can be gained through aerial surveys. The fast identification of areas with water stress or over-irrigation by drones makes it possible to make precision irrigation adjustments. By utilizing real-time data, informed decisions can be made about when, where, and how much water to apply, leading to resource conservation and improved yields. The detection of leaks and inefficiencies in irrigation systems can be enhanced by drones, which can improve overall water management. Their integration is an approach that is sustainable and data-driven, resulting in judicious water use and agricultural productivity (Cancela *et al.,* 19). Drone technology will likely play a more significant role in water management as it advances, presenting more advanced solutions for water scarcity and efficient practices. Uneven water absorption by land and variable equipment performance can leave areas under or over-irrigated. Crop wilting points and irrigation leaks can be identified using thermography and spectroscopy. Additionally, 3D maps are created from aerial photos using software and airborne laser scanning, which show catchments and water flow directions. This helps with topography evaluation and addresses soil erosion and crop health. Monitoring water stress is complex due to various factors, with thermal imaging used to detect stress. Conditions and crop genotypes influence the variation of regression equations and thresholds (Kolhe and Munde, 2020).

The paragraph discusses various methods and technologies that enable water management in agriculture. It underscores the irregular absorption of water by the soil and the possible flaws of irrigation equipment in distributing water uniformly. Identifying dry areas where crops might wilt is achieved through techniques such as thermography and spectroscopy, and leaks in irrigation systems can be detected through imaging (Bhushan *et al.,* 2014). Detailed 3D maps of their land are created by orchard growers using software and airborne laser scanning, which reveal topographical features, catchments and water flow patterns that influence soil erosion and crop health. Detecting minute temperature changes that indicate drought stress on crops is achieved through thermal imaging to monitor water stress. Genetic and environmental factors that affect plant transpiration rates and canopy temperatures can cause the applicability of this approach to vary (Debangshi and Sadhukhan, 2023).

Weed Management

Weeds are among the many challenges faced by orchards. Competition for resources like water, space, and nutrients with the main crop leads to up to 50 per cent yield losses. The most effective way to control weeds is using herbicides, which can also uproot weeds at the initial stages of growth. Excessive use of herbicides can lead to herbicide-resistant weeds, impacting the growth and yield of primary crops. Drone imaging data can accurately delineate weed-infested areas, optimizing herbicide application. Drone-based herbicide spraying is beneficial for pre- and post-emergence weed control, enabling operation in diverse field conditions and remote settings, thereby promoting safety. Incorporating drone technology for herbicide spraying and weed identification marks a significant advancement in modern agriculture. Equipped with cameras, drones improve weed detection efficiency and offer a more adaptable and systematic approach. According to Gokool *et al.* (30), this integration enhances operational efficiency by 60 times compared to conventional methods.

Yield Estimation

Drones equipped with geospatial technology provide a groundbreaking method for predicting crop yields. These unmanned aerial vehicles (UAVs) are flown over fields to capture detailed data, offering a comprehensive view of crop health and development. Essential technologies employed on these drones include geospatial imaging, multispectral and hyperspectral sensors, and Light Detection and Ranging (LiDAR) technology. These advancements enable drones to gather precise information on various crop aspects, such as plant health indicators, canopy coverage, and fruit maturity. By leveraging these technologies, farmers can accurately estimate crop yields and make informed irrigation, fertilization, and pest management decisions.

The integration of geospatial technology into drone operations enhances the accuracy and dependability of crop yield predictions, paving the way for a revolutionary future in agriculture. Geospatial imaging allows drones to capture high-resolution aerial photographs, which are then analyzed using specialized software to create maps and models of the field. Multispectral and hyperpectral sensors on drones collect data across different wavelengths beyond the visible spectrum, revealing additional insights into plant health and stress levels. LiDAR technology further enhances this capability by providing detailed 3D mapping of the terrain and crop structure, assisting in precise volume and biomass estimation. The data collected by drones equipped with these advanced technologies are pivotal for modern agriculture, empowering farmers to optimize resource allocation and maximize productivity while minimizing environmental impact. This real-time monitoring and analysis provide actionable insights that support sustainable farming practices and enhance overall efficiency in crop management. Looking forward, the ongoing development of drone technology promises even greater advancements in precision agriculture. As drones become more advanced and accessible, they are poised to revolutionize horticulture and agriculture, offering innovative solutions for predicting yields, monitoring crops, and managing agricultural operations. The future of agriculture is bright, with drones leading the way in precision and efficiency.

1. **Multispectral and Hyperspectral imaging**: These sensors capture data beyond the visible spectrum, revealing insights into plant health that are invisible to the human eye. By analyzing how different wavelengths of light are reflected, these sensors can detect subtle variations indicating stress, nutrient deficiencies, or the presence of diseases (Thomas *et al.*, 2018).
2. **LiDAR (Light detection and ranging)**: LiDAR technology utilizes laser pulses to generate detailed 3D terrain maps. This enables precise plant height measurements, canopy coverage, and overall field topography. Such data is crucial for understanding variations in sunlight exposure, water drainage patterns, and potential microclimates within the field (Debnath *et al.*, 2020).

The integration of drones and geospatial technology offers numerous advantages for predicting crop yields:

- **Enhanced accuracy and early warning**: Drone-based predictions provide significantly higher accuracy than conventional methods. Real-time, high-resolution data enables early detection of potential issues, allowing farmers to implement corrective measures before adversely affecting yield.
- **Targeted resource management**: Precise yield forecasts empower farmers to allocate resources such as fertilizers and water more efficiently. This targeted approach optimizes resource use, minimizes waste, and improves farm profitability.
- **Mitigation of crop losses**: Early identification of crop stress or diseases through drone-based predictions enables timely interventions. This proactive approach can effectively reduce crop losses and ensure more consistent harvests.

Drones equipped with advanced geospatial technologies are revolutionizing crop yield prediction. By offering a detailed perspective of field conditions and harnessing robust data analysis capabilities, this technology empowers farmers to enhance their methods and achieve maximum yields.

The Adoption of Orchard Drones in India

India stands as one of the world's foremost fruit producers, yet it grapples with the realities of a significant population of small-scale farmers. These farmers often contend with water scarcity, pests, and diseases. Drones hold promise in mitigating these challenges and boosting agricultural efficiency and productivity, although their adoption in India is still nascent. Several obstacles hinder their widespread application in Indian agriculture, limiting their potential benefits to small and marginal farmers. Addressing these challenges with supportive government policies could make this technology accessible and beneficial to these farmers. As technology progresses, drone production costs are expected to decrease, further aiding their integration into agricultural practices. The recent launch of the Kisan Drone scheme by the Government of India aims to bolster drone usage in the farm sector, signaling a potential increase in their deployment and impact.

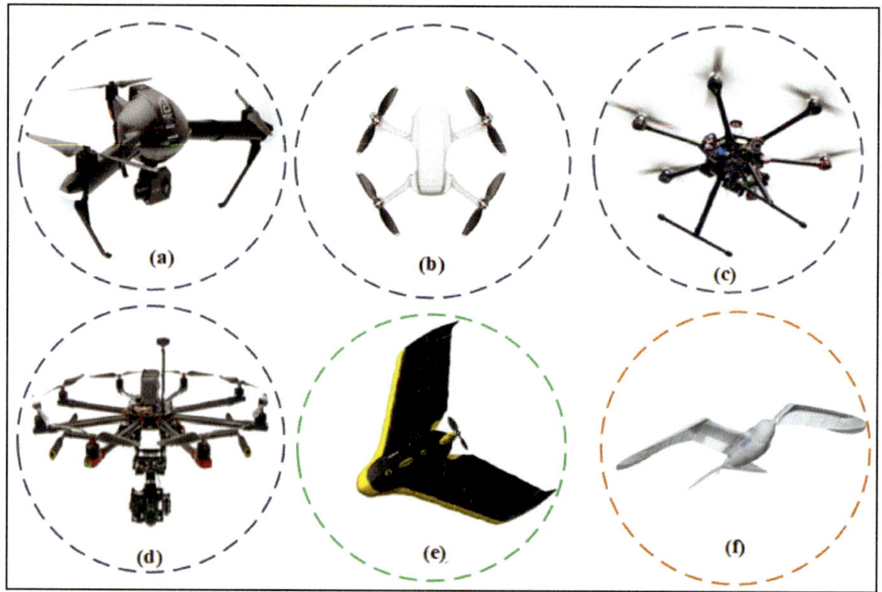

Figure 18.6: The Popular Drone Types Used in Agriculture:

(a) Rotary wing (tri-copter), (b) Rotary wing (quadcopter), (c) Rotary wing (hexacopter), (d) Rotary wing (octocopter), (e) Fixed-wing (eBeeTMTM) and (f) Flapping wing (SmartBird) (Cai *et al.*, 2013).

Challenges and Considerations

While drones offer numerous benefits in orchard management, several challenges must be addressed to achieve widespread adoption:

1. **Regulatory compliance**: Drones are subject to regulations governing flight restrictions, privacy concerns, and licensing requirements.
2. **Data management**: Managing large volumes of data drones generates demands for robust storage solutions and analytical tools.
3. **Cost**: Initial investment costs for purchasing drones and associated equipment can be substantial, though long-term savings in labour and inputs may offset these expenses.
4. **Technical expertise**: Operating drones and interpreting data require specialized training and skills, which may present barriers to adoption for some farmers.

Future Directions

The future of drones in orchard management shows promise, driven by ongoing advancements in sensor technology, artificial intelligence, and automation. Key areas of development include:

1. Autonomous Operations: Integrating drones with AI algorithms enables autonomous flight planning, data analysis, and decision-making.
2. Sensor Integration: Enhancing sensor accuracy and miniaturization to improve data collection capabilities.
3. Interoperability: Developing standardized data formats and interoperable platforms to facilitate seamless data sharing and integration with other farm management systems.

Conclusion

In conclusion, drones have not just improved but transformed orchard management. They have provided farmers with unprecedented insights into crop health, pest management, and overall farm productivity. While challenges exist, the continued technological progress and supportive regulatory frameworks are expected to accelerate drone adoption in agriculture. This evolution promises sustainable and efficient orchard management practices for the future.

REFERENCES

Abbas N, Abbas Z, Liu X, Khan SS, Foster ED and Larkin S, 2023. A survey: Future smart cities based on advance control of Unmanned Aerial Vehicles (UAVs). *Appl. Sci.,***17**: 9881.

Abdul-Jabbar TS, Ziboon AT and Albayati MM. 2023. January. Crop yield estimation using different remote sensing data: Literature review. In *IOP Conference Series: Earth and Environmental Science*, IOP Publishing. **1129**. pp. 12004.

Abdulridha J, Batuman O and Ampatzidis Y, 2019. UAV-based remote sensing technique to detect citrus canker disease utilizing hyperspectral imaging and machine learning. *Remote Sens.* **11**: 1–22.

Adamo F, Attivissimo F, Di Nisio A, Ragolia MA and Scarpetta M, 2021. June.A new processing method to segment olive trees and detect *Xylella fastidiosa* in UAVs multispectral images, In *IEEE International Instrumentation and Measurement Technology Conference proceedings.* IEEE International.**5**. pp. 2642.

Adhikari A, Kumar M, Agrawal S and Raghavendra S, 2021. An integrated object and machine learning approach for tree canopy extraction from UAV datasets. *J. Indian Soc. Remote Sens.* **49**: 471–478.

Aggelopoulou KD, Fountas S, Gemtos TA, Nanos GD and Wulfsohn D January, 2006. Precision farming in small apple fields of Greece. In *Proceedings of the 8th International Conference on Precision Agriculture.* Precision Agriculture press. **5**. pp.538-556.

Aiswarya L, Siddharam Rajesh GM, Gaddikeri V, Jatav MS, Rajput J and Asha KR, 2024. Smart Irrigation Management Through Unmanned Aerial Vehicles (UAVs). **In**: *Artificial Intelligence and Smart Agriculture: Technology and Applications,* Kusum, P. (Ed.), Springer Nature Singapore, Singapore, pp. 293-313.

Ameer SA, Alkhafaji MA, Jaffer Z and Al-Farouni M June, 2024. Empowering Farmers with IoT, UAVs, and Deep Learning in Smart Agriculture. In*E3S Web of Conferences,* EDP Sciences, pp. 4007.

Ampatzidis Y, Partel V, Meyering B and Albrecht U, 2019. Citrus rootstock evaluation utilizing UAV-based remote sensing and artificial intelligence. *Comput. Electron. Agric.* **164**: 1–10.

Anand A and Deb C, 2024. The potential of remote sensing and GIS in urban building energy modelling. *Energy Built Environ.***5**: 957-969.

Anushi SJ, Krishnamoorthi A and Singh SK, 2023. Cultivating Tomorrow: Precision Agriculture and Sustainable Crop Production. *Agric. and Food: e-Newsl.* **5**: 558-562.

Bajpeyi MM, Arun K, Ashutosh SA, Ashutosh S and Pramod KM, 2023. Use of Drone Technology in Precision Farming. In *Research Trends in Multidisciplinary Research and Development,* S. Fedorov (Ed.),Chandra Shekhar Azad University of Agriculture and Technology, Kanpur, India, pp. 123-142.

Ballester C, Brinkhoff J, Quayle WC and Hornbuckle J, 2019. Monitoring the effects of water stress in cotton using the green red vegetation index and red edge ratio. *Remote Sens,* **11**: 873.

Barmpoutis P, Kamperidou V and Stathaki T, 2019. Estimation of extent of trees and biomass infestation of the suburban forest of Thessaloniki (Seich Sou) using UAV imagery and combining R-CNNs and multichannel texture analysis. *In*:

Proceedings of Twelfth International Conference on Machine Vision (Vol. 2019, No. 3), Amsterdam, Netherlands. 114333.

Bhushan M, Kumar P and Bhusan BS, 2023. *Advancements in Precision Irrigation Techniques for Sustainable Water Management.***In**: *Farming for the Future: Smart Agriculture Innovations,* Santosh, D.T. (Ed.), Renu Publishers, New Delhi. pp. 43-60.

Cai G, Dias J and Seneviratne L, 2014. A survey of small-scale unmanned aerial vehicles: Recent advances and future development trends. *Unmanned Syst.* **2**: 175-199.

Calderon R, Navas-Cortes JA, Lucena C and Zarco-Tejada PJ, 2013. High-resolution airborne hyperspectral and thermal imagery for early detection of Verticillium wilt of olive using fluorescence, temperature and narrow-band spectral indices. *Remote Sens. Environ.* **139**: 231-245.

Cancela JJ, González XP, Vilanova M and Mirás-Avalos JM, 2019. Water management using drones and satellites in agriculture. *Water,* **11**: 874.

Cheng Z, Qi L, Cheng Y, Wu Y and Zhang H, 2020. Interlacing orchard canopy separation and assessment using UAV images. *Remote Sens.* **12**: 767.

De Luis-Ruiz JM, Sedano-Cibrian J, Perez-Alvarez R, Pereda-Garcia R and Ruiz-Fernández C, 2024. UAV and Laser Scanner Point Clouds Integration. **In**: *Advances in Design Engineering IV.* Manchado del Val, C., Suffo Pino, M., Miralbes Buil, R., Moreno Sanchez, D., Moreno Nieto, D. (Ed.), Springer, Switzerland. pp. 424-434.

Debangshi U and Sadhukhan A, 2023. Crop water stress monitoring through precision technologies: A review. *Pharm. Innov.***12**: 303-308.

Debnath S, Paul M and Debnath T, 2023. Applications of LiDAR in agriculture and future research directions. *J. Imaging.***9**: 57.

Deng X, Zhu Z, Yang J, Zheng Z, Huang Z, Yin X, Wei S and Lan Y, 2020. Detection of citrus huanglongbing based on multi-input neural network model of UAV hyperspectral remote sensing. Remote Sens. **12**: 2678.

Dong X, Zhang Z, Yu R, Tian Q and Zhu X, 2020. Extraction of information about individual trees from high-spatial-resolution UAV-acquired images of an orchard. *Remote Sens.***12**: 133.

Duarte A, Acevedo-Munoz L, Goncalves CI, Mota L, Sarmento A and Silva M, 2020. Detection of Long horned Borer attack and assessment in eucalyptus plantations using UAV imagery. *Remote Sens.***12**: 3153.

Dutta G and Goswami P, 2020. Application of drone in agriculture: A review. *Int. J. Chem. Stud,* **8**: 181-187.

f Ameer J, Abbasi E, Karamidehkordi E, Parmehr EG and Canavari M, 2024. Dos and Don'ts of using drone technology in the crop fields. *Technol. Soc.***76**: 102456.

Gaddikeri V, Asha KR, Ajay S, Malkhan SJ and Rajput D, 2023. UAVs in Orchard Management: Need, Challenges, and Applications. *Just Agriculture*.**5**: 1-7.

Garcia-Ruiz F, Sankaran S, Maja JM, Lee WS, Rasmussen J and Ehsani R, 2013. Comparison of two aerial imaging platforms for identification of Huanglongbing-infected citrus trees. *Comput. Electron. Agric.***91**: 106-115.

Gokool S, Mahomed M, Clulow A, Sibanda M, Kunz R, Naiken V and Mabhaudhi T, 2024. Exploring the Potential of Remote Sensing to Facilitate Integrated Weed Management in Smallholder Farms: A Scoping Review. *Drones*, **8**: 81.

Hani N, 2020. Minne Apple: a benchmark dataset for apple detection and segmentation. *IEEE robot. autom. lett.* **5**: 852–858.

Horstrand P, Guerra R, Rodrýguez A, Dýaz M, Lopez S and Lopez JF, 2019. A UAV platform based on a hyperspectral sensor for image capturing and on-board processing. *IEEE Access 7*, pp. 66919–66938.

Horton R, Cano E, Bulanon D and Fallahi E, 2017. Peach flower monitoring using aerial multispectral imaging. *J. Imaging*.**3**: 1–10.

Hu X and Li D, 2020. Research on a single-tree point cloud segmentation method based on UAV tilt photography and deep learning algorithm. *IEEE J. Sel. Top. Appl. Earth Obs. Remote Sens.* **13**: 4111–4120.

Hulens D, Vandersteegen M and Goedeme T, 2017. Real-time vision-based UAV navigation in fruit orchards, In *Proceedings of 12th International Joint Conference on Computer Vision, Imaging and Computer Graphics Theory and Applications.* Science and Technology Publications. **4**. pp. 617-622.

Jain S, Bhujel S, Shrivastava U, Mohapatra A and Mishra G, 2023. Advancements in drone technology for fruit crop management: A comprehensive review. *Int. J. Environ. Clim.***13**: 4367-78.

Jimenez-Brenes FM, Lopez-Granados F, De Castro AI, Torres-Sanchez J, Serrano N and Pena JM, 2017. Quantifying pruning impacts on olive tree architecture and annual canopy growth by using UAV-based 3D modelling. *Plant methods*, **13**: 1-15.

Johansen K, Raharjo T and McCabe MF, 2018. Using multi-spectral UAV imagery to extract tree crop structural properties and assess pruning effects. *Remote Sens.***10**: 854.

Ju C, Kim J, Seol J and Son H, 2022. A review on multirobot systems in agriculture. *Comput. Electron. Agric.* **202**: 107336.

Kalischuk M, Paret ML, Freeman JH, Raj D, Da Silva S, Eubanks S, Wiggins DJ, Lollar M, Marois JJ, Mellinger HC and Das J, 2019. An improved crop scouting technique incorporating unmanned aerial vehicle–assisted multispectral crop imaging into conventional scouting practice for gummy stem blight in watermelon. *Plant Dis.* **103**: 1642-1650.

Kolhe P and Munde TN, 2020. Water management made easy through drone-a case study. **In**: *International Symposium on Artificial Intelligence Based Future Technologies in Agriculture, Pune.*

Kumar A, Rani M, Aishwarya and Kumar P, 2022. Drone Technology in Sustainable Agriculture: The Future of Farming Is Precision Agriculture and Mapping. **In**: *Agriculture, Livestock Production and Aquaculture: Advances for Smallholder Farming Systems,* Kumar, A., Kumar, P., Singh, S.S., Trisasongko, B.H., Rani, M. (Ed.), Springer International Publishing, pp. 3-12.

Lawrence ID, Agnishwar J and Vijayakumar R, 2024. Autonomous Agriculture: The Role of IoT, AI, and ML in Drone Operations. **In**: *Sustainable Development in AI, Blockchain, and E-Governance Applications. IGI Global*, pp. 14-34.

Lobo Torres D, Queiroz Feitosa R, Nigri Happ P, Elena CueLa Rosa L, Marcato Junior J and Martins J, 2020. Applying fully convolutional architectures for semantic segmentation of a single tree species in urban environment on high resolution UAV optical imagery. *Sens.* **20**: 563.

Lou Z, Xin F, Han X, Lan Y, Duan T and Fu W, 2018. Effect of unmanned aerial vehicle flight height on droplet distribution, drift and control of cotton aphids and spider mites. *Agron.* **8**: 187.

Lv X and Wang Z, 2019. An improved RF detection algorithm using EMD-based WT. *KSII T. Internet Inf. Systems.* **13**: 3862-3879.

Maes WH and Steppe K, 2019. Perspectives for remote sensing with unmanned aerial vehicles in precision agriculture. *Trends Plant Sci.* **24**: 152-164.

Mahesh B, Dhanush S, Rakshita C, Raghavendra KR and Geetha Priya M, 2023. Antarctic Ice Sheet Surface Mass Balance Using UAV-Based Digital Elevation Model. In*International Conference on Cognitive Computing and Information Processing*. Springer Nature, Switzerland, **12**. pp. 39-49.

Meesaragandla S, Jagtap MP, Khatri N, Madan H and Vadduri AA, 2024. Herbicide spraying and weed identification using drone technology in modern farms: A comprehensive review.*Results Eng*. 101870.

Mekala MS, Dhiman G, Viriyasitavat W, Park JH and Jung HY, 2023. Efficient lidar-trajectory affinity model for autonomous vehicle orchestration. *IEEE* Trans. *Intell. Transp. Syst.***25**: 2708 – 2718.

Menshchikov A, Shadrin D, Prutyanov V, Lopatkin D, Sosnin S, Tsykunov E, Iakovlev E and Somov A, 2021. Real-time detection of hogweed: UAV platform empowered by deep learning. *IEEE Trans. Comput.* **70**: 1175-1188.

Mesas-Carrascosa FJ, Perez-Porras F, Merono de Larriva JE, Mena Frau C, Aguera-Vega F and Carvajal-Ramýrez F, 2018. Drift correction of lightweight microbolometer thermal sensors on-board Unmanned Aerial Vehicles. *Remote Sens.* **10**: 615.

Mohsan SAH, Othman NQH, Li Y, Alsharif MH and Khan MA, 2023. Unmanned aerial vehicles (UAVs): Practical aspects, applications, open challenges, security issues, and future trends. *Intell. Serv. Robot.* **16**: 109-137.

Molin JP, Mascarin LS, Povh FP, Amaral JR and Oliveira AS, October, 2007. *Proceedings of the 6th European Conference on Precision Agriculture*, Wageningen Academic Publications. **29**. pp. 371.

Montes-Osuna N and Mercado-Blanco J, 2020. Verticillium wilt of olive and its control: what did we learn during the last decade?. *Plant.* **9**: 735.

Mu Y, Fujii Y, Takata D, Zheng B, Noshita K, Honda K, Ninomiya S and Guo W, 2018. Characterization of peach tree crown by using high-resolution images from an unmanned aerial vehicle. *Hortic. Res.* 5.

Muhamad A, Panjaitan SD and Yacoub RR, 2024. Design and development of flight controller for quadcopter drone control. *TELECTRICAL.* **1**: 279-291.

Nevalainen O, Honkavaara E, Tuominen S, Viljanen N, Hakala T and Yu X, 2017. Individual tree detection and classification with UAV-based photogrammetric point clouds and hyperspectral imaging. *Remote Sens.* **9**: 185.

Nguyen HT, 2022. Evaluation and identification of bark beetle-induced forest degradation using RGB-acquired UAV images and machine learning techniques. *Ph. D. Thesis,*Yamagata University, Faculty of Agriculture, Yamagata.

Ok AO, Ozdarici-Ok A, 2018. Combining orientation symmetry and LM cues for the detection of citrus trees in orchards from a digital surface model. *IEEE Geoscience and Remote Sensing Letters.* **15**: 1817-1821.

Padua L, Adao T, Hruska J, Guimaraes N, Marques P and Peres E, 2020. Vineyard classification using machine learning techniques applied to RGB-UAV imagery In *Proceedings of IEEE International Geoscience and Remote Sensing Symposium. IEEE.***15**. pp. 6309-6312.

Pederi YA and Cheporniuk HS, 2015. Unmanned Aerial Vehicles and new technological methods of monitoring and crop protection in precision agriculture, In *Proceedings IEEE International Conference Actual Problems of Unmanned Aerial Vehicles DevelopmentsSymposium. IEEE.* pp. 298-301.

Picanco Filho MC, Lima E, Carmo DDGD, Pallini A, Walerius AH, da Silva RS, Sant Ana LCDS, Lopes PHQ and Picanco MC, 2024. Economic Injury Levels and Economic Thresholds for *Leucoptera coffeella* as a Function of Insecticide Application Technology in Organic and Conventional Coffee (*Coffea arabica*), Farms. *Plant.* **13**: 585.

Popescu D, Ichim L and Stoican F, 2023. Orchard monitoring based on unmanned aeial vehicles and image processing by artificial neural networks: a systematic review. *Front. Plant Sci.* **14**: 1237695.

Rejeb A, Abdollahi A, Rejeb K and Treiblmaier H, 2022. Drones in agriculture: A review and bibliometric analysis. *Comput. Electron. Agric.*, **198**: 107017.

Rejesus RM and Hornbaker RH, 1999. Economic and environmental evaluation of alternative pollution-reducing nitrogen management practices in central Illinois. *Agric. Ecosyst. Environ.* **75**: 41–53

Schiefer F, Kattenborn T, Frick A, Frey J, Schall P and Koch B, 2020. Mapping forest tree species in high resolution UAV-based RGB-imagery by means of convolutional neural networks. *ISPRS J. Photogramm. Remote Sens.***170**: 205–215.

Shah IA, Laraib A, Ashraf H and Hussain F, 2024. Drone Technology: Current Challenges and Opportunities. **In**: *Cybersecurity Issues and Challenges in the Drone Industry*, IGI Global, Pennsylvania, USA, pp. 343-361.

Singh P and Singh P, 2023. Drones in Indian Agriculture: Trends, Challenges, and Policy Implications.

Tahar KN, 2012. Aerial terrain mapping using unmanned aerial vehicle approach. Remote Sensing and Spatial Information Sciences, *Int. Arch. Photogramm.***39**: 493-498.

Thomas S, Kuska MT, Bohnenkamp D, Brugger A, Alisaac E, Wahabzada M, Behmann J and Mahlein AK, 2018. Benefits of hyperspectral imaging for plant disease detection and plant protection: a technical perspective. *J. Plant Dis. Protect.***125**: 5-20.

Torres-Sanchez J, de Castro AI, Pena JM, Jimenez-Brenes FM, Arquero O and Lovera M, 2018. Mapping the 3D structure of almond trees using UAV acquired photogrammetric point clouds and object-based image analysis. *Biosyst. Eng.***176**: 172–184.

Torres-Sanchez J, Lopez-Granados F, Borra-Serrano I and Pena JM, 2018. Assessing UAV-collected image overlap influence on computation time and digital surface model accuracy in olive orchards. *Precis. Agric.***19**: 115-133.

Torres-Sanchez J, Lopez-Granados F, Serrano N, Arquero O and Pena JM, 2015. High-throughput 3-D monitoring of agricultural-tree plantations with unmanned aerial vehicle (UAV) technology. *PLoS One.* **10**: 130479.

Toscano F, Fiorentino C, Capece N, Erra U, Travascia D, Scopa A, Drosos M and D'Antonio P, 2024. Unmanned Aerial Vehicle for Precision Agriculture: A Review. *IEEE Access.* **12**: 69188-69205.

Venkatesh J, Ramasamy KK, Aruna M, Rao KPK, Sasikala N and Nasani K, 2022. Smart agriculture monitoring scheme using machine learning strategies. In *International Conference on Innovative Computing, Intelligent Communication and Smart Electrical Systems.* **30**. pp. 1-9.

Wang D, Prato T, Qiu Z, Kitchen N and Sudduth K, 2003. Economic and environmental evaluation of variable rate nitrogen and lime application for claypan soil fields. *Precis. Agric.* **4**: 35–52.

Wang Y, Chen Y and Wang D, 2022. Convolution network enlightened transformer for regional crop disease classification. *Electron*, **11**: 3174.

Whitley KM, Davenport JR and Manley SR, 2000. Differences in nitrate leaching under variable and conventional nitrogen fertilizer management in irrigated potato systems. In: *Proceedings of the 5th International Conference on Precision Agriculture,* P. C. Robert, R. H. Rust, W. E. Larson (Ed.),American Society of Agronomy, Minnesota, USA, pp. 11-19.

Zhang C, Valente J, Kooistra L, Guo L and Wang W, 2019. Opportunities of UAVs in orchard management. *Int. Arch. Photogramm. Remote Sens. Spatial. Inf. Sci.* **12**: 673–680.

Zhang D, Xu F, Pun CM, Yang Y, Lan R, Wang L, Li Y and Gao H, 2021. Virtual reality aided high-quality 3D reconstruction by remote drones. *ACM T. Internet. Techn.* **22**: 1-20.

Zortea M, Macedo M, Mattos AB, Bernardo R and Gemignani BH, 2018. Automatic citrus tree detection from UAV images based on convolutional neural networks, In: *Proceedings of 31st Sibgrap/WIA- Conference on Graphics, Patterns and Images.* IBM Research, pp. 14-21.

Chapter 19

Fruit Based Agroforestry Systems, Carbon Sequestration and Policy Perspective

Sanjeev K. Chauhan

Directorate of Research, Dr. Y.S. Parmar University of Horticulture and Forestry Nauni, Solan–173230, Himachal Pradesh

Agroforestry is emerging as one of the diversification options for the farmers. It is a refined concept, where land units are deliberately so managed under trees and crops, with or without animals, that the system is scientifically sound, practically feasible, economically viable, socially acceptable and ecologically desirable/ sustainable. Lesser availability of land, low returns from traditional crops and the ever-increasing demand for fuel, fodder and timber are the main reasons that compel farmers to integrate multipurpose tree species into the farmland. Agroforestry is a resource-conserving, not depleting system compared with the existing land management systems involving few crops like rice, wheat, sugarcane, cotton, *etc.*, which are extremely resource-exhaustive, be it in terms of natural, financial or human resources. The most common crop rotation (rice-wheat) is over exploiting the water resources, nutritional pool and environmental health.

Experiments across the country have shown soil improvements under agroforestry to a significant level. Carbon sequestration and back transfer of nutrients through twig/litter fall, *etc.* help the system positively. Thus, agroforestry is sustaining the system.

In the coming years, the tree-based direct needs will exclusively be met from farm forestry or agroforestry. At present, poplar (*Populus deltoides*), eucalyptus (*Eucalyptus tereticornis*), dek (*Melia azaderach* and *M. composita*), willow (*Salix* spp.), ghamar (*Gmelina arborea*), maharukh/ardu (*Ailanthus excelsa*), jor tor (*Casuarina* spp), Chandan (*Santalum album*), mulberry (*Morus alba*), leucaena (*Leucaena leucocephala*), kadam (*Anthocephalaus cadamba*), teak (*Tectona grandis*), among other trees, are grown by farmers on a commercial scale.

Poplar has become the most preferred tree for farmlands in Punjab and its adjoining states. Almost any crop (cereals, pulses, vegetables, cash crops, forage crops, fruit crops *etc.*) can be grown with it during the *rabi* season. It is one of the world's fastest-growing industrial soft woods (can be harvested within a reasonably short period of 4-5 years). Poplar intercropping is a highly profitable venture as much as poplar growing is a highly lucrative business since market for its products are readily available.

Eucalyptus, another fast-growing species, has also gained ground because wood prices of poplar have fallen.

Farmers prefer it to other trees because it has self-pruning habit, aside from its light canopy cover, fast growth, multiple uses, straight bole, better survival under minimum management. It is quite hardy in nature, but being evergreen, intercropping is not profitable but sole planting is more profitable than intercropping – harvested at the age of 3.5 years.

Farmers have recently realized the benefits of growing indigenous/exotic, fast-growing dek, Casuarina, Ghamar, Maharukh, willow, kadam, *etc.* under agroforestry systems. The species can yield high economic returns within a 10–12 year rotation and can also partly provide the raw material requirements of the industry, if raised scientifically under agroforestry systems. These species are mainly used in making furniture and panels; plywood, paper, Packaging, sports industry. However, under Punjab conditions, kadam is less remunerative than poplar because of its slow growth, evergreen nature and lower price in the market.

Leucaena (*leucaena leucocephala*), a fast-growing, thornless, leguminous woody perennial has been found very useful under sloping lands. It is specifically raised under silvi-pastoral models. A number of new leucaena varieties have been introduced for their good quality fodder value and low anti-factors value. Teak/ sagwan (*Tectona grandis*), chandan (*Santalum album*), and Mahogany (*Swietenia* spp), shisham (*Dalbergia sissoo*), Australian acacia (*Acacia* spp.), though encouraged by the private nursery growers in the agroforestry systems (agri-silvicultural and horti-silvicultural models), will not be profitable and adopted because of their long rotation period.

The farmers are experimenting with various agroforestry models with the aims of productivity, sustainability, stability and adoptability. Their concept of agroforestry is aimed at conservation and optimum use of land, natural resources and other climatic resources for production of food, fodder, fuel, timber and wood.

It is in agroforestry where the vast solar energy available, nutrients present at different soil depths and water available in the sub-surface soil are efficiently utilized and environment is ameliorated.

Scientists are working to provide farmers the latest technologies related to the different agroforestry systems. They have been concentrating more on wood technology, biophysical and socioeconomic aspects of the system. Still, there is ample vacuum as regards biophysical and socio-economic processes involved in the component.

Horti-Silvicultural System

The tremendous increase in the costs of production and stagnated crop productivity, warrant not only emergent actions to sustain productivity but also measures to revive the sagging economy of the farmers. The stagnant agricultural economy of the farmers has come under concern as it has an overall impact on the economic growth of the country. After facing high growth, the agriculture sector is showing signs of fatigue for almost a decade. The agro-ecosystem has depleted very deleteriously under the pressure to increase the stagnating production and industrial revolution. Soils have become highly deficient in organic carbon and in most of micro and macro nutrients. Water table is falling at an alarming rate, which is the concern of each and every one.

Figure 19.1: Inter-Cultivation of Agricultural Crops is Normally Followed and Essential in Natural Farming.

Furthermore, the excessive use of agro-chemicals in the soil and crops has further worsened the condition of human, animal, soil and environmental health. A prudent approach in rationalizing the use of our natural resources is long due.

Moreover, under the present regime, farmers have to look for the market and change their mindset towards commercial agriculture rather than traditional agriculture. There is a mismatch between what is being produced and what the buyer market is demanding. Farmers have tried alternatives to rice-wheat or other traditional rotation like other agricultural crops, poultry, dairy, fishery, apiculture, sericulture, mushroom cultivation, *etc.* but much success has not been achieved. The only option which finds some ray of hope is integration of crops for higher productivity with differential outputs, and more specifically fruit tree species with fast growing timber species on the farm land. Integrating fruit trees with annual crops is a common feature in many variations of tree-crop combination, geometry, sequence, *etc.* It is becoming a popular intervention in natural farming with preference for nitrogen fixing crops.

Poplar Block Plantation with Guava

The horti-silvicultural model (integration of fruit trees and short rotation forest trees) remains the most appropriate alternative option for sustainable management of land unit, which will improve the economy of the farmers, besides taking care of the natural resources (soil, water and air). Fruit trees (mango, citrus, litchi, ber, pear, guava, *etc.*) based agroforestry models are becoming popular to make use of space between fruit trees (minimum 6m spacing between trees) during the initial juvenile period, as the forest tree component will be harvested when the fruit trees start commercial bearing of fruits. It has caught the attention of the farmers all over the country to exploit the national and *Poplar block plantation with mango* international fruit and wood markets. Short-rotation species like poplar, eucalypts, kadam, Sagwan, Ghamar, Subabul, casuarina, sohanjana, dek, *etc.* find a comfortable place for inclusion under agroforestry systems and the *Poplar block plantation with mango* farmers have accepted these species for growing under the different combinations and geometries. Progressive farmers have earned handsome income through the adoption of horti-silvicultural interventions. The economics of such systems is better than the traditional agricultural rotation and cereal/pulses based agroforestry models (Table 19.1).

The objective of farmers in adopting agroforestry in general and horti-silvicultural models in particular are simply stated in terms of socio-economic issues. The continuous research efforts are important to find out appropriate solutions to the problems and challenges in present day agriculture. Agricultural crops and fruit tree combination is commonly followed (Table 19.2), though inter-cultivation in orchards is not recommended to avoid fruit tree root disturbance and avoid alternate host for the insect-pest diseases. In order to realize the benefits of the horti-silvicultural models, a complete package of technology have to be developed for adoption in various agro-climatic zones after taking into account the needs, customs and tradition of the local people. On-farm survey indicated that fruit based agroforestry practices are more suitable for the farmers with large land holdings. The inability of the small farmers to meet the food security limit them to

Table 19.1: Economic Evaluation of different Cropping Cystems (₹/acre)

Cropping System	*Crop Income under Poplar Trees*		*Average Annual Income*	*Poplar Income**	*Total*	*Average Approx. Costs*	*Net Return*
	Three Years Old Tree	*Four Years Old Tree*					
I. Wheat + Paddy	39370.0	45000.00	42185.00	-	42185.00	10350	31835.00
II. Peach + Garlic + Poplar	23387.9	42523.02	32955.46	40000	72955.46	24250	48705.46
Plum + Garlic + Poplar	23085.4	26191.42	24638.41	40000	64638.41	24250	40388.41
Guava + Garlic + Poplar	22973.1	43713.42	33343.26	40000	73343.26	23750	49593.26
Kinnow +Garlic + Poplar	33705.9	43713.42	38709.66	40000	78709.66	29750	48959.66
III. Peach +Onion + Poplar	40485.5	56240.19	48362.85	40000	88362.85	22250	66112.85
Plum + Onion + Poplar	40183.0	39908.59	40045.80	40000	80045.80	22250	57795.80
Guava + Onion + Poplar	40070.7	57430.59	48750.65	40000	88750.65	21750	67000.65
Kinnow + Onion + Poplar	50803.5	57430.59	54117.05	40000	94117.05	27750	66367.05

* Average projected annual income.

Table 19.2: A Comprehensive List of Agri-Horticulture Systems in India

North India	**Himachal Pradesh**	*Malus pumila, Pistacia vera, Prunus amygdalus, Prunus armeniaca, Emblica officinalis, Citrus* spp.	*Amaranthus spp., Hordeum vulgare, Fagopyrum esculentum, Solanum tuberosum, Triticum aestivum, Pisum sativum, Solanum esculentum, Zea mays*	Verma *et al.* (2007); Goswami *et al.* (2015)
	Kashmir valley, Jammu and Kashmir	*Juglans regia, Prunus amygadalus, Prunus domestica, Prunus avium, Prunus armeniaca*	*Allium sativum, Avena sativa, Brassica campestris, Brassica oleracea, Pisum sativum, Raphanus sativus, Solanum lycopersicum, Zea mays*	Islam *et al.* (2017)
	Punjab	*Psidium guajava, Mangifera indica, Citrus* spp., *Prunus amygdalus, Prunus armeniaca*	*Curcuma longa, Allium cepa, A. sativum*	Chauhan *et al.* (2012,2013, 2021), Dhillon *et al.* (2013, 2010a,b, 2011), Dhillon and Chauhan (2012), Sangwan *et al.* (2015),
	Uttarakhand	*Mangifera indica, Punica granatum, Musa paradisiaca, Tectona grandis*	*Colocasia esculenta, Curcuma longa, Oryza sativa, Triticum aestivum, Zingiber officinale*	Karki *et al.* (2021)
North-east India	**Meghalaya**	*Ananas squennsa, Citrus limon, Citrus reticulata*	*Camellia sinensis, Piper nigrum*	Majumdar *et al.* (2004)
	Arunachal Pradesh	*Psidium guajava, Mangifera indica, Citrus reticulata, Musa* sp., *Ananus comosus, Carica papaya, Areca catechu, Pyrus communis, Prunus domestica, Vitis Vinifera, Citrus reticulata*	*Cajanus cajan, Solanum melongena, Zea mays Phaseolus vulgaris, Ipomea batatas, Brassica oleracea var. botrytis, Daucus carota, Coriandrum sativum, Capsicum annum, Cucurmis sativus, Oryza sativa, Eleusine corcana*	Pangging and Singh (2015); Deb (2020)

	Assam	*Artocarpus heterophyllus, Mangifera indica, Citrus eticulata, Musa* sp., *Ananus comosus, Zizyphus jujuba, Cocos nuccifera, Areca catechu, Citrus limon*	*Brassica sativus, Cucus carota, Coriandrum sativum, solanum melongena, Capsicum annum, Cucurbita moschata, Momordica charantia, Lycopersicum esculentum, Luffa cylindrica, Cucurmis sativus, Oryza sativa, Zea mays*	
	Tripura	*Areca catechu, Cocos nucifera, Mangifera indica, Artocarpus heterophyllus*	*Colocasia esculenta*	Sarkar *et al.* (2021)
West India	**Rajasthan**	*Ziziphus mauritiana* (L.), *Cordia myxa* (Forster), *Emblica officinalis* (Gaertn)	*Triticum aestivum, Pennisetum typhoides, Cymopsis tetragonaloba, Vigna mungo, Vigna aconitifolius*	Singh *et al.* (2014); Korwar and Pratibha (2005)
	Maharashtra	*Mangifera indica, guava*	*Capsicum annum, Manihot esculenta, Coriandrum sativum, Brassica oleraceae,Hellianthus annuus*	Bhoyar *et al.* (2016)
	Gujarat	*Mangifera indica*	*Solanum melongena, Abelmoschus esculentus*	Gunaga (2017)
Central India	**Uttar Pradesh**	*Carica papaya*	*Abelmoschus esculentus, Citrullus lanatus, Ipomoea batatas, Solanum gilo, Solanum lycopersicum*	Kumar *et al.* (2000)
South India	**Kerala/ Tamilnadu**	*Cocos nucifera*	*Amorphophalus companulatus, Coleus parviflorus, Cumuma longa, Ipomoea batatas, Manihot esculenta, Zingiber officinale*	Nair (1993)

adopt the fruit-timber interventions. The lack of appropriate research and extension approach also harm the interest of the farmers and intensive efforts are required for appropriate combinations validation and their dissemination at the farm.

Figure 19.2: During the Juvenile Phase of Fruit Trees, Integrating Fast Growing Timber Trees are Remunerative.

Long rotation trees are suggested not to be include in the system to avoid loss in yield to fruit plants due to competition for resources. Short rotation timber tree intervention is only for juvenile period. Even the delay in commercial fruiting due to inter-planting is compensated by short rotation trees. Usually there is no difference in behaviour of crop physiology excepting some time shifting (peach, plum and guava fruit physiological behaviour is reflected in Figure 19.3a-c). Since the fruit plants are primary component, therefore, appropriate growing conditions should be made available to realize the full potential of fruit crop rather than the secondary inter-cultivated crop/trees. Inter-cultivation as mentioned above is to make use of blank space rather than accommodating both the components (timber-fruit plants) in the same pit. In the second rotation, instead of block planting, timber trees should be accommodated on the boundary for income, and serve as wind break/shelterbelt for creating microenvironment conducive for the fruit plants.

Climate Change and Mitigation through Plantation Systems

The earth's climate system has demonstrably changed on both global and regional scales since the pre-industrial era. Land resources have been under tremendous pressure since the very beginning of civilization for food and forest products. Anthropogenic activities have known to affect the biosphere through

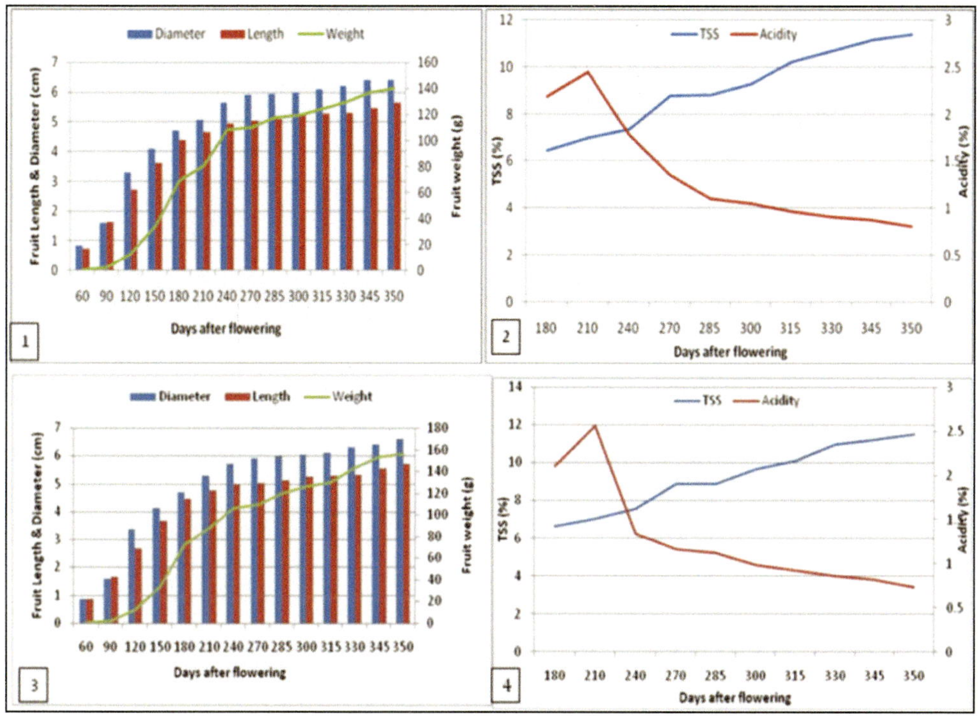

a: Guava under open (1 and 2) and under poplar canopy (3 and 4).

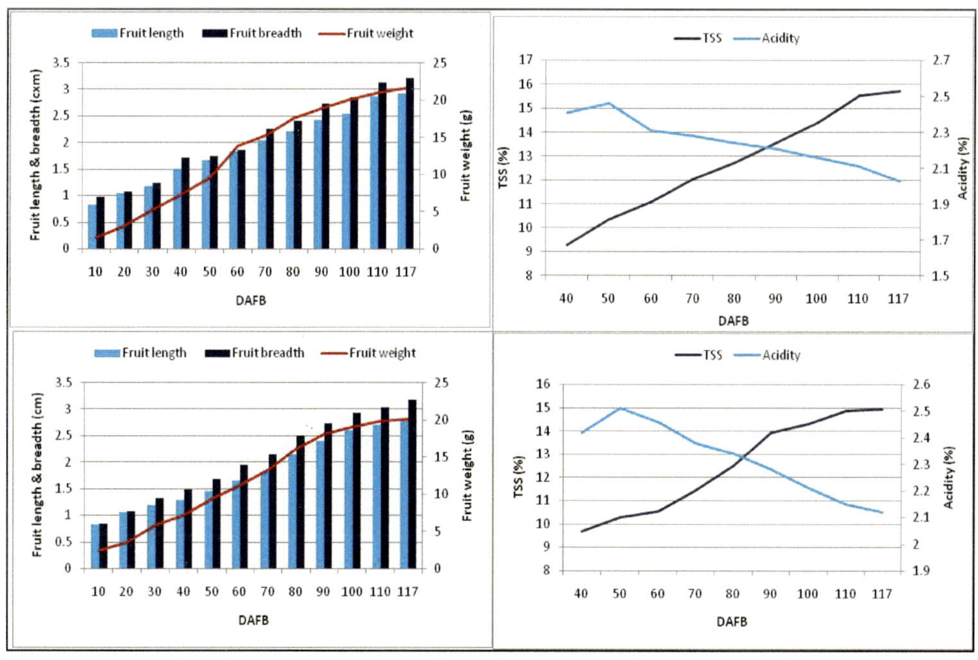

b: Peach under open (1 and 2) and under poplar canopy (3 and 4).

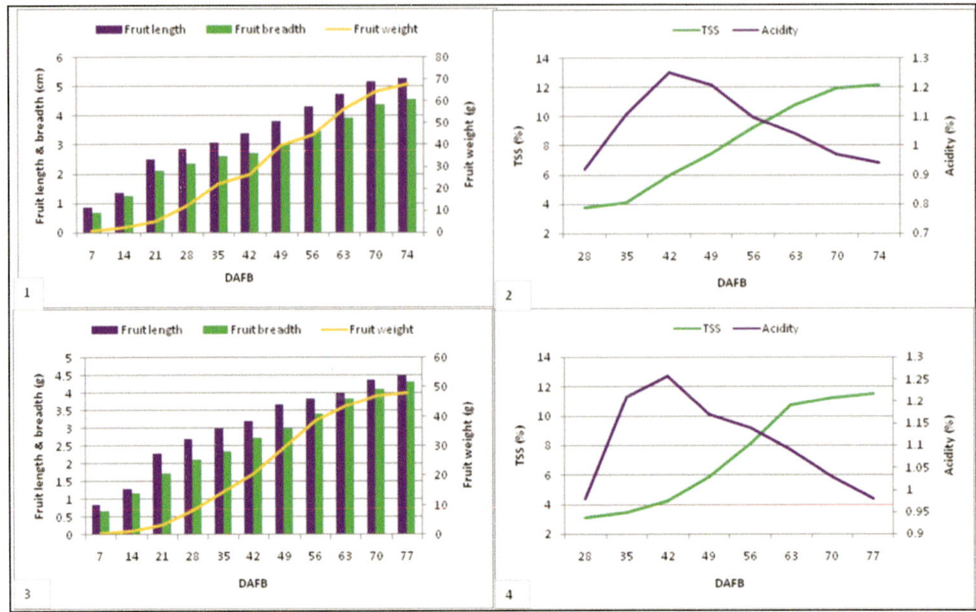

c: Plum under open (1 and 2) and under poplar canopy (3 and 4)

Figure 19.3a-c: Behaviour Remains Almost same in Open and Shaded Environment Excepting Shift in Time.

Figure 19.4: Interplanting Long Rotation Species in Orchards is not Appropriate.

Figure 19.5: After the Establishment of Orchard, Inter-cultivation in Block is not Suggestive Rather on the Boundary.

Figure 19.6: Planting Fruit Plant and Timber Tree in same Pit Rather than in between Trees/Rows is Inappropriate.

changes in the land use and forest management activities, thus altering the natural balance of greenhouse gases in the atmosphere. The environmental security thus, has become a serious concern across the globe. The meteorological records have shown significant increase in the environmental temperature referred to as 'Global warming'. Human activities have increased green house gases and raised global temperature by 0.5 degree Celsius over the past 100 years, which has been predicted to increase by 1.4-5.8°C during the next century (Bhadwal and Singh, 2002).

Figure 19.7: High Density of Timber Trees is not Desirable.

Evidence of this global crisis include the rise in global sea levels by 0.09-0.88m through the melting of the world's snow peaks and glaciers, which poses a threat to millions living in coastal areas. Each 1°C rise in temperature will displace the limits of tolerance of land species some 125 Km towards the poles, or 150 m vertically on mountains. Many species will not be able to redistribute themselves fast enough to keep up with the projected changes, and the considerable alterations in ecosystem structure and functions. A drastic effect on splitting of huge glacier in the Antarctica has already been registered, which has thrown a challenge to save the earth, lest the very survival of the mankind will be jeopardized. Climate change is one of the most important global environmental challenges facing humanity with implications for food production, natural ecosystems, freshwater supply, health, *etc.*

There is an increasing awareness, the world over about the adverse impacts of greenhouse gas emissions and the consequent climate change. The concentration of CO_2 and other greenhouse gases (GHGs) in the atmosphere has considerably increased over the last century and is set to rise further, the largest proportion of which resulting from the burning of fossil fuels and the conversion of tropical forests to agricultural land (Paustian *et al.,* 2000). Since 1750, atmospheric carbon dioxide (CO_2) concentration has risen by 30 per cent, with a steep increase observed during the last 50 years (IPCC, 2001). With an annual increase of 0.5 per cent or 3.6GT carbon yr^{-1}, atmospheric CO_2 concentration is expected to double until the mid to late 21st century (Bouwman, 1990). So, there is a growing international scientific effort to develop ways to slow down the addition of carbon dioxide to the atmosphere. One way to manage carbon is to use energy more efficiently to reduce our need for a major energy and carbon source-fossil fuel combustion. Another way is to increase our use of low–carbon and carbon free fuels through suitable technologies (nuclear power and renewable sources such as solar energy, wind power, bio-fuels, *etc.*). The third and newest way to manage carbon is through 'Carbon Sequestration'. It refers to provision of long term storage of carbon in the terrestrial biosphere, underground, or the oceans, so that the build up of carbon dioxide (the principal green house gas) concentration in the atmosphere will reduce.

Carbon emissions are strongly, though not entirely, related to economic growth and standard of living, and given the fact that developing countries whose participation is being sought are amongst the poorest countries in the world who need the 'maximum economic space' and 'environmental space' to grow and for combating climate change, these issues become very critical. At the current stage of development, India cannot afford to reduce its energy consumption. A potential option for mitigating the accumulation of CO_2 in the atmosphere is the enhanced cycling of carbon by the terrestrial biosphere through massive reforestation or sustainable afforestation programmes.

The investment in forestry sector to store carbon in the trees and forests is one of the viable options for offsetting the gases released by fossil fuel burning and for mitigating the potential effects of global warming. Perennial vegetation, notably forests have an important role to play in regional, national and international greenhouse gas balances. Agroforestry systems (including horti-silviculture) are commonly considered to be C sinks because the integration of trees results in greater CO_2 sequestration from the atmosphere and thus enhanced carbon storage in permanent tree components.

Agroforestry is one key path to prosperity for poor people suffering from hunger, malnutrition, abject poverty and the deterioration of the environment in the areas that have been bypassed by the *Green Revolution*. A key advantage of agroforestry, however, is that it provides benefits for poor farmers in developing countries and global environment payoffs. The focus on trees, as opposed to other vegetation is because trees sequester a relatively large quantity of carbon for longer period. This is accomplished further harvesting from forests and by accumulating biomass and soil carbon. Protecting existing forests and woodlands, and establishing new ones, will help us limit greenhouse gas emissions while at the same time assisting biodiversity conservation, addressing issues such as land degradation and salinity, and in some cases generating income from timber and/or carbon 'rights.' Promoting plantations outside forests help in saving in the traditional forests through meeting the forest product resources outside forests.

Carbon Sequestration - The Need of the Hour

Following industrial revolutions across the globe especially in United States, Europe, Japan, China and Russia coupled with deforestation to meet the priority requirements of increasing population in developing countries mainly in African and South – East Asian countries, the distribution of Carbon in its natural pool and in atmosphere has recorded alarming change. Major change is to make the doormant pool to an active pool – below the earth to the atmosphere. Owing to its various implications through global warming, radiation effect on plant growth, *etc.*, this issue has caught the attention of almost every nation to put their heads together. Scientific evidence suggests that increased atmospheric carbon dioxide could have some positive effects such as improved plant productivity (Keutgen and Chen, 2001). However, negative changes in the global climate (rising temperatures, higher frequency of droughts and floods) are often the most consequential processes associated with an increased concentration of CO_2 (USDA NRCS, 2000).

The environmental security has become a serious concern across the globe. The recent meteorological records have shown significant increase in the environmental temperature referred to as 'Global warming'. A drastic effect on splitting of huge glacier in the Antarctica has already been registered, which has thrown a challenge to save the earth. In the last one thousand years, the 20th century was the warmest and the last decade of the 20th century was warmest decade (nine out of ten years were warmest in recorded history and this trend is continuing in the 21st century as well. The glaciers in the Himalayas are depleting at an alarming rate causing concern among each and every one. Of the six GHGs (Carbon dioxide, methane, nitrous oxide, chlorofluorocarbon, hydrofluorocarbon and perfluorocarbons), CO_2 accounts 63, methane 24, nitrous oxide 10 per cent and other remaining 3 per cent of the carbon equivalent emissions. Thus, it is important to direct efforts to carbon mitigation in addition to methane and nitrous oxide, though accounted in terms of carbon dioxide.

Carbon dioxide is the largest single green house gas currently trapping about half of the total heat contributing to global warming. The average annual emissions of carbon were estimated to be 7.1 thousand million tons C per year during the 1980s. Global carbon emissions from fossil fuel alone are estimated to be 5.5 thousand million tons per year and increased by 61 per cent by 2015 compared to the base figures of 1990 (IPCC, 1996). The contribution of India to the cumulative global CO_2 emission during 2003 was 4.07 per cent but on per person basis, it is very small when compared with those of US, Europe and developing Asian Countries like Japan, China, *etc.* Although the global temperature is expected to increase by only about 6°C by 2100, the impacts of this on 'global climate change' is likely to be very serious. This may result into catastrophic changes in climate with added evaporation and precipitation, and vegetation movement towards poles by 200 to 300 km. Therefore, the global powers are emphasizing to make every possible effort to contain it's increase by 1.5 to 2°C.

Plants – Viable Option to Sequester Carbon

Plants grow through the well-known process of *photosynthesis*, utilizing the energy of sunlight to convert water from the soil and carbon dioxide from the air into sugar, starches, and cellulose, the *carbohydrates* that are the foundations of the entire food chain. CO_2 enters a plant through its leaves, the greater atmospheric concentrations tend to increase the difference in partial pressure between the air outside and inside the plant leaves, and as a result more CO_2 is absorbed and converted to carbohydrates. Plant absorbs 3.67 unit of CO_2 during photosynthesis to produce one unit of carbon stored in plant tissues. Crop species vary in their response to CO_2. Wheat, rice, and soybeans belong to a physiological class (C3 plants) that responds readily to increased CO_2 levels. Corn, sorghum, sugarcane, and millet (C4 plants) that follow a different pathway. The latter, though more efficient photosynthetically than C3 crops at present levels of CO_2, tend to be less responsive to enriched concentrations.

Analysis of the impact of climate change on forest productivity based on experimental research and modeling, would suggest that forest productivity may increase under elevated carbon dioxide, but the local conditions of moisture stress and nutrient availability will strongly temper any response (USDA, 2000). New field experiments on vegetation and recent meta-analysis studies of the accumulated experimental data demonstrate that elevated carbon dioxide has been shown to increase plant growth by 25-50 per cent, with experimental results varying widely. However, when the models are implemented across the geographically diverse landscapes of the United States, the model responses to elevated carbon dioxide are substantially lower, less than 20 per cent by ecosystem type reflecting local conditions of moisture stress and nutrient availability. Joyce and Nungesser (2000) compared the magnitude of these modeled responses to forest growth responses from timber management. They concluded that the growth response to timber management is larger (25 to 39%) than the projected productivity response from changes in carbon dioxide or climate (8 to 29%). Burning oil and coal releases carbon into the atmosphere far more rapidly than it is being removed, and this imbalance causes atmospheric carbon dioxide concentrations to increase. In addition, by clearing forests, we reduce the ability of photosynthesis to remove CO_2 from the atmosphere, also resulting in a net increase.

In middle and higher latitudes, global warming will extend the length of the potential growing season, allowing earlier planting of crops in the spring, earlier maturation and harvesting, and the possibility of completing two or more cropping cycles during the same season. Crop-producing areas may expand poleward in countries such as Canada and Russia, although yield in higher latitudes will likely be lower due to less fertile soils. Many crops have become adapted to the growing-season day lengths of the middle and lower latitudes and may not respond well to the much longer days of the high latitude summers. In warmer, lower latitude regions, increased temperatures may accelerate the rate at which plants release CO_2 in the process of *respiration*, resulting in less than optimal conditions for net growth. When temperatures exceed the optimal for biological processes, crops often respond negatively with a steep drop in net growth and yield. If night-time temperature minima rise more than do day-time maxima-as is expected from greenhouse warming projections, heat stress during the day may be less severe than otherwise, but increased night-time respiration may also reduce potential yields. Another important effect of high temperature is accelerated physiological development, resulting in hastened maturation and reduced yield. Climate change will modify rainfall, evaporation, runoff, and soil moisture storage. With increased temperatures and longer growing seasons, vegetation will transpire more water, thus making less water available for irrigation or domestic uses.

The demand for water for irrigation is projected to rise in a warmer climate, bringing increased competition between agriculture (already the largest consumer of water resources in semiarid regions) and urban as well as industrial users. Falling water tables and the resulting increase in the energy needed to pump water will make the practice of irrigation more expensive, particularly when with drier conditions more water will be required per acre. Peak irrigation demands are also predicted to rise due to more severe heat waves. Additional investment for dams, reservoirs, canals, wells, pumps, and piping may be needed to develop irrigation networks in new locations. Finally, intensified evaporation will increase the hazard of salt accumulation in the soil, thus rendering the land out of cultivation. Conditions are more favorable for the proliferation of insect pests in warmer climates. The possible increases in pest infestations may bring about greater use of chemical pesticides to control them, a situation that will require the further development and application of integrated pest management techniques. Such situations will again deteriorate the environment leading to chemical pollution.

Global warming is predicted to lead to thermal expansion of sea water, along with partial melting of land-based glaciers and sea-ice, resulting in a rise of sea level, which may range from 0.1 to 0.5 meters (4 to 20 inches) by the middle of the next century. According to present estimates of the Intergovernmental Panel on Climate Change (IPCC), sea level will rise 9-88cm by 2100. Such a rise could pose a threat to agriculture in low- lying coastal areas, where impeded drainage of surface water and of groundwater, as well as intrusion of sea water into estuaries and aquifers, might take place. In parts of Egypt, Bangladesh, Maldives, Srilanka, Indonesia, China, the Netherlands, Florida, and other low-lying coastal areas already suffering from poor drainage, agriculture is likely to become increasingly difficult to sustain. Some island states are particularly at risk and they have to migrate to new areas for their sustenance.

Carbon sequestration is one of the important components of the entire carbon spectrum and is an effect being institutionalized through bilateral treaties with the sole objective of managing the excess release to the atmosphere. The continued increase in concentration of GHGs in the atmosphere is likely to lead to climatic change resulting in large changes in ecosystem leading to possible catastrophic disruptions, livelihood, economic activities, living conditions and above all human health. Carbon sequestration especially sponging up carbon dioxide emissions with trees and other plants appears to be a simple, common sense approach to the climate crisis. It refers to the provision of long term storage of carbon in the terrestrial biosphere, underground, or the oceans so that the buildup of carbon dioxide (the principal greenhouse gas) concentration in the atmosphere will reduce or slow. The Intergovernmental Panel on Climate Change defines it as the "amount of carbon absorbed from the atmosphere by plants (equivalent to net photosynthesis), but which can then be lost in part to the atmosphere through subsequent decomposition of dead residues". It plays an important role in the global carbon cycle. The fact that we have temporarily thrown the global cycle out of balance suggests that we can also rebalance the system by improving vegetation management and natural sequestration of atmospheric carbon through photosynthesis. Obviously, the more trees and plants we have on earth, the more carbon sinks we have. Measuring and predicting sequestration is fundamental to providing the quantifiable unit of carbon that is traded or claimed as offsets against emissions. Estimation of sequestration generally will have to be accurate (Including a measure of uncertainty), easy to perform, cost effective and verifiable.

Measurement needs to be at regular intervals, simple to ensure that costs are minimised, and transparent to facilitate independent verification. That is, it must be based upon a methodology that is acceptable to all parties concerned. Often the best approach to fulfill these multiple needs will be a verifiable modeling approach, based upon best available knowledge and data. Carbon accounting techniques should be flexible and be a process of continuous improvement, enabling information and model prediction to be continually updated as new knowledge and real-time measurements of carbon in forests come to hand.

Carbon Sink in Agroforestry System

Removing atmospheric carbon (C) and storing it in the terrestrial biosphere is one of the options, which have been proposed to compensate greenhouse gas (GHG) emissions. Agricultural lands are reported to be source of GHG but also believed to be a major potential sink and could absorb large quantities of C if trees are reintroduced to these systems and judiciously managed together with crops and/or animals. Thus, the importance of agroforestry as a land-use system is receiving wider recognition not only in terms of agricultural sustainability but also in issues related to climate change. It is a dynamic, natural resource management practice that, through the integration of trees on farms and in the agricultural landscape, diversifies and sustains production for increased social, economic and environmental benefits. Approximately 1.2 billion people–20 per cent of the world's population–depend to a large extent on agroforestry products and services for their survival (Anon., 2000). Agroforestry is different from forestry and from conventional agriculture. The discipline of agroforestry focuses on growing trees on farms while the discipline of forestry focuses on trees in forests and commercial tree plantations.

Agroforestry is a form of agriculture that involves planting trees in crop (Figure 19.1) and livestock farming systems. Although the term is fairly new, the practice of agroforestry is based on a vast store of indigenous knowledge developed by farmers since the drawn of agriculture. Agroforestry is now providing powerful technological and policy innovations that are rapidly spreading in Africa, Asia, Latin America, and more recently, in several developed countries also. It is one key path to prosperity for poor people suffering from hunger, malnutrition, abject poverty and the deterioration of the environment in the areas that have been bypassed by the Green revolution. A key advantage of agroforestry, however is that it provides private benefits for poor farmers in developing countries and global environment payoffs.

Agroforestry is a form of natural resource management, incorporating trees into farming systems leads to greater prosperity at the farm level. Trees provide farmers with marketable products – such as lumber, poles, firewood, animal fodder, fruits and medicines – all of which earn extra income. They improve soil fertility by fixing nitrogen from the air and recycling nutrients from the soil, making use of sub-surface water, thereby helping to increase crop yields and helping to ensure stability of future production. Trees on farms help hold moisture where it is needed, reduce soil erosion and keep valuable topsoil in place, reduce the intensity of downstream flooding and maintain watershed building materials. They serve as live fences in semi-arid regions, protecting vegetable and cereal crops that would otherwise be overrun by livestock. Trees increase ecosystem biodiversity above and below ground and help ameliorate global climate change by sequestering carbon – in their live biomass as well as in the soil – that otherwise would be added to the earth's atmosphere.

Several studies have shown that the inclusion of trees in the agricultural landscapes often improves the productivity of systems while providing opportunities to create C sinks. The amount of carbon sequestered largely depends on the agroforestry put in places, the structure and function, which to a great extent are determined by environmental and socio-economic factors. The carbon sequestration potential of afforestation/reforestation is specific to the species, rate and management involved, and it is, therefore, variable. The carbon sequestration potential for agroforestry practices is even more variable, depending on the planting density, component in place and production objective of the system. The prominent role of forestry and agroforestry systems in carbon sequestration has increased global interest in these land use options to stabilize green house gases emission. Agroforestry system especially agri-silvicultural/horti-silviculture system can be carbon sinks and temporarily store carbon, while other systems (ruminant based) are purely green house gases emission systems. If the entire area of feasible land is used for forestry, the mean estimate of carbon sequestration by natural forests along with newly afforested plantation is around 153 TG (tera gram) carbon per year by 2030, equivalent to the 1990 fossil fuel emission of India. In one of the studies in Uttar Pradesh, approximately 20 million tons of carbon has been estimated to be sequestered by the farm forestry plantations (Singh *et al.,* 2000).

Actually, the carbon storage in plant biomass is only feasible in the perennial agroforestry systems (perennial-crop combinations, agroforestry, windbreaks, hedgerow inter-cropping, *etc.*), which allow full time tree growth where the wood component represents an important part of the total biomass. The carbon sequestration in agroforestry systems is a dynamic process. At the initial stage, the carbon and the nitrogen from the vegetation and the soil is lost, then there is quick accumulation during fast growth phase and at maturity or harvesting when the system returns back to agriculture or new plantation, some part of the carbon is released back though major portion is stored in the stem and the root system. The root system in general contributes to the soil carbon and the major share of stem wood carbon goes to the durable products and on an average 50 per cent of the stem wood carbon is converted to products like lumber and plywood that have a long life span. The short rotation species most preferred in the agroforestry interventions are very less suited to long term locking of carbon. For example, the productivity of eucalypt is more than the *shisham* but its carbon sink is less because of its shorter rotation and the relatively low storage time of the wood products. In timber-fruit tree combination, both the components help in carbon storage.

Agroforestry systems sustainably managed have substantial economic and carbon sequestration potential. The principal barrier to agroforestry management appears to be socio-political rather than economic. Because forest management requires long term investment, good planning, selection of approximate species, management technique, technical guidance, legal constraints and also the marketing structure need to be taken care-off. System assessment needs changes in four main carbon pools (above ground biomass, below ground biomass, soils and inter-cultivated crops). Carbon monitoring, however, requires specialized equipments and trained personnel's. It requires baseline determination of pre-project carbon pool in biomass, soils and standing litter crop, establishment of permanent sample plots for periodic measurement of changes in carbon pool, measure carbon stored in non-project area and areas with sparse vegetation, net differences in carbon accumulated in project and non-project land uses, partitioning of roots, wood and foliage for species and changes in vegetation pattern (natural forest, forest plantation, agroforestry, agriculture ecosystem, *etc.*).

Winrock International has conducted lot of studies during nineties on carbon sequestration in Australia, USA, Brazil, Guetamala, Philippines, Newzealand, *etc.* All the fast growing species proportionately conserve carbon in the system, and in India in general, Poplar, *Eucalyptus, Leucaena, Acacia, Melia, Dalbergia, Anthocephalus,* bamboo, neem, ghamari, simal, *etc.* can be incorporated in the agroforestry system to sequester the carbon in the system and interplanting of fruit trees will add to the value. Agroforestry as such can be justified in many other ways as increasing soil carbon, greatly benefit agricultural productivity and sustainability. However, the cost of carbon sequestered through agroforestry appears to be much lower through other CO_2 mitigation options. Gera *et al.* (2006) reported 115, 64 and 56 tons/ha carbon sequestration potential under poplar block, poplar boundary and *Eucalyptus* boundary plantations, respectively, under irrigated agro-ecosystem on farmers' fields. The area under both the species is increasing every year because of huge industrial wood demand. Peichl *et al.* (2006); Swamy and Puri (2005);

Thevathasan and Gordon (2004), Zhou and Wang (1997); Dabas and Bhatia (1996) also suggested great potential of tree-based intercropping systems in reducing the atmospheric CO_2 concentration compared to sole cropping systems. The agroforestry interventions have indirect carbon sequestration benefit. The surplus biomass, in the form of lops and tops, from agroforestry plantations are a net addition to the existing fuelwood supplies and can go a long way to divert some of the pressure from the natural forest. However, agroforestry alone cannot solve the current climatic problems but can only be one among a range of strategies. The international panel on climate change (IPCC) has recommended a catalogue of remedial measures to mitigate increasing CO_2 emissions. Among these remedial measures, such as reforestation, the conversion of agricultural land into agri-silvicultural/horti- systems has also been included. There is however still more work to be done to improve our understanding of carbon sequestration and green house gases mitigation through agroforestry interventions (block, boundary and hedge row plantations).

Carbon Sequestration

Agricultural or farmland soil can be a source or sink of greenhouse gases. This potentiality can be seriously diminished or depleted by intensive cultivation, poor soil management; improper agricultural practice and land use have strong impact on biotic processes of carbon sequestration (Lal 2004). Montagnini and Nair (2004) substantiates the potential of agroforestry in carbon sequestration, with their argument that increased amount of carbon can be sequestered in woody biomass in lands devoted to agriculture, while still allowing for growing of food crops. Dixon (1995) estimated that, at a global scale, agroforestry implementation on 585 to 1275 x 10^6 ha of technically suitable land, about 128 to 228 $MgCha^{-1}$ can be sequestered. Sharrow and Ismail (2004) based on their study at western Oregon, USA supported the proposition that agroforests are more efficient at accumulation of carbon and nitrogen (12% of Carbon and 2% of Nitrogen) above ground compared to sole plantations (9% of carbon; 1% of nitrogen) and pasture (<1% of N and C).

The sporadic studies in agroforestry systems mostly report a derivative of biomass productivity rather than C sequestration in soils. The value of C sequestered in wood however, depends on the end use of the wood and other products. Nevertheless, it is an issue which is much debatable and under investigation. Agroforestry has indirect effect of C sequestration also, when it helps in decreasing pressure on natural forests, which are the largest sink of terrestrial carbon. Another indirect avenue of carbon sequestration is through the use of agroforestry technologies for soil conservation, which could enhance carbon storage in trees and soils. The estimates of annual carbon emission (gross) from deforestation for India are about 42Mt. The standing biomass (above and below ground) is estimated to be 8357 Mt. The corresponding C storage is 4179 (assuming 50% of woody biomass as carbon). The total carbon stored in forests, including soil is estimated to be 9578Mt

(Ravindernath and Somashekhar 1995). When the net annual C emission in forest ecosystem is considered (uptake-release) for India, the carbon emission is nearly offset by sequestration in forests under succession and in tree plantations (63.4 Mt of carbon emission v/s 68.8Mt of carbon uptake). Horti-silvicultural, a ual tree component based system even has more potential than agri-silvicultural system.

Carbon Trading

Carbon trading (also known as emission trading) is the buying and selling of 'allowances' to emit CO_2 and its equivalents. In other words, the credits generated from sequestered carbon can be traded in a carbon trading market to achieve cost-effective greenhouse gas emissions. Under an emission trading system, emitters of greenhouse gases could be required to have a sufficient number of emission permits to cover their actual emissions. Alternatively, emitter could buy carbon credits for quantities of carbon sequestered in trees to offset their emissions. Carbon trading can occur in a number of ways *e.g.* informally between countries, over the counter and centralized through an institutional financial exchange. All the three approaches would seem to be able to operate effectively and coexist for emissions trading. Buyers and sellers would use the approach that best suited them.

Following international understanding on issues concerning global environmental security and eventual development of the concept of carbon trading as envisaged under Kyoto Protocol, the initiative has already been exhibited in this direction in many countries. Historically, developed countries have been responsible for most Green House Gas (GHG) emissions and therefore bear much of the responsibility for the problems associated with climate change. The US Acid Rain Project, the US Emission Credit Trading Programme for criteria pollutants, the US Lead phase down and the Hunter River Salinity Trading Scheme of Australia, trading programmes for emitters of greenhouse gases in UK, Dutch Forests Absorbing CO_2 Emissions (Stuart and Sekhran, 1966), Costa Rica Office for Joint Implementation to facilitate private companies in GHG mitigation (Tattenbach, 1996; Van Der Gaast, 1996) Denmark, Greenhouse Gas Emissions Trading Market in Chicago, National Carbon Accounting System in Australia, *etc.* are some of the initiatives in this direction.

Any activity that results in the improvement of the environment in any of its expressed or perceived form with positive effect should be assigned some economic value, which has been almost missing till date. As a generality, only the direct benefits accruing in the form of materials and not in the form of services were assigned values and eventually these physical commodities were marketed. Afforestation programme for example, has been valued only with respect to the timber return. The resulting environmental amelioration practically was never assigned any value. Carbon-trading deals in forestry/agroforestry could sharply reduce poverty among the rural poor, while also providing businesses with an inexpensive way to "off-set" their carbon emissions. Carbon trading allows industries in developed countries to off-set their emissions of carbon dioxide by investing in reforestation and clean energy projects in developing countries.

Trees can reduce greenhouse gases in the atmosphere by sequestering carbon, but many fear carbon-trading deals will spawn vast tree plantations with monocultures of non-native tree species on lands already claimed by local people. According to the researchers, there are 126 million hectares (311 million acres) of low-yielding crop and pastureland. Converting some of these areas to higher-yielding agroforestry would sequester five to 50 tons of carbon per hectare per year. Rehabilitating dry forests in India could double sequestration from 27 to 55 tons of carbon yearly for every hectare of dry forest improved. The agroforestry system with perennial crops may be important carbon sinks, while intensely managed agroforestry intervention with annual crops is more or less similar to conventional agriculture. The carbon sequestration payments will encourage the land holders to adopt less intensive practices since net revenue will be higher with carbon payments. Some of the Indian states (Punjab, Uttar Pradesh, Tamilnadu, *etc.*) have started realizing the benefits of carbon sink in agroforestry through the voluntary marketing system of C credits. The industries however are ready to pump funds in afforestation/reforestation project to harness the green initiatives but industries with directly related to the end use from the said activity have restrictions to invest under corporate social responsibility fund.

Challenges and Critical Issues

Agroforestry including fruit based interventions, a sustainable landuse management system requires careful planning to address the challenges being faced by the farmers and industry in the changing climate scenario. The choice of trees for agroforestry system depends upon the purpose of the farmer whether to grow them for personal use or industrial use (ecological or economic). Trees for industrial uses catch the attention of farmers to grow economically because of sustainable management and green felling restrictions from the forest area. There are around 890 paper mills and 3300 plywood units (majority in unorganized sector) and around 90 per cent raw material to run these units is sourced from outside forests, three to four per cent from Govt. forests and six to seven per cent from timber import. However, other wood based industries (furniture, construction, packaging, sports, agriculture implements, *etc.*) are equally important to project the importance of adopting and extending agroforestry area. The SWOT analysis at state level (due to differential capacities of agroforestry adoption and wood based industries) would be appropriate so that the appropriate steps are taken at state/regional scale with National projections. At the moment poor adoption of agroforestry is constraint by the following multiple (social, economic and ecological) factors:

- ☆ Agroforestry as an adaptation measure in changing climate scenario could not gain impetus and provision of payment for environmental services could be counter-productive.

- Standardization of best agronomic/management practices under different agroforestry system to increase the productivity of the system and their testing at different agro-climatic zones for higher productivity, sustainability and adoptability. Intensive efforts are required to identify the suitable species/crops/genotypes for agroforestry with marketing acceptability. Diagnosis of exiting status should be a periodic activity to refine the designs with new models for adoption.
- Below ground studies are lacking to justify the natural resource exploitation (co-existence of functional rooting strategies).
- Non-availability of superior quality planting stock, no provision of accreditation of nurseries in agroforestry species is essential but lacking (provision exists in fruit plants).
- Marketing system is not regularized, which results in malpractice in market system and farmers feel cheated. Dedicated marketing intelligence system is not in place.
- Permit system of harvesting and interstate/within the state transit is still an issue of concern for adopters of agroforestry though the provisions exist in National Agroforestry Policy.
- Industrial units for wood processing are distributed at limited scale and they are not ready to extend helping hand to the farmers for arranging raw material at affordable prices.
- Extension activities are lacking; benefits of research are not reaching properly to the poor farmers due to reluctance of line departments.
- Financial support in the form of subsidies/loan/support price including institutional credit facilities are reluctantly available on plantation projects.
- Capacity building of the farmers on some important issues *i.e.*, planting technology including geometry of planting, pruning schedule, age specific cropping sequence, resource depletion, marketing channels, *etc.* are insufficiently.
- Export of nutrition with agroforestry produce from the cultivated land, *etc.* have to be compensated for sustainability.
- Legal disputes because of common field boundaries are increasing, which did not receive attention from the policy planners.
- National/international collaboration in agroforestry research is essentially required and farmers/scientists must be trained for this programme.
- Farmers have a fixed mind that only the prices of wood govern the prices of processed product (paper, plywood, *etc.*), industries have their own issues including treating wood based industries as green initiative.

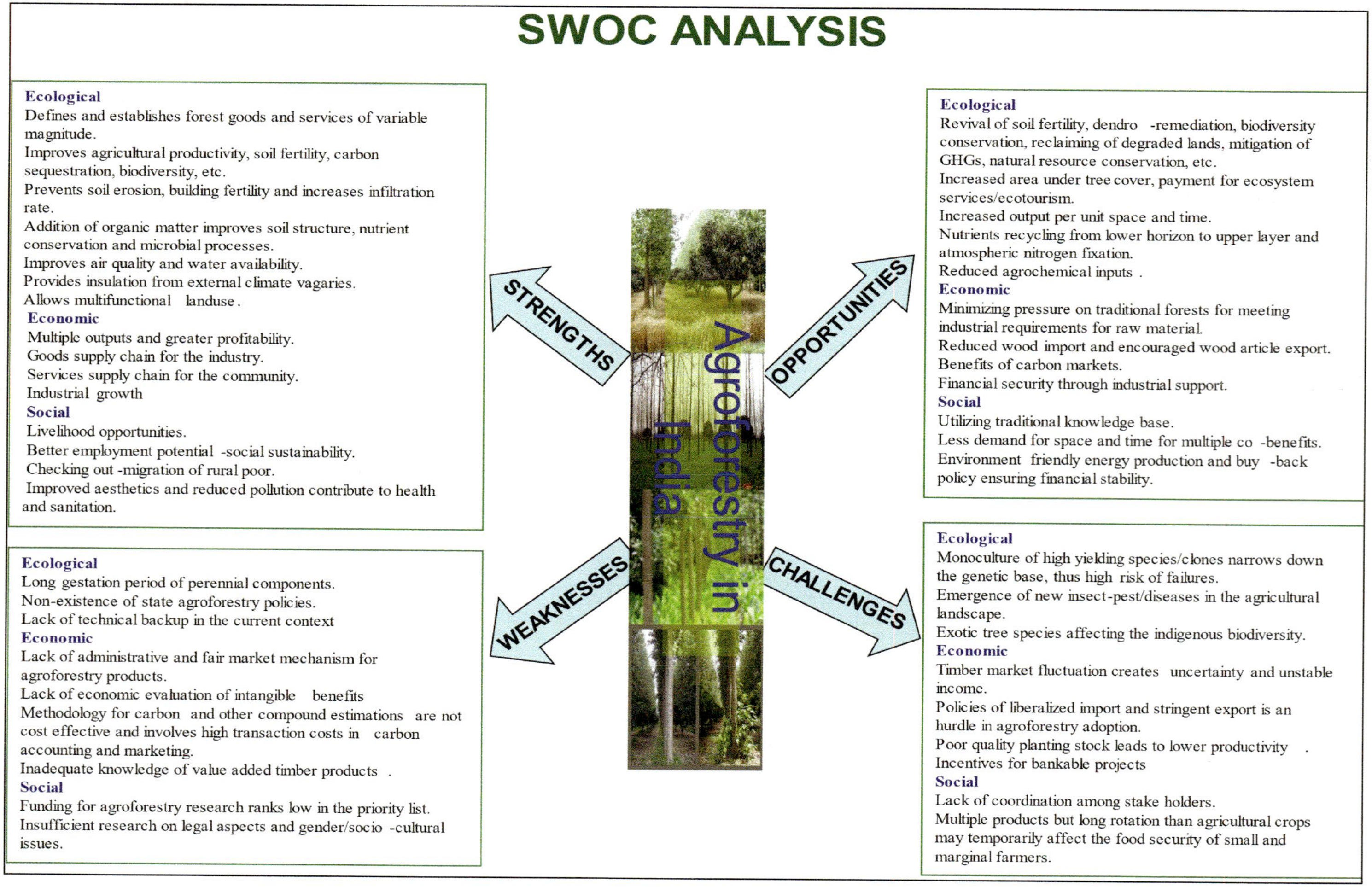

SWOC ANALYSIS

Agroforestry in India

STRENGTHS

Ecological

Defines and establishes forest goods and services of variable magnitude.
Improves agricultural productivity, soil fertility, carbon sequestration, biodiversity, etc.
Prevents soil erosion, building fertility and increases infiltration rate.
Addition of organic matter improves soil structure, nutrient conservation and microbial processes.
Improves air quality and water availability.
Provides insulation from external climate vagaries.
Allows multifunctional landuse.

Economic

Multiple outputs and greater profitability.
Goods supply chain for the industry.
Services supply chain for the community.
Industrial growth

Social

Livelihood opportunities.
Better employment potential -social sustainability.
Checking out -migration of rural poor.
Improved aesthetics and reduced pollution contribute to health and sanitation.

WEAKNESSES

Ecological

Long gestation period of perennial components.
Non-existence of state agroforestry policies.
Lack of technical backup in the current context

Economic

Lack of administrative and fair market mechanism for agroforestry products.
Lack of economic evaluation of intangible benefits
Methodology for carbon and other compound estimations are not cost effective and involves high transaction costs in carbon accounting and marketing.
Inadequate knowledge of value added timber products .

Social

Funding for agroforestry research ranks low in the priority list.
Insufficient research on legal aspects and gender/socio -cultural issues.

OPPORTUNITIES

Ecological

Revival of soil fertility, dendro -remediation, biodiversity conservation, reclaiming of degraded lands, mitigation of GHGs, natural resource conservation, etc.
Increased area under tree cover, payment for ecosystem services/ecotourism.
Increased output per unit space and time.
Nutrients recycling from lower horizon to upper layer and atmospheric nitrogen fixation.
Reduced agrochemical inputs .

Economic

Minimizing pressure on traditional forests for meeting industrial requirements for raw material.
Reduced wood import and encouraged wood article export.
Benefits of carbon markets.
Financial security through industrial support.

Social

Utilizing traditional knowledge base.
Less demand for space and time for multiple co -benefits.
Environment friendly energy production and buy -back policy ensuring financial stability.

CHALLENGES

Ecological

Monoculture of high yielding species/clones narrows down the genetic base, thus high risk of failures.
Emergence of new insect-pest/diseases in the agricultural landscape.
Exotic tree species affecting the indigenous biodiversity.

Economic

Timber market fluctuation creates uncertainty and unstable income.
Policies of liberalized import and stringent export is an hurdle in agroforestry adoption.
Poor quality planting stock leads to lower productivity .
Incentives for bankable projects

Social

Lack of coordination among stake holders.
Multiple products but long rotation than agricultural crops may temporarily affect the food security of small and marginal farmers.

Figure 19.8

Majorities of the above mentioned constraints have to be addressed appropriately. Still there is need for way forward strategies to achieve the objective of extending area under agroforestry for diversification of traditional crop rotations, enhanced farmers income, conservation of natural farming (soil/water/biodiversity), ameliorate deteriorating environment, mitigate/adapt to changing climate, conserve of traditional forests, meet the targets of Nationally Determined Contribution of CO_2 sequestration/clean energy under Paris convention, employment generation, *etc.* SWOC analysis is a continuous process to follow so that the befits are realized in true spirit.

It has been rightly emphasized, that, second green revolution should focus on tree crops, agroforestry and mixed livestock pastures and crops. Farm forestry has a greater future role for ensuring environmental sustainability and meeting sociological and economic objectives for meeting sustainable development goals. Moreover, crop diversification is need of the hour to meet the challenges of surplus food grain production, higher costs of inputs, natural resources depletion, financial burden of MSP on government, poor economic health, marketing of produce, promoting green technologies, *etc.* The objective of increased income and amelioration of environment will remain a distant dream, unless, suitable measures are adopted for ensuring sustainable and proper linkages between sellers and buyers, and rational protection to the farmers.

The sustainable afforestation efforts under agroforestry can continue only if the raw wood produced by primary producers is absorbed by the market at reasonable profits and returns. This also requires incentives for establishment of more wood based industries, wood use diversification, production of quality wood based finished products at competitive prices, strengthening of linkages between farmers and buyers, decentralized value addition, *etc.* The market structure should be properly developed including market intelligence and information. There should be inbuilt checks and regulatory mechanisms for timely interventions and minimizing exploitation of primary producers. Farmers should get reasonable returns for their produce in terms of productivity and economics.

More than 85 per cent timber is sourced from outside forests (private land) for wood based industries when all the forests are under the control of the Govt. Productivity of forest plantations is very low (4-6m^3 ha^{-1} yr^{-1}) in comparison to the agroforestry plantations (20-25m^3 ha^{-1} yr^{-1}), therefore it is appropriate to encourage and invest in agroforestry to meet the domestic/industrial requirements directly, conserve the traditional forests indirectly for environmental services including conservation of natural resources, and generate employment in the complete channel from nursery, plantation management to processing and marketing.

Farmers have made committed efforts for the viability of the wood-based industries on one hand and conservation of forests on the other but their efforts have not been recognized appropriately. Dedicated ward to honour the individual farmer/association would encourage the others to realize the importance of adopting trees. Even supporting farmers during initial lean period would be counter-productive. Some of the issues have been identified (not in the order of priority/importance) for consideration to boost the agroforestry adoption. These issues need discussion for clarification and narrowing down the priorities of major stakeholders (farmers and industry).

- Regular monitoring of state of agroforestry area and timber demand/supply data is inappropriate to address the sustainable future production/utilization projections. Effective area and demand/supply figures are essential to put efforts for appropriate planning, which is still missing and require concerted efforts.
- Many states have liberalized tree harvesting and transit rules for timber, NAP 2014 has candidly addressed these issues but still the practical problems exist on ground, which discourages farmers from adoption of tree cultivation.
- Monoculture in species/genotypes, export of nutrients, depletion of water resources, invasive species, competition with inter-cultivated crops, *etc.* are words of caution and need proper technical attention.
- Identifying niche area for agroforestry plantations would facilitate aggregation of wood based industries – Bihar and eastern UP have potential (raised poplar with the support of ICFRE) but no buyer in the area resulted to abort the tree growing. On the other hand, the Yamunanagar - Jagadhri, Haryana has transformed the region as a hub for wood based industries. There are 273 plywood manufacturing units, 297 peeling units and 331 saw mills in Yamuna Nagar district. Yamuna Nagar city has emerged as the biggest market of farm-grown timber in the country as about 35-40 lakh tonnes of timber worth about ₹ 150 -180 million is being traded annually, of which about ₹ 75-90 million is going back to farmers. Yamuna Nagar city now has heavy concentration of wood-based units, which manufacture wood products worth ₹500-600 million annually and provide direct and indirect employment to about 1 lakh people. Surprisingly, Yamunanagar district alone produces about 60 per cent plywood of the country, which has helped this city in earning the reputation of being called as the 'plywood capital of the country'. Hoshiarpur district in Punjab also has aggregation of wood based industries, resulting in large scale agroforestry plantations in Hoshiarpur and adjoining districts of Nawanshahr, Ropar, Ludhiana and Jalandhar.
- Marketing at present is unregulated/non-transparent with inefficient supply chain. Tree growers and wood based industries need complete harmony and there is complementary relationship between them. Both are not mutually exclusive rather dependent on progress of each other. Therefore, incentive for the farm grown wood based industry is equally important and these industries should be treated at par with agro-industries/green industry. Compulsory industrial support under CSR

for research, technology transfer and farmers support, *etc.* should be mandated for the wood based industry.

- Value addition at farmers level would generate subsidiary occupation – debarking, chipping, peeling/veneer, toys and sports articles, mementos, sculptures, pre-finished products, *etc.* decentralized biomass based power units (residue based) at village level, processing residue for ethanol, *etc.*
- Market intelligence and information system is essential but in poor state of affairs, resulting in direct loss to the farmers and handsome earning for the middlemen. Financial institution's limited support on plantation projects - credit facility and insurance cover for plantation projects is also discouraging the farmers for their adoption because of initial juvenile phase of tree harvesting. Bankable projects are missing and small farmers cannot afford to shift to agroforestry interventions.
- Ecotourism and provision of payment of environmental services - the enhanced uptake of carbon dioxide in agroforestry system and reduced emissions of GHGs and locking of CO_2 through non-destructive use of wood have significant potential to remove CO_2 from the atmosphere. Evidences have emerged that agroforestry systems increase above ground and soil organic carbon to mitigate GHGs. The payment of environmental services through onfarm grown trees could be additional incentive income to the farmers, the states like Punjab, UP, Tamil Nâdu, *etc.* have started sharing the benefits under voluntary marketing system of carbon trading.
- State forest development corporation can operate wood based industry [Tamilnadu Newsprint and Paper Ltd. (TNPL) – a public service and Government agency] to regulate the marketing, which otherwise is controlled by private players and timber prices of short rotation trees keep on fluctuating. Mechanism on market based support system on the pattern of Quad-partite agreement in Consortium of Industrial Agroforestry (Tamilnadu), Tri-partite agreement of ITC, or Haryana State Forest Corporation Model or WIMCO Model to popularize poplar or Non-timber forest produce as in tribal affairs ministry (agroforestry is not only timber but other products as well) is appropriate and workable. Selling prices of timber are almost same what existed in 2001-02 but cost of cultivation has increased substantially, thus posing major challenge for expansion in area (even though major industrial timber demand is fulfilled from agroforestry plantations). Present timber rates (poplar-eucalyptus) have shown substantial rise but these rates still keep on fluctuating, which shakes the confidence of farmers in adopting tree cultivation. Stability in prices will build the confidence of the farmers to grow more.
- Legal dispute of raising trees on common boundary- there is no directive/ recommendation on this important aspect - NAP 2014 is also silent on the issue.

- Planting stock and timber certification are important issues and often discussed but with little progress (procedure exist in fruit plants for the accreditation/grading of nurseries by National Horticulture Board) to support the intervention.
- Skill development - KVKs and VVKs should extend technical support – provision of an exclusive agroforestry cadre for capacity building, which is lacking at present. ICFRE can extend support at regional scale or in identified districts with positive adoption potential of onfarm trees.
- A dedicated cadre in the state to boost agroforestry – State forest dept. field functionaries are trained for trees, whereas, Agriculture Department technocrats are trained for crops, resulting that for long time agroforestry remained no body baby. A dedicated cadre in either of the department (state Agriculture or Forest) to boost onfarm plantation and intensive inter-cultivation is an essential requirement.
- Provision of establishment of an Agroforestry Board at par with other development boards (National horticulture Board, Jute Board, Rubber Board, Coffee Board, Tobacco Board, Tea Board, Silk Board, Spices Board, *etc.*) to address the issues of concern on ecology, economics and social aspects. Sub-mission on agroforestry should be upgraded as Mission to enhance the impetus to the on farm tree adoption, however recently support under RKVY has been strengthened for the research, technology transfer and adoption by the farmers.
- Addition of trees on agricultural landscape limit the resources (above/below ground) for the crops but very little emphasis has been placed on the management of trees rather major emphasis has been placed on suitability of understorey crops. Focus must shift to best management practices of trees for tree-crop interface. No focused emphasis has been placed on developing shade tolerant varieties.
- Digital agroforestry - periodic assessment of area, demand-supply, productivity, digitization of weight/volume tables to know the quantity of timber produce. Decision support system for best cultural practices, market information and trading, traceability of timber to avoid complication in free transit would be an added advantage.
- Upgrading Education, research and extension capacities and collaboration between ICAR and ICFRE - National School on Agroforestry under ICFRE at par with IIT in Engineering would attract good talent in the discipline including international students. Regional training centers for trainers training/stakeholders capacity building will also strengthen the support base. Only limited number of species (poplar, eucalyptus, dek, casuarina, subabul, bamboo, chandan, ailanthus, sagwan, kopok, ebony, sambal, kadam, ghamar, rubber, *etc.*) are being adopted at present, which needs diversification with a greater number of species and superior genotypes with respect to their endues.

- Policy provision for diverting 3-5 per cent cultivated area by each farmer for any tree (timber/fruit/avenue) would meet the aspiration of honorable Prime Minister for Har Med Pe Ped/Natural Farming/increased tree cover/enhanced income.
- Timber Import/export? - Govt. is cautious on the issue and farmers are mainly adopting short rotation species and import is targeting cheap and large size logs/lumbers, therefore posing no competition with the farmers.
- Targeting future projections – Future expansion of on farm plantations and wood based industries requires action plan with minimum area coverage annually under agroforestry (say minimum 5000-10000ha). It is possible and unutilized land can be leased out to the industries/NGOs/ Corporations or a dedicated unit, Plantation Corporation/Govt. owned timber based industries.
- Changing climate is an opportunity for mitigation and adaptation but challenge for the short rotation trees as well. The administrative decision in J and K regarding felling of poplar trees and a ban on new plantations from Kashmir valley is an eye opener. Similar PIL have been filed for eucalyptus as well, which is required to be defended scientifically to serve the economic interests of the farmers and the wood based industries.

It has been rightly emphasized, that, second green revolution should focus on tree crops, agroforestry and mixed livestock pastures and crops. By judicious use of tree-crop combinations, every piece of land can be made to produce maximum goods and services required for increasing human and cattle population. The objective of farmers in adopting agroforestry is simply stated in terms of socio-economic issues as productivity and sustainability. By introducing trees, the soil agro-ecosystem like physical, chemical and biological properties of soil are improved. The water holding capacity, improvement of permeability, aggregate stability and rhizospheric micro-environment is improved.

Agroforestry system modifies the microclimate of the place and offers environmental benefits including resilience against the changing climatic conditions. In order to realize the benefits of agroforestry, a complete package of technology have to be adopted after taking into account the needs, customs and tradition of the local people. Agroforestry research and development is emerging as a major agenda for environmental services. Policy reforms, institutional innovations and market/industry support will generate win-win situation for the farmers and resolve conflicts among the stakeholders. The country will witness the real green revolution only when countryside is clothed with greenery. The present administrative and political environment is also in favour of green technologies. The National Agroforestry Policy initiative in 2014 was a step forward but present report offers number of positive steps to capture greater domestic recognition to make multifunctional agroforestry adoption and expansion a reality in the country. New socio-economic policies have brought forestry and agriculture closer than ever. Support from different donor agencies to strengthen the research and extension base is encouraging farmers to try agroforestry. National agroforestry policy 2014 has opened up new avenues with continuous support system under Rashtriya Krishi Vikas Yojana.

REFERENCES

Anonymous, 2000. Path to prosperity through agroforestry: ICRAF's corporate strategy 2001-2010. ICRAF Publication, Nairobi, Kenya. 43p.

Anonymous, 2001. *State of the World's forests.* FAO, Rome. 181p.

Bhadwal S and Singh R, 2002. Carbon sequestration estimates for forestry options under different land use scenarios in India. *Current Science* **83**: 1380-1386.

Bhoyar S, Deshmukh H, Mahajan R, Sharma N, Bohra B and Nalgel DN, 2016. Traditional agroforestry systems practiced in lower hills of Melghat region, Chikhaldara tehsil, Maharashtra, India. *Biodiversity* **27**: 131-140.

Bouwman AF, 1990. Exchange of greenhouse gases between terrestrial ecosystems and the atmosphere. pp: 61-127 In: Bouwman A.F. (ed.), Soils and the Greenhouse Effect. John Willey and Sons, New York, USA, 575 pp.

Chauhan S., Chauhan R and Dhillon WS, 2012. Status of intercropping in poplar based agroforestry in India. *ENVIS Forestry Bulletin* **12**(1): 49-67.

Chauhan SK, Dhillon WS, Singh N and Sharma R, 2013. Physiological behaviour and yield evaluation of agronomic crops under agri-horti-silviculture system. *International Journal of Plant Research* **3**(1): 1-8.

Chauhan SK, Dhillon WS, Gupta N, Panwar P and Rajni, 2021. Effect of wind breakers using eucalyptus on cultivation of kinnow mandarin. *Ind J Horti.* **78**(3): 268-272.

Colmenares OM, Brindis RC, Verduzco CV, Grajales MP and Gomez MU, 2020. Hoticulture agroforestry systems recommendatioed for climate change adaptation: A review. *Agricultural Reviews* **41**(1): 14-24.

Dabas M and Bhatia S, 1996. Carbon sequestration through afforestation: role of tropical industrial plantation. *Ambio* **25**: 327-330.

Dixon RK, 1995. Agroforestry systems : sources or sinks of greenhouse gases? *Agroforestry Systems* **32:** 99-116.

Deb S, 2020. Traditional agroforestry systems of Northeast India. In: Roy N, Roychoudhury S, Nautiyal S, Agarwal S, Baksi S (eds) Socio-economic and eco-biological dimensions in resource use and conservation. *Environmental science and engineering*. Springer, Cham. https: //doi.org/10.1007/978-3-030-32463-6_5.

Dhillon WS and Chauhan SK, 2012. Crop yield and economic evaluation of an agri-horti- silvicultural system. *Indian Journal of Ecology* **39**(2): 310-312.

Dhillon WS, Srinidhi HV, Chauhan SK, Singh C, Singh N and Jabeen N, 2010a. Micro-envoriment and physiology of turmeric cultivated under poplar tree canopy. *Indian Journal of Agroforestry* **12**(2): 23-37.

Dhillon WS, Chauhan SK, Singh N, Rattan CS and Singh D, 2010b. Pre-bearing behaviour of some fruit crops under horti-silvicultural system. *Ind. J. Horti.* **67**(3): 311-314.

Dhillon WS, Chauhan SK and Jabeen N, 2011. Micro-climatic amelioration and yield potential in horti-silvicultural system. *Indian Journal of Ecology* 38: 124-128.

Dhillon WS, Chauhan SK, Jabeen N, Singh C and Singh N, 2012. Growth performance of intercropping system components and nutrient status of soil under horti-silvicultural system. *International Journal of Environment and Resource* **1**(1): 31-38.

Gera M, Mohan G, Bisht NS and Gera N, 2006. Carbon sequestration potential under agroforestry in Roopnagar District of Punjab. *Ind. For.* **132**: 543-555.

Goswami S, Verma KS and Kaushal R, 2014. Biomass and carbon sequestration in different agroforestry systems of a Western Himalayan watershed. *Biological Agricultural and Horticulture* **30**: 88-96.

Gunaga R, 2017. Crop biomass and yield patterns of dominant agroforestry systems of Navsari district Gujarat, India. *Indian J of Agroforestry* **19**: 72-78.

IPCC, 1996. The Science of Climate Change. Contribution of Working Group I to the Second Assessment Report of the Intergovernmental Panel on Climate Change [Houghton, J.J., L.G. Meiro Filho, B.A. Callander, N. Harris, A. Kattenberg and K.Maskell (eds.)]. Cambridge University Press, Cambridge and New York, pp.572

IPCC, 2001. Climate change 2001: The scientific basis. In: Houghton J.T., Ding Y., Griggs D.J., Noguer M., Van der Linden P.J., Dai X., Mashell K. and Johnson C.A. (eds), Contribution of working group I to the Third Assessment report of the International Panel on Climate change. Cambridge University Press, Cambridge, UK, 881 pp.

Islam MA, Qaisar KN and Bhat GM, 2017. Indigenous knowledge in traditional agroforestry sys tems of Kashmir valley: current challenges and future opportunities. *International Journal of Forestry and Crop Improvement* **8**(1): 68–77.

Joyce LA and Nungesser M, 2000. Ecosystem productivity and the impact of climate change. The impact of climate change on America's forests: A technical document supporting the 2000 USDA forest services RPA assessment, Joyce,L.A. and Birdsey, R. (eds.). Gen. Tech. Rep. RMRS-GTR-59. Fort Collins US, pp.45-68.

Keutgen N and Chen K, 2001. Responses of citrus leaf photosynthesis, chlorophyll fluorescence, macronutrient and carbohydrate contents to elevated CO_2. *J. Plant Physiol.* **158**: 1307-1316.

Karki H, Bargali K and Bargali SS, 2021. Spatial and temporal trends in soil N-mineralization rates under the agroforestry systems in Bhabhar belt of Kumaun Himalaya, India. *Agroforestry System.* DOI: 10.1007/s10457-021-00669-9.

Korwar GR and Pratibha G, 2005. Existing and improved agroforestry systems in low rainfall areas of India. Proc. *Sustainable Agriculture System*. Drylands. Niamey, Niger. 3-16.

Kumar S, Swaminathan V and Sathiamoorthy S, 2000. Effect of spacing, nutrition and intercrops on yield and quality of papaya (*Carica papaya*). *Research Crops* **1**: 58-62.

Lal R, 2004. Soil Carbon Sequestration Impacts on Global Climate Change and Food Security. *Science* **304:** 1623-1627.

Majumdar B, Kumar K, Venkatesh MS, Bhatt BP and Patiram, 2004. Effect of different agroforestry systems on soil properties in acidic Alfisols of Meghalaya. *Journal of Hill Research* **17**: 1–5.

Montagnini F and Nair PKR, 2004. Carbon sequestration: an under exploited environmental benefit of agroforestry system. *Agroforestry Systems* **61:** 281-295.

Nair PKR, 1993. An introduction to Agroforestry. Kluwer Academic Publishers, ICRAF, Nairobi, Kenya. 243.

Pangging G and Singh SL, 2015. Traditional agroforestry system in Eastern Himalaya: a case study of fringe villages of Doimukh forest range, Arunachal Pradesh, India. *International Journal of Current Research in Biosciences and Plant Biology* **2**(10): 23-28.

Paustian K, Six J, Elliott ET and Hunt HW, 2000. Management options for reducing CO_2 emissions from agricultural soils. *Biogeochemistry* **48**: 147-163.

Peichl M, Thevathasan NV, Gordon AM, Huss J and Abohassan RA, 2006. Carbon sequestration potential in temperate tree-based intercropping system, southern Ontario, Canada. *Agrof. Systems* **66**: 243-257.

Ravindranath NH and Somashekhra BS, 1995. Potential and economics of forestry options for carbon sequestration in India. *Biomass and Bioenergy* **8:** 323-336.

Sangwan AK, Dhillon WS, Singh H, Singh NP, Chauhan SK and Gill PPS, 2015. Influence of horti-silviculture combinations on pre-bearing growth and physiological parameters of pear. *Indian Journal of Horticulture* **72**(1): 21- 27.

Sarkar PK, Sarkar P, Kumar A, Pala NA and Kumar M, 2021. Carbon Storage Potential of a Waterlogged Agroforestry System of Tripura, India. *Water Air Soil Pollution* 232, 151. https: //doi. org/10.1007/s11270-021-05098-z.

Sharrow SH and Ismail S, 2004. Carbon and nitrogen storage in agroforests, tree plantations and pastures in Western Oregon, USA. *Agroforestry Systems* **60**: 123-130.

Singh B, Bishnoi M, Baloch MR and Singh G, 2014. Tree biomass, resource use and crop productivity in agri-horti-silvicultural systems in the dry region of Rajasthan, India. *Archives of Agronomy and Soil Science* **60**: 1031-1049.

Singh TP, Varalakshmi and Ahluwalia SK, 2000. Carbon sequestration through farm forestry: case from India. *Ind. For.* **126**: 1257-1264.

Stuart M and Sekhran N, 1996. Developing externally financed greenhouse gas mitigation projects in Papua New Guinea's forestry sector: A review of concepts, opportunities and links to biodiversity conservation. PNG biodiversity conservation and resource management program/UNDP Port Moresby.

Swamy SL and Puri S, 2005. Biomass production and C-sequestration of *Gmelina arborea* in plantation and agroforestry system in India. *Agrof. Systems* **64**: 181-195.

Tattenbach F, 1996. Certifiable, tradable offsets in Costa rica. Joint Implementation Quarterly. 2

Thevathasan NV and Gordon AM, 2004. Ecology of tree intercropping systems in the North temperate region: experiences from southern Ontario, Canada. *Agrof Systems* **61**: 257-268.

USDA Forest Service Gen. Tech. Rep, 2000. RMRS-GTR-59.

USDA NRCS (United States Department of Agriculture Natural Resources Conservation service), 2000. Growing carbon: a new crop that helps agricultural producers and the climate too. Available from the Soil and Water Conservation Society. *http.//www.unl.edu/nac.*

Van Der Gaast W, 1996. JI/AIJ initiatives during the Pilot Phase. The Netherlands: Joint Implementation Network.

Verma KS, Bhardwaj DR and Chand K, 2007. Agroforestry in Himachal Pradesh. In: Puri S., and Panwar P. (Eds.) Agroforestry systems and practices. New India Publishing Agency, Pitam Pura. 67-93.

WRI, 1994. Worls resources 1994-95. A report by the worlds resources institute in collaboration with UNEP and UNDP. Oxford University Press, New York and Oxford.

Zhou L and Wang H, 1997. A simulation study on CO_2 assimilation and crop growth in agroforest ecosystems in the East China Plain. *J. Environ. Sci.* **9**: 463-547.

Chapter 20

Reimagining a Paradigm Shift in Indian Agriculture through Natural Farming

Rajeshwar Singh Chandel and Sudhir Verma*

*Vice Chancellor, *Principal Scientist (Soil Science)*
Dr. Y.S. Parmar University of Horticulture and Forestry, Nauni, Solan–173230, Himachal Pradesh

The current biosphere emergency, fuelled by climate change and habitat loss, necessitates a re-evaluation of food production systems. Critical challenges in contemporary agriculture, including monoculture cropping and climate change, are examined. The chapter examines the evolution of natural farming in response to crises, delving into traditional practices, indigenous knowledge, and exploring traditional food systems. Natural farming, characterized by locally resources minimal inputs has been presented as a sustainable method to ensure farmers' livelihood through food and nutritional security. This farming system is showcased for its positive impact on soil health and its ability to counteract land degradation, insect pest and disease management and post-harvest quality of crop produce.

The Green Revolution that was initiated during the 1960s helped to cross the edge of famines and ensure food security in most developing countries including India. The global agricultural development strategy has thereafter, shifted from a system of extensification (putting more land under cultivation) to a more preferable approach of intensification (a system of using more inputs per unit of land to enhance resource use efficiency). Presently, India not only fulfills its own food requirements but also maintains substantial food grain buffer reserves and holds the distinction of being the world's largest exporter of rice (Varaprasad and Kumari, 2022).

Various attempts to improve the livelihood opportunity of smallholding farmers have brought fundamental productivity changes after 1960 and continued till 1990s through establishment of different international and national research and education institutes/universities to build local research capacities. These institutions, mandated to generate technological advances for countries that underinvest in agricultural research and education, coupled with government policy interventions thereby leading to the productivity changes and termed as 'Green Revolution' (Pingali 2012). The Green Revolution was an endeavour initiated by Dr Norman Borlaug in 1960s which later led to winning the Nobel Peace Prize for his work in developing HYVs of wheat and is often credited with having transformed India from 'a begging bowl to a bread basket'. The Green Revolution initiatives helped yield augmentation of staple cereals (rice, maize, and wheat) to many folds and substantial reductions in food insecurity and poverty in many countries of Asia and Latin America (Sharma and Singhvi, 2017).

The picture, however, remains no longer rosy. The consequences of the Green Revolution have now come under constant global scrutiny. Several adverse effects of the Green Revolution approaches including soil degradation, groundwater depletion, loss of biodiversity, water pollution (Pretty and Bharucha 2014), human health (Sharma and Singhvi, 2017) and above all loss of ecosystem productivity, became visible over time. The use of high-cost agrochemicals not only degraded the soil, polluted the environment, and affected people's health adversely, but placed farmers in debt.

Though the green revolution made food available to many and provided increased calorie consumption, it failed to provide a diverse diet. The focus on a few dominant cereals adversely affected the production of nutritionally important crops contributing to malnutrition among smallholder farmers and beyond (Pingali, 2012). After the Green Revolution, significant changes in food production affected the consumption practices of Indians. The Food and Agriculture Organization (FAO) has recorded a significant decrease in the production of millets and an increase in the production of rice between the years 1961–2017 (FAO 2019). Further, majority of the farmers use pesticides without personal protective measures, such as safety masks, impermeable gloves, long sleeve shirts *etc.*, as there is low level awareness about the deleterious effects of pesticides. Pesticides, applied over the plants, can also enter the human body, and the concentration of nitrate in the blood can immobilize hemoglobin in the blood. Extended exposure to organophosphates can also develop cancer (Xavier *et al.,* 2004).

The current ecosphere plight, fueled by climate change and habitat loss, entails a re-evaluation of current food production systems (Varaprasad and Rao, 2024). It is now high time that policy initiatives show urgency on the fact that the farmers and particularly smallholders lives matter. The immediate call of the day is to reduce cultivation costs, reduce risks, enhance biodiversity and increase returns.

Agroecology

Agroecology is a holistic approach to agriculture that integrates ecological principles into farming systems, aiming for sustainability, resilience, and social justice. By harmonizing farming with natural ecosystems, agroecology offers a sustainable alternative to industrialized agriculture, focusing on biodiversity, resource efficiency, and the empowerment of local communities. According to Gliesmann (2007), a system is in line with basic agroecological principles if Level 3 has been reached (Figure 20.1). Level 4 and 5 are going beyond production and focus on socio-economic aspects such as re-establishing connections between growers and eaters and responsible governance.

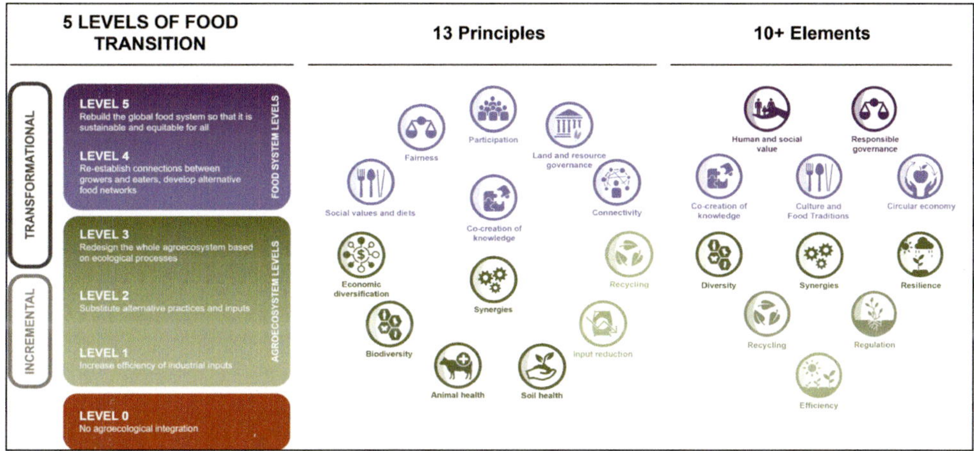

Figure 20.1: The Principles and Elements of Agroecology Arranged According to the Five Level of Food Transition (Adapted from Gliesmann, 2007).

One of the core principles of agroecology is **biodiversity**, which emphasizes the need for variety at all levels of agricultural production. This includes cultivating multiple crops, incorporating a diversity of animal species, and supporting rich microbial life in the soil. Diverse systems are more resilient to pests, diseases, and environmental stresses like climate change. Biodiversity helps reduce dependency on chemical pesticides and fertilizers, creating a more self-sustaining agricultural system. **Synergies** are also central to agroecological practices. The integration of different farming components, such as crops, livestock, and natural ecosystems, maximizes productivity while minimizing waste. This interconnectedness of various agricultural elements increases overall efficiency and sustainability. **Recycling** resources within the farm system is another important agroecological principle. This involves the reuse of nutrients, biomass, and water, reducing the need for synthetic inputs like chemical fertilizers. Agroecology also focuses on **resilience**, ensuring that farming systems can withstand and adapt to environmental stresses like climate change, pests, and disease outbreaks. Resilience is achieved through biodiversity and ecological balance, as diverse systems are better equipped to handle shocks than monocultures, which are vulnerable to insect-pest and disease invasions and or extreme weather events. **Social equity** is a key component of agroecology. The approach emphasizes fair labour practices, equitable access to resources, and the empowerment of small-scale farmers and rural communities. Agroecology values the contributions of these communities, ensuring that local smallholding farmers have a voice in decision-making processes related to agricultural policy and practice. This approach fosters food sovereignty and improves the livelihoods of small and marginalized farmers. Aiming for **efficiency**, agroecology strives to make the most of available natural resources like sunlight, water, nutrients and without resorting to industrial synthetic inputs. By working with nature, agroecology optimizes resource use and minimizes environmental degradation.

The co-creation of knowledge is also a vital principle. Agroecology promotes collaboration between scientists and local farmers, blending traditional knowledge farmers' and their time to time innovations with modern scientific insights. This participatory approach ensures that agricultural practices are context-specific and adapted to the unique conditions of each region. The principle of the **circular economy** advocates minimizing waste and reusing resources within the system. Practices like permaculture and agroforestry mimic natural ecosystems, which reduce external inputs and promote long-term sustainability. Finally, agroecology promotes **human and social values**, placing emphasis on community well-being, food traditions, and cultural heritage. It values food sovereignty, supporting local food systems, traditional agricultural knowledge and innovations that foster resilience and preserve biodiversity.

By integrating these principles, agroecology not only provides a framework for sustainable agriculture but also aims to empower farmers and communities, improve food systems, and promote ecological stewardship at every level. By adhering to these principles, agroecological systems aim to harmonize agricultural practices with the environment, ensuring long-term productivity and resilience and sustenance.

Organic Farming

Organic farming is a production system that prohibits the use of synthetically produced agri-inputs. Instead, it relies on organic material such as crop and animal residues, legumes, bio-pesticides and bio fertilizers for maintaining soil fertility and productivity and managing insect-pests and diseases under conditions of sustainable natural resources and a healthy environment (Ramanjaneyulu *et al.,* 2020). The National Organic Standards Board of the United States Department of Agriculture (USDA) defines it as 'an ecological production management system that promotes and enhances biodiversity, biological cycles, and soil biological activity. It is based on the minimal use of off-farm inputs and management practices that restore and enhance ecological harmony'(USDANOSB 1995). As a system, it entails several components with specific and integrated functions. The use of cover crops, green manures, animal manures, crop rotations, mixed farming, organic manure, and compost is encouraged to improve soil fertility, maximize biological activity, and maintain long-term soil health. Similarly, weeds are managed by using crop rotations, good crop husbandry, and biological controls. Insect- pests and diseases are managed using diversity, bio-intensive pest management, and other biological controls. Finally, it emphasizes reducing external off-farm inputs and focusing on conserving the biodiversity of the agricultural system and the surrounding environment (Kumar *et al.,* 2019). However, organic farming has several disadvantages *viz.* higher production costs, lower yields, higher overhead costs, and more labour intensive. Organic fertilizers are also more expensive than synthetic fertilizers, and organic farms must have to go through a rigorous certification process. As a result, the organic farming practices are still away from the reach of majority small holder farmers of the country.

Natural Farming

Crop management practices affect the relationship between soil processes and agro-ecosystem function to a great extent, and thus affect the sustainability of agricultural production systems (Jernigan *et al.,* 2020). Indiscriminate use of chemical inputs has adversely affected the soil quality and crop ecosystem, and threatens future food production by reducing biodiversity, declining factor productivity and contributing to environmental degradation of the existing system (Kotschi 2015; Chandini *et al.,* 2019). In the present scenario, ensuring food security, producing more with less resources and building the resilience of resource poor farmers are important in creating a food and nutrition secure future. However, the ever-increasing input costs, even in organic production systems, are a concern to the sustainability of the production systems. The subsequent focus on developing sustainable and equitable approaches to agriculture underpin the Natural Farming (NF) approach, which aims to address both environmental and socio-economic concerns within the farm sector.

NF is an agroecological approach that can potentially address multiple agricultural crises attributed to the industrial farming model (NRAS 2020). It emphasizes sustainable, ecological, and holistic methods in agriculture and encompasses minimal soil disturbance, use of locally available organic inputs, diversity, cover crops, water conservation, animal integration, natural pest control, application of local and traditional knowledge, resilience of agricultural systems and it is in harmony with nature. Indian farmers increasingly find themselves in a vicious cycle of debt, because of the high production costs, high interest rates for credit, the volatile market prices of crops, the rising costs of fossil fuel based inputs, and private seeds. Under such conditions, 'Zero Budget Natural Farming (ZBNF)' developed by Mr. Subhash Palekar, a Padam Shree awardee, promises to end a reliance on loans and drastically cut production costs, ending the debt cycle for desperate farmers (Campesina, 2016).

History of Natural Farming

It is well documented in literature that Indian agricultural history was rich, diverse and high yielding before the advent of the Green Revolution, through the ages. Krishi-Parashar systematically wrote the first textbook on Agriculture in the world (c. 400 BCE, AAHF Classic Bulletin) which deals with prediction of rainfall models on the basis of movement and position of planets, rainfall and its distribution, indicators of drought, management of farming and cattle, nutrient management, seed collection and preservation, agriculture tools and plough mechanics, agronomic practices, *etc.* (Sadhale, 1999). Vrikshayurveda (c. 1000 CE) has description of a unique fermented liquid fertiliser-cum-plant protection material called Kunapajala, which happens to be the first fermented natural liquid manure in the world. Further, important information is given about plant nutrition, plant diseases, and plant protection with natural products/formulations, construction of gardens, miracles related to agriculture and horticulture, use of plant species as an indicator of crops and animal production and description of religious plants.

Inscriptions in Chola Temple in Ramnad (Tamil Nadu) indicate rice yields of 6.6 MT/ha during 10th-13th Century, and great biodiversity in cultivation from millets, paddy, lentils, vegetables and fruits have been reported (Macrae, 2016 and Srinivasan, 2016). Edinburgh review in 19th Century reported that the land productivity in India was 3 times that of England (Watt, 1889). It reported rice yields of 6 MT/ha, as per revenue records of Madras Presidency Coimbatore in 1807, and bengal gram production of 1104 kg/ha in Gujarat. Watt (1889) has also mentioned that the yield of native cotton variety of India was substantially higher in the 19th century as compared to the national average in 2017-18.

There is also a broad tradition of NF, propounded by advocates such as Narayana Reddy (Karnataka), Shripad Dabholkar (Maharashtra), G. Nammalvar (Tamil Nadu), Partap C. Aggarwal (Madhya Pradesh) and Bhaskar Save, popularly referred to as the 'Gandhi of Natural Farming' in Gujarat. The present NF model was originally known as Zero Budget Natural Farming (ZBNF) or Subhash Palekar Natural Farming (SPNF). It is being widely accepted by farmers (Khadse and Rosset 2019; Neelam and Kadian 2016; Fitzpatrick *et al.,* 2022; Behl *et al.,* 2023). This innovative farming system was developed by Mr. Subhash Palekar during 1980s after a period of self-study of the *Vedas* (the oldest of the Hindu scriptures), organic farming and conventional agricultural science, and evaluating different methods on his own farm. The phrase 'Zero Budget' was not referred to 'zero costs' but to achieve dramatic cuts in production costs by ending dependence on external synthetic inputs and agricultural credit. Instead, it is meant that 'the need for external *financing* is zero, and that any costs incurred can be offset by a diversified short duration intercrop's (Khadse and Rosset, 2019).

The NF, which is based on principles of agroecology, can potentially address multiple agricultural crises attributed to the modern day chemical input based farming model. In India, NF is being adopted for its potential to regenerate the soil, increase farmer incomes, enhance food security and enhance resilience to climate change. It offers a suitable perspective on multiple fronts from production to consumption while addressing wholesome aspects of health (ecological, human and economic). This indigenous cow based holistic farming aims to reduce crop production costs and market dependency for farm inputs by utilizing cheap and locally sourced inputs, and increase farm carrying capacity to sustain the vital livelihood source of resource-poor farmers (Figure 20.2). It relies more on soil biology than soil chemistry by encouraging multicropping, round-the-year soil cover, addition of formulation made up of cow dung and urines to trigger the microorganisms in the soil system, and creation of soil moisture- air in the ration of 50:50 in the soil. Himachal Pradesh and Andhra Pradesh have set out the aim of transforming NF to all their farmers through state-led programmes of training, research and extension.

Figure 20.2: Natural Farming Practices (intercropping and tree paste, *paudhlep*) in Himachal Pradesh.

Similarities and Key differences between Natural Farming and Organic Farming

Both natural and organic farming are chemical/poison free farming methods, which discourage farmers from using chemical fertilizers, pesticides *etc.* on plants and in all agricultural practices, and promote nonchemical pest control methods. Both of these farming methods encourage farmers to use local breeds of seeds, and native varieties of vegetables, grains, pulses and other crops. Under organic farming, farmers can use organic fertilizers and manures like compost, vermi-compost, cow dung manure including those sources from external sources. However, such external inputs (even organic) are not added under NF. In NF, decomposition of organic matter by microbes and earthworms is encouraged right in the field itself, which gradually improves soil fertility.

Natural Farming–Present Status

NF addresses the root causes of farmers' problems in an integrated way by providing climate resilient and holistic long-term solutions, which includes a precise focus on social and economic dimensions of food systems. Even before institutional support to NF, as it exists today in various states, farmers have been practicing these tenets for decades led by various innovators. The NF practices, as propounded by Mr. Subhash Palekar, a *Padam Shri* Awardee by the Government of India (Palekar, 2013) and promoted by Acharya Dev Vrat, former Governor of Himachal Pradesh (HP) and present Governor of Gujarat, enables farmers to improve soil health, drastically reduce costs and risks, reduce irrigation requirements and increase yields (NITI Aayog 2024, Barakzai *et al.,* 2021; Chandel *et al.,* 2021). The four principles of NF being employed by the farmers to enhance productivity at low cost are *Beejamrit* (Seed treatments using local cow dung and urine); *Jeevamrit* or *Ghanjeevamrit* (soil inoculant made of cow dung and urine); *Aachhadan i.e.* Mulching (covering the whole soil surface with crop residue or live intercrops to create favourable microclimate in soil); and *Whapsa* (soil aeration *i.e.* creating a condition of moisture and air in 50:50 proportion) (Figure 20.3). Some other associated pillars are simultaneous growing of short duration intercrops, including atleast one legume, and use of local resource-based decoctions for plant protection.

Beejamrit Seed treatment Jeevamrit Whapsa line Aachhadan

Figure 20.3: Basic Tenets of Natural Farming.

Natural Farming and Soil Health

Healthy soil is the foundation of productive, profitable and environment friendly agricultural systems. Sustainability of a farming system is related to its effect on changes in soil quality over time (Karlen *et al.,* 1997). The main bio-formulation that are applied in NF for soil health improvement are Jeevamrit and Ghanjeevamrit (Figures 20.3 and 20.4). **Jeevamrit** is a fermented microbial culture. It provides nutrients, but most importantly, acts as a catalytic agent or culture that promotes the activity of microorganisms in the soil to synthesize/to make bio-available plant nutrients in situ, increases the population of native earthworms and protect against pathogens. NF argues that the dung of indigenous cows and undisturbed soil from the field has a huge number of diverse microorganisms which help in increasing the bioavailability of nutrients to the plants. Soil is a complex ecosystem hosting bacteria, fungi, plants, and animals. Soil microbes metabolise recalcitrant forms of soil-borne nutrients to liberate these elements for plant nutrition. In natural ecosystems, most nutrients such as N, P, and S are bound in organic molecules and are therefore minimally bioavailable for plants. To access these nutrients, plants are dependent on the growth of soil microbes such as bacteria and fungi, which possess the metabolic machinery to depolymerize and mineralize organic forms of N, P, and S (Jacoby *et al.,* 2017). Lu *et al.* (2014) isolated 219 bacterial strains from cow dung, among which 59 isolates displayed nematicidal activity against >90 percent of the tested nematodes.

Figure 20.3: Use of Jeevamrit in NF.

Ghanjeevamrit is the dry formulation of jeevamrit. It helps the soil to activate the available nutrients, microorganisms to make them available for the crop sown in that particular area. It increases the count of earthworms in soil which is beneficial for soil fertility. It ensures higher yield by enhancing the availability of nutrients through faster decomposition of bulky organic manures by boosting the microbial activity in the soil. Many of these formulations are rich in beneficial micro flora and can act as efficient plant growth promoters. Saharan *et al.* (2023) investigated the effect of *Jeevamrit* on microbial properties under ZBNF coupled with metagenomic analysis. Whole genome metagenomic analysis revealed the presence of proteobacteria, bacterial phyla including *Bacillus, Pseudomonas, Rhizobium,* and *Panibacillus*. On the other hand, Ascomycota was the dominating fungal phyla present in the soil samples. Further, functional analysis showed a high representation of genes/enzymes involved in amino acids and carbohydrate metabolism contributing to soil fertility, plant growth, defence, and development.

Figure 20.4: Preparation of Ghanjeevamrit by the Farmers.

Aachhadan (mulching) is defined as covering of soil surface using either live crops as short duration intercrops or crop residues like straw (dead plant biomass) (Figure 20.5). It protects soil from direct exposure from sunlight, produces humus, conserves top soil, increases water retention and transmission, moderates hydro-thermal regime, encourages soil fauna, and prevents weeds. Evidence suggests that crop residues are good sources of plant nutrients and can increase yield and water use efficiency, while decreasing weed pressure. Long-term studies of the residue recycling have indicated improvements in physical, chemical and biological health of soil (Singh and Sidhu, 2014). The NF initiatives show promise for transformation at scale and open up a new frontier in thinking about agriculture with efficient utilisation of soil and water resources. Aachhadan, associated with NF reduces soil evaporation and transpiration, and helps mitigating drought conditions.

Figure 20.5: Mulching/Aachhadan under Natural Farming.

Whapsa, is a condition where there is a presence of both air molecules and water molecules in the soil in equal proportion. Soil aeration, a result of jeevamrit and aachhadan- represents water management through improved soil structure and humus content. This helps in reducing irrigation requirements and promotes water availability to the crops through soil capillary actions, increase water use efficiency and resilience to drought (Figure 20.6). The advocates of NF counter the over-reliance on irrigation in green revolution farming. The *whapsa* condition reduces the occurrence of soil and root borne disease. As per the Centre for Study of Science, Technology and Policy (CSTEP) report based on a study conducted in Andhra Pradesh in 2020, NF requires 50 to 60 percent less water and electricity when compared to conventional farming practices. NF practices also prevent over-extraction of groundwater, enable aquifer recharge, and eventually contribute to increasing water table levels (Khurana *et al.,* 2022).

Figure 20.6: Whapsa Lines for Irrigation under Natural Farming.

Changes in farming practices are foremost reflected in the changes in biological properties such as microbial populations and soil enzymatic activity. Soil enzymes have been suggested as one of the important indicator of soil quality, and for evaluating the degree of alteration and assessing the effect of different cropping systems on nutrient dynamics and soil quality (Bandick and Dick 1999). The soil microorganisms perform various biogeochemical functions and help in replenishing soil fertility, as they are involved in nutrient cycling (Jacoby *et al.,* 2017). However, the full range of microbes associated with plants and their potential to replace synthetic agricultural inputs has only recently started to be uncovered. Various researchers have studied effect of individual inputs on crop yields, but information is scarce on the effect of NF package as a whole. NF has positively affected soil quality and microbial community composition within sustainable farming systems (Liao *et al.,* 2019; Verma *et al.,* 2018; Smith *et al.,* 2020; Rana *et al.,* 2021). Yankit *et al.* (2024) observed a significant increase in the soil microbial (bacteria, fungi and actinomycetes) population under tomato-based NF system in comparison to organic (OF) and conventional (CF) farming systems.

Figure 20.7: Earthworms and their Casts under Natural Farming.

Soil organisms like earthworms and micro-arthropods remain under-represented in soil processes. Earthworms are a major component of soil faunal communities in the NF ecosystem. Soil fauna is crucial to soil formation, litter decomposition, nutrient cycling, biotic regulation, and for promoting plant growth. Soil microarthropods have been found to be sensitive to changes in land management practices (Parisi *et al.,* 2005) and are thus being used as indicators of soil quality. NF has also shown increase in soil faunal communities (earthworms and microarthropods). Barakzai *et al.* (2021) reported an increase in soil microarthropods from Hymenoptera, Hemiptera, Coleoptera, Isoptera, and Isopoda orders under NF. They also reported order Geomorpha under NF, which was not found under CF. Total soil microarthropod population increased by more than 3 times under NF in comparison to CF. The abundance of soil micro-arthropods has been observed to be positively correlated with soil C and N (Wang *et al.,* 2015). Earthworms move the soil around and creates tunnels that alter soil structure. Earthworm tunnels bring in oxygen, drainwater and create space for plant roots. The natural feeding habits of earthworms involve ingestion of small amounts of soil through their bodies and then they excrete it in the form of earthworm casts, which improve soil fertility status (Avant, 2017). Tillage, chemical fertilization, and pesticide usage regularly influence earthworm populations. However, NF practices allow earthworms to proliferate. Cabbage intercropped with fenugreek, pea and coriander under SPNF had higher population count of earthworms (183.33 m^{-2}) as compared to CF (41.67 m^{-2}) (cabbage as sole crop). The result showed that the application of cow urine and dung based inputs help in promoting earthworm activity in soil. Similarly, earthworm cast weight was also higher in SPNF (57.23 g m^{-2}) than CF system (14.87±0.56 g m^{-2}) during the rainy season (Vipul 2021). A systematic comparison between NF and non-NF fields conducted in Andhra Pradesh reveals that the NF fields host average 232 earthworms per square meter compared with just 32 on non–NF fields (Bharucha *et al.*, 2020). Earthworms affect the SOM dynamics. There is sharp increase of mineralization during digestion.

Studies have shown an increase in soil organic carbon (SOC) under NF systems. The amount of SOC is greatly influenced by the application of *jeevamrit, ghanjeevamrit,* and other NF practices (Saharan *et al.*, 2023). NF involves addition of microbial formulations (based on cow urine and dung), which enhance the activity of soil microbes, earthworms and other soil fauna. Zhu *et al.* (2020) reported changed perception about contribution of microbes towards SOC pools.

Increased microbial activity contributes to nutrient availability in soil. Saharan *et al.* (2023) investigated the effect of *Jeevamrit* on soil chemical properties of the ZBNF field and the economics of ZBNF. The percent increase in soil properties, such as organic carbon, available phosphorus, and available potassium, was recorded up to 46, 439 and 142 percent, respectively, while micronutrients, such as Zn, Fe, Cu, and Mn, also increased up to 98, 23 per cent 62, and 55 per cent, respectively, from 2017 to 2019. Ladha *et al.* (2016) constructed a top-down global N budget for maize, rice, and wheat for a 50 year period (1961 to 2010). They reported that non-symbiotic nitrogen fixation appears to be the major source for crop N uptake. An estimated 48 per cent (737Tg) of crop N, equal to 29, 38, and 25 kg ha-per ha per year for maize, rice, and wheat, respectively, was contributed by sources other than fertilizer- or soil N. Smith *et al.* (2020) reported that NF is likely to reduce soil degradation and could provide yield benefits for low-input farmers. Nitrogen fixation, either by free-living nitrogen fixers in soil or symbiotic nitrogen fixers in legumes, is likely to provide the major portion of nitrogen available to crops. They also worked out maximum potential nitrogen fixation and release and reported that 52–80 percent of the national average nitrogen applied as fertilizer is expected to be supplied by biological fixation.

Higher SOC and microbial diversity (fungal hyphae or bacteria) helps in the formation of porous spaces, thereby leading to increase in the absorption of water. Therefore, water adheres to the surface of particles or organic matters leading to water infiltration and increase in the holding capacity. NF practices enhance the soil microbial diversity and soil fauna, and thus lead to improved soil structure and porosity, which leads to better soil physical environment and thus improvement in soil-water-air relations. Study in dry temperate zone of HP under Apple based NF system showed higher soil moisture content (SPNF 2021). The straw mulch under SPNF maintained higher soil moisture (up to 4.9 per cent, w/w) throughout the season, in comparison to that in CF without mulch. NF practices also moderated the soil temperature. Soil temperature was lower in the afternoon under SPNF in comparison to the plots under CF without mulch, and thus maintained a favourable hydro-thermal regime in the root zone.

Insect-Pest and Disease Management in Natural Farming

There is a rich heritage of plant protection in NF (Nene 2003, 2012). The use of natural materials available with farmers to control crop disorders has been mentioned since Kautilya's Artha-sastra (c. 300 BC). Knowledge of plant protection generated by our ancient and medieval scholars are applicable to the crop production even in modern India, especially at the small farmers level. These reports provide references to plant protection found in Vedas, Buddhist literature, Kautilya's Artha-sastra, Krishi-Parashara, Sangam literature of Tamils, Agnipurana, Brhat Samhita of Varahamihira, Surapala's Vrikshayurveda, Viswavallabha of Chakrapani Misra, and some documents of the medieval and pre-modern period. These also cover the plant protection and treatment practices recommended in Vrikshayurveda.

NF practices provide immunity to plants. Seed treatment with beejamrit, application of ghanjeevamrit, jeevamrit, aachhadan (mulch) and whapsa, all established tanets of NF help in increase in beneficial microorganisms, increase earthworms activity and increase soil organic carbon which cause resistance in crop plants. Therefore menance of insect-pests is reduced. However, when infestation occurs, the preparations of various decoctions *viz.* agniaster, neemaster, drakaster, brahamaster and dashparni ark from locally available plants (Figure 20.8) in cow urine helps to control the insect-pests (Palekar, 2006). Multispecies cropping system under NF is considered as the application of ecological principles based on biodiversity, plant interactions and other natural regulation mechanisms (Vandermeer, 1989). The highest number of parasitoids and predators was recorded under guava orchard intercropped with cowpea as compared to that in the sole cropped field. Mealybug-specific parasitoids *viz., Anagyrus dactylopii* (Howard) and *Coccido xenoides* sp. were recorded in the guava orchard intercropped with cowpea. *Anagyrus dactylopii* (Howard) and *Coccidoxenoides perminutus* (Girault) as efficient native parasitoids of grapevine mealybug, *M. hirsutus*, were reported by Amala *et al.* (2013). Yankit *et al.* (2019) cultivated tomato under Natural Farming (NF), Organic Farming (OF) and Conventional Farming (CF) systems and observed that incidence of invasive leaf miner, *Tuta absoluta* (Meyrick) was significantly less in NF system as compared to organic farming (OF) and conventional farming (CF).They further observed that *T. absoluta* appeared in NF plots 4 weeks later than conventional farming. The SPNF system had 7 species of natural enemies, viz; *Coccinella septempunctata* (L.), *Hippodamia variegate* (Goeze), *Episyrphus balteatus* (De Geer), *E. frequens* (Matsmura), *Metasyrphus confrator* (wiedemann), *Diadegma semiclausum* (Hellen), and *Diaeretiella rapae* (Mc Intosh), whereas, in the CF, only 5 species of natural enemies were recorded (Vipul 2021). The Shannon index described the SPNF system as a more diverse ecosystem in terms of natural enemies. Barakzai *et al.* (2022) intercropped cauliflower with pea, coriander and mustard as trap crop and reported ZBNF system harboured with less insect-pest diversity and more natural enemies as compared to the CF system.

Natural enemies' diversity indices revealed that the NF based system was more diverse than CF system. Vishwajeet (2020) recorded 5 species of insect-pests *viz.* jassid, *Amrasca biguttula biguttula* (Ishida), brinjal shoot and fruit borer, *Leucinodes orbonalis* (Guen), green house whitefly, *Trialeurodes vaporariorum* (Westwood), cotton aphid, *Aphis gossypii* (Glover), serpentine leaf miner, *Liriomyza trifolii* (Burgess) and blister beetle, *Mylabris pustulata* (Latreille) in both *i.e.* Subhash Palekar Natural Farming (SPNF) as well as Conventional Farming (CF) systems. Among natural enemies *C. septempunctata, H. variegata, E. balteatus*), *Ischiodon scutellaris* (F), were present in CF, whereas *C. septempunctata, H. variegata, Oenopia sexareata* (Mulsant), *E. balteatus, I. scutellaris* were present in SPNF system. In another study, the SPNF and CF had the same insect-pest diversity, but delayed their incidence by 1-3 weeks in SPNF as compared to CF and attracted relatively more natural enemies (Rana *et al.,* 2021). In Spiti Valley of HP, Natural farming practices successfully controlled woolly apple aphid, *Eriosoma lanigerum* (Hausmann) by using low- cost locally sources inputs when compared to conventional farming methods (Vashisth *et al.,* 2023). Bakshi *et al.* (2023) reported 11 species of natural enemies in apple orchards maintained under SPNF system and out of these 11 species, 4 species of coccinellids, 5 of syrphids, 1 of green lacewing and 1 species of Chalcidoid wasp and in CF system only 8 species of natural enemies in apple ecosystem were recorded.

Figure 20.8: Various Locally Available Farm Based Inputs Used for Plant Protection Formulations.

Tatsumi *et al.* (2023) predicted the higher microbial diversity and abundance of nutrient cycling microbes in soils from natural farming fields and also more microbial interaction between the microbes due to the absence of chemical stress. On being attacked by certain pathogens, plants cry for help by releasing volatile organic compounds (VOCs) as a part of their natural defence mechanisms. These VOCs act as stimulants for the beneficial (specific) microbes around the plant to come for rescue (Liu, 2019). In NF, the culture of the beneficial microbes is routinely added to the soil. Thus, the rescue microbial forces are present nearby to immediately provide defence services to the plants. In cotton plants, volatile organic compounds isolated from rhizobacteria (*Bacillus cereus, B. aryabhatta, B. tequilensis*) were effective in reducing mycelial growth and lessens the extremity of root rot (Nagrale *et al.,* 2022). Upperi *et al.*(2009) reported lowering of leaf spot incidence in groundnut after jeevamrit application, that could be attributed to the elemental and phenolic compounds present in cow dung and urine which led to retardation and multiplication of pathogen and restricted the additional spread of pathogen. Mycorrhizal fungi are another group of symbiotic fungal species which have been found to increase in numbers in the soil under natural farming conditions. Under NF system, plant health and productivity mostly rely on the root microbiome (Sinong *et al.,* 2021). When infected with *Fusarium*, the damaged pepper plants attract beneficial bacterial species that defend the host plant (Gao *et al.,* 2021). Jeevamrit had a substantial impact on the bell pepper plant's growth and yield characteristics, resulting in an increase in yield that was 82.4 per cent above that of control and a decrease in the prevalence of insect pests and diseases (Hameedi *et al.,* 2018).

Mixed cropping or Multiple cropping, one of the main components of natural farming, results in complex relationships between the different plants, both above- and below-ground. The benefits of mixed cropping are lower weeds population density, insect- pests infestation and diseases incidence. It could be due to increase in population of natural enemies and due to presence of non-preferential hosts and thus difficulties in detecting host. A significantly less severe Ascochyta blight (*Mycosphaerella pinodes*) attacks on peas were reported in pea-cereal intercrop than in pea monocrop. The decline in disease occurrence could be partially explained by a modification of the microclimate within the intercrop's canopy, specifically a reduction in leaf wetness duration during and after flowering (Schoeny *et al.,* 2010). In China, chilli peppers cultivated alongside maize exhibited lesser incidence of Phytophthora blight than peppers grown on their own (Yan *et al.,* 2006). Marzani (2023) investigated the impact of mixed cropping on *Ascochyta rabiei* in chickpea and interpreted that mixed cropping combinations (Chickpea-oilseed-wheat-faba bean) showed lesser disease severities in comparison to mono-cropping system. Moreover, multiple cropping systems may contain species that produce biochemical cues that disrupt the development of diseases, insect- pests, root parasitic nematodes and weeds.

Foliar sprays of sour butter milk (*Khatti Lassi*) are routinely used to check diseases on crop plants under NF system. A large number of yeasts are found in the fermented milk. Cow urine and butter milk retarded the growth of *Rhizoctonia* and *Fusarium* in soybean (Sapre and Varma 2006). Rice foliar spray of *P. fluorescens* and buttermilk (0.5%) reduced the incidence of false smut, sheath rot and grain discoloration (Jeyalakshmi *et al.,* 2021). *Sounthaster,* a concoction prepared by boiling dried ginger with milk, has been found to be very good fungicide and bactericide under NF. Ginger contains various biomolecules such as monoterpenoids, sesquiterpenoids and phenolic compounds and their derivatives provide a broad antimicrobial spectrum against different microorganisms making it an interesting alternative to synthetic antimicrobials. Phenolic compounds such as eugenol, shogaols, zingerone, gingerdiols, gingerols, *etc.* and their synergistic relationship with other compounds such as β-sesquiphellandrene, cis-caryophillene, zingiberene, α- farnesene, α- and β-bisabolene, are mainly responsible for the antimicrobial activity found in ginger essential oil, extracts, and oleoresins (Jacob, 2016). Most of these compounds are insoluble in water; thus, aqueous extracts exhibit lower antimicrobial activity than those mixed with milk. In this sense, ginger showed a high fungicide and antibacterial effect due to its high eugenol concentration. Under NF, *whapsa* condition not only results in the increased root network but also improves the health of plants. As the plants do not come in contact with water directly and high soil moisture conditions are not there, the incidence of diseases, especially soil borne will always be lower.

Quality of Natural Farming Produce

There is growing awareness and acceptability of chemical free, naturally grown food commodities with higher demand from health conscious consumers with comparatively higher nutritional quality and safety compared to chemically grown crops. NF produce has been reported to be of better quality with a higher shelf life. Studies have been carried out on the post-harvest physiology, quality retention, shelf-life extension and value addition of the produce grown under natural farming. The investigations carried in the laboratory reveals significant differences in the quality attributes of the crops *viz.* tomato, broccoli, cabbage, cauliflower and bell pepper. The crops were analysed for various physico-chemicals, nutritional and sensory parameters *viz.* total solids, ash content, TSS, titratable acidity, total sugars, reducing sugars, ascorbic acid, total phenols, spoilage index and their antioxidant activity. Sharma *et al.* (2021) studied the effect of modified atmosphere packaging (MAP) and storage conditions on the shelf-life and quality of broccoli and cabbage during 2019–20.

Shelf-life comparison was done on the basis of physiological loss in weight (PLW) at every alternate day up to 28 (broccoli) and 34 days (cabbage). Results revealed that broccoli and cabbage packed in perforated LDPE pouches and stored under low temperature storage retained higher sensory acceptability for 14-20 days with record extended shelf-life for market purpose up to 18 and 22 days, respectively with minimum PLW (<10%). Both broccoli and cabbage grown under SPNF system contained a good amount of total sugars (4.40, 6.59%), reducing sugars (2.53, 5.34%), ash (8.80, 1.28%) and ascorbic acid (86.20, 45.40 mg/100g), respectively. Sharma *et al.* (2023) studied the effect of modified atmospheric packaging (MAP), storage conditions (ambient, zero energy cool chamber and refrigerated) as well as storage period on shelf-life and quality of tomatoes produced through Subhash Palekar natural farming (SPNF) and chemical farming (CHEM) systems. The shelf-life comparison was done at every two days' interval up to 24 days based on physiological loss in weight (PLW) and spoilage percentage. SPNF and CHEM tomatoes were compared on the basis of quality attributes (firmness, acidity, TSS, sugars, total phenols, antioxidant activity *etc.*), mineral content and sensory characteristics. Results revealed that tomato stored in refrigerated condition had lowest PLW (p Â0.05) and spoilage incidence and highest sensory acceptability up to 24 days and was at par with zero energy cool chamber (ZECC) stored tomatoes. Quality evaluation indicated that tomatoes grown under SPNF system contained a higher amount of total solids, total sugars, reducing sugars, ascorbic acid, total phenols and ash. The results showed that tomatoes with MAP and stored under refrigerated storage condition retained maximum quality up to 24 days followed by tomatoes stored under ZECC (18 days) compared to 12 days at ambient conditions (Figure 20.9).

Tomatoes after 24 days

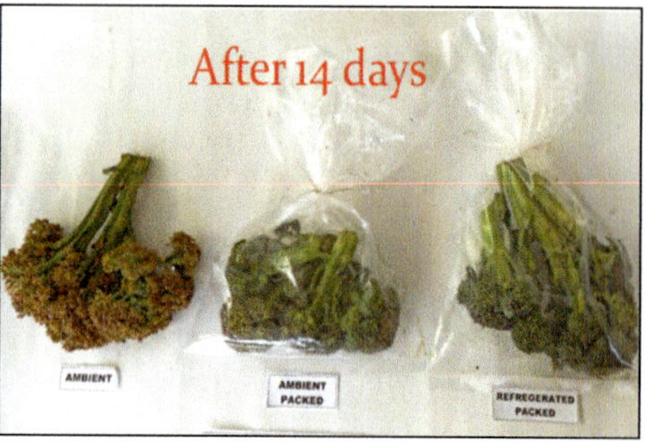

Broccoli after 14 days

Figure 20.9: Effect of MAP and Storage Conditions on Shelf Life and Quality of Tomato and Broccoli grown under Natural Farming System (SPNF).

Natural Farming-Sustaining Livelihoods

Multi-cropping is an integral component of NF systems. Up to 09 crops are being grown concurrently by the NF farmers leading to crop intensification and thus increasing biodiversity. Dr YS Parmar University of Horticulture and Forestry (UHF), Nauni, Solan HP has developed package of practices for different vegetable based NF systems (tomato+ french bean+ brinjal; capsicum + french bean + brinjal; pea+ spinach+ fenugreek; cabbage+ fenugreek+ coriander) and standardized complete set of high-density apple-based practices based on this low cost, local resource based, non-chemical and climate resilient farming model. Orchardists of Himachal Pradesh have successfully demonstrated the worth of NF practices for producing good quality apples, and apprehensions cast against the efficacy of NF practices in sustaining the apple productivity are overhyped. Several studies have reported that there is no yield penalty and decrease in income under NF (Duddigan *et al.,* 2022; Laishram *et al.,* 2022). Studies have shown that cost of cultivation has decreased and net income increased under NF. Laishram *et al.* (2022) studied the impact of NF System on rural households in Solan District of HP. Their study focused mainly on the different cropping systems of NF and comparing the economics of NF with CF systems. Study shows that farmers adopted five major crop combinations under

NF system including vegetables-based cropping system (*e.g.*, tomato + beans + cucumber and cauliflower + pea + radish) and vegetables-cereals-based cropping systems. The results indicated that vegetable-based cropping system had 19.68 percent more net return in *kharif* season and 24.64 per cent more net return in *rabi* season as compared to sole vegetable cultivated under CF. NF maximizes land use and reduces the chance of crop yield loss. NF has resulted in increased returns especially in the vegetable cropping system, where reduction in cost was 30.73 per cent (*kharif*) and 11.88 per cent (rabi) across all crop combinations in comparison to CF. There was cost savings from eliminating chemical fertilizers and pesticides as well as higher benefit from intercrops under NF. Chandel *et al.* (2021) compared the yield potential, input cost and net returns in major crop combinations under SPNF during 2018-19 and 2019-20 with CF. The cereals-pulses, cereals-vegetables, fruit+pulses-vegetables and vegetables-pulses were found as the 4 major crop combinations followed under SPNF on an average area of 0.26 ha, in HP. The fruit based SPNF farming combination was the most popular (40.6%) and profitable (REE=21.44%). A reduction of 14.34- 45.55 per cent in cost of cultivation and an increase of 11.8-21.55 per cent in the net returns, over CF, validate the superiority of SPNF over the CF.

Recent studies by National Institute of Agricultural Extension Management (MANAGE), Hyderabad and Academy of Management Studies (AMS), Lucknow on the impact evaluation of Prakritik Kheti Khushhal Kisan Yojana (PK3Y), flagship programme of GoHP on Natural Farming, have shown positive results of the scheme. The studies shows significant reduction in the use of chemical pesticides and fertilizers, and thus input costs. Almost 45 per cent beneficiaries have adopted NF fully and adoption has percolated to non-beneficiary farmers also. The keeping quality/shelf life has increased substantially and the practice supports the risk resilience in agriculture. According to a survey conducted by the university, farmers have adopted NF for a number of reasons, including family wellbeing, food self-sufficiency, environmental issues, and cost-cutting. Around 99.1 per cent of the farmers conveyed that they are getting better drought resistance in crops and the quality of produce by using NF practices. This clearly shows that NF can be an effective tool in the hands of farmers against otherwise inevitable ill impacts of climate change. Around 99.1 per cent of the respondents reported a better taste of the food produced through NF practices. Critics have expressed major apprehension about windfall decline in crop productivity under NF. But 59.1 per cent respondents reported improvements in yield levels by turning to NF practices. Saving on account of reduced cultivation cost under NF was confirmed by 89.1 per cent apple orchardists.

Agro-ecological Services and SDGs

Reduction in use of synthetic inputs under NF can improve the financial stability of smallholder farmers. By focusing on low-cost, locally available resources, NF can create more resilient and sustainable livelihoods. In addition, NF reduces the risk of chemical residues in food, leading to better health outcomes. Healthier soils and reduced chemical runoff contribute to cleaner water sources and better overall environmental health. NF promotes sustainable agricultural practices that minimize waste and use resources efficiently. By reducing reliance on synthetic chemicals, NF contributes to more responsible production methods, enhance soil health and promote carbon sequestration, which can help mitigate climate change effects. Practices like crop diversification and organic soil management improve resilience to climate variability and extreme weather events. Thus, NF practices are a step to meet the Sustainable Development Goals (SDGs). Of the 17 SDGs the most important goals for Agriculture are 'End poverty in all its forms everywhere' (SDG 1), 'End hunger, achieve food security and improved nutrition and promote sustainable agriculture' (SDG 2), 'Ensure availability and sustainable management of water and sanitation for all' (SDG 6), 'Promote sustained, inclusive and sustainable economic growth, full and productive employment and decent work for all' (SDG 8), 'Build resilient infrastructure, promote inclusive and sustainable industrialization and foster innovation' (SDG 9), 'Ensure sustainable consumption and production patterns' (SDG 12) and 'Protect, restore and promote sustainable use of terrestrial ecosystems, sustainably manage forests, combat desertification, and halt and reverse land degradation and halt biodiversity loss' (SDG 15). Moving towards a progressive vision, the NF programme envisions mapping these 7 SDG Goals, 15 Targets and 18 indicators through creation of a Sustainable Food Systems Platform for Natural Farming (SuSPNF). Thus, as per holistic metrics of sustainability, NF endeavours to meet all criteria and aspects of sustainability of farming systems and mitigating the effects of climate change that CF fails to deliver for the future.

New Initiatives

The government of HP has put in place various policies and programs to support NF practices and certify agricultural produce. These initiatives are aimed at promoting sustainable agriculture, improving soil health, and ensuring the quality and authenticity of farm products. It is among the few pioneering states in India, where the State Government has made a quantum leap in developing systems to bring NF to the grass-root level. UHF has developed package of practices for different vegetable and high-density apple-based NF cropping systems. HP has initiated steps towards establishment of circular economy model through standardization of SuSPNF model (Figure 20.10). This will create loops required to regenerate, recycle and reduce through the value chains for social, ecological and business benefits. The UHF has been working in close coordination and collaboration with NABARD/ Department of Agriculture, GoHP to establish SuSPNF model. Under SuSPNF, university is setting up and hand-holding Farmers Producer's Organizations (FPO) with NF farmers.

Work is also being initiated on Sustainable Agriculture Systems (SAS) of Chaupal Naturals, Shimla and Gram Disha Trust, Mandi in collaboration with researchers from 11 European countries under the European Union funded

ACROPICS project. All economic activity is aligned to ensure that overall systems health is rebuilt.

Figure 20.10: HP Government Initiatives for Marketing of NF Produce.

Therefore for a farm, this also means all activity which brings loops of regenerating the farm ecology, recycling natural biomass available and providing the correct or 'true' economic benefits for the farmers *e.g.* considerations in True Cost Accounting. Circular Farm Economy addresses Net-Carbon Neutrality in all operations of goods and service supply. It involves three aspects *viz.* regenerating Farm Systems to sequester carbon in soil and produce, products resulting as output from agriculture are reused, recycled or reduced (optimized) to allow net energy savings, and designing value chains from the products to ensure there is no loss or wastage, and reduction in GHG emissions over time. UHF is acting as Producer Organisation Promoting Institute (POPI) to establish and hand-hold the FPOs and executing blended financing and facilitating the venture based activities in the campus (Figure 20.11). In this endeavour, UHF has established a model outlet for selling the NF produce, and demonstrate the supply chain model from farm to fork.

Another innovation is for certification of NF produce. Existing third-party certification has some challenges as it is expensive for the individual farmer. Moreover, it certifies the product and not the producer and therefore requires extensive planning and documentation for initial as well as any revisions in crops sown and grown – thus making it difficult for farmers to apply and avail certification. Under PGS certification, individual farmers cannot apply unless the group of farmers are living in a similar geographical area and Regional Councils (RCs) require funds for data collection, and management and the absence of funding have led to many RCs becoming defunct and non-operational. Although, PGS certificate is free to obtain and the documents needed are basic, its operation and procedure are difficult to be used by an individual farmer/group of farmers as it does not provide any local language support. This is highlighted by the fact that governments usually rely upon services from Implementation Agencies (IAs) which are tasked to register farmers and groups on the PGS-India platform.

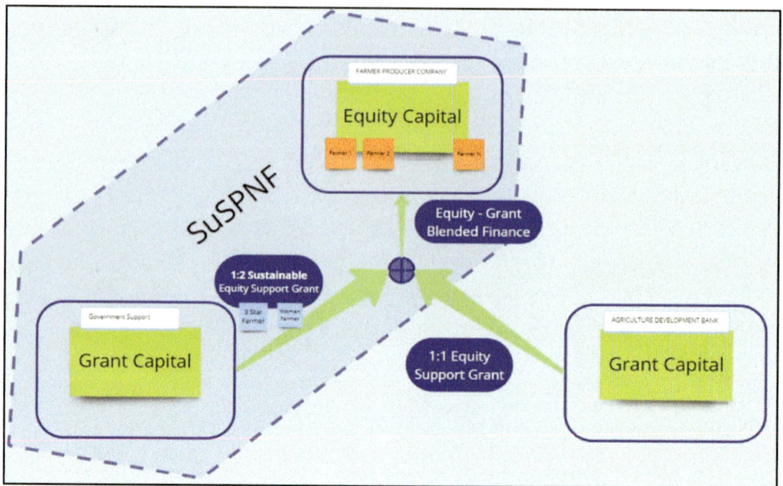

Figure 20.11: Blended Financing Model for Supporting NF FPCs under SuSPNF.

Thus, a unique and Innovative Self-Assessed Certification mechanism (CETARA-NF) accessible through a Web portal and mobile application is also being put in place for NF produce (Chandel *et al.,* 2022). The ultimate purpose is to provide an inexpensive and easy process for farmers to certify and for consumers to gain confidence on the food they are consuming. Nearly 1,40,000 farmers have registered for the CETARA Certificate on the web portal till July 2024 (Figure 20.12) (www.spnfhp.in).

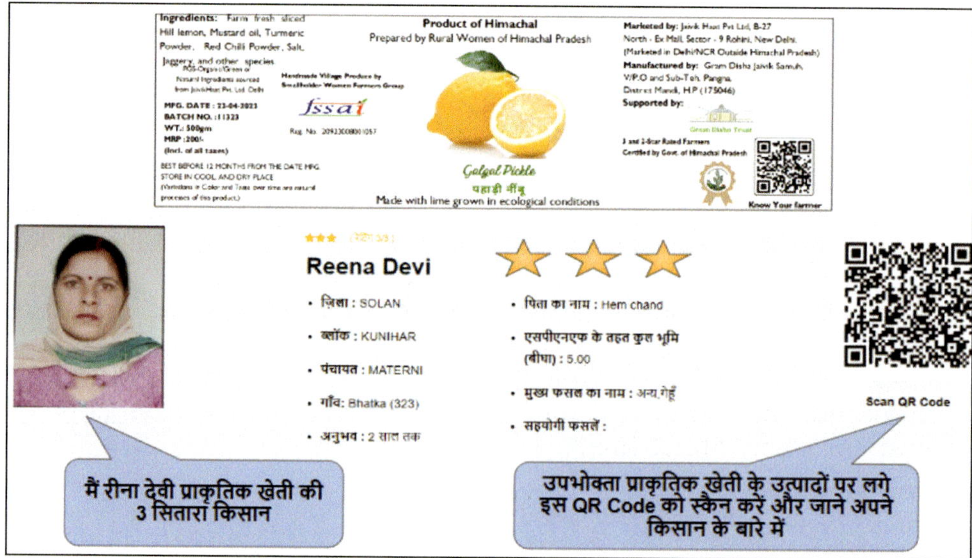

Figure 20.12: CETARA Certification of NF Produce.

Conclusion

Natural farming promotes climate resilience in harmony with nature and reduces environmental deterioration. It minimizes dependency on external inputs and is a step towards developing carbon neutral production systems. NF supports ecological integrity, climate resilience, food security, and rural development, hence, making it a viable solution to the complex societal issue of sustainable food production. The global inclination towards NF has been increasing as individuals, communities and governments recognize the need for more sustainable agricultural practices. External evaluation of NF initiatives in HP has already shown the promising impacts of NF. Furthering the research and development on climate resilient NF practices will support the movement of NF through generation of scientific evidences. The establishment of circular economy through standardized SuSPNF model, which includes entrepreneurship development through FPCs and involving and integrating transparency and traceability system embedded in CETARA certification will help in resolving the sustainability challenge in the future. This will create loops required to regenerate, recycle and reduce through the value chains for social, ecological and business benefits. This will lead to up-scaling and out-scaling of sustainable food systems and will be a stepping stone towards meeting SDGs. Furthermore, by implementing Nutri Gardens under institutional framework, schools will create a vibrant learning environment that benefits students understanding the concept sustainable food systems. Besides, the strengthening of FPOs will be a viable option to enhance the capability of individual NF farmers, improve their market access and contribute to the overall sustainability, reliability and resilience of the food system pathway by connecting producers and consumers following the principles of transparency and traceability.

REFERENCES

Amala U, Yadav DS and Bhosale AM, 2013. Studies on parasitoid complex of mealybug infesting grapes in Maharashtra. *Journal of Applied Horticulture* **15**: 117–119.

Avant S, 2017. Earthworms work wonders for soils. https: //www.usda.gov/media/ blog/2017/04/21/earthworms-work-wonders-soils#: ~: text=Earthworm per cent 20tunnels per cent 20bring per cent 20in per cent 20 oxygen,in per cent 20 comes per cent 20out per cent 20much per cent 20better!

Bakshi D, Chandel RS, Verma SC, Chandel VGS, Katna S, Bharat N and Sharma U, 2023. Impact of natural farming practices on the diversity of natural enemies in apple. In: National Conference on Natural and Organic Farming for Ecological, Economical and Nutritional Security held at Palampur, Himachal Pradesh, India w.e.f 7-9th June, 2023. 318-320 pp.

Bandick AK and Dick RP, 1999. Field management effects on soil enzyme activities. *Soil Biology and Biochemistry* **31**: 1471–1479.

Barakzai AW, Chandel RS, Sharma PL, Verma SC, Singh MP and Yankit P, 2021. Effect of farming systems on diversity and seasonal abundance of insect pests and their natural enemies in cauliflower. *Indian Journal of Entomology* 84(3): 535-540. https: //doi: 10.55446/IJE.2021.60.

Barakzai AW, Chandel RS, Verma S, Sharma PL, Bharat NK, Singh MP and Yankit P, 2021. Effect of zero budget natural farming and conventional farming systems on biological properties of soil. International *Journal of Current Microbiology and Applied Sciences* **10**(02): 1122-1129.

Behl P, Osbahr H and Cardey S, 2023. New possibilities for women's empowerment through agroecology in Himachal Pradesh, India. *Sustainability* 16(1): 140. https: //doi.org/10.3390/su16010140

Bharucha ZP, Mitjans SB, Pretty J, 2020. Towards redesign at scale through zero budget natural farming in Andhra Pradesh, India. *International Journal of Agricultural Sustainability* **18**(1): 1-20.

Campesina, 2016. Agroecology Knowledge Hub. FAO-UN. https: //www.fao.org/agroecology/detail/en/c/443712/.

Chandel RS, Gupta M and Gupta A, 2022. Certified Evaluation Tool for Agriculture Resource Analysis-Natural Farming (CETARA-NF) (SuSPNF (PK3Y) Digital Farmer Self Assessed Certified-Evaluation System) A Handbook- Vision, Approach and Methodology. SPIU, Department of Agriculture, Govt of HimachalPradesh, p26. https: //spnfhp.in/Upload/MediaGallery/PDF/3/2023-Jan-05-18-44-47-348.pdf.

Chandel RS, Gupta M, Sharma S, Sharma PL, Verma S and Chandel A, 2021. Impact of Palekar's natural farming on farmers' economy in Himachal Pradesh. *Indian Journal of Ecology* **48**(3): 872-877.

Chandini RK, Kumar R and Prakash O, 2019. The impact of chemical fertilizers on our environment and ecosystem. In: Research Trends in Environmental Sciences, 2nd ed, 69-86. https: //www. researchgate. net/publication/331132826.

Duddigan S, Collins CD, Hussain Z, Osbahr H, Shaw LJ, Sinclair F, Sizmur T, Thallam V and Ann Winowiecki L, 2022. Impact of zero budget natural farming on crop yields in Andhra Pradesh, SE India. *Sustainability* 14(3): 1689. https: //doi.org/10.3390/su14031689.

FAO, 2019. FAOSTAT: Food and agricultural organisation. http: //www.fao.org/faostat/en/#data/QC/visualize.

Fitzpatrick IC, Millner N, Ginn F, 2022. Governing the soil: natural farming and bionationalism in India. *Agriculture and Human Values* **39**(4): 1391-406.

Gao M, Chao X, Cheng Gao, Clement KM Tsui, Meng-Meng Wang, Xin Zhou, Ai-Min Zhang, and Lei Ca, 2021. Disease-induced changes in plant microbiome assembly and functional adaptation. *Microbiome* **9**: 1-18.

Gliessman SR, 2007. Agroecology: The ecology of sustainable food systems, 2nd ed., Boca Raton, FL : CRC Press, Taylor and Francis Group.

Hameedi A, Thakur KS, Kansal S., Mehta DK, Yousafzai A. and Mohammad MH, 2018. Effect of organic nutrient sources on growth, yield and quality of bell pepper (Capsicum annuum L.) under mid hill condition of Himachal Pradesh. *International Journal of Multidisciplinary Research and Development* **5**(1): 135-138.

Jacob JN, 2016. Comparative studies in relation to the structure and biochemical properties of the active compounds in the volatile and nonvolatile fractions of turmeric (*C. longa*) and ginger (*Z. officinale*). *Studies in Natural Products Chemistry* **48**: 101-135

Jacoby R, Peukert M, Succurro A, Koprivova A and Kopriva S, 2017. The role of soil microorganisms in plant mineral nutrition–current knowledge and future directions. *Frontiers in Plant Science* **8**: 1617. https: //doi.org/10.3389/fpls.2017.01617

Jernigan AB, Wickings K, Mohler C L, Caldwell BA, Pelzer CJ, Wayman S and Ryan MR, 2020. Legacy effects of contrasting organic grain cropping systems on soil health indicators, soil invertebrates, weeds, and crop yield. *Agricultural Systems* **177**: 102719.

Jeyalakshmi C, Renukaand R and Ramanadane T, 2021. Evaluation of foliar sprays of biocontrol agents, botanicals and fungicides for the production of healthy rice seed in the coastal ecosystem. *Journal of Cereal Research* 13 (Spl-1): 77-83. http: //doi.org/10.25174/2582- 2675/2021/115310

Karlen DL, Mausbach MJ, Doran JW, Cline RG, Harris RF and Schuman GE, 1997. Soil quality: a concept, definition, and framework for evaluation (A Guest Editorial). *Soil Science Society of America Journal* **61**(1): 4 10. https: //doi: 10.2136/sssaj1997.03615995006100010001x.

Khadse A and Rosset PM, 2019. Zero budget natural farming in India: From inception to institutionalization. *Agroecology and Sustainable Food Systems* **43**(7–8): 848–871. doi: 10.1080/21683565.2019.1608349

Khurana A, Halim MA and Singh AK, 2022. Evidence (2004-20) on holistic benefits of organic and natural farming in India, Centre for Science and Environment, New Delhi.

Kotschi J, 2015. A soiled reputation: Adverse impacts of mineral fertilizers in tropical agriculture. Commissioned by World Wildlife Fund (Germany) to Heinrich Böll Stiftung.

Kumar SR, Spandana P, Ramanjaneyulu AV and Srinivas A, 2019. Components of organic farming. In: Organic Farming (Eds. Gopinath, KA and Ramanjaneyulu, A.V.). Daya Publishing House. A division of Astral International Pvt. Lid., New Delhi.

Ladha JK, Tirol-Padre A, Reddy CK, Cassman KG, Verma S, Powlson DS, van Kessel C, de B. Richter D, Chakraborty D and Pathak H, 2016. Global nitrogen budgets in cereals: A 50-year assessment for maize, rice, and wheat production systems. *Scientific Reports* **6**: 19355.

Laishram C, Vashishat RK, Sharma S, Rajkumari B, Mishra N, Barwal P, Vaidya MK, Sharma R, Chandel RS and Chandel A, 2022. Impact of natural farming cropping system on rural households–evidence from Solan district of Himachal Pradesh, India. *Frontiers in Sustainable Food Systems* **6**: 878015.

Liao J, Xu Q, Xu H and Huang D, 2019. Natural farming improves soil quality and alters microbial diversity in a cabbage field in Japan. *Sustainability* **11**(11): 3131. https: //doi.org/10.3390/su11113131

Liu H, Macdonald CA, Cook J, Anderson IC, Singh BK, 2019. An ecological loop: host microbiomes across multitrophic interactions. *Trends in Ecology and Evolution* 34(12): 1118-30. https: //doi.org/10.1016/j.tree.2019.07.011

Lu H, Wang X, Zhang K, Xu Y, Zhou L, Li G, 2014. Identification and nematicidal activity of bacteria isolated from cow dung. *Annals of Microbiology* **64**: 407-11. https: //doi.org/10.1007/s13213-013-0660-7

Macrae S, 2016. Identifying Agrarian Low-Density Urbanism Amongst the Charter States of the Chola and Sinhalese. *Trent University Occasional Papers in Anthropology* **18**: 105-127.

Marzani QA, 2023. Management of ascochyta blight (Ascochyta rabiei (Pass.) Labrouse) using chickpea-based mixed cropping systems through different combinations of companion crops. *Australasian Plant Pathology* **52**(3): 215-223.

Nagrale DT, Gawande SP, Shah V, Verma P, Hiremani NS, Prabhulinga T and Waghmare VN, 2022. Biocontrol potential of volatile organic compounds (VOCs) produced by cotton endophytic rhizobacteria against *Macrophomina phaseolina*. *European Journal of Plant Pathology* **163**(2): 467-482.

Neelam HSC and Kadian KS, 2016. Cow based natural farming practice for poor and small land holding farmers: a case study from Andhra Pradesh, India. *Agricultural Science Digest* - A Research Journal **36**(4): 282-86.

Nene YL, 2003. Crop disease management practices in ancient, medieval, and pre-modern India. *Asian Agri-History* **7**(3): 185-201.

Nene YL, 2012. Potential of Some Methods Described in Vrikshayurveda in Crop Yield Increase and Disease Management. *Asian Agri-History* **16**(1): 45-54.

NITI Aayog, 2024. Significant Contributors- Natural Farming: NITI Initiative.

NRAS, 2020. State of rural and agrarian india report 2020. Available online:

Palekar S, 2006. The principles of spiritual farming II. 2nd ed. Amravati: Zero budget natural farming research, development and extension movement, Amravati, Maharashtra, India.

Palekar S, 2013. Subhash Palekar Krishi - vegetable crops (Part II). Zero Budget Spiritual Farming Research Development and Extension Movement. Amravati, Maharashtra, India.

Parisi V, Menta C, Gardi C, Jacomini C and Mozzanica, 2005. Microarthropod communities as a tool to assess soil quality and biodiversity: a new approach in Italy. *Agriculture Ecosystems and Environment* **105**(1-2): 323-33. https: // doi: 10.1016/j.agee.2004.02.002.

Pingali PL, 2012. Green revolution: impacts, limits, and the path ahead. *Proceedings of the National Academy of Sciences* 109(31): 12302-308. https: //doi. org/10.1073/pnas.0912953109

Pretty J and Bharucha ZP, 2014. Sustainable intensification in agricultural systems. *Annals of Botany* **114**(8): 1571-96.

Ramanjaneyulu AV, Sainath N, Srinivas M, 2020. Promotion of organic farming: roles of key players. *Biotica Research Today* **2**(8): 731-4.

Rana A, Chandel RS, Sharma PL, Yankit P, Verma S, Verma SC and Sharma P, 2021. Insect-pests, natural enemies and soil microflora in cabbage grown under Subhash Palekar natural and conventional farming systems. *Indian Journal of Ecology* **48**: 1442-8.

Sadhale N, 1999. Krishi-Parashara (Agriculture by Parashara). Agri-History Bulletin No. 2. Agri-History Foundation, Secunderabad 500009, India. 94 pp.

Saharan BS, Tyagi S, Kumar R, Vijay, Om H, Mandal BS, Duhan JS, 2023. Application of jeevamrit improves soil properties in zero budget natural farming fields. *Agriculture* **13**(1): 196.

Sapre JK and Varma RK, 2006. Influence of cow urine and butter milk on the activity of lytic enzymes secreted by three soil-borne pathogens of soybean. *Journal of Mycopathological Research* **44**(2): 279-283.

Schoeny A, Jumel S, Rouault F, Lemarchand E, Tivoli B, 2010. Effect and underlying mechanisms of pea-cereal intercropping on the epidemic development of ascochyta blight. *European Journal of Plant Pathology* **126**(3): 317-31.

Sharma N and Singhvi R, 2017. Effects of chemical fertilizers and pesticides on human health and environment: A review. *International Journal of Agriculture, Environment and Biotechnology* **10**(6): 675-680. https: //doi: 10.5958/2230-732X.2017.00083.3.

Sharma R, Gautam S, Hamid H, 2021. Effect of MAP on shelf-life of broccoli and cabbage grown under natural farming. *Indian Journal of Agricultural Sciences* **91**(9): 1353-7.

Sharma R, Gautam S, Verma SC, Sood G, Bharat NK, Thakur KS, Chandel A, Thakur US, Bishist R, Sharma S, Kumar P, 2023. Influence of modified atmospheric packaging on shelf-life and quality of naturally grown tomato (*Solanum lycopersicum* L.) stored under different conditions. Brazilian Archives of Biology and Technology **66**: e23220466.

Singh Y and Sidhu HS, 2014. Management of cereal crop residues for sustainable rice-wheat production system in the Indo-Gangetic plains of India. *Proceedings of the Indian National Science Academy* **80**(1): 95-114.

Sinong GF, Yasuda M, Nara Y, Lee CG, Dastogeer KM, Tabuchi H, Nakai H, Djedidi S, Okazaki S, 2021. Distinct root microbial communities in nature farming rice harbor bacterial strains with plant growth-promoting traits. *Frontiers in Sustainable Food Systems* **4**: 629942.

Smith J, Yeluripati J, Smith P and Nayak DR, 2020. Potential yield challenges to scale-up of zero budget natural farming. *Nature Sustainability* **3**: 247-52.

SPNF, 2021. Annual report - SPNF project. Dr YS Parmar University of Horticulture and Forestry, Nauni, Solan (HP) India

Srinivasan TM, 2016. Agricultural practices as gleaned from the Tamil literature of the Sangam age. *Indian Journal of History of Science* **51**(1): 167-89.

Tatsumi C, Lin J, Ishiguro M and Uchida Y, 2023. Natural farming diversifies resource-utilisation patterns and increases network complexity in soil microbial community of paddy fields. Agriculture, *Ecosystems and Environment* **356**: 108618.

Upperi SN, Lokesh BK, Maraddi GN and Agnal MB, 2009. Jeevamrutha, a new organic approach for disease management and crop production in pomegranate and groundnut. *Environment and Ecology* **27**(1): 202-204.

USDANOSB, 1995. Definition of "Organic." United States Department of Agriculture National Organic Standards Board. United States Department of Agriculture, Washington, D.C.

Vandermeer JH, 1989. The ecology of intercropping. Cambridge University Press, Cambridge, UK.

Varaprasad K and Rao TV, 2024. Perspective chapter: agroecology-based natural farming in India. https: //www.intechopen.com/chapters/88975

Varaprasad KS and Kumari VS, 2022. Agroecology-based biodiversity management. *Indian Journal of Plant Genetic Resources* **35**(3): 46-49. https: //doi: 10.5958/0976-1926.2022.00040.7

Vashisth S, Verma S, Verma SC, Dev I and Chandel RS, 2023. Woolly apple aphid management in dry temperate zone with Natural Farming. *Indian Horticulture* **69**(2): 16-18.

Verma S, Chandel R S, Kaushal R, Yankit P and Sharma P, 2018. Soil quality management through zero budget natural farming. *ICAR News* **24**(3): 8-9.

Vipul, 2021. Comparative effect of SPNF and Chemical Farming systems on insect-pests, natural enemies and soil microflora in cabbage. M.Sc. Thesis, Department of Entomology, Dr YS Parmar University of Horticulture and Forestry, Nauni, Solan.

Vishwajeet, 2020. Population dynamics of insect-pests of brinjal and their natural enemies under Natural Farming System and Conventional Farming System. M.Sc. Thesis, Dr YS Parmar University of Horticulture and Forestry, Nauni, Solan.

Wang S, Tan Y, Fan H, Ruan H and Zheng A, 2015. Responses of soil microarthropods to inorganic and organic fertilizers in a poplar plantation in a coastal area of eastern China. *Applied Soil Ecology* **89**: 69-75. doi: 10.1016/j.apsoil.2015.01.004.

Watt G, 1889. A dictionary of the economic products of India. Superintendent of Government Printing. Central Secretariat Library.

Xavier R, Rekha K and Bairy KL, 2004. Health perspective of pesticide exposure and dietary management. *Malaysian Journal of Nutrition* **10**(1): 39-51.

Yan S, Tianfu Z, Yunyue W, Jianbin C, Xiahong H, Chengyun L, Youyong Z, 2006. Effect of intercropping on disease management and yield of chilli pepper and maize. *Acta Horticulturae Sinica* **33**(5): 995-1000.

Yankit P, Chandel RS, Verma S, Sharma PL, Verma SC, Gaikwad MB, Sharma P, Chauhan S, Keshava and Gautam US, 2024. Insights on soil biological properties and crop yields under natural farming in western Himalaya. *Indian Journal of Agricultural Sciences* **94**(3-S1): 089–094

Yankit P, Chandel RS, Sharma PL and Verma SC, 2019. Zero budget natural farming for the management of invasive leaf miner (*Tuta absoluta*) in tomatoes. *ICAR News*, **25**(1): 11-12.

Zhu X, Jackson RD, DeLucia EH, Tiedje JM, Liang C, 2020. The soil microbial carbon pump: From conceptual insights to empirical assessments. *Global Change Biology* **26**(11): 6032-9.

Index

D

E

F

G

O

P

Q

R

S